A1 C1

Mechanics and Usage

1. Do punctuation and spelling follow standard practice?
2. Is word choice accurate and precise?
3. Does grammatical usage follow standard practice?
4. Is each sentence effective in structure and emphasis?
5. Is the content free of typos and careless errors?

Graphics and Other Visuals

1. Do the visuals support the project's purpose and the needs of its readers?
2. Is each visual placed appropriately?
3. Is each visual neat and attractive?
4. Does each visual enhance the verbal information?
5. Is each visual numbered and captioned?
6. Is each visual referred to in the text?
7. Are all visuals, either borrowed or containing borrowed information, documented?

Headings

1. Do headings reflect the organization?
2. Does the wording of each heading reflect the information that follows?
3. Are headings of different levels distinguished by techniques such as point size, style, and color and by placement on the page?
4. Are major headings preceded by a triple space and followed by a double space?

Layout and Design

1. Is the material arranged usefully and pleasingly?
2. Does the design invite a positive audience response?
   - Does ample white space on each page make the text easy to read?
   - Is the typeface of the text readable and large enough to suit reader needs?
   - Are special fonts (such as Old English, Classic Greek, and script) and styles (such as all capitals, boldface, and italics) used sparingly?
3. Is the communication project designed so that readers can comprehend it in full and easily locate individual parts?

TECHNICAL ENGLISH

EIGHTH EDITION

TECHNICAL ENGLISH

WRITING, READING, AND SPEAKING

Nell Ann Pickett
Hinds Community College

Ann A. Laster
Hinds Community College

Katherine E. Staples
Austin Community College

New York Boston San Francisco
London Toronto Sydney Singapore Madrid
Mexico City Munich Paris Cape Town Hong Kong Montreal

Editor in Chief: Joseph Terry
Development Manager: Arlene Bessenoff
Development Editor: Carol Hollar-Zwick
Senior Marketing Manager: Carlise Paulson
Supplements Editor: Donna Campion
Production Manager: Patti Brecht
Project Coordination, Text Design, and Electronic Page Makeup: Nesbitt Graphics, Inc.
Cover Design Manager: John Callahan
Cover Designer: Kay Petronio
Senior Manufacturing Buyer: Dennis J. Para
Printer and Binder: R. R. Donnelley & Sons
Cover Printer: Coral Graphic Services, Inc.

For permission to use copyrighted material, grateful acknowledgment is made to the copyright holders on pp. 683–686, which are hereby made part of this copyright page.

Library of Congress Cataloging-in-Publication Data

Pickett, Nell Ann.
Technical English : writing, reading, and speaking.–8th ed. / Nell Ann Pickett, Ann A. Laster, Katherine E. Staples.
p. cm
Includes bibliographical references and index.
ISBN 0-321-00352-7
1. Readers–Technology. 2. Technology–Problems, exercises, etc. 3. English language–Technical English. 4. Technical writing. I. Laster, Ann A. II. Staples, Katherine. Ill. Title.

PE1127.T37 P5 2000
808′.0666–dc21 00-033182

Please visit our website at http://www.awl.com/englishpages

ISBN 0-321-00352-7

1 2 3 4 5 6 7 8 9 10—DOC—03 02 01 00

To our students, our colleagues, and our friends

In Memoriam

HARRY J. PARTIN

26 March 1927 10 July 1999

Electronics Technology Teacher, Hinds Community College

1961–1999

Husband of Nell Ann Pickett

BRIEF CONTENTS

DETAILED CONTENTS

PART II

The Technical Communicator's Tools 27

PART V

Oral Communication 469

REFERENCE SECTIONS

PREFACE

Today's workplace is defined by constant change. Technological change affects the nature and practice of every career and the work of every professional, broadening workplace knowledge and resources and widening the marketplace for goods, services, and information. In the face of such change, every successful professional is both a learner and a teacher. Workplace change requires flexibility, active learning, problem solving and critical thinking, social awareness, and the ability to apply all of these to workplace communication.

Effective communication is integral to individual career success. In the workplace community, professionals must support clients and serve as members of diverse and even geographically distant teams. To communicate successfully, workplace professionals must consider the needs and values of persons whose knowledge and experience may differ vastly from their own. Workplace communicators must therefore develop information for multilevel audiences and use new, often technological, means of information delivery. Technology shapes communication tools and research options, providing more choices for locating, shaping, and delivering information.

Technical English: Writing, Reading, and Speaking, Eighth Edition, prepares workplace communicators for professional responsibilities and for change. This text provides learners with a range of practical strategies for communication choices and conventions, tools, and media.

This edition of *Technical English* is another milestone on a journey that began many years ago. Our first edition evolved out of questions we asked as we worked to create an effective communication course. What communication skills do professionals need to succeed in such diverse fields as electronics, drafting, health sciences, commercial art, marketing, computer science, paralegal studies, and agriculture? How can we best present material that students can read, apply, and adapt to their individual career needs?

To answer these questions, we explored workplaces as well as classrooms. We worked closely with technical instructors, we interviewed professionals from many fields, and we collected samples of technical writing wherever we went. We read, assimilated, and read some more. And we talked with technical communication instructors. Our research for each edition has identified and refined a proven body of knowledge and skills that people need in order to be effective communicators. We decided that the best way for learners to develop workplace communication skills is to write documents and give oral presentations on subjects relevant to their chosen careers. This approach not only develops communication skills; it applies knowledge from students' majors and prepares future professionals by asking them to write within their disciplines.

Our goal then remains our goal now: to prepare students to meet the communication demands of their chosen careers. Effective communication skills are increasingly important for entry-level employment and for career advancement. We want the users of this book to become informed communication problem solvers, and we want this eighth edition of *Technical English* to be as useful on the job as it is in the classroom.

Technical English provides an introductory textbook in technical communication written accessibly for students and teachers in college and in on-the-job training. We believe that effective communication skills are vital for success in a global workplace. This textbook is designed to develop visual, written, and spoken communication and research skills. We are humbled and pleased by the continued interest in this textbook and by the comments we receive from students, teachers, and workplace professionals.

WHAT'S NEW IN THIS EDITION

The eighth edition of *Technical English* reflects new directions in technical writing instruction and new developments in workplace communication. It addresses the concerns of a changing workplace that increasingly depends on technology to serve diverse, even global, audiences. To help learners become active and responsible workplace communicators, the text emphasizes ethical decision making through discussion, readings, and case studies. *Technical English*, Eighth Edition, also incorporates practical approaches to document design, visual communication, and internationalism.

The eighth edition of *Technical English* offers a new sequence of instruction and many new elements. We have devoted more discussion to the roles, communication tools, and composing processes of workplace authors, while continuing to emphasize typical information products in the writing assignments. Many new examples appear throughout the text. We have streamlined the plan sheet of earlier editions to encourage responsible decision making and audience analysis. We have incorporated readings within chapters to encourage critical reading.

In this edition, we are pleased to welcome Katherine Staples as collaborator and partner. She brings to *Technical English* her experience as technical communication teacher, author, and consultant.

THE ORGANIZATION OF THIS TEXTBOOK

This textbook is arranged to help readers understand and use the elements of workplace communication, to help them apply the writing process, and to prompt them to use these skills and strategies in workplace contexts. Rather than appearing in a separate section, readings are integrated in the activities at the end of each chapter. Since the ability to articulate values and to apply them to problem solving is important in the workplace, we include short case studies that encourage students to analyze situations, to respond, and to consider the implications of their responses. We have reorganized *Technical English*, Eighth Edition, into five parts: Getting Started with Technical Communication, The Technical Communicator's Tools, The Technical Writing Process, Situations and Strategies for Technical Communication, and Oral Communication. Also included are three reference sections: The Technical Communicator's Guide to Research, Appendix 1 The Search for Employment, and Appendix 2 The Search for Standard English Usage.

Part I Getting Started with Technical Communication explains the contexts and tools that technical communicators use in developing solutions to workplace communication problems. Chapter 1 provides a full definition of technical communication, while Chapter 2 explains the wide range of workplace audiences. Each chapter in Part I includes individual and collaborative activities, a reading, and a case study to encourage students to discuss the role of technical communicator and to apply it to their own career goals.

Part II The Technical Communicator's Tools introduces effective language, information design, visuals, and communication technologies. Each chapter in Part II includes examples, individual and collaborative activities, and readings that reinforce the principles and skills presented.

Part III The Technical Writing Process explains the steps for developing a workplace communication project, from generating ideas to planning, arranging, drafting, and revising, both individually and collaboratively. The text emphasizes efficient, exact, and clear organization and composing skills that apply to any communication project. Each chapter in Part III includes individual and collaborative activities to help learners become efficient and effective communicators.

Part IV Situations and Strategies for Technical Commmunication introduces typical forms of and occasions for workplace communication—procedures, descriptions, definitions, summaries, proposals, reports, and correspondence. Each chapter in Part IV includes plan sheets, examples, individual and collaborative activities, a reading, and case study, all of which focus directly on the type of workplace communication presented.

Part V Oral Communication discusses the development and delivery of oral presentations. The chapter addresses the use of visuals and offers suggestions for coping with nervousness. The single chapter in Part V includes individual and collaborative activities, a reading, and a case study.

The Technical Communicator's Guide to Research provides practical information, reference sources, and documentation models valuable for research projects. To help students conduct such projects, this guide includes a student research report and an interview with its author.

Appendix 1 The Search for Employment explains the ways in which effective communication can help job seekers identify their goals and abilities and present themselves honestly and credibly to employers. Appendix 1 also suggests ways to locate job openings and to apply for them.

Appendix 2 The Search for Standard English Usage is an easy-to-use reference guide with sections on usage, conventions in language usage, and common errors in writing. It is designed so that readers can quickly and easily find answers to specific questions.

APPRECIATION

Every edition of *Technical English* has been influenced by many individuals. One of the rewards of writing a textbook is the people we meet and learn from. Many students, teachers, and workplace communicators have been generous with their support. In particular, we wish to thank the colleagues whose thoughtful reviews and suggestions have enhanced this eighth edition of *Technical English*. Each of them contributed comments from a specific area of expertise:

- Lee Brasseur, Illinois State University
- Pamela Ecker, Cincinnati State Technical and Community College
- Sherry Little, San Diego State University
- Ann Neville, The University of Texas at Austin
- Karen Schriver, KSA Document Design and Research
- Emily Thrush, Memphis State University

We also wish to thank the Society for Technical Communication for encouragement. In particular, we wish to acknowledge Maurice Martin, editor of *Intercom*; Anita Dosik, Publications Director; and Bill Stolgitis, Executive Director.

We are grateful to our manuscript reviewers and consultants for their helpful suggestions: Jerry Carr, Marjorie Morris, and McLendon Library Staff, Hinds Community College; Roger Bacon, Northern Arizona University; Marian G. Barchilon, Arizona State University East; Crystal Brantley, Vance-Granville Community College; Pat Cearley, South Plains College; Roger Fox, Garland County Community College; Michelle Holt, North Idaho College; Theresa M. Jackson, West Iowa Tech Community College; Becky Kamm, Northeast Iowa Community College—Calmar; Frank Masiello, New York City Technical College (CUNY); John Metz, University of Toledo; Anne Schoolfield, University of North Texas.

We hope that *Technical English: Writing, Reading, and Speaking,* Eighth Edition, helps students to become active and successful communicators in their careers and in their workplace communities. We welcome comments and suggestions. Please e-mail us.

Nell Ann Pickett
picketthcc@aol.com

Ann A. Laster
aalaster@aol.com

Katherine E. Staples
kstaples@bga.com

ANCILLARIES

PRINT RESOURCES FOR STUDENTS AND INSTRUCTORS

An updated Instructor's Manual is available to adopters of this edition.

NEW! *Researching Online,* Fourth Edition, by David Munger and Shireen Campbell, is an indispensable media guide that gives students detailed step-by-step instructions for performing electronic searches; using e-mail, listservs, Usenet newsgroups, IRC's, and MUD's; and assessing the validity of electronic sources.

Visual Communication, by Susan Hilligoss (Clemson University), introduces document design principles and features, practical discussions of space, type, organization, pattern, graphic elements, and visuals.

MEDIA RESOURCES FOR STUDENTS AND INSTRUCTORS

Daedalus Online is the next generation of the highly awarded Daedalus Integrated Writing Environment (DIWE), uniting a peer-facilitated writing pedagogy with the inherently cooperative tools of the World Wide Web. This writing environment offers students prewriting strategies and prompts, computer-mediated conferencing, peer collaboration and review, comprehensive writing support, and secure 24-hour availability. *Daedalus Online* also offers instructors a suite of interactive management tools and linking assignments to facilitate a heuristic approach to writing instruction. For more information, visit **http://www.awl.com/daedalus.**

NEW! *The English Pages* at **http://www.awl.com/englishpages.** The completely revised *English Pages* Web site encourages students and teachers to explore the multiple uses of traditional and online resources for writing. For instructors, Teaching Perspectives—interactive Webs started by experienced faculty on pragmatic teaching topics—provide fresh teaching ideas and an online forum for collaboration. The Syllabus Sharing Project allows you to discover how other instructors are teaching from your Longman textbook, and Online Casebooks provide the raw materials for creating essay assignments from credible online sources. For students, interactive tutorials offer specific guidelines for learning important writing processes, a lively e-zine provides a publication medium for students' writing projects, and more than 25 student papers with layered annotations demonstrate the features of effective and ineffective essays in every area of the English curriculum.

INTRODUCTION

This introduction provides an overview of the subject and goals of *Technical English,* Eighth Edition, a text designed to help you learn and apply workplace communication skills.

THIS TEXT, TECHNICAL COMMUNICATION, AND YOU

The rate of workplace change is increasing, and everyone touched by change depends on some form of technical communication—written, spoken, or visual—to learn, to adapt, or simply to keep up. On the job, you will need to provide practical information for a wide range of readers and listeners. These readers and listeners—your audience—may be beginners or experts, distant or nearby. You may often communicate with people whose needs, knowledge, and expectations are unlike your own. Your audience may have information to share with you as well, and you can learn by reading, listening, and responding thoughtfully. As an effective workplace communicator, you will become an active and responsible member of your workplace community, supporting customers and team members, learning from them—and with them.

This text assumes that communication will be a necessary part of your career. It introduces common types of workplace messages with suggestions to help you present practical information in writing and orally, visually and verbally. The text then asks you to develop and prepare workplace communication projects, making choices about subject, audience, purpose, content, medium, page layout, document design, and visuals. To help make your choices informed and efficient, *Technical English* discusses workplace communication knowledge, tools, and media. And since learning to learn is essential to your career, the text discusses strategies for research, showing you how to locate information and how to make sure that what you locate is credible and current. Understanding resources and using them wisely can help you to make your workplace communication successful.

Since people do not work, learn, or communicate in a vacuum, effective workplace communication depends in part on understanding others. Every decision you make about a technical communication project depends on a thoughtful response to four related questions:

- Who is your audience (reader, listener, or viewer)?
- What practical problem can your communication help to solve?

- How can you best plan, present, and deliver information?
- How can you assure that the information and the way you present it meet your audience's needs?

The chapters in this text remind you to ask and answer these four questions before you begin any communication project and as you work to develop and revise it. The plan sheets that accompany discussion of each type of workplace communication will help you understand your audiences and develop sound strategies. To test and strengthen these strategies, you'll be encouraged to work collaboratively.

Understanding workplace audiences, mastering communication skills and techniques, and collaboration are essential to careers. Also essential are critical thinking and an awareness of your responsibility to others. To develop critical thinking skills, chapters in *Technical English* provide readings that consider workplace communication and allow you to draw conclusions. In addition, case studies present workplace situations and invite you to respond. While there are no right or wrong answers in these discussions, your responses will help you better understand various views and values.

PART I

Getting Started with Technical Communication

Part I of *Technical English* introduces you to a different kind of communication, one that can make a difference in the workplace and beyond. But what kind of communication is this, and how does it differ from personal and academic communication? What makes it special, and what makes it effective? **Chapter 1 Technical Communication: What Is It?** and **Chapter 2 Defining Workplace Readers: Who Are They?** provide answers to these questions.

In every career, professionals regularly communicate for a variety of reasons. However, the workplace calls for communication only when there is a need—to inform readers or to solve problems. The audiences for workplace communication vary widely. The needs of a reader who will use a new camera, for example, differ from the needs of the technician who will repair the camera. The new owner will need easy instructions for loading film and taking photographs; the technician will need far more technical information in the repair manual.

The effective technical communicator must present information to meet the needs of specific audiences. Chapter 1 defines technical communication, explaining the workplace communicator's role, responsibilities, and choices.

Chapter 2 explains the importance of understanding different audiences and emphasizes the importance of sensitivity to every audience's needs.

As you read these chapters, consider the ways in which effective communication can contribute to your career and your workplace. Who are your intended audiences? What kinds of information can best help them?

CHAPTER 1

Technical Communication: What Is It?

CHAPTER GOALS

This chapter:

- Defines the nature of technical communication in the workplace
- Explains the central importance of audience in technical communication
- Describes the nature of workplace collaboration and peer editing
- Explores the responsibilities of the technical communicator

INTRODUCTION

When most people think of technical communication, they think of a highly specialized document—a report, a Web page, user instructions—written about a highly technical subject. However, technical communication also means the broader, more important process of developing usable information to teach technical and nontechnical audiences about an expanding range of products and services. In your career, you will be developing information as well as reading it, using it, and responding to it. Making a sale, documenting problems, providing instructions, reporting on progress, proposing purchases—workplace communicators daily create information for many purposes and audiences. Technical communication touches nearly every workplace and every person.

How can such communication be most effective? Skillful workplace communicators conduct business, share new ideas, and solve problems by combining their career and academic knowledge with their knowledge of people and their understanding of communication. This chapter introduces you to technical communication and to the knowledge and skills that you will need as your career and your on-the-job communication responsibilities change and grow.

WHAT IS TECHNICAL COMMUNICATION?

Technical communication is a field in which professionals develop information to guide readers, listeners, and viewers in solving practical problems. At one time, these communicators were highly specialized technicians, engineers, or scientists who developed advanced, sometimes theoretical, information for technically educated readers. However, as every profession becomes more specialized and as more people rely on technologies in the workplace and at home, an increasing number of readers need and use technical information. The ability to communicate openly, honestly, and clearly is important to every professional. Although the communication work of full-time technical communicators will differ from that of professionals who communicate as part of other careers in other fields, the same knowledge, skills, and ethical concerns apply to communicators in every workplace.

Technical writing is only one kind of technical communication, although an important one. Workplace communicators use writing, speaking, images, and even the medium and physical design of information to reach their audiences. Consider this example: A large store distributes, sells, and services electronic and video equipment, including a popular and sophisticated VCR manufactured by a major international company. However, customers who buy the VCR often complain that they cannot use it because the manufacturer's instructions are impossible to understand or follow. To solve this problem, the store manager decides to develop a set of easy-to-use instructions for customers and assigns the service manager to prepare them. As the service manager plans the project, she asks questions like these:

- Will customers read all or some of the instructions I write?
- Do I know enough about the VCR to write a clear set of instructions for its users?
- If not, where can I find the most current and reliable information?
- How much of what I learn about the VCR do I really need to tell my readers?
- How do I learn the kinds of problems customers are having with the manufacturer's instructions?
- What kind of language can make the instructions easy and unthreatening for nontechnical readers?
- How can I arrange information and design pages so that readers can look up information as well as read and follow instructions?
- What kind of visuals would be helpful? Where should they be placed?
- How can I test to be sure that my instructions are helpful and effective?

Easy, readable, and reliable VCR instructions for nontechnical readers are a good example of a successful technical communication project. Such instructions meet the needs of the people who read and use them, and they reflect current information, thoughtful arrangement and design, and clear writing. As you study this textbook, you will learn to ask questions to define your audience and its needs and to use technical communication resources to solve workplace problems efficiently and responsibly.

TECHNICAL COMMUNICATION AND AUDIENCE

The way an audience reads or hears, comprehends, and responds to information depends to a great extent on content, arrangement, and design. It also depends on other factors, such as the audience's background, knowledge, attitude, or special skills—as well as on the audience's motivation. The informed communicator employs a variety of visual and verbal techniques to make information clear and usable. Since some of these communication techniques may be new to you, Part I introduces you to visual, verbal, and information-finding techniques and to the communication technologies most readily available to you in the workplace.

Also new to you may be the concept of audience.

- Who will read what you write or listen to what you say?
- What does that audience already know about the subject?
- What attitude does the audience likely have toward it?
- What knowledge, experience, or education does the audience have?
- Why does the audience need the information, and how will the audience locate, learn, and apply it?

Since workplace audiences are becoming more diverse, practical ways to define and meet their needs for workplace information are discussed in Chapter 2.

WORKPLACE COMMUNICATION AND ACADEMIC WRITING

Many elements of technical communication reflect the conventions and practice of good academic writing. As a student, you are part of a learning community, and writing is an important part of your role as a learner. The academic emphasis on analytical thinking, logic, organization, and writing conventions helps you define and express your own ideas and respond to new ones. Effective academic writing helps you learn how to learn.

In the workplace, you join another learning community, and the skills and knowledge of academic writing will help you as you continue to learn throughout your career. However, the goals and audiences of workplace communication are more diverse than those in academic communication. In workplace communication, you can be the teacher, the partner, and the collaborator as well as the learner. The knowledge you have to share can help others, but you'll also respond to new information and take part in group projects. The added dimensions of graphics and other visuals, layout, information design, and choice of medium make the communication process more effective for you and for your audiences. Through your workplace communication, you will continue learning about other people and about your chosen field. As you advance in knowledge and responsibility, you will find yourself providing a wider range of information to more people—as a manager, an innovator, and a problem solver.

The rewards of the workplace learning community differ from those of the academic one. Communication aids your learning as a student, and excellent grades acknowledge and reward intellectual growth and the ability to master new knowledge and skills. In the workplace, communication allows you to grow not only intellectually, but professionally and personally as well. The workplace acknowledges and rewards professional growth with new responsibilities, new opportunities to learn, and career advancement. However, communication in both learning communities—in academia and on the job—requires active curiosity, discipline, commitment to critical thinking, responsiveness, intellectual honesty, respect for others, and knowledge of writing conventions.

COMMUNICATING ON THE JOB

In technical communication, the subject, medium, and genre (or type) of presentation are often defined for you. That is, you will be assigned a document to prepare or a subject to research; perhaps the need for a document will arise within your own organization. For instance, you may be asked to analyze an existing process to determine if a change can increase productivity. Or you may see a need for new equipment and write a proposal to justify its purchase.

No one jumps out of bed and says, "I believe I'll write a report today." You might, however, arrive at work one day to meet your supervisor, who says, "Sales in the southeast are down for the last quarter. Contact and interview all the sales managers in this area to come up with an explanation for the decrease in sales and recommendations for solving the problem. Kristin will have the sales figures. And get Jason to develop some good visuals. I need a written report on my desk by Friday so that I can use it to make a presentation to our sales division. I'll need charts and tables to support my talk."

Will the finished report inform readers or persuade them that your recommendation is the right course of action? Actually, it will do both. By presenting factual information about sales records, you are informing your audience about existing problems. By supporting your recommendations with factual material and providing ample evidence for your conclusions, you are also showing your audience that your conclusions are correct. Even your effort to get the latest information, to present it logically, and to design the report and graphics and other visuals clearly and attractively persuades your audience that you have prepared your response professionally and credibly. Workplace communication is always factual, even when it persuades.

Collaboration

The report assignment above illustrates another important aspect of technical communication: collaboration. You may be only one of several communicators gathering, analyzing, and presenting needed information. As a member of a team—a collaborative group—you may be responsible for the project, but the information for the report and work on it may come from a number of people. In some instances, you may be responsible for only one section. In others, you may be assigned to perform only one task—researching, editing, or planning and conducting interviews. In collaborative work, you will likely be guided by a group leader who directs others and makes decisions so that the finished project will accomplish its purpose. Or you yourself may be the group leader. A project may be carried out through group meetings, telephone conferences, or a networked system. Computer networking has decided advantages, allowing team members to see each other's work in progress and to interact during revision.

In any group task, some members may find themselves doing extra work, and others may not fulfill their responsibilities. The bottom line, nevertheless, is that the job must be completed on time, and the information must meet the reader's needs. A responsible collaborator does his or her share, meets deadlines, participates actively, and keeps other team members informed.

Peer Review

Rarely does even the most experienced writer complete a document in one attempt. Good writing takes hard work and revision, and workplace communicators value the suggestions of their peer reviewers to make their collaborative work successful. Peer review, in which team members edit and comment on each other's work, helps technical communicators keep their readers' needs in mind. Because writers tend to overlook slips and omissions in their own work, thorough and tactful peer reviewers improve a project and the performance of all those who work on it.

THE TECHNICAL COMMUNICATOR'S RESPONSIBILITIES

Technical communicators, like other professionals, must consider their responsibility to others in the workplace community: team members, managers, customers, and the people who will use technical information long after the communicators who created it have moved on. The individual communicator is responsible for his or her own workplace performance, conduct, safety, and integrity and is concerned with the well-being of others. The Society for Technical Communication (STC), the largest organization for technical communication professionals, outlines such responsibilities in its Ethical Guidelines for Technical Communicators. The guidelines read, in part, as follows:

> **Legality**
> We observe the laws and regulations governing our professional activities in the workplace . . . .
>
> **Honesty**
> We seek to promote the public good in our activities. To the best of our ability, we provide truthful and accurate communications. We dedicate ourselves to conciseness, clarity, coherence, and creativity, striving to address the needs of those who use our products. We alert our clients and employers when we believe material is ambiguous. Before using another person's work, we obtain permission . . . .
>
> **Fairness**
> We respect cultural variety and other aspects of diversity in our clients, employers, development teams, and audiences. We serve the business interests of our clients and employers, so long as such loyalty does not interfere with the public good . . . .

Professionalism
We seek candid evaluations of our professional performance from clients and employers. We also provide candid evaluations of communication products and services. We advance the technical communication profession through our integrity, standards, and performance.

(Published by permission of the Society for Technical Communication.)

These ethical guidelines highlight three important issues in workplace communication. First, readers rely on information, which means that workplace communicators have an obligation to present it truthfully and accurately. Second, standards of legality, honesty, and professional competence should be upheld in all workplace communication. Third, since so many people depend on information for their safety, security, and employment, communicators have a responsibility to society as a whole.

GENERAL PRINCIPLES for Technical Communication

- Technical communication is functional, always solving a problem by providing information to help meet a practical need.
- Technical communication is sensitive to audience. The technical communicator asks: Who will read, see, or hear this material? Why will the audience need it? What kind of details can be most useful to the reader? What approach to the material and its arrangement is appropriate for the user?
- Technical communication is based on a communicator's ethical and professional commitment to audience and to keeping up with changes in subject matter.
- Technical communication involves choices in medium (the means for communicating a message) and genre (the type of message and its characteristic format).
- In technical communication, logical organization guides and directs those who use the information.
- In technical communication, accurate, precise terminology is essential. Technical communicators use exact, concise, concrete language.
- In technical communication, conventional standards of writing—such as grammar, usage, spelling, and punctuation—are observed. Technical communicators avoid any distracting inconsistencies in their use of writing conventions.
- Technical communication requires the ability to think critically, objectively, thoroughly, and creatively.
- In the workplace, communicators plan, draft, revise, design, and even edit or deliver information through the use of technical tools and equipment.
- Technical communication is rarely the work of a single individual. Instead, workplace communicators usually work in teams, sharing expertise and making sure that projects meet the needs of diverse audiences.

CHAPTER SUMMARY

TECHNICAL COMMUNICATION

Workplace communication is practical, user-centered, and an important part of every career. Information must be adapted for changing audiences with different needs, and can be delivered electronically, in person, or in print. Workplace communicators must consider visual and spoken communication as well as writing, and must constantly adapt to changes in their careers, in their audiences, and in the nature of communication itself. Adapting to these changes means learning to work cooperatively with people who are very different from themselves. Most important of all, professionals, whatever their chosen careers, must consider the ethics of their communication roles.

ACTIVITIES

INDIVIDUAL AND COLLABORATIVE ACTIVITIES

1.1. Using such career research sources as the *Occupational Outlook Handbook, The Dictionary of Occupational Titles,* or *Encyclopedia of Careers,* or computer software such as *Challenge* and *Discover,* locate a job title and description that matches your career goals. First, name your occupational choice (such as computer programmer, engineer, CAD operator, veterinarian, electronics technician, editor, or health care professional). Then research the duties, skills, educational preparation, and work settings for your career. Write an analysis of what you learn about your chosen career.

1.2. Use research sources to write a brief report on the kinds of communication skills you will need on the job in your chosen career. Who will be the audiences for your communication projects? How will your projects support your audiences' needs? Who will be your collaborators in developing and preparing projects?

1.3. In teams of two or three, plan and conduct short telephone interviews with professionals from different fields to determine how much and what kinds of communication these professionals routinely perform. As a team, organize and present your findings in a short report or as an oral presentation to your class.

1.4. Locate different kinds of technical communication materials you find at home. For example, consider the nutritional analysis of prepared foods, warnings on appliances, or assembly instructions. How are these written, designed, and placed for easy everyday use? Are they effective? How could they be improved?

READING

1.5. The following job description from the U.S. Department of Labor's *Dictionary of Occupational Titles* lists the many roles and duties of a technical communicator.

U.S. Department of Labor, Job Description of a Technical Communicator

WRITER, TECHNICAL PUBLICATIONS (profess. & kin.)

Develops, writes, and edits material for reports, manuals, briefs, proposals, instruction books, catalogs, and related technical and administrative publications concerned with work methods and procedures, and installation, operation, and maintenance of machinery and other equipment. Receives assignment from supervisor. Observes production, developmental, and experimental activities to determine operating procedure and detail. Interviews production and engineering personnel and reads journals, reports, and other material to become familiar with product technologies and production methods. Reviews manufacturer's and trade catalogs, drawings and other data relative to operation, maintenance, and service of equipment. Studies blueprints, sketches, drawings, parts lists, specifications, mock ups, and product samples to integrate and delineate technology, operating procedure, and production sequence and detail. Organizes material and completes writing assignment according to set standards regarding order, clarity, conciseness, style, and terminology. Reviews published materials and recommends revisions or changes in scope, format, content, and methods of reproduction and binding. May maintain records and files of work and revisions. May select photographs, drawings, sketches, diagrams, and charts to illustrate material. May assist in laying out material for publication. May arrange for typing, duplication, and distribution of material. May write speeches, articles, and public or employee relations releases.

From *Dictionary of Occupational Titles,* Vol. I., 4th ed. (1991 rev.) U.S. Department of Labor.

QUESTIONS FOR DISCUSSION

a. In addition to writing, what kinds of tasks can a technical communicator perform on the job? What other kinds of communication do these tasks require? How important is spoken and visual communication in this career?
b. What kinds of technical communication responsibilities outlined in the job description could apply to other careers?

c. What kinds of responsibilities to others do the activities of a technical communicator involve?
d. How many of the roles listed in the job description could involve collaboration?
e. What sort of personal qualities would help a professional communicator be successful in the range of duties specified in the job description?

READING

1.5. In the following poem, poet, professor, and technical writer Chick Wallace asks questions about the nature of technical writing and personal life.

A Technical Writer Considers His Latest Love Object

CHICK WALLACE

1. How can "X" be described?
2. What is the essential function of "X"?
3. What are the components of "X"?
4. How is "X" made or done?
5. How should "X" be made or done?
6. What are the types of "X"?
7. How does "X" compare to "Y"?
8. What are the causes of "X"?
9. What are the consequences of "X"?
10. What kind of representative (*enter species here*) is "X"?
11. What is my memory of "X"?
12. What is the efficiency ratio of the performance of "X"?
13. Is "X" profitable in the long or short term?
14. What is my personal response to "X"?
15. What is the present status of "X"?
16. How should "X" be interpreted?
17. What is the present value of "X"?
18. What case can be made for and against "X"?
19. Conclusion: Hello, darling. I've been thinking. . . .
20. Results: Wedding scheduled 4:00 p.m. 6/27. Take the weekend off.

(Reprinted by permission of Chick Wallace.)

QUESTIONS FOR DISCUSSION

a. In the title of the poem, what attitude does the phrase "Latest Love Object" suggest?
b. If you were considering marriage, what questions would you ask yourself? Would your questions be logical? How would your questions differ from those in the poem?

c. What is implied about dedication to work in the closing sentence, "Take the weekend off"?
d. In groups of three or four, compare and contrast the qualities of technical communication implied in the *Dictionary of Occupational Titles* job description on page 11 with the qualities of technical communication presented in the poem.
e. In what ways does "A Technical Writer Considers His Latest Love Object" present its subject *nontechnically?* How is the writer's personality presented? How does the author use humor? What is the poem's purpose?

CASE STUDY

1.6. In a group discussion or a short written assignment, respond to the following case study by identifying the ways in which the case illustrates the nature of technical communication in the workplace.

VCR Blues

Gary has worked in the service area of Smithville Electronic Sales and Service for three years as head customer service representative. Smithville Electronic Sales and Service sells many VCRs, among which are the popular (and widely advertised) ones manufactured by Konyo. However, over the last six months, Gary has had over 33 calls and letters from customers who wish to return their Konyo VCRs. The customers complain that the manufacturer's instructions are so difficult, complicated, and technical that all of the great features they read about in advertisements are impossible to use on their Konyo equipment. Even though the store has a no-return policy on working VCRs, the callers are getting angrier and more numerous. Gary decides to discuss the problem with Cheryl, the sales manager, and Charlie, the service manager.

After Gary explains the complaints and shows Cheryl and Charlie copies of the VCR user's manual, Cheryl says, "This isn't our problem, is it? Our sales staff explained how to use the VCRs, and that meets the customers' needs. They all get a free Konyo manual, don't they? Besides, we're not in the writing business. Our job is to sell good equipment to customers who want it. And there's nothing wrong with any of those VCRs. The problem is the customers!"

Charlie adds, "Well, it's true that even our service technicians find the manual a little hard to use. The print is so tiny, and the diagrams really are pretty bad. It's almost impossible to find any information you need for reference or even to read that stuff. But why do people who use VCRs need to know all the technical information? Wouldn't we be saving money if we just warned customers to pay close attention during the sales talk?"

QUESTIONS FOR DISCUSSION

a. How would *you* respond to Cheryl and Charlie?
b. How can Smithville Electronic Sales and Service best resolve customer complaints about VCR instructions?
c. What obligation does Smithville Electronic Sales and Service have to its VCR customers? Would providing information that is helpful to customers be good for the company? For its employees? In what ways?
d. What have Cheryl and Charlie misunderstood about technical communication? Is it always technical in its treatment of a subject? Have they considered the real audience for the VCR instructions?
e. Would developing clear VCR user instructions for customers be a good investment of Smithville Electronic Sales and Service's time and money? Why or why not?
f. Should Smithville Electronic Sales and Service share the customers' complaints with Konyo? Should they refer angry customers to the manufacturer?
g. What would *you* suggest to solve the problem? Why?

CHAPTER 2

Defining Workplace Readers: Who Are They?

CHAPTER GOALS

This chapter:

- Classifies types of workplace readers
- Outlines factors that complicate audience analysis
- Explains the importance of meeting reader needs in the reader's context
- Describes the diverse and international nature of workplace readers
- Shows how the needs and makeup of audience affect the workplace communicator's choices in subject, organization, language, information design, and graphics and other visuals

INTRODUCTION

Thoughtful and effective communication requires a sensitive understanding of a project's audience since the knowledge level and expectations of those who need information can vary widely. Some workplace readers study documents, taking notes in order to apply information or make decisions. Others need to locate and use only selected information. They will not read any more than necessary, and they resent wasting time searching for what they need. Change in the workplace means that the needs, interests, and expectations of information users are more diverse than ever before.

For workplace communicators, a clear understanding of audience determines which type (or genre) of written document will best meet the reader's needs. Audience affects the communicator's decisions about content, page layout, information design, language, and medium. However, making assumptions about readers can be risky. Many audiences will not share the communicator's expertise, background, cultural assumptions, organizational culture, or primary language. Even readers with the same kind of knowledge and interests can differ widely in their needs.

For all of these reasons, workplace communicators must be sensitive and responsive to others. Communicators need strategies to recognize different audiences and flexibility to respond appropriately. This chapter describes the widening range of workplace readers, explains the importance of respecting diverse audiences, and provides strategies for analyzing audiences in order to meet their needs successfully.

READERS AND EXPERTISE

Workplace information is always targeted for a specific audience. It often classifies readers by expertise, or the degree of knowledge and experience the communicator expects the intended audience to have. A typical classification system appears on page 17.

Category	*Characteristics*
Experts	Have advanced knowledge and skills in their field. Understand technical information and language within that field. Handle theory and practical application with ease.
Technicians	Understand technical information and language within their field. Handle practical application with ease.
Professionals (nonexpert)	Have the education and ability to read and understand difficult and technical information (although such reading may require study and more than one reading).
Lay (general) audience	Have no specialized education but need practical information. May or may not be highly motivated to read information in full.

An Environmental Protection Agency manual describing standards for automotive emission, for example, would be targeted for an audience of technicians, while the maintenance guide that comes with a new car would be targeted for lay readers. A technician (with technical expertise) has the background knowledge and skill to apply the technical information to automotive repair and inspection. The driver of the new car (a lay reader) only wants to learn how to adjust the new air conditioner—and to learn about the car's other features quickly and easily.

It is possible to classify different types of workplace readers. However, any classification system is at best no more than a workable way to describe the readers' degree of technical knowledge. Terms such as "expert" and "professional" can be helpful in planning a project's content. Real workplace readers, however, are often harder to classify because a workplace audience may fit into more than one category.

What do readers expect, and how will they use or apply the information? Answers to such questions determine a project's medium, or means of delivery—written, spoken, or electronic. How can readers most easily and conveniently read, locate, and use information? What kinds of format and medium will be most comfortable for them? These answers determine a communication project's organization, design, tone, and style.

PRIMARY AND SECONDARY AUDIENCES

Technical communicators use every possible means to recognize and meet readers' needs. But defining readers and understanding their needs can be difficult, particularly since the audience for technical communication is often made up of readers with very different needs. In planning a project that must serve more than one audience, it is useful to prioritize audiences as *primary* and *secondary. Primary audiences* typically need information to perform a task, to make a decision, or to decide a course of action. *Sec-*

ondary audiences are those whose activities will be affected by the decision. Secondary readers may read an entire document or only the portion that directly affects them.

Suppose, for instance, that a training manager is assigned to develop a repair manual for custom manufacturing equipment that her company designs and builds. The manual will be used by companies that buy and install the equipment to instruct newly hired technicians. However, the same manual must also be helpful for the experienced technicians who will use it for quick reference in emergency breakdowns. The newly hired technicians, who will read, follow, and study entire procedures, will be the manual's primary audience. The secondary audience, the experienced technicians who need to look up selected information quickly, is also important. The repair instructions must be written, organized, illustrated, and designed to serve the needs of both audiences.

A second example illustrates another kind of need, one in which the relationship between author and readers is much more direct. Members of a desktop publishing team know that new software will help them prepare publications more quickly than the programs they currently use, making their work easier. After several group meetings and some research, the team leader writes a report proposing and justifying the software purchase. The primary readers of the proposal are the supervisor, who will recommend the purchase, and the department head, who can approve or deny funding. Secondary readers are the publishing team members, the managers and authors who submit manuscripts to the group, and the upper administration who will be pleased about plans for a smooth and efficient flow of information.

UNDERSTANDING THE READER'S CONTEXT

The two examples above illustrate another important issue for technical communicators. Audiences understand, interpret, and use technical information in their own organizational, professional, or personal contexts. These contexts may differ widely from the context in which the technical communicator creates the information.

In the case of the repair manual, for example, the technical author will be familiar with the equipment and with repair procedures, but will likely not have performed repairs on a noisy, busy manufacturing floor. This author is certainly not an inexperienced new employee, eager to perform duties well and worried about failure, nor will the author be the one to resolve the crisis and end the expense caused by downtime when the equipment is out of order. However, the readers of this technical author's repair manual depend on it—and on the author's understanding of their situation—to solve a workplace problem by showing them how to find information, learn, and act quickly to make repairs.

Likewise, while the author of the proposal to purchase new software for the publishing team is part of the organization in which the proposal will be read and acted on, that author must remember that the proposal represents more than the short-term interests of the group that submits it. To be successful, such a proposal must demonstrate that the new software will meet the needs of all the primary and secondary readers it addresses: the supervisor, the department head, the publishing team, and the company's upper administration. In addition, whether the proposal is accepted or rejected, the author and all of its readers will need to continue to work cooperatively.

Technical communicators can easily become so involved in their own contexts that they lose sight of their intended audiences' contexts. It is important for communicators to consider the circumstances in which readers will see, understand, and act on the information provided.

MOTIVATING READERS

Workplace authors may assume that because readers need information, they will have to read what the authors produce. This is not automatically the case. The communicator's choices of language, organization, information design, or medium can directly or indirectly influence readers to read, ignore, value, or disregard a piece of communication.

- Is the subject matter too technical?
- Does it *seem* or *look* too difficult?
- Does the communicator reinforce a negative assumption?
- Is the information hard to see or follow?

A close examination of an example illustrates the positive effects of careful communication.

Most people dislike paying taxes, and the Internal Revenue Service (IRS) knows it. To make the problem even worse, tax law is difficult, technical, complicated, and constantly changing. To make it easier for taxpayers to understand and calculate their taxes correctly, the IRS is attempting to help taxpayers understand rules, avoid mistakes, and pay no more than what they owe. The IRS mission states that the agency's goal is to

> provide America's taxpayers top quality service by helping them understand and meet their tax responsibilities and by applying the tax law with fairness and integrity to all.

To meet this goal, the IRS has established a wide range of services, including service centers with materials and free advice, toll-free help lines, and, most important, a series of forms, instructions, and publications that attempt to make a complicated subject, tax law, as easy as possible for readers to understand. One of the first federal agencies to initiate a plain English policy, the IRS considers clear writing a major responsibility to the public.

Why does the IRS spend valuable time revising tax forms? The agency certainly has the authority to penalize those who evade paying taxes, even those who make honest mistakes. However, threats, audits, and even legal action do not encourage cooperation and compliance with the law. It is more costly both for the IRS and for taxpayers if the agency must conduct audits, correct mistakes, and collect back taxes and penalties. Both the IRS and taxpayers benefit if the agency helps the public understand tax rules and follow them correctly. The goal of the IRS clear writing policy and of its services is to save money by collecting the correct amount of tax from those who owe it. The taxpayers' goal in accepting advice and reading tax documents is to save their own money. These goals—that of the agency and that of the audience of taxpayers—overlap. By making written documents as clear as possible and by offering help, the IRS creates cooperation, persuading taxpayers by providing clear information. People will never enjoy paying taxes, but they can ask for free help to be certain that they are paying the correct amount and following the rules. When an author's goal and the audience's goal overlap, communication can create more than a means to provide information. It can also build a deserved trust.

DIVERSITY AND THE WORKPLACE AUDIENCE

Today's workforce is changing rapidly. Jobs traditionally held almost exclusively by men or by women are no longer gender specific. A diverse workforce means increased hiring and promotion of minorities and numbers of employees for whom English may be a second language. Employees are increasingly aware of their own rights to be free from discrimination on the basis of age, gender, ethnicity, disability, or belief.

For a communicator, workplace diversity calls for sensitivity to differences and a deliberate effort to avoid assumptions. For example, language can be a form of discrimination when jokes, discussions, or generalizations exclude, stereotype, or belittle anyone. Individuals or organizations who allow such language, spoken or written, risk legal action. Such language also creates an organizational climate of hostility, division, and distrust.

The increasingly international nature of the workplace is a closely related issue. Because communication technologies make international business easy and fast, members of nearly every organization communicate with cultures distant from their own. To be effective, such communication calls for education and tact. It requires that communicators examine their own assumptions from the viewpoint of those with different customs or workplace practices. While Americans appreciate direct communication that saves time and makes a quick transaction, other cultures may find such an approach hasty or even offensive. Direct communication may be considered bad business and bad manners, which can ruin any working relationship. In international communication, it's important to learn about and understand the values of other cultures. Only this kind of knowledge, sensitivity, and willingness to learn about one another can make communication positive on both sides.

Sensitivity to workplace diversity allows communicators to work cooperatively with people of different backgrounds. Willingness to consider common goals and needs of all employees and to model open, accessible, inclusive communication means a productive workplace and a wide market for all of the goods and services that communication promotes.

GENERAL PRINCIPLES for Defining Workplace Readers

- The audience for a communication project is the people who will read or hear it.
- Communicators must consider the needs, expertise, and contexts of their audiences.
- The audience for a project determines its subject matter, design, language, and medium.
- Effective workplace communicators are sensitive to audiences whose needs, cultures, and languages differ from their own.

CHAPTER SUMMARY

DEFINING WORKPLACE READERS

Awareness of readers and determination to meet reader needs with integrity, sensitivity, and skill are essential for the effective technical communicator. This sensitivity to audience requires a willingness to keep learning. Communicators must avoid assumptions about a reader's knowledge, needs, and workplace context and ignore stereotypes about those different from themselves. In a changing and diverse workplace, technical communicators must be willing to learn about new audiences with varied needs. When the goals of communicators and the goals of readers overlap, communicators earn the trust of their audiences.

ACTIVITIES

INDIVIDUAL AND COLLABORATIVE ACTIVITIES

2.1. Find and review a short IRS form and the instructions that accompany it. In what ways do the writing and design of these materials reflect the IRS mission as stated on page 19. How could the forms and instructions be improved?

2.2. In a small group of two or three, analyze assembly instructions for a toy or operating instructions for a small appliance.
  a. Who is the intended audience for these instructions?
  b. How many times should the audience have to use them?
  c. Are visuals effective for this audience? Is the writing too technical? Are the steps easy to see and to follow on the page?

  Discuss your findings in a memo to the manufacturer, listing the strengths and weaknesses of the instructions for the audience.

2.3. What makes a piece of communication convincing? Consider the following questions as you analyze two different workplace documents or campus brochures for organizations or services.
  a. What assumptions about audience do you see in these materials?
  b. How are these assumptions evident in the way the materials are written?
  c. How easy or difficult are they for readers to see, read, or follow?
  d. How clear and fair is the discussion of the subject?

  List the elements that make each document effective or ineffective.

2.4. In small groups of two or three, list the audiences for a college catalog.
  a. What kinds of information does such a catalog provide?
  b. Who are the catalog's readers, and how do they use different kinds of information?
  c. Does a college catalog provide legal information?
  d. How can it be used and by whom?

  Consider distant audiences and audiences who will use the catalog as a record in later years. How does the catalog meet the needs of its audiences?

READING

2.5. The following casual interview with communications professor Jack Hunter appears in Studs Terkel's book *Working*. In the interview, Hunter discusses the importance of communicators in every field and the responsibility of communicators to use their power in ways that are ethical, respecting their audiences.

Interview from Working

STUDS TERKEL

Jack Hunter

I'm a college professor. As a communications specialist, I train students to become more sensitive and aware of interpersonal communication—symbolic behavior, use of words, as well as nonverbal behavior. I try to ignite symbols in your mind, so we can come to a point of agreement on language. This is an invisible industry. Since the Second World War we've had phenomenal growth. There are seven-thousand-plus strong teachers in this discipline.

I'm high on the work because this is the way life is going to be—persuading people. We're communicating animals. We're persuadeable animals. It's not an unethical thing. It's not the black mustache and the black greasy hair bit. There is an unethnical way—we're cognizant of the ways of demagogic persuasion—but we train students in the ethical way. Business communication is a very important field in our industry. We train people so they can humanize the spirit of both parties, the interviewer and the interviewee. In the first ten minutes of an interview, the interviewer has usually made up his mind. We find out the reasons. Through our kind of research we tell business: what you're doing is productive or counterproductive.

I'm talking about specialists, that we're accustomed to in the movie world. One guy blew up bridges, that's all he could do. Here's a guy who's an oral specialist or writing or print or electronics. We're all part of the family. Nobody has a corner on communication.

Many Ph.D's in the field of speech are now in business as personnel directors. I have good friends who are religious communicators. I had the opportunity to go with a bank in a Southern state as director of information. I would have overseen all the interoffice and intraoffice communication behavior—all the written behavior—to get the whole system smoother. And what happens? Profit. Happiness in job behavior. Getting what's deep down from them, getting their trust . . . .

Communications specialists do have a sense of power. People will argue it's a misuse of power. When a person has so much control over behavior, we're distrustful. We must learn how to become humane at the same time.

QUESTIONS FOR DISCUSSION

a. Hunter calls human beings "persuadeable animals." In what ways does technical communication persuade its readers, even if its primary purpose is to inform them?
b. How do communicators "humanize the spirit of both parties," communicators and audience? In what ways can technical communicators help to make this happen?
c. What sort of communication specialists do you see in the workplace? What is their role within the workplace? What is their responsibility to audiences?
d. What power do communicators have over others? How can that power be used ethically? How can it be abused? Give some examples of each kind of communication—ethical and unethical.

CASE STUDY

2.6. In a group discussion or in a short written assignment, discuss the ways in which the following case study suggests the importance of considering diverse audiences with different needs and different levels of knowledge.

Usual and Customary

As the jumbo jet nears JFK International Airport on its red-eye flight from Zurich, the attendants, who have distributed copies of the United States Customs Declaration (Form 6059B) to all of the passengers with their morning coffee, notice a commotion in the center of the cabin. Two of them, Henri and Buffy, hurry to the cabin to help.

"What's the trouble?" Buffy asks. "How can we help?"

"It's this blessed form," one of the passengers replies. "I'm an American citizen, I travel outside the country regularly, and I still can't understand it! What on earth is a 'monetary instrument'? And who exactly is the responsible family member? I have been trying to explain it to Mr. Patek, here, and I just can't. He's frustrated, and so am I!"

"That's quite true," says Mr. Patek. "I am here on vacation with my family. There are fourteen of us traveling, and we are all related, some rather distantly. But who is the head of the family? Does more than one person fill out a form? We are bringing money for our relatives, but we are not sure if we are all one family. If so, we must declare our money. If not, we do not have to. We are afraid that the American Customs will seize our money! What should we do? The form is so unclear, and it is so worrisome."

"That's not all," adds another passenger. "I am from Singapore. In my country, we welcome people from many cultures, and one of the ways we do this is by making information easy to follow in more than one language. I read English fluently, unlike my business associate, who depends on me for translation. I cannot understand what this form is asking. How very inconsiderate of visitors!"

"I'm from Australia, and we have one of the most delicate ecosystems in the world. We have to be very careful about what people bring into our country. We certainly let them know clearly what's what before they come into the country! Just what do you Yanks consider an 'agricultural product,' anyway?"

"Humph," snaps another traveler. "I'm a citizen, and I spent an hour filling out a long list of everything I bought on my vacation in Europe. Then I put on my glasses and was able to read the fine print. I didn't have to make a list unless I brought in more than $1,400! This form isn't just hard to understand. It's hard to read!"

Buffy and Henri sigh. They have heard all of these complaints before.

APHIS/FWS USE ONLY

WELCOME TO THE UNITED STATES

CUSTOMS USE ONLY

DEPARTMENT OF THE TREASURY
UNITED STATES CUSTOMS SERVICE

FORM APPROVED
OMB NO. 1515-0041

CUSTOMS DECLARATION

19 CFR 122.27, 148.12, 148.13, 148.110, 148.111

Each arriving traveler or responsible family member must provide the following information (only ONE written declaration per family is required):

1. Family Name

2. First (Given) Name | 3. Middle Initial(s) | 4. Birth Date *(day/mo/yr)*

5. Airline/Flight No. or Vessel Name or Vehicle License No. | 6. Number of Family Members Traveling With You

7. (a) Country of Citizenship | 7. (b) Country of Residence

8. (a) U.S. Address *(Street Number/Hotel/Mailing Address in U.S.)*

8. (b) U.S. Address *(City)* | 8. (c) U.S. Address *(State)*

9. Countries visited on this trip prior to U.S. arrival

a. b.

c. d.

10. The purpose of my (our) trip is or was. *(Check one or both boxes, if applicable)* ☐ Business ☐ Personal

11. I am (We are) bringing fruits, plants, meats, food, soil, birds, snails, other live animals, wildlife products, farm products; or, have been on a farm or ranch outside the U.S.: ☐ Yes ☐ No

12. I am (We are) carrying currency or monetary instruments over $10,000 U.S., or foreign equivalent: ☐ Yes ☐ No

13. I have (We have) commercial merchandise, U.S. or foreign. *(Check one box only)* ☐ Yes ☐ No

14. The total value of all goods, including commercial merchandise, I/we have purchased or acquired abroad and am/are bringing to the U.S. is: ▷ $ *(U.S. Dollars)*

(See the instructions on the back of this form under "MERCHANDISE" and use the space provided there to list all the items you must declare. If you have nothing to declare, write "–0–" in the space provided above.)

SIGN BELOW AFTER YOU READ NOTICE ON REVERSE

I have read the notice on the reverse and have made a truthful declaration.

X

Signature | Date *(day/month/year)*

U.S. Customs use only — Do not write below this line — U.S. Customs use only

INSPECTOR'S BADGE NUMBER | STAMP AREA

TIME COMPLETED

Customs Form 6059B (101695)

NOTICE

ALL PERSONS ARE SUBJECT TO FURTHER QUESTIONING AND THEIR PERSONS, BELONGINGS, AND CONVEYANCE ARE SUBJECT TO SEARCH. (19 CFR 162.3 - 162.8)

The unlawful importation of controlled substances (narcotics, chemicals, prescription medicines if not accompanied by a prescription, etc.) regardless of amount is a violation of U.S. law.

AGRICULTURAL AND WILDLIFE PRODUCTS

To prevent the entry of dangerous agricultural pests and prohibited wildlife, the following are restricted: Fruits, vegetables, plants, plant products, soil, meats, meat products, birds, snails, and other live animals or animal products, wildlife and wildlife products. Failure to declare all such items to a Customs/Agricultural/Wildlife officer can result in penalties and the items may be subject to seizure.

CURRENCY AND MONETARY INSTRUMENTS

The transportation of currency or monetary instruments, REGARDLESS OF AMOUNT IS LEGAL; however, if you take out of or bring into the United States more than $10,000 (U.S. or foreign equivalent, or a combination of the two) in coin, currency, traveler's checks or bearer instruments such as money orders, personal or cashier's checks, stocks or bonds, you are required BY LAW to FILE a report on Form 4790 with the U.S. Customs Service. If you have someone else carry the currency or instruments for you, you must also file the report. FAILURE TO FILE THE REQUIRED REPORT OR FAILURE TO REPORT THE TOTAL AMOUNT YOU ARE CARRYING MAY LEAD TO THE SEIZURE OF ALL THE CURRENCY OR INSTRUMENTS, AND MAY SUBJECT YOU TO CIVIL PENALTIES AND/OR CRIMINAL PROSECUTION.

MERCHANDISE

VISITORS (NON-RESIDENTS) must declare in Item 14 the total value of all articles intended for others and all items intended to be sold or left in the U.S. This includes all gifts and commercial items or samples. (EXCEPTION: *Your own* personal effects, such as clothing, personal jewelry and camera equipment, luggage, etc., need not be declared.)

U.S. RESIDENTS must declare in Item 14 the total value of ALL articles, including commercial goods and samples, they acquired abroad (whether new or used; dutiable or not; and whether obtained by purchase, received as a gift, or otherwise). Including those articles purchased in DUTY FREE STORES IN THE U.S. OR ABROAD, which are in their possession at the time of arrival. Articles which you acquired on this trip mailed from abroad, (other than articles acquired in insular possessions and various Caribbean Basin countries) are dutiable upon their arrival in the U.S.

THE AMOUNT OF DUTY TO BE PAID will be determined by a Customs officer. U.S. residents are normally entitled to a duty free exemption of $400 on those items accompanying them; non-residents are normally entitled to an exemption of $100. Duty is normally a flat rate of 10% on the first $1000 above the exemption. If the value of goods declared in Item 14 EXCEEDS $1400 PER PERSON, then list ALL articles below and show price paid *in U.S. dollars* or, for gifts, fair retail value. Please describe all articles by their common names and material. For example: MAN'S WOOL KNIT SWEATER; DIAMOND AND GOLD RING; etc. Also, please have all your receipts ready to present to the Customs officer, if requested. This will help to facilitate the inspection process.

COMMERCIAL MERCHANDISE can be defined as articles for sale, for soliciting orders, or other goods not considered personal effects of the traveler.

IF YOU HAVE ANY QUESTIONS ABOUT WHAT MUST BE REPORTED OR DECLARED ASK A CUSTOMS OFFICER

DESCRIPTION OF ARTICLES *(List must be continued on another Form 6059B)*	VALUE	CUSTOMS USE
TOTAL ▷		

Paperwork Reduction Act Notice: The information collected on this form is needed to carry out the Customs, Agriculture, and Currency laws of the United States. We need it to insure that travelers are complying with these laws and to allow us to figure and collect the right amount of duty and taxes. Your response is mandatory.

The estimated average burden associated with this collection of information is 3 minutes per respondent or recordkeeper depending on individual circumstances. Comments concerning the accuracy of this burden estimate and suggestions for reducing this burden should be directed to U.S. Customs Service, Paperwork Management Branch, Washington DC 20229. *DO NOT send completed forms to this office.*

QUESTIONS FOR DISCUSSION

a. Who are the audiences for U.S. Customs Form 6059B?

b. What are some of the technical terms the form uses? How could these terms be clearer?

c. The instructions appear on one side of the form, and the descriptive information on the other. Is this the most helpful arrangement for the form's users? How would you change the form's design?

d. The print on Form 6059B is very small. To address this problem, a revised form could have larger print with instructions on a removable sheet. How would you revise the form to make it easier to see and to read?
e. Form 6059B says that "the estimated average burden associated with the collection of information is 3 minutes per respondent or recordkeeper depending on individual circumstances." Do you agree? Why or why not?
f. Would translating Form 6059B make it any easier for international users to fill out? Why or why not?
g. Is Form 6059B a persuasive document? What message does it send to first-time visitors to the United States? To non-English speakers?
h. What suggestions or comments would you send to the U.S. Customs Service about this form?

PART II

The Technical Communicator's Tools

Workplace change and geographically and culturally distant audiences call for innovative solutions to communication problems. The technical communicator's tools are the skills and resources used to create and deliver workplace information. The chapters in Part II describe these valuable communication tools. What special tools do workplace communicators need?

Clear language is an essential tool for any communicator. Whether writing for experts or beginners, communicators always seek to use the clearest style possible. Clear, concise writing allows readers to follow ideas with ease. **Chapter 3 The Technical Communicator's Tools: Clear Language** describes the features of such writing and provides practical guidelines for achieving it.

Information design, like written language, uses a series of strategies to make a communication project easy for audiences to use. However, information design depends on the audience's ability to *see,* rather than to *read,* a text. If a text looks difficult—crowded, dense, dark, and hard to read—audiences won't find it useful or inviting. Effective information design means planning the size, shape, and structure of a document to suit the needs of readers. It also means using visual devices, such as headings, page numbering, type size and style, and symbols, so that readers can easily skim, read, or review the text. **Chapter 4 The Technical Communicator's Tools: Information Design** explains the principles of information design.

Graphics fulfill audience needs by visually displaying information. Graphics can illustrate concepts, present details, establish comparisons, and generate interest. Like the effective use of language and information design, the effective use of graphics depends on the communicator's awareness of audience. **Chapter 5 The Technical Communicator's Tools: Visuals** explains the uses of materials that can visually display and clarify information.

Technical communication tools can be used to draft, develop, and deliver many types of workplace communication. **Chapter 6 The Technical Communicator's Tools: Technology** explains different media, including print, e-mail, and recorded media such as film, video, and CD-ROM, exploring the advantages and disadvantages of each. Chapter 6 also discusses the use of some of the more popular communication technologies, including word processors, digital technologies, and graphics software.

Used wisely in combination, these tools can help the informed workplace communicator.

CHAPTER 3

The Technical Communicator's Tools: Clear Language

CHAPTER GOALS

This chapter:

- Defines *effective writing style*
- Explains the significance of clear writing style in the workplace
- Discusses accurate diction
- Provides strategies for conciseness
- Provides strategies for effective sentence patterns

INTRODUCTION

Technical communicators must write with an understanding of their audience and the audience's needs. Such writing requires an awareness of language and usage. Thus, clear written language is a basic tool for technical communicators. This chapter discusses a variety of topics to direct your thinking about clear written language. These topics include style, with an emphasis on workplace style; specific and general words, with a consideration of jargon and gobbledygook; sexist language; tone, tactfulness, and fairness; effective sentence patterns; and strategies for conciseness, simplicity, and clarity.

WHAT IS *STYLE?*

Style is the way you write, the way you craft language to serve the needs of readers. It has to do with

- choosing words (diction)
- using patterns of words and sentences to express ideas clearly
- selecting and arranging words in clear patterns to communicate meaning to a defined audience with specific needs
- creating a tone appropriate for the writing situation

In speaking you have a variety of ways to make meaning clear and to establish rapport with audience: voice, body language, pitch, facial expressions, and simple reiteration or explanation in response to questions. In written language, however, you have only one chance to explain ideas and to establish an appropriate tone.

Workplace Style

Workplace writing style is expository, providing and perhaps explaining information. Workplace writing typically requires utilitarian language with an emphasis on exactness, and it requires that you present information accurately, completely, and clearly. Such a style supports organizational patterns, including informative headings, and provides transitions to guide readers through a document (see also Chapters 4 and 7). It requires that you present information as objectively as possible, with an appropriate tone and without bias. In other words, you present information in an ethical manner by using a clear style suitable to your readers.

The President of the United States issued a memorandum (1 June 1998) declaring that the federal government's writing must be in plain language, making clear what the government is doing, what it requires, and what services it offers. He states that using plain language saves the government and the private sector time, effort, and money. The President's message underscores a basic characteristic of effective style: writing must communicate clearly, accurately, and responsibly. Whether you are writing for the federal government, a Fortune 500 company, or an international or local business, you have the same goal: to use a style appropriate for the writing situation so that readers can understand and use information as they wish or need.

One challenge for workplace communicators is making choices in language to convey information to specific audiences for specific purposes. This challenge is compounded by the fact that readers of workplace writing may represent different cultures, educational backgrounds, hierarchies within an organization, and geographical regions. Readers may speak a different language or require translation of a document.

You must be aware of readers and their reason for reading a document; this awareness leads to careful consideration of language choices and arrangement of material in a document. As a writer, you should develop a rich, diverse vocabulary, giving yourself a wide range of choices and enhancing your ability to find and use the right words for each audience and purpose. Available sources such as general and specialized dictionaries, handbooks, and encyclopedias; textbooks; manuals; and knowledgeable people can help you find accurate and appropriate words for different writing situations.

To select appropriate language that best conveys your intended meaning to an identified audience, choose specific words, avoid inappropriate jargon, avoid sexist language, and convey information in an appropriate tone with tactfulness and without bias.

SPECIFIC AND GENERAL WORDS

Words can be classified as specific or general. A specific word identifies a particular person, object, place, quality, concept, or occurrence; a general word identifies a group or a class. For example, *chief executive officer (CEO)* indicates a person of a particular rank; *officers* indicates a group. For any general word there are numerous specific words; the group identified by *officers* includes specific terms such as *president, chairperson,* and *board of directors.*

Words can also be classified as concrete or abstract. Concrete words such as *computer terminal, mannequin, T square, tractor, desk,* and *thermometer* have physical referents; they name persons or things that can be perceived through the senses. *Scheduling, production, control, courage, recovery,* and *fear* refer to qualities, ideas, conditions, or concepts; such words are abstract.

The more specific the word choice, the more likely the reader is to understand a specific meaning. When general words are replaced with specific, concrete ones, the meaning of a sentence becomes clearer.

General A computer with printer is expensive.

While this sentence makes a judgment and might be used to persuade, it lacks details to support the judgment. The following informative sentence, which includes specific details, supports the judgment.

Specific The HP Pavilion 8250 with a Hewlett Packard Desk Jet 722C costs about $1,500.

General The chart shows manufactured goods representing a large percentage of the goods produced in Manitoba.

Specific The chart shows manufactured goods representing 44 percent of the goods produced in Manitoba.

General The instrument was on the shelf.

Specific The thermometer lay on the topmost shelf of the medicine cabinet.

Language could be used more easily and communication would be much simpler if words meant the same things to all people at all times. However, meaning shifts with the user, the situation, the section of the country, and the context of a word in a sentence and in a document. Most words have multiple meanings.

Also, the meaning of a word can be interpreted in two very different ways: denotation and connotation. The denotative meaning of a word is literal and objective. The connotative meaning refers to attitudes and ideas implied by words because of certain associations. For example, the word *cheap* means inexpensive, but it also suggests poor quality or materials. Since technical writing calls for exactness and objectivity, choose words without misleading or unwanted connotations. If you need direction about connotation in word choice, consult the usage notes in a reliable dictionary.

Using general words is much easier, of course, than using specific ones; language is filled with umbrella terms with broad meanings. Using general words, however, indicates a lack of consideration for your audience. As a writer, be careful not to assume that your audience brings to the reading of a document your understanding and knowledge. Find out as much as you can about an audience. Consider the audience's educational level, experience, and interest in or need for the information. Use what you learn to determine the connotation and difficulty of technical language.

Finding words that express specific meaning requires thought. To write effectively, you must find the specific words at an appropriate level to convey your intended meaning to your audience.

Jargon and Gobbledygook

Jargon is the specialized or technical language of a trade, profession, class, or group. Used in a specialized context for an intended audience that understands the terminology, jargon is appropriate. For example, a computer programmer communicating with a computer engineer could use such terms as input, interface, menu driven, or 48MB SDRAM memory. However, if these specialized words are applied to actions or ideas not associated with computers, jargon is used inappropriately and is often offensive. If such specialized words are used extensively—even in the specialized occupational context—with an audience who does not understand the terminology, jargon is used inappropriately. Consider the following examples of inappropriate jargon.

EXAMPLE 1
Nurse to patient: A myocardial infarction is contraindicated.
(Translation: No heart attack.)

The statement would be clear if made by a doctor to a nurse.

EXAMPLE 2
TV repairer to customer: A shorted bypass capacitor removed the forward bias from the base-emitter junction of the audio transistor. (Translation: A capacitor shorted out and killed the sound.)

The statement would be clear if spoken by one TV repair expert to another.

Mechanic Dan Anderson writing in the mid-February 1999 issue of *Farm Journal* gives an example of jargon used inappropriately for an audience. The farmer who asks "Why is my bean platform bulldozing like a D-8 Caterpillar?" gets the following answer from a farm equipment mechanic: "The relay isn't getting a signal from the potentiometer to energize the header lift solenoid." Obviously, a farm equipment mechanic understands the statement, but many farmers would not understand because of the jargon. Eric Hehs writes in the April 1999 issue of *Code One*, a magazine from Lockheed Martin Tactical Aircraft Systems, "The effects of the strake-generated vortices were further enhanced by automatic variable camber wings." Jargon is appropriate here because of the specialized audience of the magazine.

The problem with inappropriate use of jargon is that the writer does not consider the audience. If you must use jargon when writing for an audience that does not understand its meaning, explain the specialized meaning simply.

Jargon enmeshed in abstract pseudotechnical or pseudoscientific words is called *gobbledygook* (from *gobble*, to sound like a turkey). Consider the following examples.

EXAMPLE 1

The optimum operational capabilities and multiple interrelationships of the facilities are contiguous on the parameters of the support systems.

Revised The facility operates best when support systems are in place.

EXAMPLE 2

Integrated output interface is the basis of the quantification.

Revised Agreement among constituents determines the quantity.

The message in both sentences is unclear because of jargon and pseudotechnical and pseudoscientific words—words that *seem* to be technical or scientific but in fact are not. Further, the words are all abstract; that is, they are general words that refer to ideas, qualities, or conditions. Not one word in either sentence above is concrete; not one word in either sentence creates an image in the reader's mind.

Some writers believe that a flowery, pretentious style with flashy words and phrases impresses a reader. More often, such style confuses and annoys the reader who must work harder to understand meaning.

SEXIST LANGUAGE

Implied sexist language is offensive to many people. Sexist language uses words that arbitrarily assign roles or characteristics to people on the basis of gender. For example, sexist language assumes that all nurses, secretaries, and clerks are female; all carpenters, physicians, and engineers are male. It assumes that the masculine pronoun *he* and the noun *man* (and compounds such as *chairman*, *salesman*, *repairman*) refer to all individuals of unidentified sex. It stereotypes jobs by gender when in fact, men and women are both included, as in *flight attendant* and *police officer*. The careful writer avoids using sexist language. For suggestions on how to avoid sexist language, see pages 669–671 in Appendix 2.

TONE, TACT, AND BIAS

Tone, tact, and nonbias play major roles in effective communication. In any writing situation, using an appropriate tone, tact, and impartiality indicates your concern for the reader.

Tone indicates your attitude as a writer toward the subject and the audience. It is what you say and how you say it. Tone can be classified broadly as informal and formal. Informal language creates a more personal tone; formal language creates a more distant one. Informal language includes slang, colloquialisms, and regional dialects; sentence fragments; contractions and personal pronouns; and other characteristics of casual spoken language. Formal language uses multisyllable words, complex sentence structure, and stylistic techniques such as complex or extended figures of speech.

Most workplace writing falls somewhere between informal and highly formal. It uses standard vocabulary, conventional sentence structure, and few if any contractions. Consider the following examples.

Informal style	Yea! We zapped the competition. Our sales staff sold a humongous amount of merchandise to go over the top of our goal.
Between informal and formal	The Atlanta sales staff reached the goal established by the district managers three weeks ahead of schedule.
Formal	In 1999 the Salyer Company received a prestigious award for pioneering innovative sales and marketing practices, inspiring company sales personnel to surpass existing sales records.

The writing situation—the subject, the purpose, and the audience—determines tone. One writing situation may call for a humorous tone, another for a serious one. Yet another situation may call for a formal tone or a conversational one. Suppose you must write a memo to fellow collaborators about a meeting time. Depending on your attitude toward the reader and the meeting, you might write one of the following sentences.

The meeting of the collaborative team takes place at 2:00 p.m.

The collaborative team meets at 2:00 p.m.

Let's get the team together at 2:00 p.m.

Suppose we chat as a team at 2:00 p.m.

Let's mull over the ideas as a team at 2:00 p.m.

Let's chew the fat with team members at 2:00 p.m.

REMINDER

Be consistent in tone throughout a document.

The first sentence is somewhat formal; the following sentences move progressively from a somewhat formal tone to a very informal one. As a writer, you must be careful that each document has the appropriate tone for the situation.

Ethical communicators are honest about readers' needs. They avoid manipulating the reader through tone. They do not use tone to come across as honest, trustworthy individuals who then use the reader for their own benefit.

Tactfulness means treating readers with respect, being concerned about and considerate of the readers' feelings. It is exercising good manners. An easy phrase to include in a letter, for example, is "thank you." Such a simple phrase generates goodwill and a positive attitude. Tactfulness is avoiding a patronizing, dictatorial, or negative tone.

NEGATIVE EXAMPLES

I told you that plan would not work.

Your department has submitted three consecutive reports filled with errors.

When I told you to prepare the surgical packages first, that is exactly what I meant.

We cannot ship the needed supplies for six months.

These sentences create a strong, negative tone. Such writing reveals the writer as a person who is inconsiderate of readers, and it projects a negative image of the writer and the company or business.

Bias is your preference for some ideas or courses of action over others. In some writing situations, your opinion is expected; in others, you feel you must speak out because of some recognized or perceived danger or ethics violation. Most writing situations, however, call for an impartial presentation of material; in such situations, keep your opinions to yourself.

We all have strong feelings on many subjects, particularly controversial ones. We have preferences for one piece of equipment over another, one brand over another, one method of performing a task over another, and one vendor over another. Keeping our feelings and opinions to ourselves can be difficult. For example, suppose your company is buying new computers. You must select the vendor. Your best friend owns a computer business, but the computers you have decided on are much more expensive at her store. Since you are in a position to sway the decision in selecting a vendor, you could easily withhold information and recommend that the company purchase the computers from your friend. To be ethical and unbiased, however, you must indicate the costs of like computers from several vendors.

To be effective and ethical, communicators must recognize and understand their own biases. Sometimes, however, recognizing bias is difficult. For example, employees often have bias toward a product or service they sell or offer. Such a bias may lead them to push the product or service without revealing both its pros and cons. In writing that calls for impartiality, you want to be fair in the presentation of information, being careful not to mislead readers through your biases and preferences. Quite simply, you want to be ethical.

EFFECTIVE SENTENCE PATTERNS

Some writers believe that flowery, verbose prose makes them seem intelligent. What they forget is that a concise, simple, and clear style helps their intended audience to understand their message and respond appropriately. Following the guidelines below can help you to write concisely, simply, and clearly.

Strategies for Conciseness

In communication, clarity is essential, and conciseness is a key element. Conciseness—saying much in a few words—means omitting nonessential words. You can be concise, however, without being brief: what is short is not necessarily concise. The essential quality in being concise is making every word count. In striving to be concise, avoid an overly simplified style that uses childish language and a condescending tone.

Avoid short, choppy sentences and telegraphic style. Telegraphic style eliminates function words that show relationships between content words (nouns, verbs, adjectives, and adverbs). Function words include

- articles (*a, an, the*)
- prepositions (*by, at, in, to*)
- relative pronouns (*which, that*)
- linking verbs (*feels, is*)

Omission of these function words causes difficulty in reading and results in ambiguous meaning.

Telegraphic style	Lift handle and turn to left. (Turn what to left?)
Revised	Lift the handle and turn it to the left.

It is misleading to suggest that all ideas can be expressed briefly. Some ideas, by their very nature, require more complexity in language and sentences. However, to express complex ideas clearly, keep language and structure as simple as possible.

Omitting Nonessential Words

Nonessential words weaken emphasis in a sentence by thoughtlessly repeating an idea or throwing in "deadwood" to fill up space. Wordiness results when padded phrases (meaningless words), redundancy (unnecessary repetition), and affectation (inflated language that sounds more important than it is) characterize style. Note the improved effectiveness in the following sentences when unnecessary words are omitted.

Wordy	The arrival of the train is at 2:30 p.m. in the afternoon.
Revised	The train arrives at 2:30 p.m.
Explanation	Since "p.m." indicates afternoon, the prepositional phrase "in the afternoon" is redundant. It adds no new information to the sentence. Note the action in the verb *arrives*.

Wordy	At this point in time we need to revise the procedures manual.
Revised	We need to revise the procedures manual now.
Explanation	"At this point in time" is a long way (padded phrases) to say "now."

A common problem with wordiness occurs with the use of expletives such as *it* or *there* to begin a sentence.

Wordy There are several reasons why Carter and Sons will not use temporary employees.

Revised Carter and Sons will not use temporary employees for several reasons.

Explanation "There are" and "why" are empty words. The meaning of the sentence begins with the words "Carter and Sons." The real action is in the verb in the revised sentence.

Wordy It is my belief that renovation of the factory building will increase productivity.

Revised Renovation of the factory building will increase productivity.

Explanation "It is my belief that" is meaningless; it is padding. Obviously if you make the statement, it is your belief. The idea lies in the verb *will increase,* the main verb in the revised sentence.

Wordiness also results when you construct sentences with empty verbs followed by weak nouns (nominalization). Verbs such as *do*, *perform*, *conduct*, and *make* can cause trouble. To restructure sentences with nominalizations, eliminate the empty verb and make the weak noun the verb in the sentence.

Wordy The team members will make an assessment of the quality of their work. (Empty verb: *make*; weak noun: *assessment*)

Revised The team members will assess the quality of their work. (Using the verb form *assess* of the noun *assessment* creates a direct, effective sentence in fewer words.)

Wordy In January 1964, Phil Oestricher conducted a flight test on the YF-16 No.1 over the Mojave Desert. (Empty verb: *conducted*; weak noun: *test*)

REMINDER

Good writers choose strong verbs.

Revised In January 1964, Phil Oestricher flight tested the YF-16 No.1 over the Mojave Desert. (Eliminating the empty verb *conducted* and changing the noun *test* to the verb *tested* creates a direct, effective sentence in fewer words.)

Avoid the temptation to use affectation as a way to impress or intimidate readers. Don't hide behind pretentious words and phrases. The example below shows that affectation results in an artificial tone and gobbledygook.

Wordy The company established a policy of providing copy machines at multiple locations throughout the building to enable employees to copy materials easily and efficiently in order to carry out company business in a timely manner; however, the established policy stipulates that the copy machines not be utilized for personal business.

Revised Do not use the company copy machines for personal business.

REMINDER
Good writers avoid wordiness.

Eliminating unnecessary, meaningless words makes writing more exact, more easily read and understood, and more economical. Careful revision weeds out the clutter of deadwood (padded phrases), needless repetition (redundancy), and inflated language that sounds more important than it is (affectation).

Strategies for Simplicity

Many sentences contain complete ideas yet are ineffective because they lack conciseness. You can focus on key ideas by combining sentence parts or entire sentences.

Reducing several words to one word	the supervisor *of the shift* the *shift* supervisor
Reducing a clause to a phrase or to a compound word	a house *that is shaped like a cube* a house *shaped like a cube* a *cube-shaped* house
Reducing a compound sentence	Mendel planted peas for experimental purposes, and from the peas he began to work out the universal laws of heredity.
to a complex sentence	As Mendel experimented with peas, he began to work out the laws of heredity.
to a simple sentence	Mendel, experimenting with peas, began to work out the laws of heredity.
Combining two short sentences	Many headaches are caused by emotional tension. Stress also causes a number of headaches.
into one sentence	Many headaches are caused by emotional tension and stress.

You can use coordination and subordination to combine ideas and to show the relationship between ideas.

RELATIONSHIP OF IDEAS UNCLEAR

1. The company did not hire her.
2. She was not qualified.

Revised The company did not hire her because she was not qualified.

Explanation Recasting the two sentences makes the relationship of ideas clearer with the addition of the subordinate conjunction *because*. The second sentence is changed to an adverb clause, *because she was not qualified*, telling why the company did not hire her.

RELATIONSHIP OF IDEAS UNCLEAR

1. Latisha completed the research.
2. She was not able to present her findings at the convention.

Revised Latisha completed her research, but she was not able to present her findings at the convention.

Explanation The addition of the coordinate conjunction *but* indicates that the idea following may show contrast or provide an alternative to the idea preceding it.

RELATIONSHIP OF IDEAS UNCLEAR

1. Magnetic lines of force can pass through any material.
2. They pass more readily through magnetic materials.
3. Some magnetic materials are iron, cobalt, and nickel.

Revised Magnetic lines of force can pass through any material, but they pass more readily through magnetic materials such as iron, cobalt, and nickel.

or

Although magnetic lines of force can pass through any material, they pass more readily through magnetic materials: iron, cobalt, and nickel.

Explanation The addition of the coordinating conjunction *but* indicates the alternative relationship between the two main ideas; sentence 3 has been reduced to *such as iron, cobalt, and nickel*. In the second revision, what was the main idea is now subordinated.

Using coordination and subordination helps eliminate short, choppy sentences.

SHORT, CHOPPY SENTENCES

1. The mission's most important decision came.
2. It was early on December 24.
3. Apollo was approaching the moon.
4. Should the spacecraft simply circle the moon and head back toward Earth?
5. Should it fire the Service Propulsion System engine and place the craft in orbit?

Revision

1. As Apollo was approaching the moon early on December 24, the mission's most important decision came.
2. Should the spacecraft simply circle the moon and head back toward Earth, or fire the Service Propulsion System engine and place the craft in orbit?

Use coordination and subordination to emphasize details: important details appear in independent clauses and less important details appear in dependent clauses and phrases. Almost any group of ideas can be combined in several ways. Choose the arrangement of ideas that best fits in with preceding and following sentences. More importantly, arrange ideas to make key ideas stand out.

SHORT, CHOPPY SENTENCES

1. Carmen Diaz is the employee of the year.
2. She has been with the company only one year.
3. She has not had special training for the job she performs.
4. She is highly regarded by her colleagues.

Possible revisions

Although Carmen Diaz, the employee of the year, has been with the company only one year and has had no special training for the job she performs, she is highly regarded by her colleagues.

Carmen Diaz is the employee of the year although she has been with the company only one year and has had no special training for the job she performs; she is highly regarded by her colleagues.

Although Carmen Diaz, highly regarded by her colleagues, is the employee of the year, she has been with the company only one year and has had no special training for the job she performs.

Explanation Each sentence contains the same information; through coordination and subordination, however, different information is emphasized.

Choices for Clarity

Voice in grammar indicates the relationship between the subject of a sentence and the action of the verb. Transitive verbs have two voices: active and passive. In a sentence with an active voice verb, the subject acts; in a sentence with a passive voice verb, the subject is acted upon. The two following examples say the same thing, but the emphasis changes in the second sentence. The emphasis in the first sentence is on the subject, *Amy Randazzo;* the emphasis in the second sentence is on the object, *the computer lab renovation proposal.*

EXAMPLES

Amy Randazzo wrote the computer lab renovation proposal. (active voice; the subject of the sentence, *Amy Randazzo,* has carried out an action: *wrote*)

The computer lab renovation proposal was written by Amy Randazzo (passive voice: the subject of the sentence, *computer lab renovation proposal,* has been acted upon: *was written*)

Note: A passive voice verb always has at least two words, a form of the verb *to be* and the past participle of the main verb. Keep in mind that passive voice sentences can be indirect, wordy, and often difficult to understand.

For your audience and purpose, choose the voice of the verb that gives the desired emphasis.

- Generally use active voice verbs. Active voice verbs are effective because the reader knows immediately the subject of the discussion; you mention first *who* or *what* is doing something.

  The machinist *values* the rule depth gauge.

  Roentgen *won* the Nobel prize for his discovery of x rays.

- Use passive voice verbs when *who* or *what* is less significant than the action or the result and when *who* or *what* is unknown (preferably unnamed), relatively insignificant, or cannot be identified.

  The lathe *is broken* again.

  The test on the three pesticides *was completed* yesterday.

  The transplant operation *was performed* by an outstanding heart surgeon.

  Light *is provided* by windows on the north wall.

GENERAL PRINCIPLES for Clear Language

- An effective style serves the readers' needs. It focuses on appropriate diction, word and sentence patterns, and tone to communicate meaning to a defined audience with specific needs.
- Workplace style typically requires utilitarian language that emphasizes exactness, and it presents information accurately, completely, and clearly.
- Workplace communicators must make language choices to convey information to specific audiences for specific purposes.
- Careful communicators avoid explicit or implied sexist language.
- Communicators who are concerned for the reader use an acceptable tone, tactfulness, and fairness. They evaluate a writing situation—the subject, the purpose, and the audience—to determine what is appropriate.
- A concise, simple, and clear style helps an audience to understand the communication and respond appropriately.

CHAPTER SUMMARY

CLEAR LANGUAGE

Every decision a writer makes affects style—choice of words, patterns of phrasing, sentences. For an appropriate style, use specific language, avoid inappropriate jargon, avoid sexist language, and convey information using an appropriate tone with tactfulness and without bias. To write effective sentences, choose a concise, simple, clear pattern. Prune excess verbiage to achieve conciseness, omit nonessential words, and combine sentence elements for simplicity. Choose the active voice unless *who* or *what* is less significant than the action or the result is unknown (preferably unnamed) or relatively insignificant. Above all, style means making thoughtful choices about language that help readers and respect their needs.

ACTIVITIES

INDIVIDUAL AND COLLABORATIVE ACTIVITIES

3.1. In a technical journal related to your major or some other area of interest, select a brief article to analyze for style. Using the questions below as a guide, write a report.
   a. Describe the likely readers. How has the writer taken these readers into account?
   b. Describe the tone of the article.
   c. Has the writer carefully selected specific words? Give examples to support good or poor word choices.
   d. Is jargon used? If so, give examples from the article. Is the jargon used appropriately or inappropriately?
   e. Does the writer use sexist language? If so, give examples of sexist language used in the article.
   f. How would you describe the tactfulness of the writing? How would you describe any bias in the article?
   g. Are sentences clear and concise? Find examples of effective and ineffective sentences.

3.2. Working in teams of three or four, find a brief article that illustrates effective writing style. Identify specific examples of effective style by analyzing language (word choice, jargon and gobbledygook, sexist language); tone, tact, and bias; concise, effective sentences; and strong verbs. Distribute copies of the article to the class. Discuss your findings with the class; have each team member lead the discussion on an aspect of style.

3.3. Read the following selection written by a computer technician to convince an office manager that computers need to be replaced. Then, in small groups, discuss the following questions:
   a. How would you describe the tone?
   b. What does the tone say about the author?
   c. Has the author considered the readers?

> It has come to my attention that it is time for us to consider replacing some of the computers that we have been using for the past several years. Of course I realize that some of the computers are still quite useful. I wonder if it would be a good idea to move these old computers into the secretarial area. I think most of the existing software can be used with these computers and therefore no new software would need to be purchased. The secretaries should be happy about this idea since they will not have to learn new software, don't you think? Now, first and foremost, how many new comput-

ers do you think we should buy? Second of all, what, in your opinion, would be the best brand of computers to purchase? Do you think the new computers should have added memory and faster response? Third, and finally, how many additional projects do you think we can expect to complete on an annual basis with the additional computers?

As a final activity, revise the paragraph to create a concise, reader-centered version.

3.4. Rewrite the following sentences to replace general words with specific words. If necessary, review the section on "Specific and General Words."

a. The computer has a great deal of power.
b. Medical personnel dressed the wound.
c. The office worker keyboarded the letter.
d. The store displayed its goods.
e. The technician repaired the device.
f. The kitchen workers prepared the banquet food.

3.5. Rewrite the following sentences to eliminate sexist language. If necessary, review the section "Sexist Language."

a. A lawyer has little time to read outside of his field.
b. The male nurse realized that he had won the patient's confidence.
c. The CEO called a meeting of his quality control men.
d. Each team member will contribute more positively if he has a voice in the decisions.
e. The average worker is worried about his salary.
f. Our department has completed a man-sized job.

3.6. Read the following sentences. If necessary, review the section "Effective Sentence Patterns." Identify the type of wordiness: padded phrases, redundancy, and affectation. Rewrite the sentences for conciseness.

a. It is my opinion that all technicians should update their certification annually.
b. The training session should end by 3:30 p.m. in the afternoon.
c. The team chairperson will set a meeting at an early date.
d. To complete the manual, join together the three sections the teams have completed: directions for use, the descriptions of the parts, and the supportive graphics.
e. Pursuant to our agreement, highly skilled painters will perform renovations to the interior and exterior of selected facilities.
f. Pharmaceutical firms spend large sums of money to conduct experiments on new drugs.

3.7. Rewrite the following sentences to change the passive voice verbs to active voice verbs. If necessary, review the section "Choices for Clarity."

a. The programming language is written easily by the computer programmer.
b. Instructions on how to install the garage door opener are explained in a pamphlet.

c. Volunteer work with Habitat for Humanity was done by the employees at Citizens Bank.
d. Dozens of police officers were reassigned by the Chief of Police.
e. The meal celebrating Mardi Gras was prepared by the French chef.
f. Three major health benefits are offered employees by the hospital.

3.8. Each of the following groups of sentences contains related information stated in simple sentences. Combine this information into one sentence or, if you think necessary, two or more sentences. How do different combinations affect emphasis? In groups of three or four, compare versions and discuss the advantages and disadvantages of each combination. If necessary, review the section "Strategies for Simplicity."

1. a. All drawings are projections.
   b. Two main types of projections are perspective and parallel projection.

2. a. The operating tasks each company must perform are affected by many factors.
   b. One factor is changes in market.
   c. Another factor is new technology.
   d. A final factor is government regulations.

3. a. Flat surfaces are measured with common tools.
   b. The steel rule and the depth gauge are common tools.
   c. Round work is measured by feel with nonprecision tools.
   d. The spring caliper and the firm-joint caliper are nonprecision tools.

4. a. Ohm's law states that the current in an electrical circuit varies directly as the voltage and inversely as the resistance.
   b. This law was developed by George Simon Ohm.
   c. He was born in Germany.
   d. He was born in 1787.
   e. He was a physicist.

5. a. Of all the microbiologists, none was so accurate as Leeuwenhoek.
   b. None was so completely honest.
   c. None had such common sense.
   d. He died in 1723.
   e. He was 91.

6. a. Every time a manager delegates work to a subordinate, three actions are either expressed or implied.
   b. She assigns duties.
   c. She grants authority.
   d. She creates an obligation.

7. a. One essential part of an air damper box is a vane.
   b. The vane is made of a thin sheet of lightweight alloy.
   c. It is stiffened by ribs stamped into it.
   d. It is stiffened by the bending of its edges.
   e. Its edges conform to the surfaces of the damping chamber.

8. a. Water from newly constructed wells will normally have a very high bacterial count.
   b. This count will not settle down for a matter of weeks or even months unless the supply is treated.
   c. This treatment reduces the bacterial population that found its way into the water during construction operations.

9. a. Landscape architects employ texture as a valuable tool.
   b. Repetition of a dominant plant texture unifies a plan.
   c. Contrast of texture at corner and focal points gives emphasis.

10. a. Any material employed to grow bacteria is called a culture medium.
    b. This medium, to be satisfactory, must have the proper moisture content.
    c. It must contain readily available food materials.
    d. It must have the correct acid-base balance.

READING

3.9. In the following article, Marvin H. Swift, a professor of communication at the General Motors Institute, analyzes the way in which a manager reworks and rethinks a memo of minor importance to point up a constant management challenge of major importance—creating a clear, accurate, well-focused message expressed in an appropriate tone.

Clear Writing Means Clear Thinking Means . . .

MARVIN H. SWIFT

If you are a manager, you constantly face the problem of putting words on paper. If you are like most managers, this is not the sort of problem you enjoy. It is hard to do, and time-consuming; and the task is doubly difficult when, as is usually the case, your words must be designed to change the behavior of others in the organization.

But the chore is there and must be done. How? Let's take a specific case.

Let's suppose that everyone at *X* Corporation, from the janitor on up to the chairman of the board, is using the office copiers for personal matters; income tax forms, church programs, children's term papers, and God knows what else are being duplicated by the gross. This minor piracy costs the company a pretty penny, both directly and in employee time, and the general manager—let's call him Sam Edwards—decides the time has come to lower the boom.

Sam lets fly by dictating the following memo to his secretary:

[Draft 1]

To: All Employees
From: Samuel Edwards, General Manager
Subject: Abuse of Copiers

It has recently been brought to my attention that many of the people who are employed by this company have taken advantage of their positions by availing themselves of the copiers. More specifically, these machines are being used for other than company business.

Obviously, such practice is contrary to company policy and must cease and desist immediately. I wish to therefore to inform all concerned—those who have abused policy or will be abusing it—that their behavior cannot and will not be tolerated. Accordingly, anyone in the future who is unable to control himself will have his employment terminated.

If there are any questions about company policy, please feel free to contact this office.

Now the memo is on his desk for his signature. He looks it over; and the more he looks the worse it reads. In fact, it's lousy. So he revises it three times, until it finally is in the form that follows:

[Final Version]

To: All Employees
From: Samuel Edwards, General Manager
Subject: Use of Copiers

We are revamping our policy on the use of copiers for personal matters. In the past we have not encouraged personnel to use them for such purposes because of the costs involved. But we also recognize, perhaps belatedly, that we can solve the problem if each of us pays for what he takes.

We are therefore putting these copiers on a pay-as-you-go basis. The details are simple enough . . . .

Samuel Edwards

This time Sam thinks the memo looks good, and it *is* good. Not only is the writing much improved, but the problem should now be solved. He therefore signs the memo, turns it over to his secretary for distribution, and goes back to other things.

FROM VERBIAGE TO INTENT

I can only speculate on what occurs in a writer's mind as he moves from a poor draft to a good revision, but it is clear that Sam went through several specific steps, mentally as well as physically, before he had created his end product.

- He eliminated wordiness.
- He modulated the tone of the memo.
- He revised the policy it stated.

Let's retrace his thinking through each of these processes.

Eliminating Wordiness

Sam's basic message is that employees are not to use the copiers for their own affairs at company expense. As he looks over his first draft, however, it seems so long that this simple message has become diffused. With the idea of trimming the memo down, he takes another look at his first paragraph.

> It has recently been brought to my attention that many of the people who are employed by this company have taken advantage of their positions by availing themselves of the copiers. More specifically, these machines are being used for other than company business.

He edits it like this:

> *Item:* "recently"
> *Comment to himself:* Of course; else why write about the problem? So delete the word.
> *Item:* "It has been brought to my attention"
> *Comment:* Naturally. Delete it.
> *Item:* "the people who are employed by this company"
> *Comment:* Assumed. Why not just "employees"?
> *Item:* "by availing themselves" and "for other than company business"
> *Comment:* Since the second sentence repeats the first, why not coalesce?

And he comes up with this:

> Employees have been using the copiers for personal matters.

He proceeds to the second paragraph. More confident of himself, he moves in broader swoops, so that the deletion process looks like this:

> Obviously, such practice is contrary to company policy ~~and must cease and desist immediately. I wish therefore to inform all concernd—those who have abused policy or will be abusing it—that their behavior cannot and will not be tolerated. Accordingly, anyone in the future who is unable to control himself will have his employment terminated.~~

The final paragraph, apart from "company policy" and "feel free," looks all right, so the total memo now reads as follows:

[Draft 2]

To: All Employees
From: Samuel Edwards, General Manager
Subject: Abuse of Copiers

Employees have been using the copiers for personal matters. Obviously, such practice is contrary to company policy and will result in dismissal.

If there are any questions, please contact this office.

Sam now examines his efforts by putting these questions to himself:

Question: Is the memo free of deadwood?
Answer: Very much so. In fact, it's good, tight prose.
Question: Is the policy stated?
Answer: Yes—sharp and clear.
Question: Will the memo achieve its intended purpose?
Answer: Yes. But it sounds foolish.
Question: Why?
Answer: The wording is too harsh; I'm not going to fire anybody over this.
Question: How should I tone the thing down?

To answer this question, Sam takes another look at the memo.

Correcting the Tone

What strikes his eye as he looks it over? Perhaps these three words:

- Abuse . . .
- Obviously . . .
- . . . dismissal . . .

The first one is easy enough to correct: he substitutes "use" for "abuse." But "obviously" poses a problem and calls for reflection. If the policy is obvious, why are the copiers being used? Is it that people are outrightly dishonest? Probably not. But that implies the policy isn't obvious; and whose fault is this? Who neglected to clarify policy? And why "dismissal" for something never publicized?

These questions impel him to revise the memo once again:

[Draft 3]

To: All Employees
From: Samuel Edwards, General Manager
Subject: Use of Copiers

Copiers are not to be used for personal matters. If there are any questions, please contact this office.

Revising the Policy Itself

The memo now seems courteous enough—at least it is not discourteous—but it is just a blank, perhaps overly simple, statement of policy. Has he really thought through the policy itself?

Reflecting on this, Sam realizes that some people will continue to use the copiers for personal business anyhow. If he seriously intends to enforce the basic policy (first sentence), he will have to police the equipment, and that raises the question of costs all over again.

Also, the memo states that he will maintain an open-door policy (second sentence)—and surely there will be some, probably a good many, who will stroll in and offer to pay for what they use. His secretary has enough to do without keeping track of affairs of that kind.

Finally, the first and second sentences are at odds with each other. The first says that personal copying is out, and the second implies that it can be arranged.

The facts of organizational life thus force Sam to clarify in his own mind exactly what his position on the use of copiers is going to be. As he sees the problem now, what he really wants to do is put the copiers on a pay-as-you-go basis. After making that decision, he begins anew:

[Final Draft]

To: All Employees
From: Samuel Edwards, General Manager
Subject: Use of copiers

We are revamping our policy on the use of copiers for personal use . . . .

This is the draft that goes into distribution and now allows him to turn his attention to other problems.

THE CHICKEN OR THE EGG?

What are we to make of all this? It seems a rather lengthy and tedious report of what, after all, is a routine writing task created by a problem of minor importance. In making this kind of analysis, have I simply belabored the obvious?

To answer this question, let's drop back to the original draft. If you read it over, you will see that Sam began with this kind of thinking:

- The employees are taking advantage of the company.
- I'm a nice guy, but now I'm going to play Dutch uncle. . . . I'll write them a memo that tells them to shape up or ship out.

In his final version, however, his thinking is quite different:

- Actually, the employees are pretty mature, responsible people. They're capable of understanding a problem.
- Company policy itself has never been crystallized. In fact, this is the first memo on the subject.
- I don't want to overdo this thing—any employee can make an error in judgment.
- . . . I'll set a reasonable policy and write a memo that explains how it ought to operate.

Sam obviously gained a lot of ground between the first draft and the final version, and this implies two things. First, if a manager is to write effectively, he needs to isolate and define, as fully as possible, all the critical variables in the writing process and scrutinize what he writes for its clarity, simplicity, tone, and the rest. Second, after he has clarified his thoughts on paper, he may find that what he has written is not what has to be said. In this sense, writing is feedback and a way for the manager to discover himself. What are his real attitudes toward the amorphous, undifferentiated gray mass of employees "out there"? Writing is a way of finding out. By objectifying his thoughts in the medium of language, he gets a chance to see what is going on in his mind.

In other words, *if the manager writes well, he will think well.* Equally, the more clearly he has thought out his message before he starts to dictate, the more likely he is to get it right on paper the first time round. In other words, *if he thinks well, he will write well.*

Hence we have a chicken-or-the-egg situation: writing and thinking go hand in hand; and when one is good, the other is likely to be good.

Revision Sharpens Thinking

More particularly, rewriting is the key to improved thinking. It demands a real open-mindedness and objectivity. It demands a willingness to cull verbiage so that ideas stand out clearly. And it demands a willingness to meet logical contradictions head on and trace them to the premises that have created them. In short, it forces a writer to get up his courage and expose his thinking process to his own intelligence.

Obviously, revising is hard work. It demands that you put yourself through the wringer, intellectually and emotionally, to squeeze out the best you can offer. Is it worth the effort? Yes, it is—if you believe you have a responsibility to think and communicate effectively.

Source: Reprinted by permission from *Harvard Business Review* (January–February 1973). Boxes and bracketed headings have been added to the drafts of the memo.

QUESTIONS FOR DISCUSSION

a. In groups of three or four, compare the qualities of style discussed in this chapter with the qualities of style presented in the article.
b. Answer the following questions about the memo.
    1. Describe the tone of Draft 1. Describe the tone of the final draft. What happens to the tone as Sam rewrites the drafts?
    2. Identify sentences with excess verbiage. What techniques does the writer use to eliminate excess verbiage as he writes the drafts of the memo?
    3. What level of word choice and phrasing is used? Is the level formal, informal, or in between?
    4. Is the writer tactful? Is the writer unbiased?
    5. What voice is used in each draft of the memo?
c. In groups of three or four, analyze what happens in the writer's thinking as he moves from Draft 1 to Draft 2 to Draft 3 to the final draft of the memo.
d. In groups of three or four, identify characteristics of writing that are implied by Swift's analysis of Sam's process of writing the drafts of the memo.

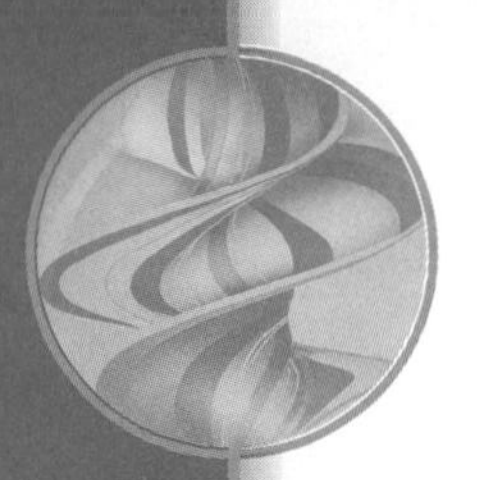

CHAPTER 4

The Technical Communicator's Tools: Information Design

CHAPTER GOALS

This chapter:

- Defines information design as an element of communication
- Shows how information design can help readers use documents in different ways
- Outlines the elements of information design that can help readers understand and navigate documents
- Explains the elements of information design for pages and screens
- Describes the features of type and white space
- Provides strategies for using effective information design

INTRODUCTION

What do readers do when a document is unappealing, the print is too small, the text is dense, and the language is confusing? They generally resist reading it. Trying to read poorly planned, poorly designed information is frustrating.

Good information design tries to make every document meet the needs of its readers by providing and integrating elements that help readers to see, understand, and learn what they need. Since different kinds of information serve different needs, information design must reflect how and why readers use the information. Also, since readers see a document before they read it, information design provides principles to make the visual elements of pages—white space, type, and visuals—as appealing as they are clear and usable.

This chapter discusses the elements of information design that can help you make strong design decisions about your communication projects.

WHAT IS INFORMATION DESIGN?

Information design is a strategy of arrangement and emphasis that helps communicators tailor documents to readers' needs. Based on principles that allow for ease of use, or usability, information design can help communicators plan so that readers can readily understand and follow a document's parts and structure. Information design also concerns the way readers see a document—from individual pages to the document as a whole. This kind of information design uses visual elements to establish hierarchies of information, to highlight ideas, and to direct the reader's attention. Information design elements include text and white space, column widths, visuals (including kind, size, and placement), spacing, color, geometric shapes, and sizes and styles of typefaces (for text, headings, and other features). By using these elements to support a document's purpose, you can help your readers to use your document more easily.

HOW PEOPLE READ DOCUMENTS

In the workplace, readers typically read two kinds of information: information that they need to *learn* (such as a trade journal article) or information that they need in order to *act* (such as a procedure). Readers read to learn in some of the ways students read to study: they skim, read steadily, highlight, and review. Readers who read to act follow a procedure step by step, whether they perform each step as they read it, skip a step or two, or read the entire procedure before beginning work. Whether they read to learn or read to act, readers may need to find material selectively. For example, a reader may want to locate and reread a specific section of a trade journal article or find and review a key step in a procedure.

Documents designed for learning differ from documents designed for action. However, both types of documents must allow readers to locate specific parts, or *navigate*, within the document as a whole. No matter how valuable the subject matter may be, if a reader can't understand the purpose and structure of the document as a whole or locate and read its parts, the document will be ineffective. Many readers resent struggling to find what they need. Effective information design makes good information better by making it more accessible to readers.

REMINDER

Consider how your audience will use your document as you plan the document's information design.

DESIGNING DOCUMENTS FOR YOUR READERS' NEEDS

When readers use a document, they need to see how the document's structure reflects its purpose and meets their own needs. In some ways, information design for an entire document is like the organization plan discussed on pages 139–140 of Chapter 7 Planning and Drafting, and appearing on the back endpaper. Information design makes the organization of the document and its parts as clear as possible for a reader to follow. Think of this kind of design as a series of strategies for helping your reader to understand your purpose and organization and also to locate parts of the document.

Helping Readers Understand

To help your readers understand a document, give them the big picture: an overview of the document's purpose, scope, and sequence. This understanding will allow your readers to see how the document's parts fit its purpose and main idea.

Previewing

Before they begin to read a document, readers need some idea of its overall content. To help readers preview the document as a whole, you can provide an overview in an introduction that discusses the document's purpose, scope, and subject and briefly explains the sequence of discussion. You may also wish to provide a preliminary summary of the report's subject matter to tie together main ideas. While these two kinds of introductory information may seem to overlap, both are new and necessary to the reader who is using them to understand the material ahead. (For more information on summaries, see Chapter 12 Summaries.)

Summing Up

Just as previews prepare a reader for what is to come, closing comments allow the reader to pull together what has been learned. This is the point at which a closing section—a summary or a conclusion—reminds the reader how newly learned specific information fits a document's main ideas. Such a closing section is equally useful for the reader who has read the entire document and the reader who wants to understand the context of only one section. In addition, reminders, review questions, and summaries that appear in margins (called *sidebars*) help readers understand how the specifics they are reading apply to larger issues.

Helping Readers Learn

People learn in pieces, or chunks, step by step, tying one piece of knowledge to another. Since reading is a way of learning, it's important to divide written information into visually separate chunks of information that are small enough for a reader to learn one chunk at a time. It's also important to label each chunk so that readers can easily recognize the subject matter of each chunk and its relationship to other chunks. Clear labeling allows readers to read, skim, and locate chunks of information selectively.

Chunking

It's hard for a writer to imagine that a reader could have difficulty in understanding a carefully planned sequence of ideas. However, the reader, unlike the writer, is new to the information and may or may not need all of it. Chunking written information helps the reader to understand and use the sequence of ideas. For example, typical chunks of a book might include chapters, sections, and paragraphs, while typical chunks of a procedure might include groups of related steps, with each step numbered in sequence. Looking at a page similar to the one below with no paragraph indentations, you can easily see how important visual chunking is in making a page appealing for readers.

Page without Paragraphing

How large or small should chunks of written information be? The answer to this question lies in the purpose, length, and complexity of a document. For example, a one-page procedure for changing a fuse could have chunks as small as a single sentence. After all, each step will be numbered, and the procedure as a whole will have few steps. However, a longer document, such as a book, could have more chunks at more levels: chapters, sections, and subsections, all related to one another. The chunks of the book will be longer, each one providing more fully developed ideas.

Long lists of short chunks are hard for readers to understand and follow. For example, in a procedure consisting of 19 short steps, your reader will have difficulty understanding how the steps fit together as a whole. To help, chunk related steps into groups of between five and seven steps each. This way, your reader will be able to see, follow, and understand how the steps and the groups fit together. The example below illustrates this principle.

Labeling

Dividing text into groups of visually defined chunks is a good beginning, but the relationship between the chunks is not clear until each chunk is *labeled*, or given a clear descriptive heading. A heading, or label, identifies the topic or subtopic of a section or chunk of information. Each heading reflects content, that is, what the information is about. The heading helps the writer to stay on the subject; the heading plus the white space marking the block of information improves the appearance of the page and the readability of the text. By giving the reader a visual impression of topics and their relation to one another, headings reflect the organization of the material. Headings also remind the reader of movement from one point to another and provide visual interest to what otherwise would be a solid page of unbroken print. The examples below illustrate the importance of labeling chunks of text with clear headings.

Text without Headings

Text with Headings

XXXXXX

XXXXXX

XXXXXX

Labels are an important way to identify and describe visual information such as drawings, charts, graphs, and tables. Identify each visual with a short descriptive heading, or *caption*. (For more information on visuals and their captions, see pages 88–89 in Chapter 5 Visuals.)

Levels of Heading

Headings show organization. The form and placement of each heading indicate its level. Uppercase letters, underlining or italics, type size, and position on the page help to differentiate levels. Headings that identify higher levels of material should look more important than those that identify lower levels.

For every written communication project, you need to develop a system of headings to reflect the project's major points and supporting points. (If your document has an outline or table of contents, the headings should correspond exactly to the headings in the body of the text.)

Following are some suggested systems of headings.

TWO LEVELS OF HEADINGS (the underlining may be omitted)

MAJOR HEADING (this heading may be centered)

Division of Major Heading

or

MAJOR HEADING

Division of Major Heading

THREE LEVELS OF HEADINGS (the underlining may be omitted)

MAJOR HEADING (this heading may be centered)

Division of Major Heading

Subdivision of division of major heading. The paragraph begins here.

FOUR LEVELS OF HEADINGS

MAJOR HEADING (this heading may be centered)

DIVISION OF MAJOR HEADING

Subdivision of Division of Major Heading

Sub-subdivision of division of major heading. The paragraph begins here.

Notice how each of the heading systems above makes the level of heading easy to see at a glance. An example of a page of text that visually differentiates headings to indicate level appears on page 60.

Text with Two Levels of Headings

Helping Readers Navigate

When readers skim, review, or look for a quick answer, they read selectively, finding and reading the information that meets their specific needs. Information design can make it easy for readers to move through, or *navigate,* text to the information they select.

In information design, *navigation* means finding a sure, quick path to an information destination. The elements that allow for easy navigation are tables of contents, indexes, and page numbering. All of these define a subject, help a reader locate it, or both.

The table of contents is a list of all of the headings that appear in a document, providing the page number on which each heading appears. The table of contents also lists all other parts of the document, including the bibliography, glossaries, and any materials that may appear in an appendix. Since a table of contents reflects the levels of headings, it is a helpful guide to the relationship between the parts of a document, especially since the table of contents appears before the first page of written text.

Like a table of contents, an index is a guide to the parts of a document. Unlike a table of contents, an index does not explain the sequence and levels of a document's headings. Instead, an index, generally the last part of a book or a long document, provides a complete list of all major topics as single words or short phrases. The topics are arranged in alphabetical order and cross-referenced. A reader can use an index to identify and locate key words and ideas. The purpose of the index is to allow the reader to identify topics and their locations, moving freely from one topic to another.

REMINDER

Consider the document's length and purpose when determining which elements to use to help readers understand a document.

If the table of contents and the index are maps to information, page numbers are the signposts that a reader must follow to arrive at a destination. By numbering each page clearly and placing the number in a consistent location on each page, information designers create a system that makes it easy for readers to move and locate selected information within a document.

DESIGN STRATEGIES FOR PAGES

So far this chapter has discussed ways to *tell* readers how to use documents. Information design makes all of these features far more useful (and far more interesting) through visual elements that *show* readers the sequence, levels, and parts of documents. These visual elements—chunks of written information, white space, emphasis markers, and such visual information as drawings, graphics, and tables—allow readers to easily read or navigate through information. An easy-to-use design also makes a favorable impression on readers.

Page design has always been important in guiding readers to the parts and highlights of written text. Before computer technologies such as word-processing and graphics software made page design easy and affordable, page design was the business of designers and printers. Authors wrote text; other people designed the text and set the type. Computer programs and document design principles have changed the way we prepare and view written information. The simplest word processing programs offer choices of many styles and sizes of type, page sizes and margins, symbols, and lines. Easy-to-use graphics programs and images from royalty-free clip art packages allow writers to create visuals that complement page design. (See Chapter 5 Visuals.)

Elements of Page Design

When people read, whether to learn or to act, they first see what is on the page. To allow the eye to move easily from one chunk of information to another, information designers balance white space and information elements of print and visuals.

White Space

Sufficient white space creates an uncluttered, inviting page, one that is easy to see and read. Indenting for paragraphs, double spacing, and allowing ample margins ($1\frac{1}{2}$ inches minimum on all four edges) are the most typical ways of allowing for plenty of white space. Additional white space providers include headings (with triple spacing before a major heading and double spacing after the heading), vertical listing (placing the items up and down rather than across the page), and columns (setting up the page with several columns of short lines, as in a newspaper, rather than a page of long lines).

Type

Type is the physical representation of letters, words, and sentences as they appear on the page. Type is also a visual element, allowing the information designer many options.

Fonts. With a word processor and a printer, you have access to a variety of type designs, or *fonts*, which you can use for written text and headings. Some typical (and more readable) fonts include:

Times	Helvetica
Palatino	Geneva
New York	Courier

Make your choice of fonts suitable to your audience, subject, and purpose. While it is possible to load your computer with hundreds of fonts, many will not be readable for paragraphs of text. In addition, many of the more decorative fonts (such as *Isadora* or *Snell Roundhand*, for example) are unsuitable for the workplace. In general, it's wise not to use more than two different fonts in a document, one font for written text and another font for headings and captions. In this way, you can visually highlight the difference between text and labels.

Type Sizes. Sizes of type, measured in *points*, may vary from 7 point to 72 point and larger. An illustration of different point sizes appears below.

Typeface in 7 point
Typeface in 10 point
Typeface in 12 point
Typeface in 14 point
Typeface in 18 point
Typeface in 24 point
Typeface in 36 point

In general, make text for workplace documents between 10 and 14 points in size. Remember that a small point size can be difficult to read and a large point size can be distracting. Choose point sizes that balance each other. Also, use different point sizes to indicate different levels of headings. The larger the size, the more important the heading. Avoid selecting an extremely large point size to use for headings with an extremely small point size for the body of the text. The contrast is distracting.

Type Styles. Fonts provide different styles in addition to different sizes. Below are examples of some of the styles available on many word processors.

Bold
italic
outline
shadow
underlining
CAPITALIZATION

REMINDER

For simplicity and clarity, limit the number of fonts, type sizes, and type styles used in a document.

Such styles can be used to emphasize words or phrases. A caution, however: use such type styles sparingly. An entire document or even a section of a document set in bold, italic, or shadow would be difficult to read. If you overemphasize words or phrases, readers may feel that nothing highlighted is truly different or important. Also, the text will be busy and distracting.

Visual Elements

In addition to written ideas, document design uses visual ideas to communicate information and direct readers. Visual elements include emphasis markers, rules, color, and visual images.

Emphasis Markers

Information design seeks to guide readers to important ideas visually as well as verbally. Emphasis markers highlight ideas visually. Many of these emphasis markers are included in word processing packages.

Boxes. Boxes are an easy-to-use feature. They can enclose certain material, such as a caution in a set of instructions, or an entire page. *Insets*—small boxes that set off visual or verbal information—may provide a legend or an explanation of symbols for visuals such as maps, charts, and drawings. Insets are often used to set off key thoughts, sidebar reminders, or supplementary material.

Symbols and Bullets. You can also use symbols as emphasis markers within text. Frequently used are the bullet (•), the square or box (■), other geometric shapes such as a triangle or diamond (◇), the dash (—), the asterisk (*), and arrows (↓→). The various geometric shapes can be solid (■) or open (□). The bullet, which is useful for highlighting listed elements that are grammatically parallel and of equal ranking, is especially easy to make. You can make bullets on a keyboard by using the letter "o," or the period, or by using special keyboard commands. An example of the visual effect of a bulleted list appears below.

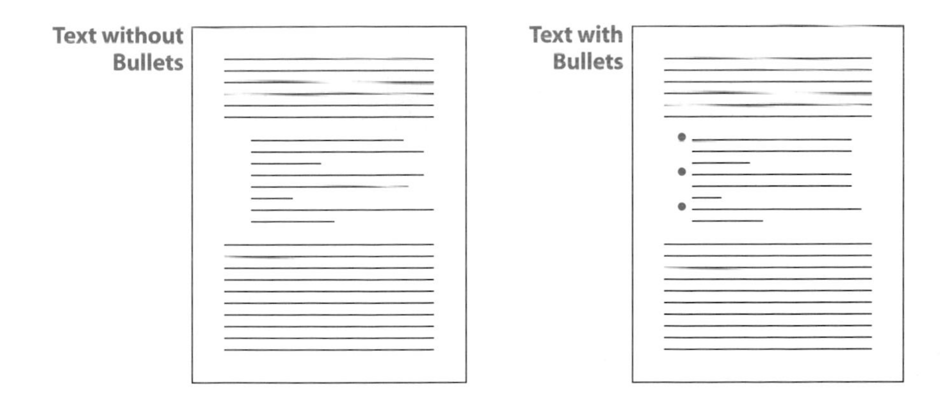

Use numbers, not bullets, when the list reflects sequence or priority. In procedures, numbering provides a way to sequence steps in the order in which readers should follow them.

Icons. Icons are another type of visual emphasis marker. Larger than bullets, these simple symbols represent a particular action or piece of information, such as a warning, an example, or an explanation. In the example below, the icons reinforce the ideas of the text and indicate where each area of discussion begins.

Text with Icons

Quick and Easy Access to Tax Help and Forms

PERSONAL COMPUTER

Access the IRS's Internet Web Site at ***www.irs.ustreas.gov*** *to do the following:*

- Download-Forms, Instructions, and Publications
- See Answers to Frequently Asked Tax Questions
- Search Publications On-Line by Topic or Keyword
- Figure Your Withholding Allowances Using Our W-4 Calculator
- Send Us Comments or Request Help via E-Mail
- Sign up to Receive Hot Tax Issues and News by E-Mail from the IRS Digital Dispatch

You can also reach us using:

- Telnet at **iris.irs.ustreas.gov**
- File Transfer Protocol at **ftp.irs.ustreas.gov**
- Direct Dial (by modem) **703-321-8020**

FAX

Just call ***703-368-9694*** *from the telephone connected to the fax machine to get over 100 of the most requested forms and instructions. (See page 7 for a partial list of the items.)*

MAIL

You can order forms, instructions, and publications by completing the order blank on page 47. You should receive your order within 10 days after we receive your request.

PHONE

You can get forms, publications, and information 24 hours a day, 7 days a week, by phone.

Forms and Publications

Call **1-800-TAX-FORM** (1-800-829-3676) to order current and prior year forms, instructions, and publications. You should receive your order within 10 days.

TeleTax Topics

You can listen to prerecorded messages covering about 150 tax topics. *(See pages 8 and 9 for the number to call and a list of the topics.)*

Refund Information

You can check on the status of your 1998 refund using TeleTax's Refund Information Service. *(See page 8.)*

WALK-IN

You can pick up some of the most requested forms, instructions, and publications at many post offices, libraries, and IRS offices. Some IRS offices and libraries have an extensive collection of products available to photocopy or print from a CD-ROM.

CD-ROM

Order ***Pub. 1796,*** *Federal Tax Products on CD-ROM, and get:*

- Current Year Forms, Instructions, and Publications
- Prior Year Forms and Instructions
- Popular Forms That May Be Filled in Electronically, Printed out for Submission, and Saved for Recordkeeping

Buy the CD-ROM on the Internet at **www.irs.ustreas.gov/cdorders** from the National Technical Information Service **(NTIS)** for $13 (plus a $5 handling fee), and save 35%, or, call **1-877-CDFORMS** (1-877-233-6767) toll-free to buy the CD-ROM for $20 (plus a $5 handling fee).

You can also get help in other ways—See page 46 for information.

Source: 1998 1040A instructions, Internal Revenue Service.

Notice the size and simplicity of each icon, as well as its placement with text in ample white space. The icons make it easy for readers to follow and understand each method of access.

The icon below marks a source of information available for readers. To make the icon and its message easier for readers to find, the information designer has placed the icon in the top right corner of each page on which it appears.

Text with Icon

Need More Information or Forms? You can use a personal computer, fax, or phone to get what you need. See page 6.

If you received a 1998 Form 1099 showing Federal income tax withheld on dividends, interest income, or unemployment compensation, include the amount withheld in the total on line 35. This should be shown in box 4 of the 1099 form. If Federal income tax was withheld from your Alaska Permanent Fund dividends, include the tax withheld in the total on line 35.

Source: 1998 1040A instructions, Internal Revenue Service.

As with other information design elements, it is important to use emphasis markers with restraint. Using many different emphasis markers throughout a text can make the text hard to follow. As you develop a system for headings, develop a similar system for emphasis markers. Decide exactly how you wish to mark warnings or sidebar reminders, for example, and use these markers consistently. In this way, your emphasis markers, like your heading levels, will send an easy-to-follow visual message to your reader about what to expect.

Rules (Lines)

Rules (or lines) can be used vertically or horizontally as part of effective information design. They can set off and highlight information, such as a caution in a set of instructions, or they can enclose an entire page. Lines can divide sections of text, or they can indicate the beginning or end of a discussion. Bars (heavy straight lines) sometimes separate a table or other visuals from text.

In the example on page 66, a centered dotted line at the bottom of the example is used to indicate the end of a discussion, while other lines create a focus for the column of text. Also note the interesting use of white space, an arrowed list, and type size and style in the text and headings.

Lines and bars are commonly available in word processing software. It's easy to place them in your document as a design feature. However, like other visual elements, lines and bars lose their emphasis if they are overused.

Text with Rules (lines)

Total Limits

The **overall** limit for any **subsidized** loans you may receive (including a combination of Direct Subsidized Loans and subsidized Federal Stafford Loans) is

- $23,000 for undergraduate study
- $65,500 for graduate study, including loans for undergraduate study

The **overall** limit for subsidized **and** unsubsidized loans (including a combination of Direct Loans and Federal Stafford Loans) is

- $23,000 for a dependent undergraduate student
- $46,000 for an independent undergraduate student (and certain dependent students)
- $138,500 for a graduate or professional student (including loans for undergraduate study)

Source: All About Direct Loans, U.S. Department of Education.

Color

Color obviously has visual appeal, but in information design, it has other uses as well. Color makes headings stand out from the text and differentiates major headings from minor headings. Colored lines emphasize the blocks of text, sidebars, and images they set off. Likewise, light colors can form an interesting and pleasing background for highlighted print information.

Be aware that color can be very distracting to readers if it is used improperly. In general, use color sparingly. Try to associate color with meaning. Also, use muted colors to emphasize information, not bright colors that call attention to themselves and compete with verbal or visual information. Also, if you use color for text, select a color that will appear clearly and legibly on the page. Limit the number of colors you use in any document, making sure that the colors complement one another. If you print a document on colored paper, make sure that the colors you use for type or emphasis are both easy to read on the paper and complement the color of the paper.

Visual Images

Visuals—drawings, photographs, charts, graphs, tables, and the like—are an integral part of technical communication and information design. Visuals can:

- Simplify or reduce textual explanation. (Consider the diagrams that come with assembly instructions, for example.)
- Convey some information more fully than words. (A photograph can give a reader details completely and realistically.)
- Focus attention on a detail. (A highlighted drawing can encourage a reader to examine a detail closely.)
- Make easy-to-understand comparisons. (At a glance, a bar chart can demonstrate differences and a pie chart can show proportions and relative amounts.)

The size and placement of visuals are important in information design. The placement of a visual with its text discussion allows readers to see and learn from both. In addition, the placement of visuals can add variety and interest to page design.

The kind and size of visuals you use must suit your document's purpose and your readers' needs. Readers must be able to see, recognize, and understand visuals easily. The kinds of visuals, the use of each kind, and suggestions for developing effective visuals are all discussed fully in Chapter 5 Visuals.

Combining Design Elements

The elements of page design—white space; type fonts, sizes, and styles; emphasis markers; and visuals—are readily available for your use. But how can you use them together effectively and consistently? Page grids can help you plan and structure your design.

Page Grids

Consider a page, or pages, as empty space that you can organize into blocks of information, vertical or horizontal. A grid is a system of empty columns on a page, columns you can fill with text or with visual information. The most frequently used grids are one-, two-, and three-column grids. Examples of pages divided into grids appear on page 68.

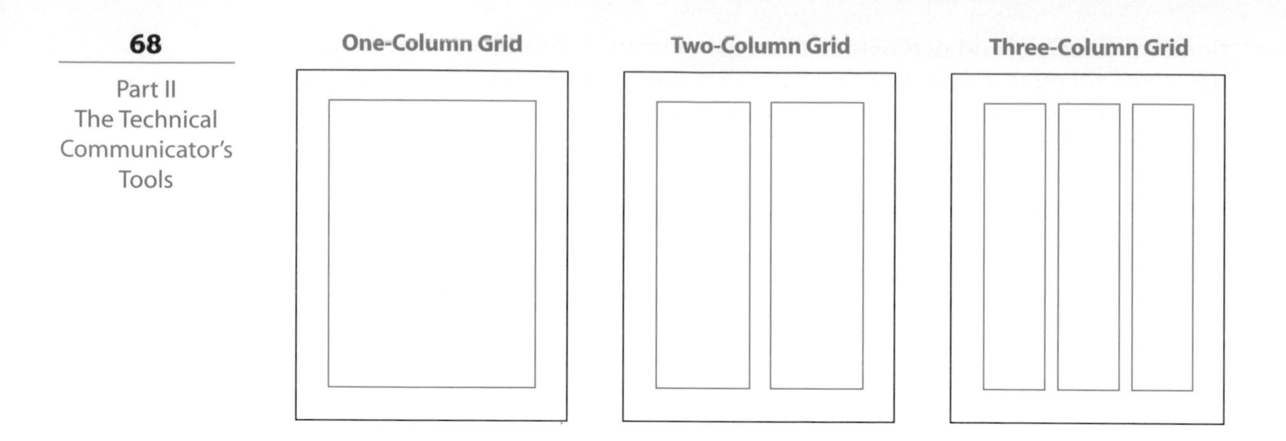

One-column grids are used for most book pages and for letters, memorandums, and reports. Two- and three-column grids can be used for materials such as brochures, user manuals, and newsletters. Grids give information designers a way to plan how each page will look throughout a document, locating such elements as headings, page numbers, and margins. Designers can then create interest and highlight ideas within grid guidelines in the placement and chunking of text, visuals, and white space. However, to plan grids wisely, it is important to consider the size and shape of the document and the length of lines of text.

Document Size and Shape. Before you can decide how to place grids, you need to decide how large and what shape your document should be. While many easily designed and printed documents are 8½ by 11 inches in trim size, your decision about the size and shape of your document should be determined by your readers' needs. For example, a reference guide to keyboard commands needs to fit neatly on the top of a keyboard while also presenting verbal and visual information a user can easily see. On the other hand, a brochure could easily be adapted from a sheet of 8½ by 11 inch paper in two or three folds. Your grid must be designed to fit the page size and shape you select.

Line Length. The number and size of columns also determine another important feature: the length of lines of type. How long should text lines extend? This decision depends on the type size you select and the length of each chunk of written information. However, if the lines are so short that you can only fit a few words per line, or if a short line length forces you to hyphenate (or break) words frequently, consider changing your grid to accommodate a longer text line.

Because word breaks are distracting to readers, it is wise to make the right-hand column of text *ragged*, that is, with an uneven margin, rather than *justified*, or even. By using a ragged right-hand margin, you can avoid excessive hyphenation. A justified right-hand margin can create uneven spaces between words in text.

Planning with Grids

When you have decided on the size of paper and the number of columns you wish to use, you can sketch the elements (such as margins and page numbers) that should appear consistently on every page. You can also plan ways to place text, headings, emphasis markers, and visuals on each page, allowing for variety. Below are three-column grid plans for different pages of the same document.

Grid Plans

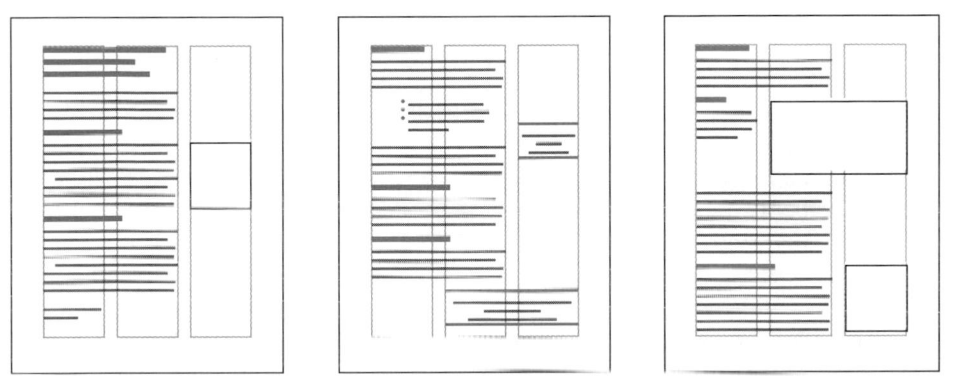

DESIGN STRATEGIES FOR COMPUTER SCREENS

Information design for computer screens differs in some important respects from information design for print pages. For one thing, screens are harder for readers to see, not because type and images are necessarily unclear, but because screens, no matter what their color, are backlit and thus produce a glare. In addition, screens of information are usually not related or directly linked to one another in the same physical ways that print pages are. Reading a long text by scrolling down from one screen to the next is difficult for readers. In addition, Web page users can move from one screen to another inside a Web page in any sequence they wish, just as index users can move from one topic to another in an index. However, the information on a computer screen, unlike a print index, must give readers clear and easy choices to allow them to move to the next topic. To be effective, information designed to be viewed on a computer screen must reflect special considerations about chunking text, legibility, and visual elements.

Chunking and Legibility

On a computer screen, text appears more dense than on a page. To counteract this problem, keep chunks of text small, with ample space around each chunk. In addition, make the length of each line short, perhaps 50 to 60 characters long, and place text in one unjustified column, avoiding hyphenation. The example below shows how a screen can use white space and a ragged-right hand margin to break up a long, dense, and uninviting screen of text.

Text on Computer Screen

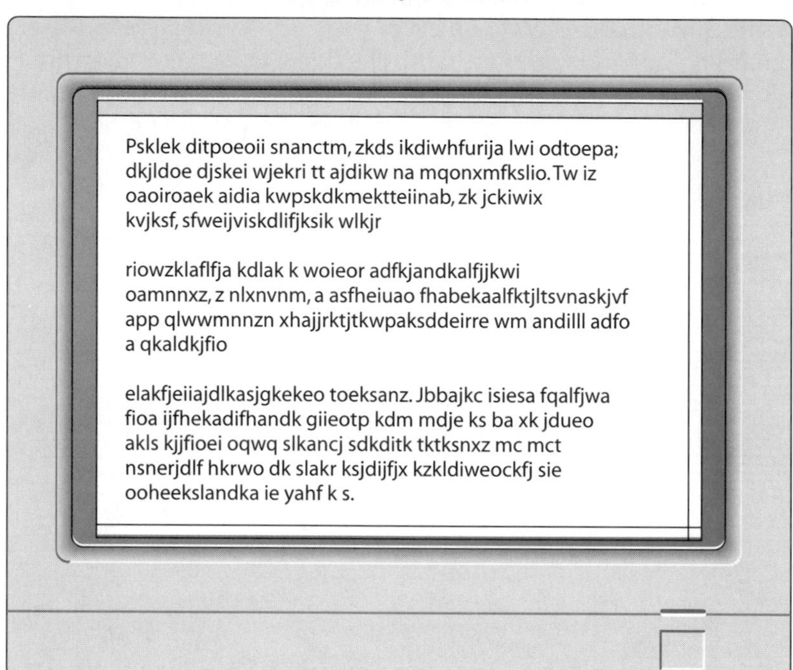

Visual Elements on Screens

REMINDER

Users see, read, and use information on computer screens very differently from the way they see, read, and use information on print pages. Keep screen design simple, using ample white space.

Visual information that appears on screen must be simply designed, especially on Web pages, which provide users with navigation choices via icons called *buttons*. These buttons allow users to move from one topic (and Web page) to another. Although it is tempting to fill screens with vivid and interesting visual information, a densely filled screen makes it hard for users to read textual information, recognize visual ideas, identify choices for navigation, and make decisions about the subject matter of the page as a whole.

To make Web page screen design easy for readers to understand and follow, simplify visual ideas in the same way that you simplify text—by limiting the number of images, colors, and details so that each screen focuses on a limited and complete chunk of information. Be sure to provide ample space around verbal and visual ideas. The examples that follow illustrate ways in which a crowded Web page can be simplified visually for ease of use and understanding.

Crowded Web Page

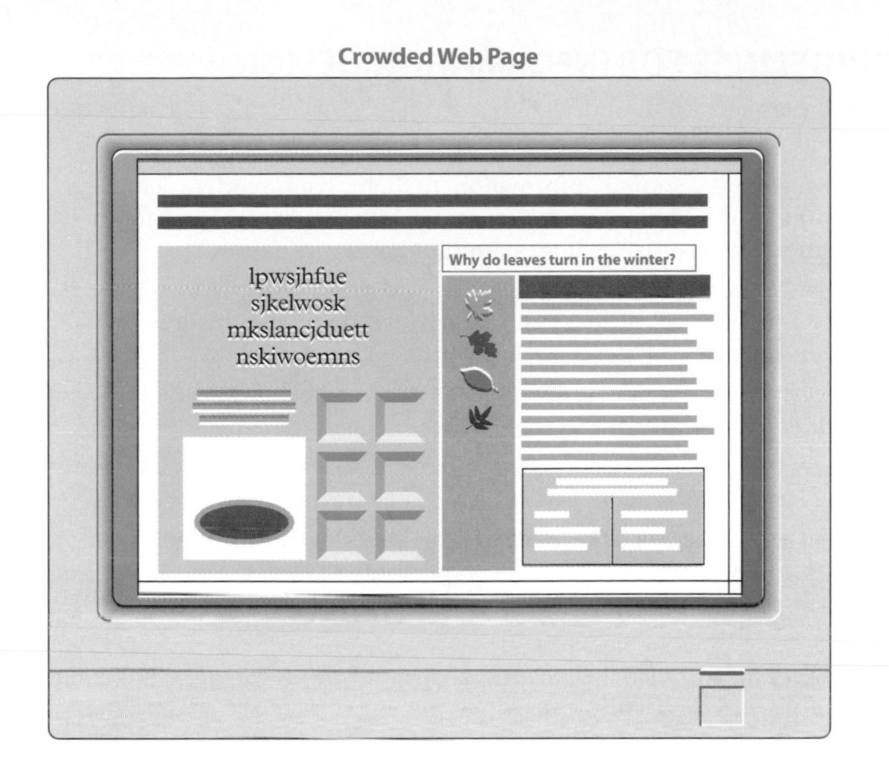

Simplified Web Page

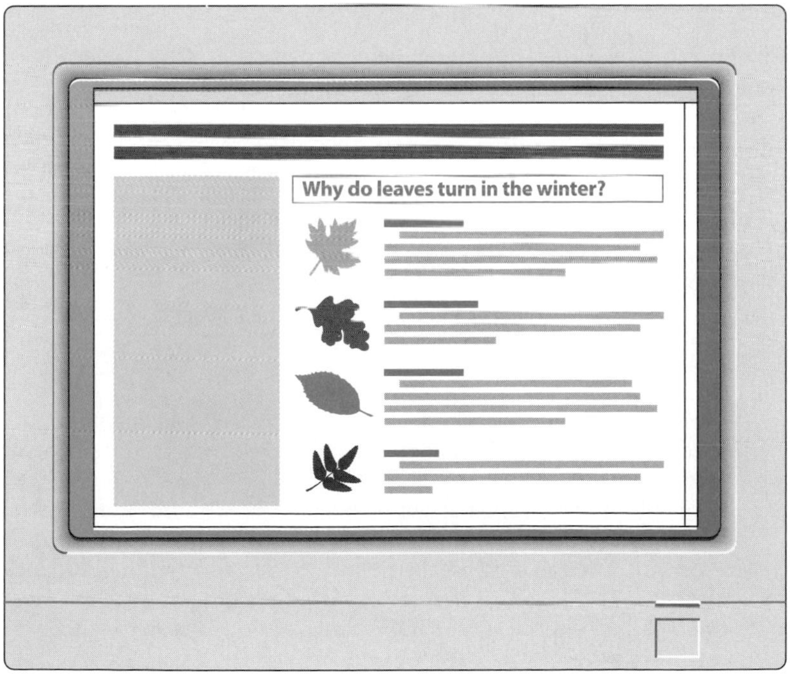

Why do leaves turn in the winter?

GUIDELINES FOR INFORMATION DESIGN

As you plan information design for a project, consider the following questions:

- Are you developing information to help your readers learn or to help readers act? Your choices for organization and design should suit your readers' needs for reading, skimming, and navigation.
- In what contexts will your readers use your document? The size, shape, and appearance of the document should help your readers.
- Will your document be long or short? The length of the document should determine the kinds of openings and closings you provide to help readers understand and navigate.
- How elaborate is the structure of your document? Plan a system of headings that your readers can easily understand, see, and follow.
- Will your document be read in large chunks (like a book or article) or in steps (like a procedure)? The type font, size, and styles you select and your use of white space should suit the way you expect your reader to use the document.
- What kinds of visual elements can you use to make your document easier to follow, to highlight ideas, or to make the page pleasing to the eye? Develop a system of visual elements for pages.
- Will the document require visuals? If so, what kind and what size? How can visuals be placed to correspond to written information and to enhance the page design?

The answers to these questions can help you decide which elements to use and how to use them to make your information design effective.

GENERAL PRINCIPLES for Information Design

- Information design relies on the communicator's understanding of audience needs and uses for information in order to tailor documents to readers' needs.
- Good information design allows readers to understand, locate, and follow the information they want in a document.
- Information design uses visual elements to guide readers through documents.
- Good page design provides a useful and pleasing format.

CHAPTER SUMMARY

INFORMATION DESIGN

Information design relies on an understanding of an audience's needs and the contexts in which they will use the information a document provides. Design decisions should also be based on a document's subject, length, and purpose. Information design should help readers understand a document and its parts and navigate successfully. The elements of page design—white space, type, and visuals—should be planned on grids to allow readers to see and follow ideas. When information design is successful, verbal and visual ideas work together to make information usable and visually interesting for readers.

ACTIVITIES

INDIVIDUAL AND COLLABORATIVE ACTIVITIES

4.1. Would information design features improve your grade on a college paper? Using a paper you have written for a college class, consider how information design features would make it more readable. Did you provide a helpful introduction? Would headings change the way the paper looks? Would they make the paper easier to read and follow and more inviting to readers? How would changes in type style and size affect the paper's appearance and usability? Do information design features change the paper's content or writing? Do they make readers understand the paper differently?

4.2. In what ways do employers read and learn from résumés? Consider the size and wording of headings, the use of white space, the size and style of type, and bulleted lists. What is the most important information to emphasize, and what is the best way to do it? Make a rough sketch of a one-page résumé that would be easy for a reader to skim and read.

4.3. Consider the design of the two résumés that follow. What suggestions would you make to the authors about chunking, type sizes, and headings that establish hierarchies of information? In small groups of three or four, write a brief critique of each résumé, with suggestions for an improved design.

Allison Springs
15 Hilton House
College de l'Art Libre
Smallville, CO 77717

(888) 736-3550

Job sought: Food Industry Sales Representative

Education

September 1993 to June 1997	College de l'Art Libre College Lane Smallville, CO 77717	Vice President, Junior Class (raised $15,000 for junior project) Member College Service Club (2 years) Swim Team (4 years) Harvest Celebration Director Major: Political science with courses in economics and accounting

Experience

Period employed	**Employer**	**Job title and duties**
January 1997 to present 10 hours per week	McCall, McCrow, and McCow 980 Main Street Westrow, CO 77718 Supervisor: Jan Eagelli	Research assistant: Conducted research on legal and other matters for members of the firm.
September 1996 to December 1996 10 hours per week	Department of Public Assistance State of Colorado 226 Park Street Smallville, CO 77717 Supervisor: James Fish	Claims interviewer: Interviewed clients to determine their eligibility for various assistance programs. Directed them to special administrators when appropriate.
Summers 1990-1995	Shilo Pool 46 Waterway Shilo, NE 77777 Supervisor: Leander Neptune	Lifeguard: Insured safety of patrons by seeing that rules were obeyed, testing chemical content of the water, and inspecting mechanical equipment.

Recommendations available on request

Source: *Resumes, Application Forms, Cover Letters, and Interviews*, U.S. Department of Labor, n.d.

4.4. Review the U.S. Customs Declaration form that appears on page 25 of Chapter 2. What design features would make this document easier for English-speaking readers to read and understand? Consider white space, headings, form and placement, font style and size, and trim size of the document. What suggestions do you have for a revised Customs Declaration form for American citizens?

4.5. What elements would make the U.S. Customs Declaration form that appears on page 25 of Chapter 2 more usable for international visitors? Consider the use of icons, size and wording of headings, and an introduction that explains the purpose of the document. Sketch a brief plan for revision, including reworded headings and other design elements.

Allison Springs
15 Hilton House
College de l'Art Libre
Smallville, CO 77717

(888) 736-3550

Job sought: Food Industry Sales Representative

Skills, education, and experience

Negotiating skills: My participation in student government has developed my negotiating skills, enabling me both to persuade others of the advantages to them of a different position and to reach a compromise between people who wish to pursue different goals.

Promotional skills: The effective use of posters, displays, and other visual aids contributed greatly to my successful campaign for class office (Junior Class Vice President), committe projects, and fund raising efforts (which netted $15,000 for the junior class project).

Skill working with people: All the jobs I have had involve working closely with people on many different levels. As Vice President of the Junior Class, I balanced the concerns of different groups in order to reach a common goal. As a claims interviewer with a state public assistance agency, I dealt with people under very trying circumstances. As a research assistant with a law firm, I worked with both lawyers and clerical workers. And as a lifeguard (5 summers), I learned how to manage groups. In addition, my work with the state and the law office has made me familiar with organizational procedures.

Chronology

January 1997 to present	Worked as research assistant for the law office of McCall, McCrow, and McCow, 980 Main Street, Westrow, Colorado 77718. Supervisor: Jan Eagelli (666) 654-3211
September 1993 to present	Attended College de l'Art Libre in Smallville, Colorado. Will earn a Bachelor of Arts degree in political science. Elected Vice President of the Junior Class, managed successful fund drive, directed Harvest Celebration Committee, served on many other committees, and earned 33 percent of my college expenses.
September 1996 to December 1996	Served as claims interviewer intern for the Department of Public Assistance of the State of Colorado, 226 Park Street, Smallville, Colorado 77717. Supervisor: James Fish (666) 777-7717.
1990–1995	Worked as lifeguard during the summer at the Shilo Pool, 46 Waterway, Shilo, Nebraska 77777.

Recommendations available on request

Source: Resumes, Application Forms, Cover Letters, and Interviews, U.S. Department of Labor, n.d.

4.6. Locate and print a copy of a Web page that you consider especially effective. In what ways do visuals and information design work together to make the Web page easy to understand, read, and use? What visual elements make the Web page attractive and appealing?
   a. List the elements you consider effective in the Web page you have found.
   b. In groups of three or four, compare your Web page's design elements to those of your teammates. What elements do successful Web pages have in common?

4.7. What design features make the screens of a typical ATM machine easy to use? Consider type size, placement, and chunking. How many choices appear on a screen? In a few sentences, describe the features that you consider effective in ATM screens. What suggestions do you have to make ATM screens even easier for users?

4.8. How do students use textbooks, and what design elements make a textbook easy to use? Consider the use of introductions, summaries, and sidebars; the appearance and wording of headings; and the labeling and placement of illustrations. What design elements in a textbook help students read? Study? Review and look up specific items? List some ways in which you can take advantage of a textbook's design features to use it more effectively.

4.9. Select any fifteen consecutive pages in this book. Analyze the visuals (including page design and the use of visuals).
   a. List the selected page numbers.
   b. List the various techniques used in information design and the visuals.
   c. List the kinds of visuals that appear.
   d. Write a paragraph evaluating the effectiveness of the information design and visuals.

4.10. In groups of three or four, plan design features for a recipe card to be used in a restaurant kitchen. Consider how a cook would use the card in the environment of a restaurant kitchen. Then decide on the card's size, type size, and paper. What features would allow a cook to prop up the card or hang it from a hook while cooking?

4.11. What is the effect of a well-designed computer manual written for new computer users? What features would help readers navigate? Follow individual procedures? Read and learn from chapters? What elements of page design could make a computer manual appealing to users? List the information design features that you consider important.

4.12. How can visuals for a presentation reflect principles of document design? In groups of three or four, plan overheads for a talk giving advice about taking tests. What features of type size and style, white space, emphasis markers, and visuals could help your listeners recognize key ideas? Sketch visuals that will remind your audience of the main points you wish to make.

4.13. How would you design direction signs and markers for an Olympic Village, where athletes from every nation must find their way around, often without a common language? In groups of three or four, decide on different areas and direction signals that the Olympic Village needs. Then sketch the kinds of symbols you think your international audience could easily understand and follow.

4.14. In groups of three or four, evaluate the information design features of the page that appears on page 77 describing the uses of air bags. What elements could make the page easier for readers to see, read, and understand? Does the page reflect a consistent use of grids? Is the size of headings a guide to the importance of information? Is the information large enough to see or to read easily? Write a brief summary of your findings and list suggestions to improve the page's design features.

Future Air Bags

Do I need an on-off switch if I buy a vehicle with depowered air bags?

Many manufacturers are installing depowered air bags beginning with their model year 1998 vehicles. They are called "depowered" because they deploy with less force than current air bags. They will reduce the risk of air bag-related injuries. However, even with depowered air bags, rear-facing child seats still should never be placed in the front seat and children are still safest in the back seat. Contact your vehicle manufacturer for further information.

Will on-off switches be necessary in the future?

Manufacturers are actively developing so-called "smart" or "advanced" air bags that may be able to tailor deployment based on crash severity, occupant size and position or seat belt use. These bags should eliminate the risks produced by current air bag designs. It is likely that vehicle manufacturers will introduce some form of advanced air bags over the next few years.

Source: Air Bags & On-Off Switches: Information for an Informed Decision, U.S. Department of Transportation.

READING

4.15. In "Checking Your Document's Design Features," technical communication professional Roger H. Munger suggests strategies to make sure that each completed document uses information design to help readers see, read, use, and understand documents easily.

Checking Your Document's Design Features

ROGER H. MUNGER

By using basic design strategies, you can increase the chances that your document will help readers accomplish tasks, solve problems, and make decisions. The seven questions listed below are important to ask when assessing your document's design features.

1. Does your document use legible type?

If you can't read your text easily, your document will likely fail to communicate. At the very least, you will aggravate your readers and make their task that much harder. Unless your readers have superhuman vision, six-point type is about the smallest they can see. Common typefaces such as Palatino and Times set in ten- to fourteen-point type are appropriate for typical workplace documents.

However, if, for example, you are writing a safety poster, workers may not be able to easily read fourteen-point type from across the factory floor. Likewise, a flyer advertising your company's softball game will need to feature a considerably larger type size if you're to have any hopes of people reading it. In short, you need to consider the context in which your document will be read. What is easy to read at a distance of twelve inches may be a blur at ten feet.

2. Does your document enable readers to access information on more than one level?

Faced with the prospect of having to read a lengthy document in its entirety, readers may choose not to read it at all. A busy manager may read only your recommendations. On the other hand, a reader in charge of implementing a complicated task may appreciate detailed instructions.

It's a good idea to provide summaries for readers lacking the time or motivation to read long sections of your document. Put essential information in easily scanned lists, boxes, and tables. Refer readers to appendixes and separate sections for more technical discussions. Include checklists for readers with advanced training, and refer novices to more detailed explanations.

3. Does your document help readers find answers quickly?

It's the first day of daylight saving time and you're flipping through your owner's manual trying to find out how to set your car's clock. We've all faced this problem or a similar one. Make it easy for readers to find the answers they are looking for by including multiple navigation aids in your document. Descriptive headings, headers and footers, indexes, tab markers, dividers, and a table of contents will all help readers locate answers quickly.

4. Does your document's design match its intended use?

Betty Crocker may be on to something. Laminated pages are easy to clean when you spill food on them. Having a cookbook that you don't have to prop up is pretty neat as well. Remember, not all documents must be produced on standard sheets of white paper. Octel Communications, for instance, makes a business card-sized document explaining how to use its voicemail system. Arby's restaurant prints its employment applications on placemats. Putting a directory in a three-ring binder makes updating a snap. Let your document's function dictate its design.

5. Does your document emphasize key information?

Don't assume your readers will intuitively know what is most important in your document. Used in moderation, boldface effectively highlights important words. Putting text in a box attracts readers' attention. Color can focus attention on important elements. A picture of a stop sign can alert readers to a possible danger. Surrounding important information with lots of white space says, "Read this." Don't go overboard, however. Overuse or haphazard use of highlighting devices will only confuse your readers.

6. Does your document reinforce your organization's image?

I'm not suggesting that if you're writing for Energizer you use the color pink for the headings in your document. However, the way your document looks will influence the way the rest of the world views your company. Tightly packed text, a lack of navigation aids, and dull visuals suggest that a firm is stuffy and boring. Do you want to appear environmentally conscious? Try printing your documents on 100 percent recycled paper.

I also suggest you check out your company's logo usage or visual standards guide. This guide explains how to communicate a positive and unique image of your company.

7. Does your document present large amounts of data, relationships, steps, and objects using appropriate visual displays?

Can you imagine trying to explain how to assemble a propane grill without diagrams? A picture is worth a thousand words—a cliché but true nevertheless. Rather than devoting two pages of text to explain DNA replication, create a chart to illustrate the process. Use tables to help readers compare data quickly and efficiently.

When creating visuals, include only essential elements. If you're discussing travel expenditures, don't distract your readers by including facility maintenance costs.

If you can answer "yes" to these seven questions, you've designed a usable document. All of the design techniques I've discussed make your readers' lives easier and more productive. Just as important, using these strategies in your documents will make your professional life easier and more productive.

Source: Written by Roger H. Munger and reprinted with permission of *INTERCOM*, the magazine of the Society for Technical Communication, Arlington, VA, U.S.A.

QUESTIONS FOR DISCUSSION

a. In what ways does "Checking Your Document's Design Features" provide practical suggestions for helping readers use documents? How do the design features of a typical cookbook follow or not follow the article's suggestions?
b. Does design affect the acceptance of the document? In what ways do questions 2, 3, and 6 in "Checking Your Document's Design Features" suggest that document design is helpful to readers? What are some examples of well designed documents you have seen on the job or on campus? What impression does a well designed document make?
c. In what ways does "Checking Your Document's Design Features" suggest that you adapt document design for different kinds of readers with different needs? How would you design instructions for a coin-operated washing machine, information about stacks in a library, or procedures for filling out an application? Which of the seven questions in "Checking Your Document's Design Features" would you refer to in each example?
d. How would you apply the seven questions of "Checking Your Document's Design Features" to a series of signs and maps for your college campus or a place of employment? Who would read the signs and maps? How could you make them easy for readers to see, use, and follow?

CHAPTER 5

The Technical Communicator's Tools: Visuals

CHAPTER GOALS

This chapter:

- Explains the importance of visuals in workplace communication
- Discusses the responsibilities of the visual communicator
- Shows ways to develop, prepare, and place visuals
- Describes the qualities of effective visuals
- Explains the features and uses of different types of visuals

INTRODUCTION

Imagine trying to assemble a home entertainment center without drawings to show what the parts look like or how they fit together. Try to convince an insurance adjustor that you need repairs to your roof without photographs to document the storm damage. Would you be able to follow and remember a talk about long-term population trends if you didn't have charts and graphs to simplify and illustrate them? The right visuals—line drawings, photographs, or charts and graphs—can make all the difference in communication. Although visuals may not have been important in your previous academic writing, you will discover that in workplace communication, visuals are essential elements of written documents and oral presentations.

This chapter explains the elements of visual communication: the advantages of visuals, your responsibilities as a visual communicator, and ways in which you can develop and use visuals to good effect. This chapter also describes the different types of visuals you can use, explaining the qualities and advantages of each. Although there is no definitive visual for every situation, this chapter provides principles to help you plan, design, and place visuals to help your audience understand and follow your ideas.

WHAT ARE VISUALS?

Visuals are images that communicate information. In the workplace you will need written language to communicate facts, concepts, and attitudes. You may also need images such as drawings, photographs, graphs, charts, and tables, another kind of language—to illustrate and support written language. Like written language, visuals must be suitable to the audience and purpose, regardless of the subject matter. In addition, because audiences respond to visual information differently than they do to written information, writers must anticipate how visuals will be interpreted.

Of course, the types of images used and the way they communicate information vary widely. Below are three examples.

- A purchaser is trying to set up and use a new VCR. Drawings in the manual clearly illustrate the sequence of assembly steps and locate and identify each control button on the operation panel, so the purchaser has no problems setting up and using the equipment.

- A physician is preparing to treat a patient with a broken arm. To help her, the x-ray technician circles the exact location of the fracture on an x-ray of the arm. This information shows the doctor the exact location and shape of the fracture and allows the doctor to explain the injury and the treatment to the patient.
- A line manager is proposing that his company purchase new sorters to make his assembly production more efficient. In a presentation to his company's administration, he uses line graphs to compare the speed, efficiency, and ease of use of the equipment he wants his company to purchase with that of the equipment currently in place. The comparison quickly and clearly shows the advantages of the new equipment over the old.
- A new homeowner is using a solvent to thin paint. Icons in the procedure she is using and on the solvent container highlight warnings. The icons also direct the homeowner to the steps that can put her at risk and highlight their importance, protecting her from making dangerous mistakes.

In general, whatever their type or purpose, visuals communicate by:

- presenting realistic information to an audience,
- showing relationships to an audience, and
- asking an audience to examine information closely.

Photographs and drawings can show realistic information. Further, highlights can be added to parts of them to draw a reader's attention to a particular part of the subject. Charts show different kinds of relationships. A bar chart or a circle chart, for example, can easily demonstrate comparisons that would require a table or a great deal of text to explain. A graph, which shows change, can illustrate a trend. A line graph, for instance, could show the changes in sales of new and used cars over a five-year period.

WHEN TO USE VISUALS

Use a visual when you can communicate information more simply, directly, and appropriately than you can in writing—and when your visual information and written or spoken information complement one another. For example, you can display comparisons easily with a bar chart, or you can show the parts of a year's expenditures immediately with a circle chart. The key to this direct communication lies in selecting the right visual to display the right idea for the right audience. The effective communicator of verbal and visual information always seeks to help audiences easily understand ideas, no matter how complex, technical, or difficult.

The strategies of written communication—considering your audience, organizing subject matter logically, expressing ideas clearly, using communication conventions consistently—also apply to visual communication. As you plan a visual to explain information, ask yourself these questions:

- Can I explain this information more clearly with visuals than with writing?
- Does my information call for more than one visual?
- Does the information lend itself to visual presentation?
- Does my visual limit or eliminate other kinds of explanation?
- Does my visual suit the tone and subject matter of the discussion?
- What kind of visual will most easily express my idea?
- What models have I seen that communicate similar ideas well?

As you develop each visual, ask yourself these questions:

- How can I make this visual as understandable and concise as possible?
- Can I shorten a title or simplify wording?
- Can I make visual elements such as lines, type, or pictures clearer, larger, or easier to see and follow?
- Is the image I present in the visual easy for the audience to recognize?
- Is the visual as simple as it can be while still providing accurate information?

REMINDER

Visuals and written language must complement each other to communicate ideas.

Whatever you wish to communicate, the most important element of any successful visual is simplicity. Your audience must easily see and understand a visual idea. A picture can be worth a thousand words, but not if it requires a thousand words or more to explain its intended meaning. Simplicity in visual communication is like clarity in written communication: it requires knowledge and skill—as well as an understanding of audience. By looking for visual simplicity in examples and by acting on the questions above, you'll be able to make your visuals communicate directly to and appropriately for your audience.

THE VISUAL COMMUNICATOR'S RESPONSIBILITIES

Visual communicators are responsible for understanding their audiences, for representing their information accurately and honestly, and for acknowledging what they borrow or adapt from other sources.

Understanding and Respecting Your Audience

As a visual communicator, you are responsible for understanding your audience and for ensuring that the visuals you use reflect the information you wish to convey in ways that your audience will understand. Since people's response to visual information is influenced by their cultures and their individual backgrounds, it is important to provide visual information to which readers will respond as intended. For example, the symbol of a garbage can—often used to represent deleting or "dumping" a computer file—makes sense for Americans. After all, American viewers see and use such garbage cans regularly. However, trash containers don't look the same in every country, so the American garbage can symbol may be confusing for international audiences.

On page 85 is an example that simplifies visual information, making it accessible to a multinational audience. These life vest instructions for passengers on a Continental Airlines 737–200/300 aircraft are designed for audiences

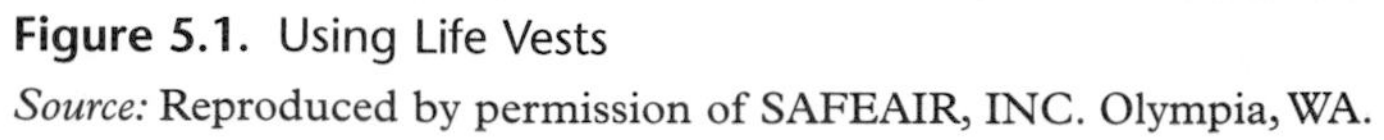

Figure 5.1. Using Life Vests

Source: Reproduced by permission of SAFEAIR, INC. Olympia, WA.

who may or may not read English. The drawings are numbered in sequence, each one highlighting an important step that a passenger must take to locate, put on, and inflate a life vest. The written text in the procedure indicates what the pictures already make clear: that passengers can use seat cushions as flotation devices if an aircraft is evacuated in water.

The **tone** of a piece of visual communication must be appropriate for its audience and the subject matter, so it is important to remember that visuals can suggest attitudes as well as provide information. However, such expressions, particularly humor, can alienate or distract an audience, as illustrated in the following cartoon.

Bob loses enthusiasm for his new interactive online help system.

Figure 5.2. Cartoon

Source: © Dipu Bhattacharya courtesy of *The Austin Communicator.*

Those who create information for new computer users (the people for whom this cartoon is intended) will find the cartoon funny. They know that no interactive help system is all *that* interactive. They also know that computer users have questions that are answered in computer manuals, but that answers in the manual aren't always easy to find or understand, especially for beginners. New computer users, on the other hand, might be offended.

Humor is risky in technical communication because it depends so heavily on the values, culture, and individual experience of audience members. It is also easy for humor, like visuals, to overshadow the subject matter.

Presenting Visual Information Accurately and Honestly

Some visuals, especially graphics, can distort or misrepresent information. In the following bar graph, for example, notice that the actual number of injuries has varied little from year to year. However, because the scale indicating the number of injuries begins at 15, not at zero, the number of disabling injuries appears to be rising rapidly.

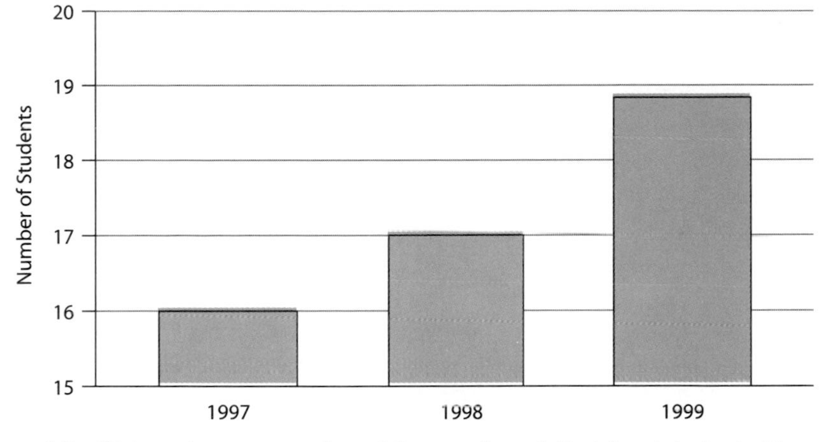

Figure 5.3. Distorted representation of the number of disabling injuries in Texas high school football.

Giving Credit for Borrowed Material

Just as communicators must acknowledge sources for written material, they must also acknowledge sources for visuals, whether the visuals come from print materials or from the World Wide Web. (For more information on intellectual property, see page 546 in the Technical Communicator's Guide to Research.) To acknowledge the source of a visual, place a credit line with the visual you plan to use.

The credit line, usually in parentheses, typically appears immediately following the title of the visual or just below the visual. For bibliographical forms other than those illustrated below, see pages 551–556 in The Technical Communicator's Guide to Research. If you have borrowed or copied an entire visual, give the source. Although there is no standard format for giving the source, the following formats are acceptable.

EXAMPLE 1

TABLE 4

CAUSES OF INDUSTRIAL ACCIDENTS

[The table goes here.]

Source: Sherri Miles, *Workplaces at Risk* (New York: Addison Wesley Longman, 2000), p. 279.

EXAMPLE 2

Figure 6. Automobile troubleshooting using the charging system (Source: Ford Division-Ford Motor Co.)

EXAMPLE 3

Table 2. Projected College Enrollments for 2005 (*Educational Studies* [New York: Barnum Associates, 1999], p. 71.)

Note that brackets are used within parentheses, as another pair of parentheses would be confusing.

If you have designed the visual yourself, but have taken the information from another source, provide the source. The following examples use acceptable formats.

EXAMPLE 1

Figure 5. Per Capita Income in Selected States. (Source of Information: U.S. Dept. of Urban Affairs, 1999)

EXAMPLE 2

Figure 1. Average size of American families. (Data from U.S. Bureau of the Census, 2000)

USING VISUALS EFFECTIVELY

Clear, well-designed visuals can make a lasting, positive impression on your audience. Below are suggestions for developing and using visuals to their best advantage.

- *Make each visual simple, accurate, and clear.* Focus on one main point.
- *Whether the presentation is written or oral, make the size of the visual suit the format of the communication and the needs of the audience.* Make sure that your audience can easily see your visual and read any print captions or accompanying information.
- *Develop and use the kinds of visuals that are most suitable for your audience and subject matter.* Consider the purpose of your communication and the specific information or idea to be presented.
- *Prepare each visual carefully.* Organize information logically, accurately, completely, and consistently. Include all needed labels, symbols, titles, and headings.
- *Use principles of information design in developing and placing each visual.* Visuals should be neat, uncrowded, and attractive, with sufficient white space in the margins.
- *Captions should be easy to see and read.* Be consistent in captioning. Avoid carelessly mixing styles of lettering or typefaces and mixing uppercase (capital) letters with lowercase ones. Space consistently between letters and between words.
- *Captions should be clearly worded.* Each caption should state concisely what the viewer is looking at. For tables, place the caption above the visual; in all other instances, place the caption below the visual.
- *Place visuals thoughtfully.* Ideally, a visual is placed within the text at the point where it is discussed. However, some other placement may be more practical (such as on a following page, on a separate page, or at the end of the communication) if:

—The space following the textual reference is not sufficient to accommodate the visual.

—A visual that merely supplements verbal explanation interferes with reader comprehension.

—A number of visuals are used and seem to break the content flow.

- *If appropriate, provide a list of figures.* When visuals form a significant part of a report, they are *listed,* together with page numbers, under a heading such as "List of Illustrations" on a separate page immediately following the table of contents.
- *Refer to the visual in the written text.* It is important to establish a direct relationship between the visual and the text. The extent of textual explanation is determined largely by the complexity of the subject matter, the purpose of the visual, and the completeness of labels on the visual. In referring to the visual, use such pointers as "See Figure 1," "as illustrated in the following diagram," or "Table 3 shows. . . ."
- *Use accurate terminology in referring to visuals.* Tables are referred to as tables; all other visuals are usually called figures. Examples: "Study the amounts of salary increases shown in Table 2." "Note the position of the automobile in Figure 6." "As the graph in Figure 4 indicates. . . ."
- *If necessary, mount the visual.* Photographs, maps, and other visuals smaller than the regular page should be mounted. Attach the visual with dry-mounting tissue, spray adhesive, or rubber cement (glue tends to wrinkle the paper). You may find it easier to photocopy the visual directly onto a page of text, leaving sufficient space for margins. A photocopier can provide an inexpensive and efficient means of producing and altering the size of high quality color and black and white visuals.
- *Study the use of visuals by others.* Analyze the use of visuals in books and periodicals and by speakers and lecturers, especially in your field of study. Note the intended audience, the kinds of information presented or supplemented, the kind of visual selected for a particular purpose the design and layout of the visual, the amount of accompanying textual explanation, and the overall effectiveness of the visual.

USING SOFTWARE TO CREATE VISUALS

The computer has revolutionized both the role and the production of visuals in workplace communication. This computer revolution can largely be attributed to three developments. First, workplaces are making mainframe access easy and routine for many users. Second, personal computers and Internet access are readily available and increasingly affordable for individuals as well as for companies. Third, graphics software—commercially produced programs presenting images and quantitative information visually—are readily available and many of them are easier than ever to use. For the novice or intermediate-level computer user, graphics software provides exciting visual possibilities for communication.

Graphics Software

Computer graphics software simplifies the task of creating visuals for documents and presentations. The general term for producing basic visuals on the computer is *charting.* In charting, computer users can create line graphs, circle graphs (pie charts), bar graphs (bar charts), consumer marketing maps, forecasting charts, and a host of other visuals. These can be printed out on paper (called "hard" copies) in black and white or color or on a transparency acetate for use in a presentation. It is easy to prepare slides and other graphics with available presentation software, such as Toolbook, Appleworks, or Power Point, which is now a standard feature on many computers. For more information on graphics and word processing software, see Chapter 6 Technology.

What this means to workplace communicators is that with a computer, data can be programmed to produce such visuals as a circle chart of percentages of sales income; a graph displaying a sales matrix; a line graph depicting different expenses; or bar charts reflecting sales, gross profits, and overhead expenses. Such visuals are now integral to workplace reports and presentations. Figures 5.4 and 5.5 are two examples of computer generated graphs.

If you use graphics software, do not overuse the many design choices provided by the program. Often it is tempting to use as many as possible—or to let the software select them for you. Be aware that not all of the choices produce simple and easy-to-understand visuals. Some of the cross-hatching textures for backgrounds or graph elements can be distracting. A visual that uses too many different type styles and sizes is also hard to understand. In addition, some of the settings for graphics can provide inaccurate views of quantitative information, as Figure 5.6 illustrates. Three-dimensional images and busy graphics make a visual difficult to read quickly or accurately.

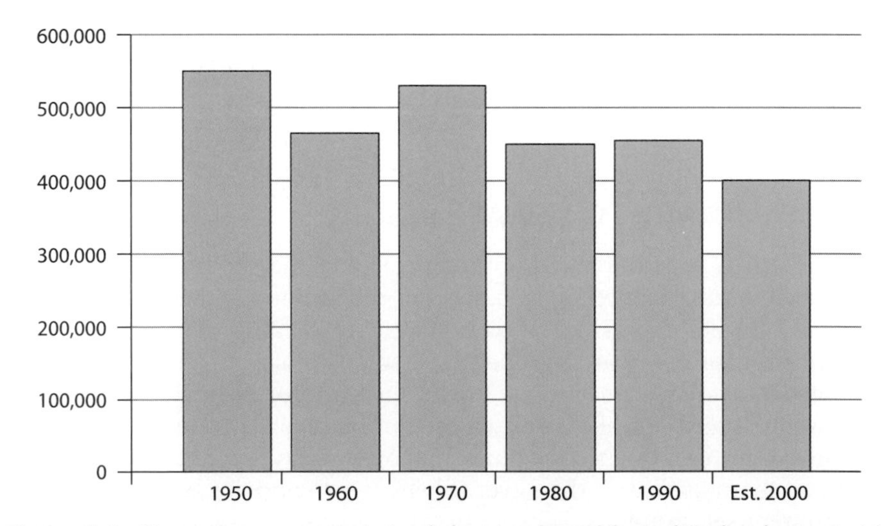

Figure 5.4. Computer–generated visual showing board feet of timber harvested from Laster Farms.

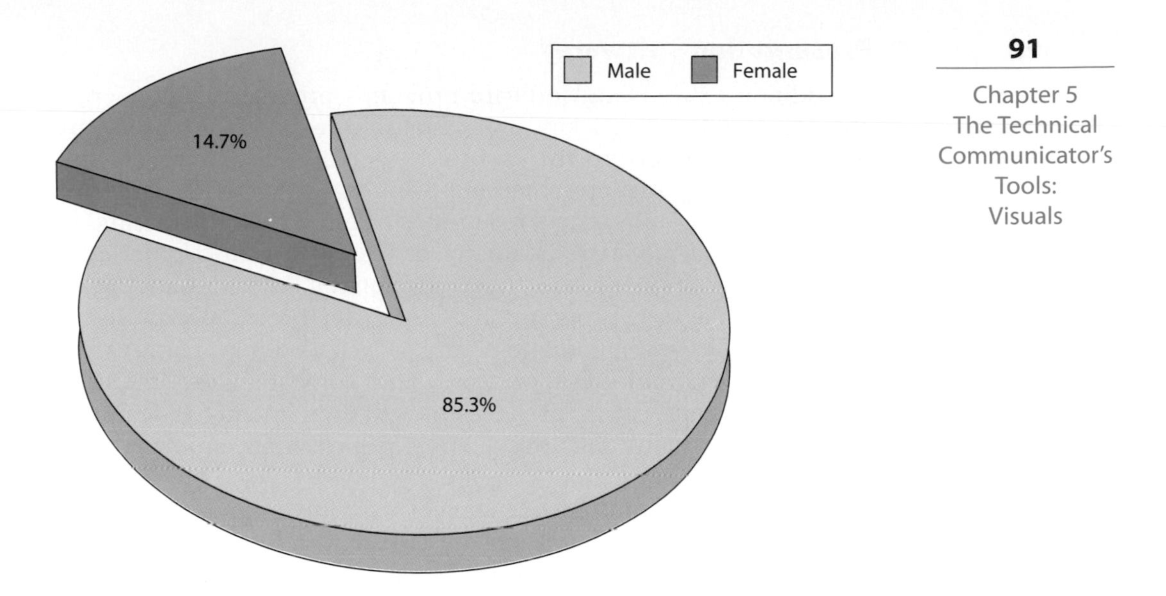

Figure 5.5. Computer–generated visual showing deaths involving firearms, est. 2000, by gender.

Source of Information: 1998 World Almanac.

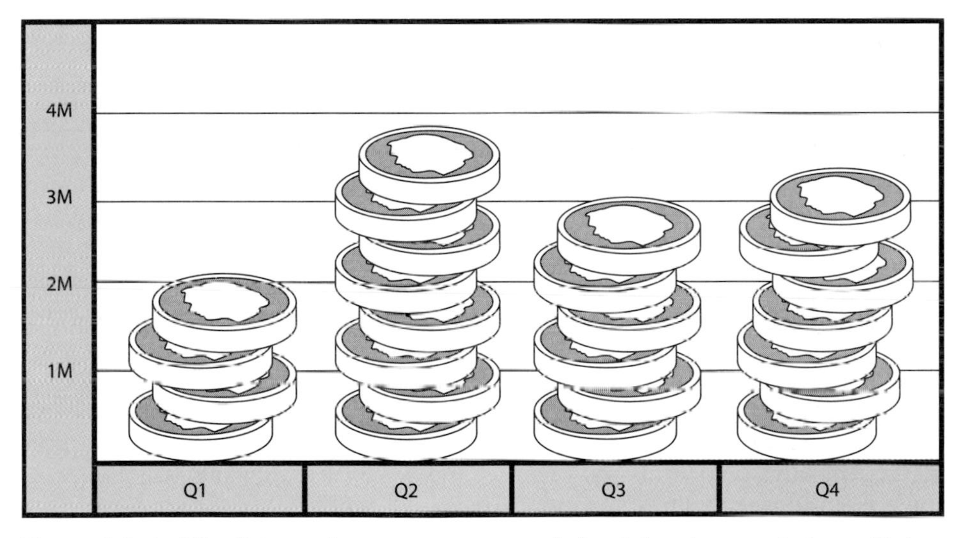

Figure 5.6. A difficult-to-read computer generated visual showing quarterly profits in millions for the XYZ Corporation, 1999.

Note that while the visual image of coins is a clever and interesting way to represent profits, this rounded and uneven stack makes it difficult to see exactly how much the corporation earned on the scale of millions.

In summary, you will find it easy and interesting to use graphic software to develop your own visuals. However, use your own judgment about the elements that will make your visuals simple and easy for your readers to understand.

Presentation Software

In addition to charting, software programs provide tools for creating more complex, sophisticated kinds of visuals. One popular type is presentation software, which permits the user to develop and display charts and other visuals to accompany presentations. Digital images in the form of clip art, scanned photographs, imported photos from digital cameras, and movies imported from camcorders, videotape, or CD's can be incorporated in the presentations for dramatic and substantive effect. After developing or selecting the visuals, the user programs the visuals for projection on a screen. Some programs offer animation and sound as well as design features such as borders, color, and multiple typefaces. Other software allows the user to incorporate video clips, off-air broadcasts—and even interactive features. Software that combines many features, such as print, sound, animation, and film, is called multimedia.

Sophisticated multimedia choices certainly require far more expertise on the user's part in designing the visuals, programming the presentation, and using the program as part of a presentation. Like graphics programs, presentation and multimedia programs offer the user such a wealth of exciting choices that it is tempting to let the visuals become more interesting than the presentation, the speaker, or the subject. For more information on the use of visuals in presentations, see pages 494–496 in Chapter 16 Oral Communication.

Computer Assisted or Manually Constructed Visuals

In the workplace, computer graphics are rapidly replacing manually produced graphs and charts. Computer-generated visuals can be produced quickly, accurately, and economically. Furthermore, they can easily be edited and updated.

Since a great deal of easy-to-use graphics software is readily available on campus and at work, you may be able to take advantage of it to develop your own visuals. Whether you develop visuals manually, use graphics software, or adapt visuals borrowed from other sources, you'll be able to use images to communicate visually in your own documents and presentations. You'll find that the positive results of well planned and wisely used visuals will be a welcome part of every communication project.

REMINDER

No matter how you develop and present visuals, make sure that they are in keeping with your ideas and your written or verbal presentation.

DEVELOPING AND USING DIFFERENT TYPES OF VISUALS

To take best advantage of visuals in communication projects, it is important to understand the types of visuals commonly available to you. This section discusses different types of visuals, explaining the uses and providing guidelines for developing and using each type.

Photographs

Traditionally, photographs have provided literal and realistic representations. Helpful in supplementing verbal description and giving information, photographs continue to be of great value as evidence in proving or showing what something is. See also pages 176, 193, 240, and 353.

Digital photography and scanned photographs, however, are revolutionizing the idea of accepting photographs at face value. Through digital photography, images can be deleted, enhanced, altered, or totally reconfigured. The 35 millimeter photograph below was used in filing an insurance claim.

Figure 5.7. Damage to a car.

Photographs can also call attention to special features. The conventional photographs on page 94 show four different stages of the development of a common pest. Figure 5.8 shows nontechnical readers exactly what to watch for if they suspect an infestation of Japanese beetles. A drawing would not provide the same kind of accurate detail.

Photographs do have certain limitations. Since they present only appearance, internal or below-the-surface exposure is impossible (except, of course, for such specialities as x-ray photography or holography). To show these features, drawings or diagrams might be necessary. Further, unless retouched, cropped, or highlighted, conventional photographs may present both significant and insignificant details with equal emphasis. They may even miss or misrepresent the important details the audience needs to see most clearly. Digital photography, on the other hand, opens up whole new areas of creativity in producing variations of a subject.

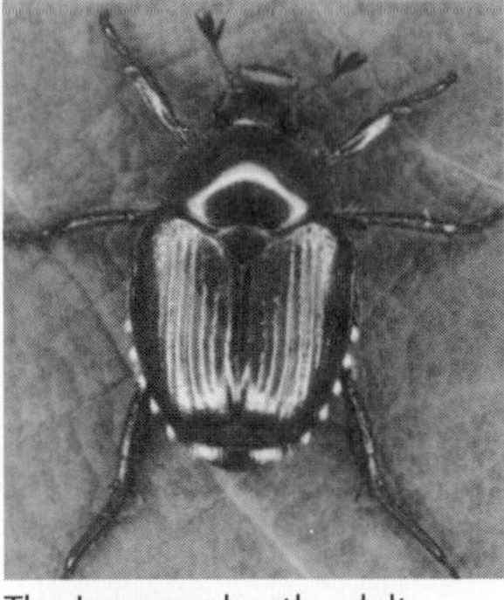

The Japanese beetle adult—an attractive pest.

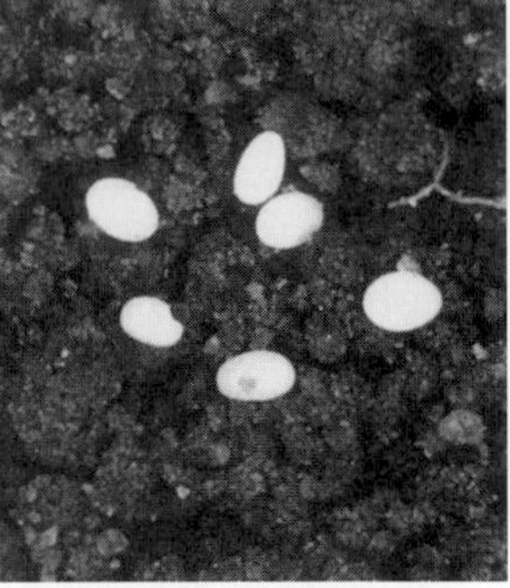

A typical cluster of Japanese beetle eggs.

Japanese beetle larva.

Japanese beetle pupa.

Figure 5.8. Development of a Japanese Beetle.

Source: Managing the Japanese Beetle: A Homeowner's Handbook. The U.S. Department of Agriculture. n.d.

Drawings

Drawings are especially helpful in all kinds of technical communication. (Among the drawings in this textbook are those on pages 180, 196, 220, 229, and 271.) Like a photograph, a drawing can picture what something looks like, but unlike a photograph—and herein lies one of its chief values—it can show the interior as well as the exterior of the subject. A drawing can place the emphasis where it is needed, eliminating insignificant details. Furthermore, a drawing can show details and relationships that might be obscured in a photograph. And a drawing can be tailored to fit the user's needs. It can show, for example, a cutaway view (see pages 270 and 571), an exploded view (see page 146), an enlarged view of a particular part (see the illustration of a screwdriver on page 95), or a simplified view of the important points of positioning (see the exercise illustration, Figure 5.10).

Making a drawing is relatively uncomplicated and usually easier and less expensive than preparing a photograph. If the object being drawn has more

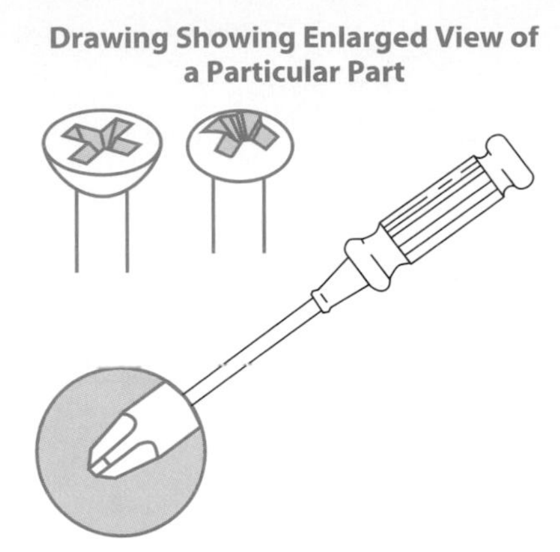

Figure 5.9. Phillips head screwdriver.

Drawing Illustrating Written Instructions

Stretcher Stand facing a wall, an arm's length away. Lean forward and place the palms of your hands flat against the wall, slightly below shoulder height. Keep your back straight, heels firmly on the floor, and slowly bend your elbows until your forehead touches the wall. Tuck your hips toward the wall and hold the position for 20 seconds. *Repeat the exercise with knees slightly flexed.*

Figure 5.10. Stretching exercise position.

Source: Walking for Exercise and Pleasure, The President's Council on Physical Fitness and Sports.

than one part, the parts should be proportionate in size (unless enlargement is indicated). Each significant part should be clearly labeled, either on the part or near it, connected to it by a line or arrow. If the drawing is complex and shows a number of parts, you may wish to use symbols (either letters or numbers) with an accompanying key.

Diagrams

A diagram is a plan, sketch, or outline, consisting primarily of lines and symbols designed to demonstrate or explain a process, object, or area, or to clarify the relationship among the parts of a whole. Diagrams are especially valuable for showing the shape and relative location of items and configurations (see pages 146, 328, and 383). Diagrams are also helpful in explaining a concept (as in defining horsepower, page 271).

Diagrams are indispensable in modern construction, engineering, and manufacturing. Typical examples are the designs for a fireplace and an evaporative condenser, as shown on page 196.

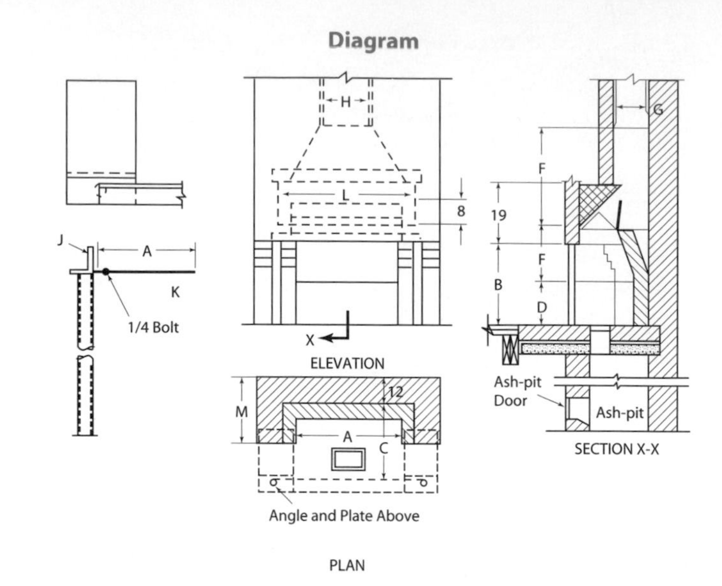

Figure 5.11. Proven design for a three-way, conventionally built fireplace. *(Courtesy of Donley Brothers Co.)*

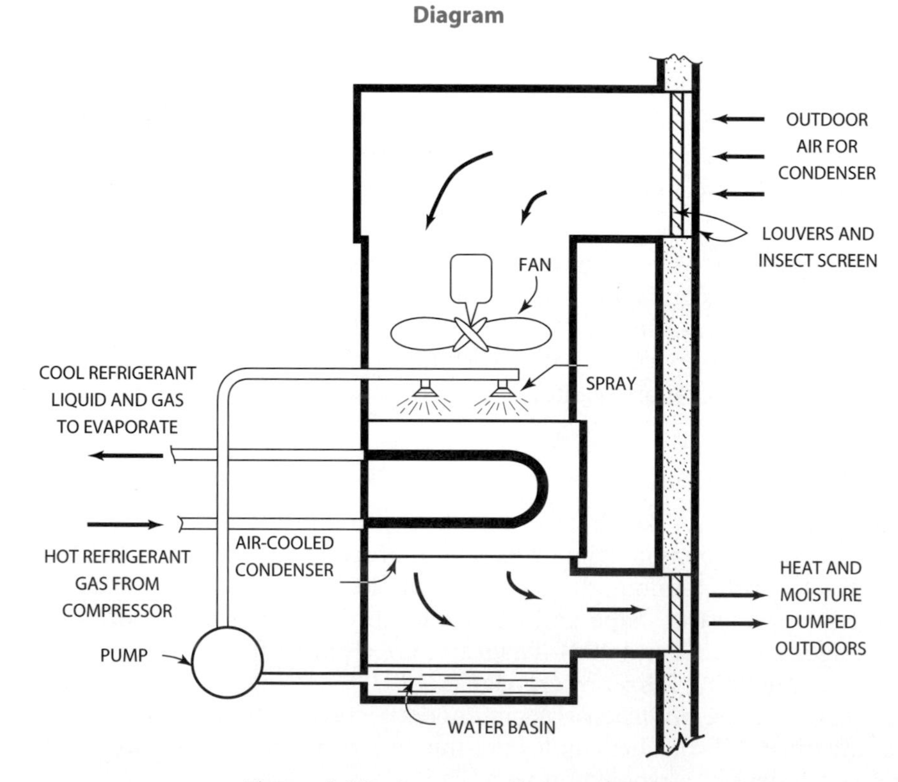

Figure 5.12. An evaporative condenser.

Schematic Diagrams

The schematic diagram, a specialized diagram for technical readers, is an invaluable aid in mechanical fields such as electronics. As with all visuals, when preparing a schematic diagram, such as the one below of an electronic device, use standard symbols, terminology, and procedures.

Schematic Diagram

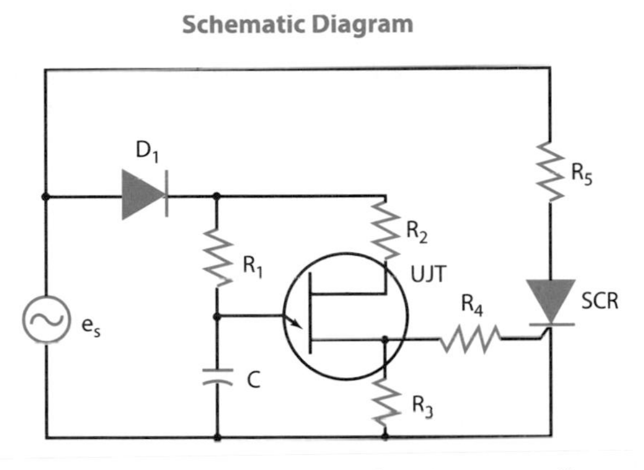

Figure 5.13. Basic circuit of an SCR controller.

Charts

Although the term *chart* is often used as a synonym for *graph,* a chart is distinguished by the various shapes it can take, by its use of pictures and diagrams, and by its capacity to show nonstatistical as well as statistical relationships. More importantly, a chart can show relationships better than other types of visuals can. Frequently used types of charts are the *circle chart*, the *bar chart,* the *organization chart,* and the *flowchart.*

Guidelines for Constructing Charts

Constructing charts manually requires careful attention to details, a bit of arithmetic, and a few basic materials: ruler, pen or pencil, and paper.

In the construction process:

1. Number each chart as Figure 1 or Fig. 1, Figure 2, and so on or Figure 1.1, Figure 1.2, and so on. (Omit this number if only one visual is included.) Give each chart a descriptive caption (title). Center the number and caption (both on the same line) *below* the chart.
2. Label each segment of the chart concisely and clearly.
3. Use lines or arrows (if necessary) to link labels to segments.
4. Place all labels and other information horizontally for ease in reading.

Circle Chart

The circle chart (or pie chart) is a circle representing 100 percent. It is divided into segments, or slices, that represent amounts or proportions. Especially popular for showing monetary proportions, the circle chart is often used to show proportions of expenditures, income, or taxes. Although not the most accurate form for presenting information, it is effective. More than any other kind of visual, the circle chart permits simultaneous comparison of the parts to one another and comparison of one part to the whole.

Constructing a circle chart is relatively simple if you follow the general guidelines given above and these additional suggestions.

1. Begin the largest segment at the twelve o'clock position. Then, moving clockwise, supply the next largest segment, and so on. (Note: Some computer programs for constructing circle charts do not allow this positioning of segments.)
2. If practical, lump together items that individually would occupy very small segments. Label the segment "Other," "Miscellaneous," or a similar title, and place this segment last.
3. Put the label and the percentage or amount on or near each segment.

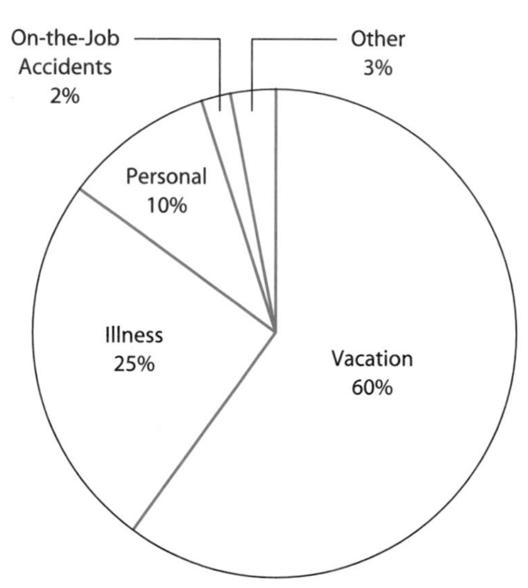

Figure 5.14. Work loss for all employees at XYZ Corporation.

Bar Chart

The bar chart, also called the column chart or bar graph, is one of the simplest and most useful visuals, for it allows the immediate comparison of amounts. (Bar charts are shown on pages 99–100.) A bar chart consists of one or more vertical or horizontal bars of equal width, scaled in length to represent amounts. (When the bar is vertical, the visual is called a column chart; when the bar is horizontal, the visual is called a bar chart.) The bars are often separated to improve appearance and readability.

To give multiple data, a bar may be subdivided, or multiple bars may be used, with cross-hatching, colors, or shading to indicate different divisions. Note the differences in the three examples of bar charts on pages 99 and 100.

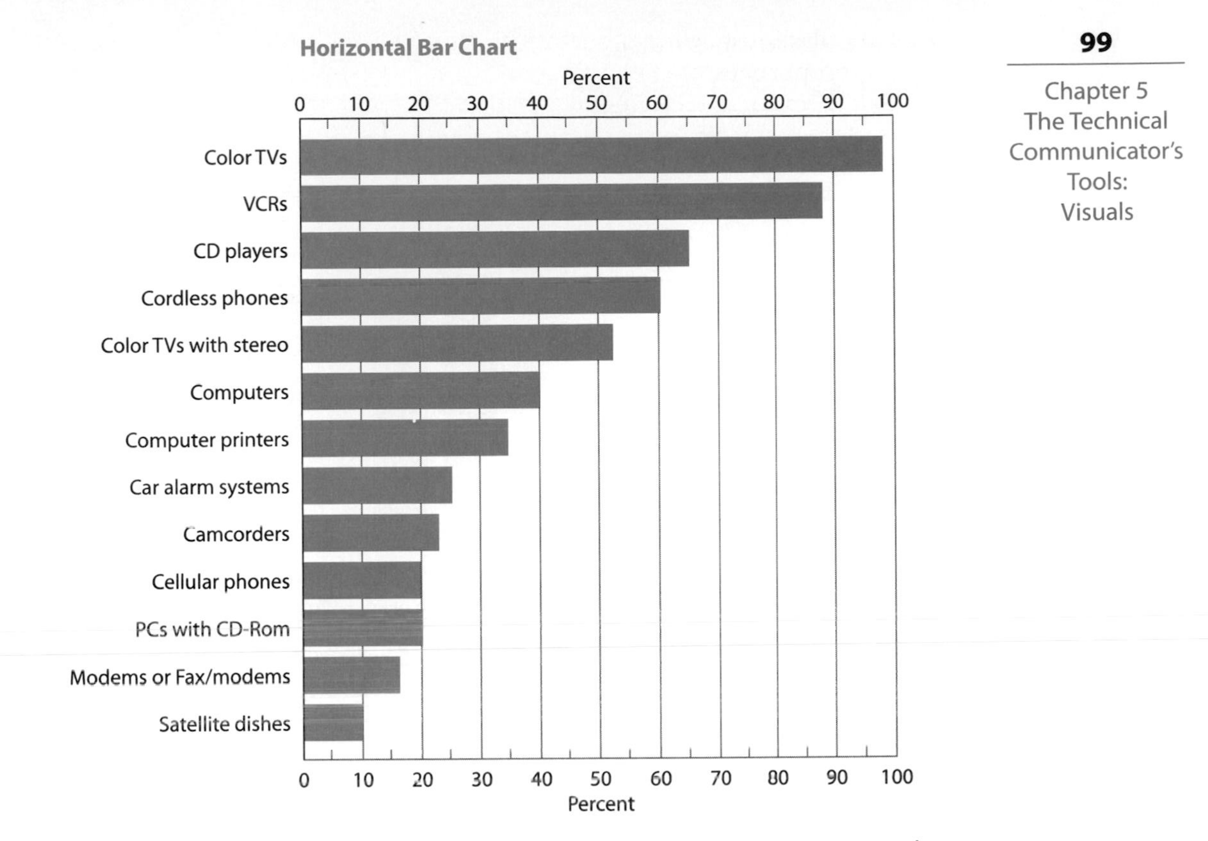

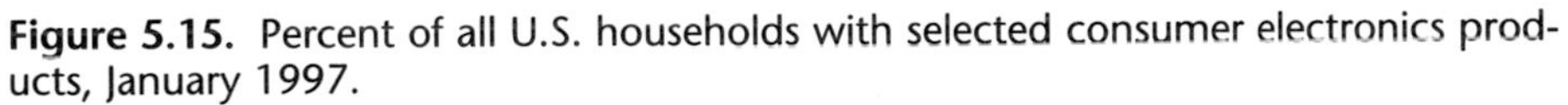

Figure 5.15. Percent of all U.S. households with selected consumer electronics products, January 1997.

Source: U.S. Department of Labor, 1999.

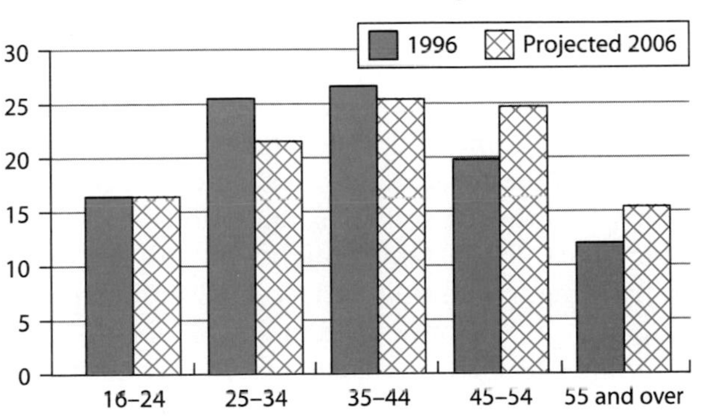

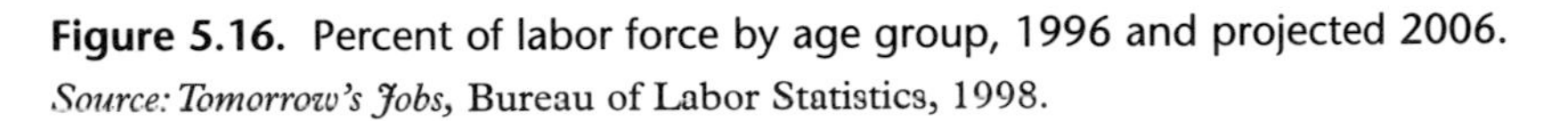

Figure 5.16. Percent of labor force by age group, 1996 and projected 2006.

Source: Tomorrow's Jobs, Bureau of Labor Statistics, 1998.

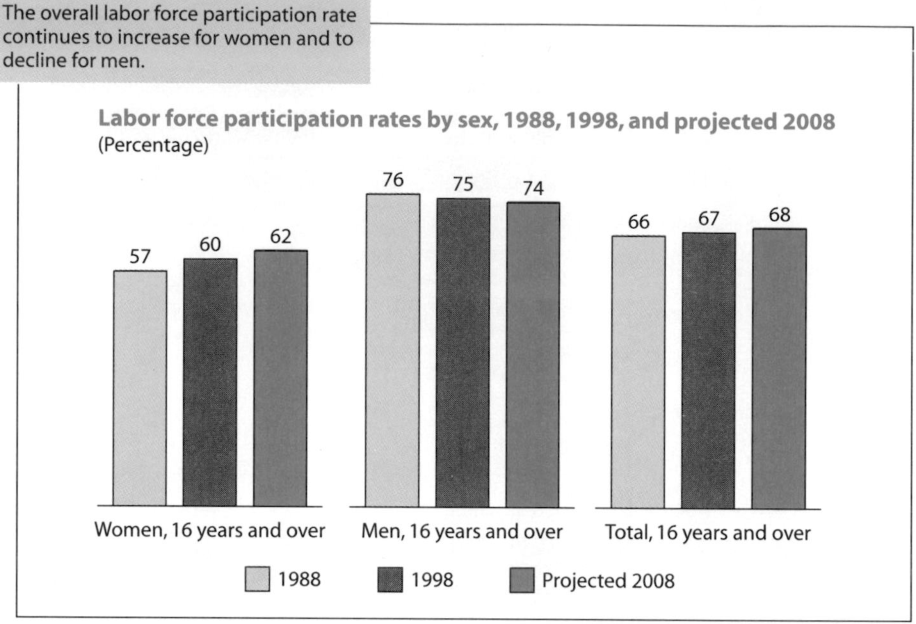

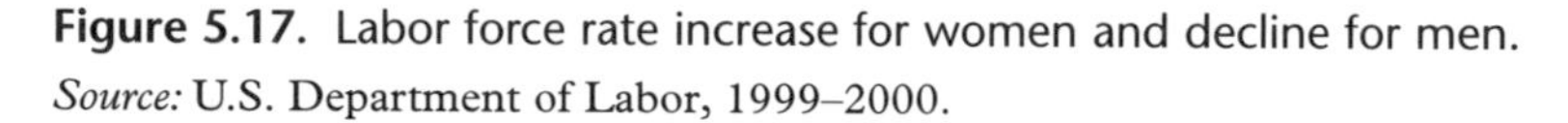
Figure 5.17. Labor force rate increase for women and decline for men.
Source: U.S. Department of Labor, 1999–2000.

Organization Chart

The organization chart is helpful in showing the structure of businesses, institutions, and governmental agencies. Unlike many other charts, the organization chart does not present statistical information. Rather, it reflects lines of authority, levels of responsibility, and kinds of working relationships. Organization charts depict the interrelationships of (1) personnel; (2) administrative units, such as offices or departments; or (3) functions, such as sales, production, and purchasing.

A staff organization chart shows the position of each individual in the organization, to whom each is responsible, over whom each has control, and the relationship to others in the same or different divisions of the organization. An administrative unit organization chart shows the various divisions and subdivisions. The administrative units of a large supermarket, for instance, include the produce department, the meat department, the grocery department, and the interrelationships of different activities, operations, and responsibilities. A college organization chart, for instance, might show the structure of the college by functions: teaching, community service, research, and the like. A bank organization chart, such as the one in Figure 5.18, shows lines of authority and responsibility among officers, managers, and other employees.

An organization chart must be internally consistent; that is, it should not jump randomly from, say, depicting personnel to depicting functions. Blocks or circles containing labels are connected by lines to indicate the organiza-

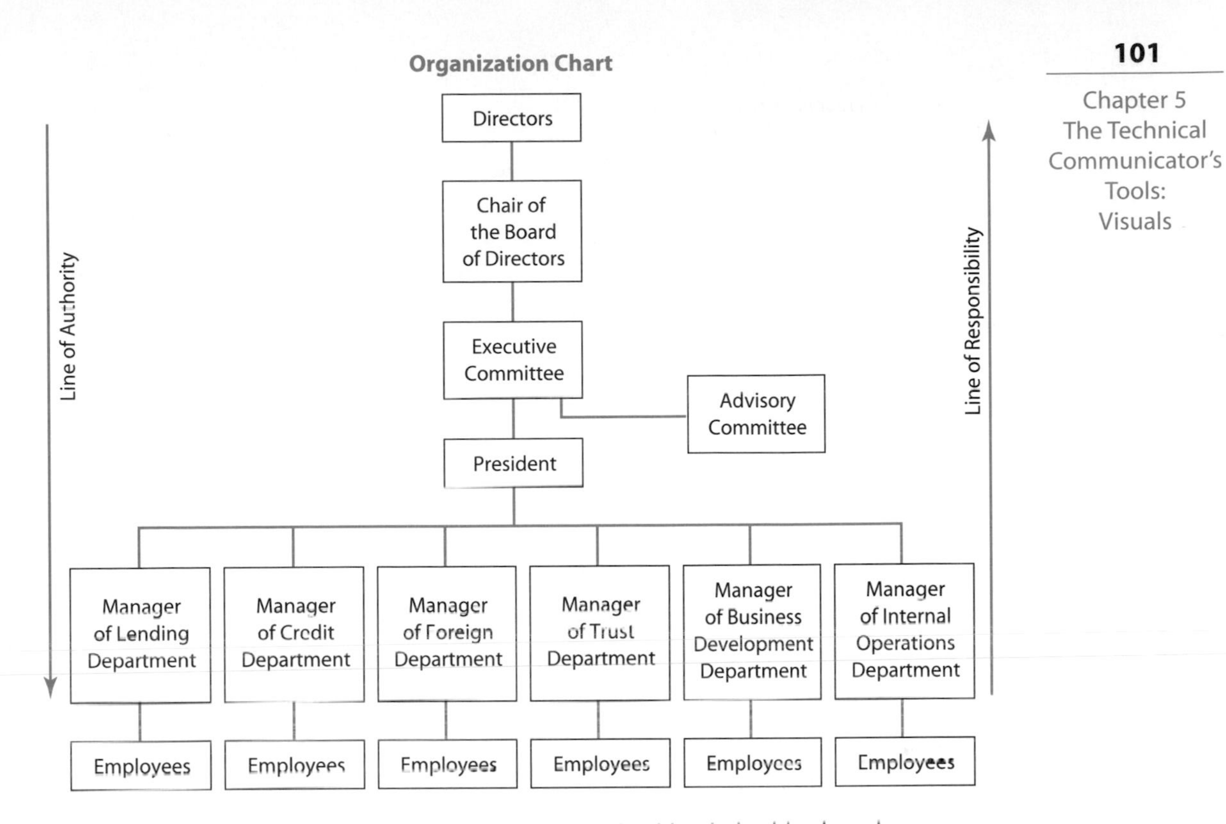

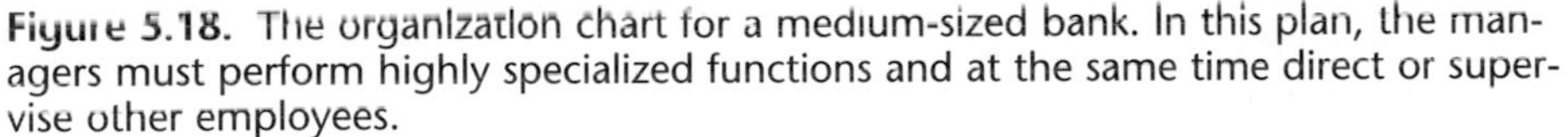

Figure 5.18. The organization chart for a medium-sized bank. In this plan, the managers must perform highly specialized functions and at the same time direct or supervise other employees.

tional arrangement. Heavier lines are often used to show chain of authority, while broken lines may show coordination, liaison, or consultation. Blocks on the same level generally indicate the same level of authority.

Flowchart

The flowchart shows the flow or sequence of related actions. It presents events, procedures, activities, or factors and shows how they are related. The flowchart can also show the path of development of a product from its beginning as raw material to its complete form, the movement of persons in a process, the steps in the execution of a computer operation. Labeled blocks, triangles, circles, and the like (or simply labels) represent the steps, although sometimes simplified drawings that suggest the actual appearance of machines and equipment may be used as well. Usually, arrows show the direction in which the activity or product moves.

The flowchart below uses labeled blocks and a circle, arrows, and screened and unscreened lettering to explain the work process in a laundry and dry cleaning plant.

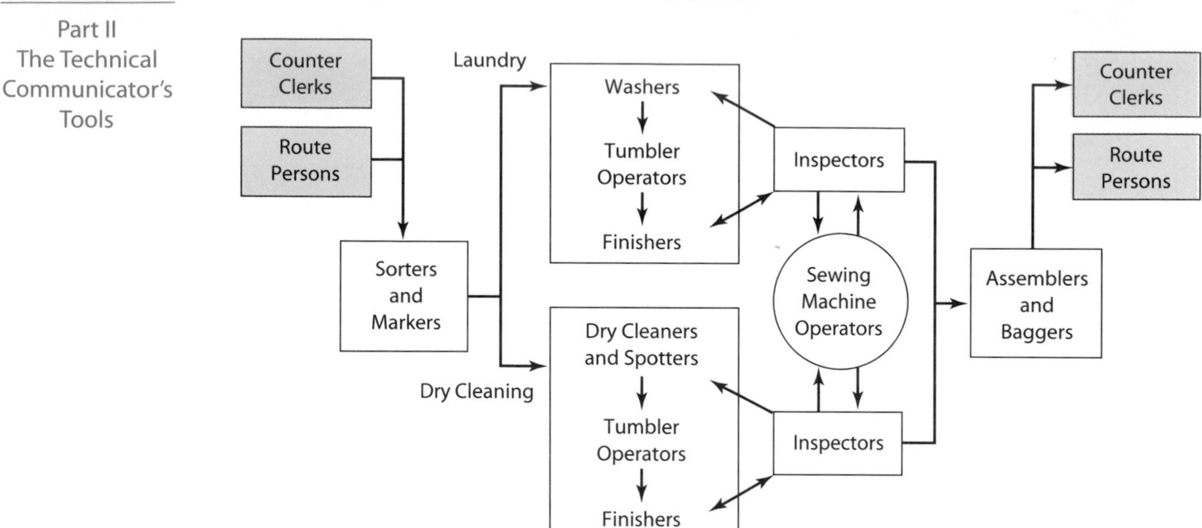

Figure 5.19. How work flows through a laundry and dry cleaning plant.
Source: Bureau of Labor Statistics. n.d.

Graphs

Graphs present numerical data in easy-to-read form. They are often essential in communicating statistical information, as a glance through a business periodical, a report, or an industry publication will show. Graphs are especially helpful in identifying trends, movements, relationships, and cycles. Production or sales graphs, temperature and rainfall curves, and fever charts are common examples. Graphs simplify data and make interpretation easier. But whatever purpose a specific graph may serve, all graphs emphasize *change* rather than actual amounts.

Consider, for instance, the graph on page 104 showing the monthly changes over a one-year period in Consumer Price Index levels for video products other than televisions. At a glance you can see the change over the year. Presented verbally or in a table, the information would be less dramatic and would require more time for study and analysis. But presented in a graph, the information is immediate and memorable for the audience.

Guidelines for Constructing Graphs

Consider the following as you prepare graphs:

1. A graph is labeled as Figure 1 or Fig. 1, Figure 2, etc., or Figure 1.1, Figure 2.1, etc. (omit the label if only one visual is included). Give each graph a descriptive caption, or title. Place the label and caption (on the same line) below the graph.

2. A graph has a horizontal and a vertical scale. The vertical scale usually appears on the left side (the same scale may also appear on the right side if the graph is large), and the horizontal scale appears below the graph.
3. Generally, the independent variable (such as time or distance) is shown on the horizontal scale; the dependent variable (such as money, temperature, or number of people) is shown on the vertical scale.
4. The horizontal scale increases from left to right. If this scale indicates a value other than time, labeling is necessary. The vertical scale increases from bottom to top; it should always be labeled. Often this scale starts at zero, but it may start at any amount appropriate to the data being presented.
5. The scales on a graph should be planned so that the line or curve creates an accurate impression that is justified by the facts. To the viewer, a sharp rise or fall in the line means a significant change. Yet the angle at which the line goes up or down is controlled by the scales. If a change is important, the line indicating that change should climb or drop sharply; if a change is unimportant, the line should climb or drop less sharply. See the multiple line graph on page 104.
6. A graph may have more than one line. The lines should be amply separated for easy recognition and distinction, yet close enough for clear comparison. Each line should be clearly identified either above or to the side of the line or in a legend. Often, variations of the solid unbroken line are used, such as a dotted line or a dot/dash line. Note the kinds of lines and their identification in the graphs below and on page 104.
7. The line connecting two plotted points may be either straight or curved. Straight lines are usually preferable if the graph shows changes that occur at stated intervals; a curved line is preferable if the graph shows changes that occur continuously.

Multiple-Line Graph

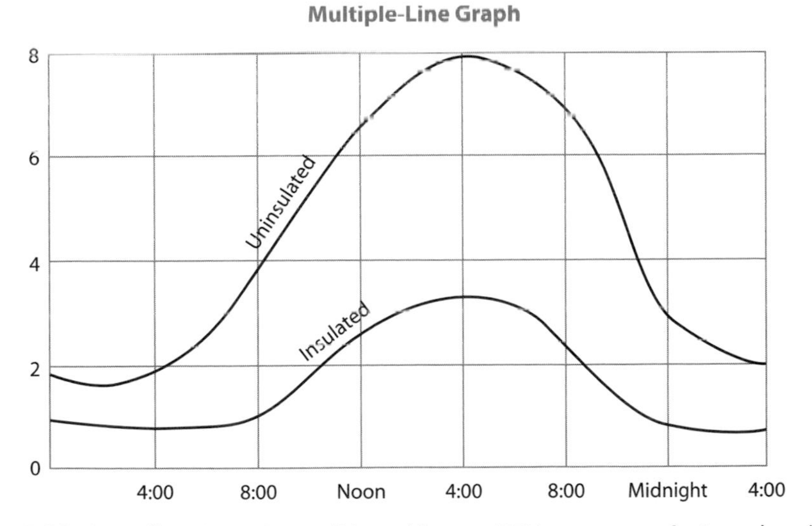

Figure 5.20. Heat flow into air-conditioned house (BTU per square foot per hour).

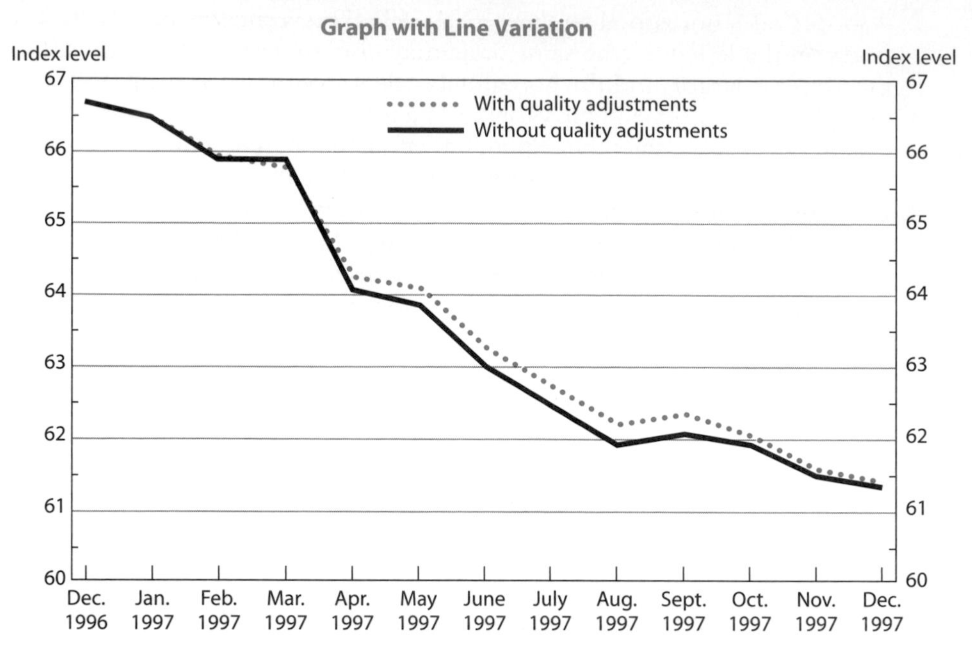

Figure 5.21. Consumer Price Index levels for video products other than televisions, with and without hedonic quality adjustments, December 1996–97.

Source: Monthly Labor Review. U.S. Department of Labor, September 1999.

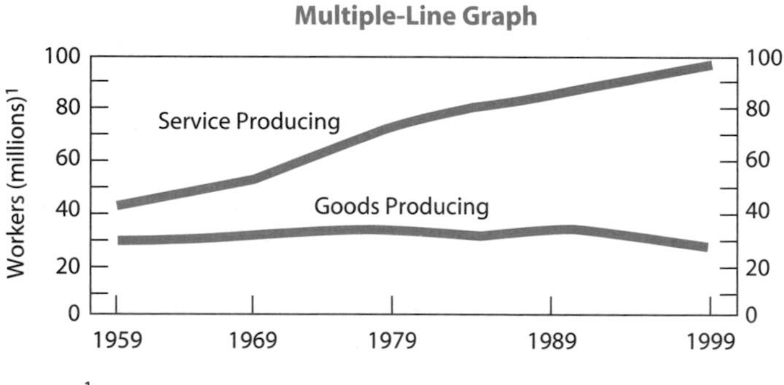

[1]Includes wage and salary workers, the self-employed, and unpaid family workers.

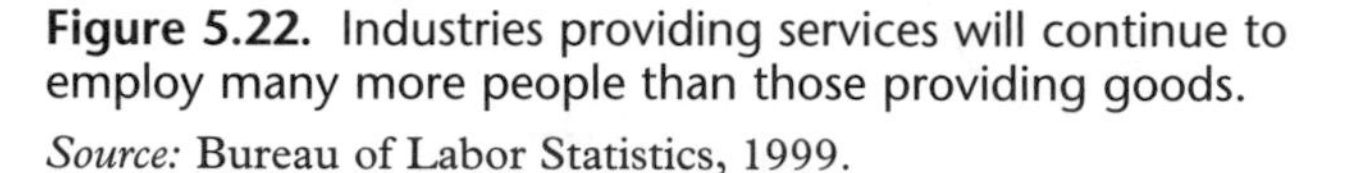

Figure 5.22. Industries providing services will continue to employ many more people than those providing goods.

Source: Bureau of Labor Statistics, 1999.

Tables

Tables are an excellent means of presenting large amounts of data concisely. (See the tables on pages 106, 108, 182, 304, 361–363, 367, and 372.) Although tables lack the visual appeal of charts or graphs, they are unbeatable as a method of organizing and depicting statistical information compiled through research. In fact, information in most other visuals showing numerical amounts and figures derives from data originally calculated in tables.

Tables may be classified as informal (such as those on pages 304, 321–324, and 361–363) or formal (pages 106 and 367). Informal tables are incorporated as integral parts of a paragraph. Thus, they are not given identifying numbers or titles. Formal tables are set up as separate entities (with identifying number and title) but are referred to and explained as needed in paragraphs of related discussion.

Guidelines for Constructing Tables

As you prepare tables, observe the following practices:

1. Number each table (the number may be omitted if only one table is included) and give it a descriptive caption (title); number and caption are omitted in informal tables. Center the number and the caption *above* the table. Often the word *table* is in all capitals, and the number is in arabic numerals, centered above the caption in all capitals.

   TABLE 2
   LEADING CAUSES OF DEATH AMONG AMERICANS

   Other acceptable practices include giving the table number in roman numerals or as a decimal sequence. Captions may also be set with each main word capitalized, as in the first example below, or with only the first word capitalized, as in the second example.

   Table II. Leading Causes of Death among Americans
   Table 2.1. Leading causes of death among Americans

   An example of a formal table appears on page 106.

2. Label each column accurately and concisely. If a column shows amounts, indicate the unit in which the amounts are expressed, such as: Wheat (in metric tons).
3. To save space, use standard symbols and abbreviations. If items need clarification, use footnotes, placed immediately below the table. (Table footnotes are separate from ordinary footnotes, which are placed at the bottom of the page.)
4. Generally, use decimals instead of fractions, unless it is customary to use fractions (as in the size of drill bits or hats).
5. Include all factors or information that affect the data. For instance, omission of wheat production in a table "Production of Chief United States Crops" would make the table misleading.
6. Use ample spacing and rules (straight lines) to enhance the clarity and readability of the table, but as a general rule, use as few lines as necessary.
7. If a table continues on another page, repeat the column headings.
8. If a table continues on another page, use the word *continued* at the bottom of each page to be continued and at the top of each continuation page. If column totals are given at the end of the table, give subtotals at the end of each page; and at the top of each continuation page, repeat the subtotals from the preceding page.

REMINDER

Use the type of visual that most clearly and accurately expresses the idea you wish to communicate.

Fatal Occupational Injuries by Event or Exposure, 1993–98

Event or exposure[1]	Fatalities			
	1993–97	1997[2]	1998	
	Average	Number	Number	Percent
Total	6,335	6,238	6,026	100
Transportation Incidents	2,611	2,605	2,630	44
Highway incident	1,334	1,393	1,431	24
Collision between vehicles, mobile equipment	652	640	701	12
Moving in same direction	109	103	118	2
Moving in opposite directions, oncoming	234	230	271	4
Moving in intersection	132	142	142	2
Vehicle struck stationary object or equipment	249	282	306	5
Noncollision incident	360	387	373	6
Jackknifed or overturned—no collision	267	298	300	5
Nonhighway (farm, industrial premises) incident	388	377	384	6
Overturned	214	216	216	4
Aircraft	315	261	223	4
Worker struck by a vehicle	373	367	413	7
Water vehicle incident	106	109	112	2
Railway	83	93	60	1
Assaults and violent acts	1,241	1,111	960	16
Homicides	995	860	709	12
Shooting	810	708	569	9
Stabbing	75	73	61	1
Other, including bombing	110	79	79	1
Self-inflicted injuries	215	216	223	4
Contact with objects and equipment	1,005	1,035	941	16
Struck by object	573	579	517	9
Struck by falling object	369	384	317	5
Struck by flying object	65	54	58	1
Caught in or compressed by equipment or objects	290	320	266	4
Caught in running equipment or machinery	153	189	129	2
Caught in or crushed in collapsing material	124	118	140	2
Falls	668	716	702	12
Fall to lower level	591	653	623	10
Fall from ladder	94	116	111	2
Fall from roof	139	154	156	3
Fall from scaffold, staging	83	87	97	2
Fall on same level	52	44	51	1
Exposure to harmful substances or environments	586	554	572	9
Contact with electric current	320	298	334	6
Contact with overhead power lines	128	138	153	3
Contact with temperature extremes	43	40	46	1
Exposure to caustic, noxious, or allergenic substances	120	123	104	2
Inhalation of substances	70	59	48	1
Oxygen deficiency	101	90	87	1
Drowning, submersion	80	72	75	1
Fires and explosions	199	196	205	3
Other events or exposures[3]	26	21	16	–

[1] Based on the 1992 BLS Occupational Injury and Illness Classification Structures.

[2] The BLS news release issued August 12, 1998, reported a total of 6,218 fatal work injuries for calendar year 1997. Since then, an additional 20 job-related fatalities were identified, bringing the total job-related fatality count for 1997 to 6,238.

[3] Includes the category "Bodily reaction and exertion."

NOTE: Totals for major categories may include subcategories not shown separately. Percentages may not add to totals because of rounding. Dash indicates less than 0.5 percent.

Figure 5.23. Formal table.

Source: Monthly Labor Review, U.S. Department of Labor, September 1999.

GENERAL PRINCIPLES for Visuals

- To communicate your message, visuals must be designed and used so that your audience can easily and quickly recognize the main ideas.
- Visuals must be appropriately developed, usefully placed, and discussed in the text in order to be effective. Visuals must be also timed to appear appropriately with spoken information in a presentation.
- Credit must be given for all borrowed material. Whether you have borrowed an entire visual or the information on which the visual is based, you must credit the source.
- Graphics software can make it easy for writers and speakers to develop their own visuals.
- Photographs can represent real or altered appearances.
- Drawings and diagrams show isolated or interior views, exact details, and relationships.
- Charts show relationships. Common types of charts are the circle chart, the bar chart, the organization chart, and the flowchart.
- Graphs show change by charting data. Graphs are helpful in identifying trends, movements, and cycles.
- Tables present specific statistical information, typically with the information in columns.

CHAPTER SUMMARY

VISUALS

Visuals, used well, have the power to communicate more immediately, more fully, and more interestingly than the written or spoken word. However, visual language, like written language, must be direct and simple to communicate. Also, since visuals convey attitudes and values in addition to ideas and since audiences perceive visual ideas in their own personal and cultural contexts, the visual communicator must be especially sensitive to the ways in which others might interpret visual messages. The right kind of visual, carefully prepared for the right audience and message, together with words, can communicate a message fully, memorably, and powerfully.

ACTIVITIES

INDIVIDUAL AND COLLABORATIVE ACTIVITIES

5.1. From documents such as periodicals, reports, brochures, and government publications, collect ten visuals pertaining to your major field of study. If necessary, make a photocopy from the original source.
   a. Write two or three sentences evaluating the effectiveness of each visual for its audience.
   b. In groups of three or four, compare the visuals you have collected and commented on above. What are the advantages and disadvantages of the visuals you and your teammates have found?

5.2. Present the following information in an appropriate visual or visuals: Government service, one of the nation's largest fields of employment, provided jobs for 22 million civilian workers in 1999—about one out of six persons employed in the United States. Nearly four-fifths of these workers were employed by state or local governments, and more than one-fifth worked for the federal government.

5.3. Select information from the table below, which describes the estimated preparation time required to complete IRS Form 1040A and its schedules. Present the selected information in a nontable visual.

5.4. In groups of three or four, discuss the advantages and disadvantages of using computer-generated graphics. Does the time spent learning to use

Estimated Preparation Time

The time needed to complete and file Form 1040A and its schedules will vary depending on individual circumstances.

The estimated average times are:

Form	Record-keeping	Learning about the law or the form	Preparing the form	Copying, assembling, and sending the form to the IRS	Totals
Form 1040A	1 hr., 17 min.	2 hr., 29 min.	4 hr., 1 min.	35 min.	8 hr., 22 min.
Sch. 1	20 min.	4 min.	10 min.	20 min.	54 min.
Sch. 2	33 min.	10 min.	35 min.	28 min.	1 hr., 46 min.
Sch. 3	13 min.	14 min.	25 min.	35 min.	1 hr., 27 min.
Sch. EIC	0 min.	2 min.	4 min.	20 min.	26 min.

Figure 5.24. Estimated preparation time for completing and filing Form 1040A and its schedules.

Source: Internal Revenue Service, 1998 1040A Instructions.

the software outweigh the advantages of creating visuals quickly? Why or why not?

5.5. What kinds of visuals can best illustrate an oral presentation? In groups of three or four, outline an oral presentation, describing the kinds of visuals you would include and indicating where they should appear in the presentation. Include a discussion of your audience. What kinds of visuals would be easiest to handle and show during a presentation?

Photographs

5.6. Find two photographs (from your own collection, from newspapers, or from periodicals) that present evidence. Write a brief comment about each photograph. In what ways does each photograph present evidence as only a photograph can?

5.7. Consider a situation or communication project in your career area in which a photograph communicates essential information. Then find or take a suitable photograph. Finally, write a paragraph to accompany the photograph that describes the information presented in the photograph.

Drawings and Diagrams

5.8. Make a drawing of a piece of equipment used in your major field of study and indicate the major parts. Then write a paragraph identifying the piece of equipment and include the drawing.

5.9. Make a diagram of a classroom building or a workplace building to show an emergency evacuation plan.

Circle Charts

5.10. Present the following information in the form of a circle chart.

FROM STEER TO STEAK

Choice steer on hoof	1,000 lbs
Dresses out 61.5%	615 lbs
Less fat, bone, and loss	183 lbs
Saleable beef	432 lbs

5.11. Prepare a circle chart depicting your expenses at college for a term. Indicate the total amount in dollars. Show the categories of expenses in percentages.

Bar Charts

5.12. Prepare a bar chart showing the total amount in dollars paid annually in Gomez's retirement plan and the individual amounts paid by Gomez, Comal County, and the state.

> Roy Gomez is a transportation maintenance supervisor for the Comal County Highway Department. His salary is $50,000 per year. Gomez participates in a retirement plan into which he annually pays 5 percent of his salary, Comal County pays 4 percent of his salary, and the state pays 2 percent of his salary.

5.13. Make a bar chart showing the sources of income and areas of expenditure for a household, for your college, or for some other organization or firm.

Organization Charts

5.14. Make an organization chart of the administrative personnel of your college or some other institution, or of the personnel in a firm.

Flowcharts

5.15. Make a flowchart of the registration procedures or some other procedure in your college or of a procedure in your place of employment.

5.16. Make a flowchart depicting the flow of a process or product in your field of study from beginning to completion.

Graphs

5.17. Construct a line graph depicting increases in minimum wage for a twenty-year period.

5.18. Prepare a line graph showing the enrollment for the past ten years in your college and in another college in your area. If you are unable to obtain the exact enrollment figures, use your own estimates.

Tables

5.19. Conduct a survey of twenty people near your age concerning their preferences in automobiles: make, color, accessories. Present your findings in a table for a car dealership.

5.20. From a bank or home finance agency, obtain the following information concerning a $50,000 house loan, a $70,000 house loan, and a $90,000 house loan:

- Interest rate
- Total monthly payment for a 30-year loan
- Total interest for 30 years
- Total cost of the house after 30 years
- Cost of insurance for 30 years
- Estimated taxes for 30 years

a. Present the information in a table.
b. In groups of three or four, compare the tables you have developed. What visual elements can make a table easiest to follow and understand?

READING

5.21. In this selection from "Pictures Please—Presenting Information Visually," William Horton, visual communication expert and fellow of the Society for Technical Communication, discusses the power of visuals in making information appealing and interesting.

From Pictures Please—Presenting Information Visually

WILLIAM HORTON

WHY USE GRAPHICS?

Graphics Record Information Concisely. Even great writers profess the concise power of graphics.

> A picture shows me at a glance what it takes dozens of pages of a book to expound.
>
> *Ivan Turgenev*

This power is not lost on business leaders.

> The higher one looks in administrative levels of business, the more one finds decisions are based on tabular or graphic formats.
>
> *Norbert Enrick*

A study by the Wharton School of Management at the University of Pennsylvania found that speakers who used visual aids were twice as successful in achieving their goals as speakers who did not use visual aids. Participants in meetings where visual aids were used retained five times as much information, and these meetings took only half as long.

Pictures Entice and Seduce Readers. Most magazines and paperback books use an attractive graphic on the cover. Even staid technical journals, such as the *Journal of the American Medical Association,* are using graphics to entice readers to pick up the publication and seduce readers into reading the articles. Similarly, graphics hold attention once people look inside a publication. Consider all the photos, charts, and illustrations in news magazines and newspapers.

Thoughts Are Visual. . . . In *The Ego and the Id,* Freud noted the visual nature of thought:

> . . . it is possible for thought processes to become conscious through a reversion to visual residues. . . . Thinking in pictures . . . approximates more closely to unconscious process than does thinking in words, and is unquestionably older than the latter both ontogenically and phylogenetically.

This importance was echoed by Albert Einstein who, in a letter quoted in *The Creative Process,* edited by Brewster Ghiselin, described his own thought process:

> The words of the language, as they are written or spoken, do not seem to play any role in my mechanism of thought. The psychical entities which seem to serve as elements in thought are certain signs and more or less clear images which can be "voluntarily" reproduced and combined. . . . The above mentioned elements are, in my case, of visual and some of muscular type. Conventional words or other signs have to be sought for laboriously only in a secondary stage, when the mentioned associative play is sufficiently established and can be reproduced at will.

. . . Even in Words. Consider, finally, the importance of vision and seeing throughout our language. We speak of someone as being "far sighted" or "visionary" or a "seer." When we agree we see "eye-to-eye." We are on the "lookout" for bargains and we relish the "sight for sore eyes." Something that gets our attention is "eye catching" or "eye opening." We trust "eyewitnesses" and hire "private eyes." After "looking into" something we develop our own "viewpoint."

In summary, graphics

- Communicate what words cannot
- Are understood more quickly than words
- Are remembered better than words
- Record information concisely
- Entice and seduce readers

Using graphics to express ideas requires as much thought and consideration as using words.

Source: From Horton, William, "Pictures Please—Presenting Information Visually," from Barnum, C. M. & Carliner, S., *Techniques for Technical Communicators,* ©1994 by Allyn & Bacon.

QUESTIONS FOR DISCUSSION

a. In what ways does "Pictures Please—Presenting Information Visually" suggest that graphics and visuals are persuasive? What kinds of graphics and visuals can best persuade readers?

b. In what ways do visual and verbal information work together? What examples of visual and written communication best reinforce each other?

c. How do Horton's principles of visual communication apply to oral communication? How can graphics or visuals influence the audience of an oral communication?

d. What are the roles of visuals in advertisements? Select a typical magazine advertisement for discussion. In what ways does the visual impress the reader? How much specific information does the text of the advertisement provide? Would this approach be ethical or appropriate in workplace communication?

e. Are visuals only about information? What kinds of visual information are important in workplace materials you have seen? How do these visuals interest you in the subject matter or encourage you to learn more?

CHAPTER 6

The Technical Communicator's Tools: Technology

CHAPTER GOALS

This chapter:

- Explains the uses of communication technologies
- Discusses the tools communicators can use to create, record, and edit information
- Discusses the types of media that can deliver information
- Evaluates the uses of different types of tools and media
- Provides practical guidelines for selecting and using technical tools and media for communication projects

INTRODUCTION

Today's communication technologies, many of which are computer-based, provide more options for writing, revising, and designing text and developing visuals than ever before. Other communication technologies allow information to be delivered faster to more people than at any time in history. While these new communication technologies may sound magical, they're not. Every new communication technology—from quill pen to computer keyboard, from newsprint to broadcast satellite—solves some communication problems but they create new ones. To make informed choices, you must be aware of the features, advantages, and drawbacks of the communication technologies you use.

This chapter introduces communication technologies, both the tools that you can use to develop information and the media that can be used to deliver information. While explaining in depth how to use these technologies is beyond the scope of this text, this chapter provides an overview of what is available for workplace communicators, explaining choices, pointing out advantages, and describing special considerations for each tool and medium.

WHAT ARE COMMUNICATION TECHNOLOGIES?

Communication technologies offer writers, editors, information designers, and visual communicators wonderful options for creating and delivering information. But what, exactly, is a technology? Technology applies science to find a solution to a practical problem. For example, consider the pocket calculator. When people solve mathematical problems manually, they often work slowly and make mistakes. The pocket calculator is a technical solution to this problem. It provides a speedy, affordable way for people to solve mathematical problems quickly and accurately. Technical solutions to such practical problems can be large or small, but when they are successful (such as the pocket calculator), they can rapidly affect work, education, government, and daily life. No matter what problem technology addresses, a technical solution involves equipment. Equipment can be extremely sophisticated, requiring extensive skills and training to use. Equipment can also be expensive.

Communication technologies provide tools that communicators can use. They also provide the means, or media, for delivering information. These media can be as direct as a telephone call, as immediate as an e-mail message, or as global as a teleconference. No matter how simple or sophisticated, communication technologies speed the process of transmitting or broaden the impact of information. To use a tool or medium well, you must recognize its features, advantages, and limitations.

TECHNICAL TOOLS FOR COMMUNICATORS

Some communication tools are easy to use and widely adaptable to many kinds of communication (such as the lead pencil). Other tools are complicated and highly specialized (such as a rapidograph). Electronic communication tools are based on the use of the computer.

The Computer as a Communication Tool

When working with the computer as a tool, you use the computer equipment (hardware) to run specialized programs (software). The keyboard and mouse are called input devices equipment (hardware) used directly to create information. You use the computer keyboard to type written information, much as you use the keyboard on a typewriter. You can use the mouse, a rounded plastic device that your hand guides to relay commands. Both the keyboard and the mouse can direct computer programs (software) to record written or visual material and to print or send it to other sources.

Output devices are pieces of computer equipment that allow you to see what you have developed. Output devices include the monitor, a screen that displays print and visuals, and the printer, which can print information onto paper, also known as hard copy. Saved information can also be stored on a disk or hard drive. The diagram below shows input and output devices for a typical computer.

Computer System

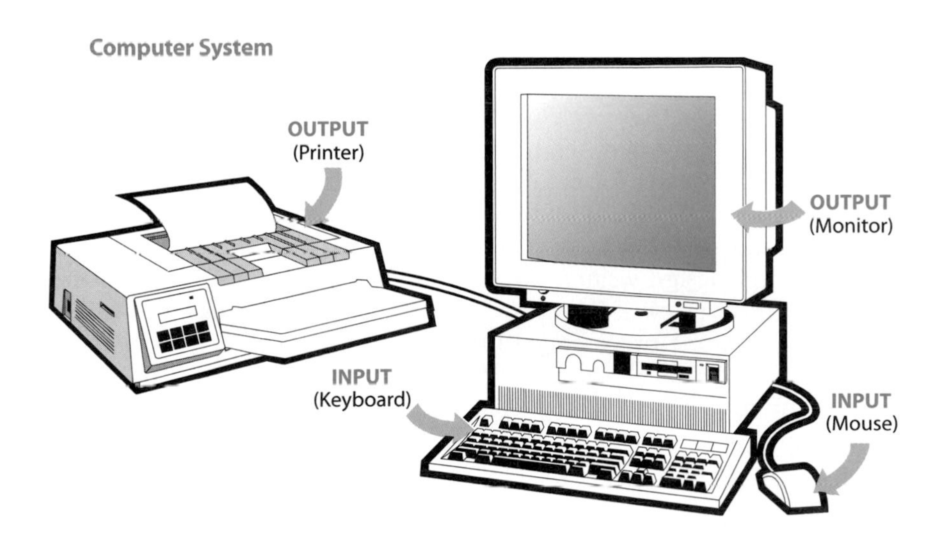

Input devices send commands to a computer's central processing unit, an electronic device that (among other things) runs computer programs, or *software*. These individual programs are the tools communicators use to write, design, create visuals, plan presentations, and develop Web pages.

Tools for Writing

Using software to create written information is called word processing. With a computer and a printer, you can draft, edit, design, print, and distribute documents. Below are some typical functions of word processors.

- **Writing and editing.** You can type (or enter) text on the keyboard and then insert, delete, or move text you have created. By reviewing a printed copy of the text, you can easily see what to change. You can then make these changes in text using the mouse or the keyboard to enter commands.
- **Page design**. Word processing programs typically offer a choice of type fonts and sizes, as well as a choice of type styles, including bold and italic. In addition, word processors allow you to set margins, center lines and pages, and control the justification of lines of type. Word processing programs make it easy to use headings.
- **Spell check.** A spell check is a regular word processing feature. You can use a spell check to review the spelling of single words or of words in an entire document. When the spelling of a word is uncertain, most spelling checkers will suggest several possible corrections. You can decide on a spelling choice, ignore the correction, or type a different correction.
- **Grammar check.** A grammar check is a common part of many word processing systems. This feature can be used to review the correctness of grammar in a document's sentences. The grammar check highlights possible errors, offering suggestions that you can act on or ignore. Some grammar checks also review for style, offering suggestions about matters such as wordiness and the use of passive voice.
- **Collaboration.** Because it is easy for others to see and comment on a text file, word processors make collaboration easy.

Word processors vary widely in the range of features they offer and in their ease of use. Some allow communicators to create lists and spreadsheets; others provide templates for common charts and graphs. Whatever their features, word processors are routine in the workplace. Access to word processors is easy, and the programs offer an increasingly affordable way to draft, edit, revise, and design written documents. Microsoft Word®, Appleworks®, and WordPerfect© are well-known word processing programs.

Relying on word processing software for editing and revision is risky. While word processors allow you to draft quickly, an unrevised rough draft is still rough, no matter how attractively it may be designed. In addition, spelling checkers and grammar checkers offer suggestions, but you must make final decisions about all changes. A spelling checker recognizes known words, but it cannot know the exact word you had in mind. In addition, a spelling checker will not highlight words that are correctly spelled. For example, if you type "ion" and meant to type "in," a spelling checker will not recognize the error. Likewise, grammar checkers often highlight correct sentences, suggesting that they are incorrect. You must be responsible for reviewing, editing, and revising documents to make sure that they are clear, accurate, and correct.

Tools for Designing

Using software to design information is called desktop publishing. Information designers can use desktop publishing programs to create high quality material for commercial printing of newsletters, brochures, reports, and even books. Typical desktop publishing features are listed below.

REMINDER

Writers are responsible for the subject matter, organization, clarity, and correctness of what they write. No matter how attractive it may look on the page, unrevised writing will still be rough. As they say in the computer world, GIGO (garbage in, garbage out).

- **Page design.** Desktop publishing offers complete information design features, including many type fonts, sizes, and styles and exact alignment on the page. In addition, desktop publishing helps you to implement page grid plans by providing columns, rules, and borders. It can also be used to number pages, chapters, and captions, providing style options for many kinds of publications. You can use or adapt any of these features.
- **Word processing.** Many desktop publishing programs provide the writing and editing features of word processors.
- **Use of visuals.** Desktop publishing allows you to take visuals such as graphs, charts, or drawings from other sources and place them with text, altering the size of the visual as necessary.
- **Accurate preview.** Desktop publishing allows you to clearly see the pages you are working with on a computer monitor.

Since you can both plan design and set type with desktop publishing software, costs of design and page layout are far lower than typesetting, and anyone who can use a desktop publishing system has the resources to create professional quality publications. QuarkXPress® and PageMaker® are popular desktop publishing programs; Word Perfect®, Appleworks®, and MSWord® also have desktop publishing features but with considerably less capability.

Desktop publishing requires skill, knowledge, and training. While desktop publishing offers far more features than word processing, it also requires that you understand what these features are and how to take advantage of them. In addition, desktop publishing software requires powerful computers to run quickly and large monitors to allow designers to see their work clearly. Communicators who understand all of the commands of the desktop publishing system but do not understand the elements of design can create cluttered and hard-to-read documents.

REMINDER

Desktop publishing requires knowledge of visual design principles and of desktop publishing programs.

Tools for Creating Visuals

Software used to create and edit visual information is called a graphics program. You can use graphics programs to create, texture, and color illustrations, charts, and graphs. You can then print these visual images or save them electronically for use in desktop publishing, presentations, or Web pages. Below are some of the many available features of graphics programs.

- **Drawing.** A drawing program allows you to create illustrations, to move and change the individual elements of the drawing, and to use a variety of lines and textures.
- **Painting.** A paint program allows you to use the mouse to simulate painting on a monitor screen. In paint programs, it is possible to control the use and appearance of color quite accurately. However, paint programs do not allow you to isolate and move parts of the design in the same way that drawing programs do.
- **Image editing.** Image editing programs allow you to scale, alter, and enhance visual images created by other programs. This software also allows you to edit images—drawings and photographs—that have been *scanned*, or converted from print to digital format. Image editing programs can also be used to edit images taken from copyright-free clip art programs or the Web.
- **Charts, graphs, and tables.** Graphics programs provide sophisticated systems for creating and adapting a variety of visuals common in workplace communication and for enhancing these visuals with color and texture.

Graphics software offers visual communicators tools they need to create vivid, interesting, and effective visuals. However, using graphics software requires familiarity with elaborate software systems, powerful computers, and large monitors, as well as an in-depth understanding of visual principles. Adobe Illustrator®, CorelDRAW®, and Freehand® are common graphics programs.

REMINDER

You should respect the intellectual property you scan or download from the Web. You must document the source and, if needed, request permission for its use.

Tools for Presentations

Presentation graphics software allows you to prepare charts, graphs, and diagrams as files to be used in an oral presentation. With a computer connected to a data projector, you can project these images onto a screen during your presentation (see page 494 Chapter 16 Oral Communication). Common features of presentation programs are listed below.

- **Backgrounds.** Presentation programs provide backgrounds and sample page designs so that you can make all of the visuals used in a speech consistent in design. These backgrounds and templates can also be used to frame and highlight imported images such as photographs, drawings, and clip art taken from other programs or sources.
- **Text.** You can write and edit text to accompany visuals or to highlight spoken ideas.

- **Visuals.** Presentation packages offer templates and formats you can use to develop graphs, charts, and diagrams to accompany oral presentations. These templates are similar to the ones that accompany some word processing programs.
- **Sequencing.** You can use presentation software to provide visual sequence and transition between images in much the same way that an edited film or video does.
- **Media.** Some presentation software allows you to incorporate sound and video into your presentation.

Presentation graphics offer many options for planning and creating text, visuals, and media to make presentations memorable and interesting. More elaborate multimedia features require more skill, but basic presentation programs are designed to be usable. Presentation programs include Harvard Graphics®, Freelance Graphics®, PowerPoint®, and Charisma®.

Most presentation software is designed to be easy to learn and to use. However, to display the materials created with presentation software, you must use a computer, a data projector, and an overhead projector during your presentation. This requires you to manage the programs while making your presentation. Presentation packages offer exciting ways to enhance an oral presentation, but if projected materials are more interesting and exciting than the oral presentation itself, you risk losing your audience.

REMINDER

The features of presentation software should enhance, not compete with, an oral presentation.

Tools for Creating Web Pages

Web authoring software allows you to create Web pages. Some Web authoring software automatically encodes a computer file with HTML, or Hypertext Markup Language. The HTML code tells the Web browser software how to display the file on a monitor screen. A Web file can contain text, graphics, and links. Authoring systems allow you to link individual pages to one another in the series of pages, called Web sites, you create. You can use the software to see individual pages or the structure of the entire Web site.

To use Web authoring software, you must understand the Web as a technical medium. Readers cannot use, see, or read information on a Web page in the same way they would a page of printed text (see pages 56–60 Chapter 4 Information Design).

REMINDER

Web-page authoring requires an understanding of medium and design as well as knowledge of how to use the software.

TECHNOLOGY THAT DELIVERS COMMUNICATION

The means by which people deliver and receive information is called a medium. Your choice of medium can influence the ways in which information is perceived. For example, print on paper is a medium that requires a more active response than a television broadcast or film. A telephone conversation, on the other hand, requires active participation from both parties. The appeal to different senses, the degree of involvement or distance of the person who receives the message, and the speed of delivery are important factors to consider for different media.

Print on the Page

Text on paper is a time-honored medium for conveying a wide range of verbal and visual information. The printed page provides a tangible record of a message, and its parts and pages can be arranged so that users can read, skim, or easily locate information. The printed page provides a wide range of sizes, shapes, and designs that can be adapted for many needs (see pages 60–69 Chapter 4 Information Design).

The drawback of printed information is that, despite the best efforts of information designers, it can require time and concentration for readers to understand. In addition, paper copies cannot be delivered instantly to readers. To allow print materials to arrive with the speed of telephone calls, a fax (facsimile) machine allows you to scan a print image and send it to distant locations over telephone wires. Sending a fax requires that the sender and the receiver of a fax message use special fax machines to scan, transmit, receive, and print copies. While fax copies provide a paper record of communication, their quality is generally less clear than that of an original paper document. In addition, a fax transmission can be more costly to send than a printed copy.

Speaking and Listening: The Telephone

To allow people to conduct conversations at a distance, the telephone provides rapid worldwide access. Thanks to communication satellites, telephone reception between continents can be clear and audible, allowing two people to speak and respond to one another readily. To open the conversation to more than one person, it is possible to set up teleconferences that link distant individuals to one connection.

For all their immediacy, telephone conversations cannot provide a record of exactly what was said. In addition, because people cannot see one another during a telephone conversation, they cannot respond to the visual cues that make one-on-one communication easier and more complete. This lack of visual cues also makes it difficult for all participants to follow and take part in a teleconference.

A telephone call or teleconference has the advantage of immediacy. However, when a call takes place between people in different time zones, the caller must take the time differences into account. A 9:00 a.m. business call from New York City, for example, will be answered in San Francisco at 6:00 a.m. International teleconference calls, with even greater differences between participant time zones, should be planned for the convenience of all participants.

Because the telephone is distancing as a medium, it is important for those who use it to be particularly attentive to one another, using good manners.

Callers should identify themselves fully, respond appropriately to one another, and listen to one another without interrupting. In a teleconference, a moderator can establish an agenda and direct the conversation, making sure that everyone has an opportunity to speak and that no one is interrupted.

Since a telephone call is expected to be immediate, being put on hold can be frustrating. If putting a caller on hold is necessary, let the caller know that he or she is being put on hold and ask if this is acceptable. Likewise, while telephone answering machines postpone a conversation, they should still reflect a willingness to communicate. Recorded messages in workplace telephone systems should courteously encourage leaving a message and promise a reply.

REMINDER

A rude or insensitive telephone conversation leaves a strongly negative impression.

Seeing and Hearing: Videoconferences

To allow people in distant locations to see and hear one another in a conversation, it is possible to set up real-time videoconferences. However, such conferences, often used in distance education, can be less spontaneous than teleconferences because the video camera can transmit only one image from one location at a time. In addition, such conferences require a great deal of equipment at each meeting site, including cameras, telephone links, and microphones. It may also be necessary for technicians to manage the audio, visual, and broadcast features of the event. In addition, such conferences still have the kinds of problems with time zones that teleconferences do, and transmission can be slow, making the conversation even less immediate for participants.

Written Conversations: E-Mail

E-mail, which transmits written messages over the Internet, combines features of the written page with features of the telephone. Like the written page, e-mail uses text to convey ideas. Like the telephone, e-mail gives the illusion of immediacy. However, since e-mail is written on a keyboard and read on a monitor screen, some of the issues of readability and legibility, true of Web page screens, also hold true for e-mail (see pages 69–70 Chapter 4 Information Design). Long messages in large blocks of text are hard to read, particularly when they require the reader to scroll down the page.

Because e-mail messages are easy to write and send instantly, they can create other risks for authors. While e-mail can feel as informal and private as conversation, it isn't. E-mail messages are easy to dash off hastily, even emotionally, particularly since the conventions of tone and format that apply to letters and memorandums do not apply. Haste in drafting can also mean errors in writing, particularly since some e-mail systems do not provide spell checks. In general, it is unwise to be in a hurry to press "send," particularly if you are annoyed. Since e-mails are part of a network, they can be forwarded to anyone with e-mail, making a private message extremely public.

REMINDER

The tone of a hasty e-mail may send a more unpleasant message than you intend.

Despite these problems, e-mail provides some important advantages. E-mail users can send messages at any hour, and recipients can open messages and reply at their convenience. In addition, it is easy to attach computer files of text or graphics, transmitting documents quickly and easily. Since the documents arrive as files, recipients can print the files or edit and comment on them. In this way, e-mail complements collaboration.

Written Conferences: Listservs

Listservs involve many people in an e-mail conversation by sending out messages to people who are part of mailing lists through automated e-mail systems. Recipients can respond to listserv mail messages, or they can simply read and follow the conversation. To join a listserv, users send an e-mail subscription request to the listserv manager, who uses a listserv program to maintain the membership. Once subscribed, users receive all messages sent out over the listserv.

Listservs provide a valuable way for subscribers to keep up with new professional ideas. Thousands of listservs link individuals with interests in every career area. However, listservs are even more public than e-mail. Listserv subscribers should respond courteously to one another, perhaps following the listserv conversation to learn more about the conversation and the people in it before joining in with an e-mail comment. More importantly, they should avoid sending personal messages to a listserv address, since literally thousands of people could be reading a private communication. Since it is easy to confuse a listserv address with that of an individual listserv subscriber, it is easy to send a personal message to an entire listserv membership, wasting the listserv subscribers' time and embarrassing the sender.

Watching and Hearing: Media for Oral Presentations

Recorded information in film, videotape, CD-ROM, audio, and presentation software provides sources that you can use to enhance your presentations. (See Chapter 16 Oral Communication, especially page 494.)

Projected materials are uniquely effective in oral presentations. Among the most common projected materials are films and videos, filmstrips, slides, and transparencies. These are projected onto a screen or some other appropriate surface such as a wall. A slide projector, a data projector, and an overhead projector, a ½-inch or a ¾-inch tape player, or similar equipment is required. Using files created with graphics presentation software, you can project onto a screen a wide variety of materials including pictures, animation, and sound. Presentation software has made the use of other media, including slides, less common since it is easier to use and more flexible. Presentation software also allows a speaker to project images without darkening a room. However, other media are still available to accompany oral presentations.

Films (more accurately, motion pictures) and videos can portray action, sound, and movement. They present a sense of continuity and logical progression. Since the preparation of films and videos requires specialized knowledge and is expensive in terms of time, equipment, and materials, it may be easier to borrow, rent, or purchase commercially prepared films and videos. A projector is needed to show films and videotapes and it may be necessary to darken the room so that the audience can easily see the projected image.

A filmstrip is a series of photographs arranged in sequence on 35-millimeter film. Filmstrips may be supplemented by captions on the frames (pictures), recorded narration, or script reading. Filmstrips are compact, easily handled, and, unlike slides, always in proper sequence. Although rather difficult to prepare locally, filmstrips from commercial producers are inexpensive and cover a wide array of topics. To show a filmstrip, a projector and screen are needed.

Slides, usually 2 inches by 2 inches, are taken with a 35-millimeter camera; they provide colorful, realistic reproductions of original subjects. Exposed film is sent to a processing laboratory, which returns the slides mounted and ready for projection. Also, sets of slides on particular topics can be purchased; these may be accompanied by taped narrations. Like filmstrips, slides require a projector and a screen. Slides are quite flexible. They are easily rearranged, revised, handled, and stored; and automatic and remotely controlled projectors are available for greater efficiency and effectiveness. If handled individually, however, slides can get out of order or be misplaced. Filmstrips and slides have been largely replaced by modern equipment that is easier to use and more flexible. For example, slides can be scanned or received from film processors as digitized images on disk or CD and used with computer projection devices.

Transparencies are easy and inexpensive to prepare and use, and the projector is simple to operate. The overhead projector permits you to stand facing an audience and project transparencies (sheets of acetate, usually $8\frac{1}{2}$ inches by 10 inches) onto a screen behind you. The projection may be enlarged to fit the screen so you can more easily point to features or mark on the projected image. Room light is kept at a moderate level. To prepare a transparency, you can write on the acetate with a grease pencil or special markers, or you can photocopy images onto the acetates. Transparencies can also be made by printing from computers to laser or inkjet printers onto acetate film.

REMINDER

In selecting media to enhance talks, you should choose media best suited to the style, subject, budget, and convenience.

Reading, Hearing, and Seeing: The World Wide Web

With potential for audio, video, animation, film, and print, the World Wide Web has changed the way people find, seek, and use information. However, using the Web as a medium requires special consideration to make information clear, easy to see, and easy to navigate.

While it is tempting to use all of the media available to make a Web site exciting, too much information is confusing for users who need to understand what each Web page says, what it shows, and what choices it provides. In addition, files with visuals can take a long time to download and appear on a screen, which annoys users.

It is important to remember that a single Web page rarely provides all of the information a user needs. Instead, users need to be able to understand available information choices, move quickly to another page, and read each page easily. On a Web site, each Web page must act as a separate, small information chunk that can lead readers to other chunks in the order they select. For this reason, it is especially important for Web site designers to consider the needs of their users carefully. What information do users need? In what order will they typically seek it? What key words will they recognize? Web designers who try to say too much and give too many directions confuse the users.

REMINDER

Keep Web sites and Web pages easy to see and navigate.

Ideally, a Web page will have a few clearly labeled links that help users understand the relationships between ideas and give them a way to retrace their steps. A simple and consistent visual design that makes choices easy to identify visually will help users to navigate the Web site. Guidelines for effective page design also apply to screen design: type elements, visuals, color, and headings. Bright and flashy colors and visuals that do not enhance the subject matter will distract Web page users, not guide them (see pages 70–71 in Chapter 4 Information Design).

GUIDELINES FOR SELECTING COMMUNICATION TOOLS AND MEDIA

As you decide which tools and media to use for your communication projects, consider the answers to the following questions.

- What communication tools are readily available to you? Are they sufficient to develop the kinds of information projects you work on?
- What is your budget for communication tools? How much will it cost to purchase and upgrade new tools as they appear?
- What features for writing, editing, designing, and using visuals do you need? How much time are you able and willing to spend in learning to use them well?
- Who are your readers? How will they use the information your project provides?
- What medium can your readers access and use most easily?
- How can you use the features of a communication tool to make a medium useful and interesting for your audience?
- What experience have you had in using new media? How much time can you spend investigating before you use it?

Don't feel that a well-prepared, designed, or presented project is necessarily better for the use of sophisticated tools or media. Base your choices on your audience, your experience, and your ability.

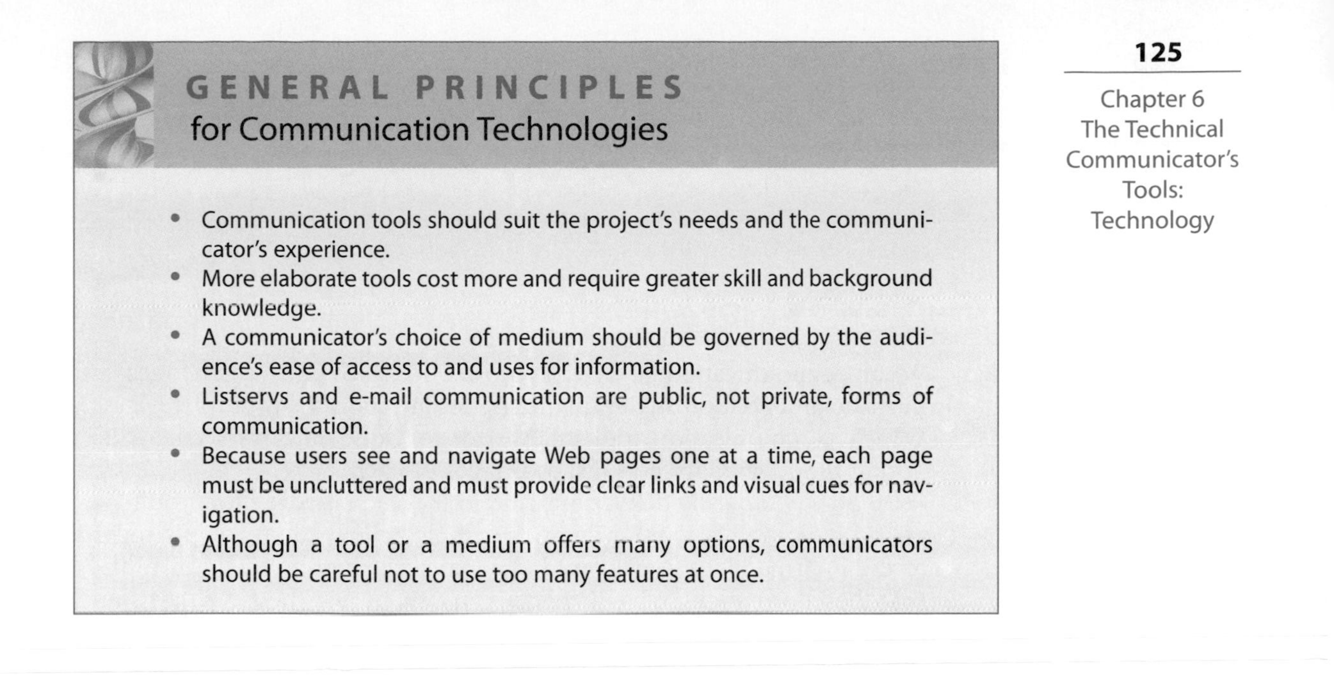

GENERAL PRINCIPLES for Communication Technologies

- Communication tools should suit the project's needs and the communicator's experience.
- More elaborate tools cost more and require greater skill and background knowledge.
- A communicator's choice of medium should be governed by the audience's ease of access to and uses for information.
- Listservs and e-mail communication are public, not private, forms of communication.
- Because users see and navigate Web pages one at a time, each page must be uncluttered and must provide clear links and visual cues for navigation.
- Although a tool or a medium offers many options, communicators should be careful not to use too many features at once.

CHAPTER SUMMARY

TECHNOLOGY

Communication tools and media, many of which are computer-based, offer authors, designers, visual communicators, and presenters many ways to develop, design, and deliver information. Communicators' creativity can be enhanced by the capabilities of word processing, desktop publishing, graphics, and Web authoring software. As you develop skills in using such software, the possibilities of applications increase. However, you must be cautious about getting caught up in the technology, forgetting audience and purpose.

In delivering communication, you have options such as the printed page, the telephone, videoconferencing, e-mail, listservs, and the World Wide Web. Through the use of film, videotape, CD-ROM, audio, and presentation software, you can enhance oral presentations. Base your choices on your audience, your purpose, your experience, and your ability.

ACTIVITIES

INDIVIDUAL AND COLLABORATIVE ACTIVITIES

6.1. What communication tools do you use most often? Which ones are most helpful for drafting and which ones are more helpful for revising? Which communication tools allow you to work quickly? Make a list, noting the advantages and disadvantages of each.

6.2. How do you typically deliver information? Which media allow you to respond most readily to your audience? Which allow you to keep a record of the information you deliver? How expensive is each medium to use? List the information delivery media you use, noting the advantages and disadvantages of each.

6.3. In groups of three or four, discuss the design options of a typical word processing system. How do elements such as different type fonts and sizes and graphics programs allow you to take advantage of effective page design? How can you decide which features to use and which to disregard?

6.4. What media would you use for an interactive four-hour conference between nine individuals located in distant states and countries? What issues of culture and physical distance would you need to consider in deciding on which media to use? Write a summary of your media choices, indicating your reasons for selecting them.

6.5. In what ways have new technologies overloaded people with information? In groups of three or four, describe the kinds of promotional and personal information that new technologies have made inexpensive and easy to produce, distribute, and even personalize. How can communicators with real messages for readers use tools and media to counteract print and electronic junk mail?

6.6. Discuss the use of technology to accompany a presentation. What kinds have you used? How complicated are they? Is learning and preparing the technology worth the benefits it offers? List the kinds of technologies that can be used to accompany an oral presentation, commenting briefly about the advantages, disadvantages, and costs of each.

6.7. What are the advantages of two-way communication (such as telephone or e-mail) over one-way communication? What are the risks and disadvantages? In groups of three or four, compare two-way to one-way communication and evaluate the media used for two-way communication.

6.8. In what ways do applying and interviewing for a job by long distance differ from applying for and interviewing for a job in person? (See pages 593, 607–612 of Appendix 1 The Search for Employment for additional information.) How would you prepare your credentials and your interview for each situation? Write a summary of your preparation for both types of application and interview.

6.9. In groups of three or four, discuss the uses of e-mail in the workplace. In what ways does e-mail differ from other kinds of correspondence? What sorts of conventions (such as letter format) should e-mail follow in a workplace setting? In what ways is e-mail a potential risk for workplace users? How can you avoid such risks?

6.10. Although communication technologies make global conversation between distant organizations quick and easy, it does not necessarily promote better understanding or cooperation. If you were moderator of an English-speaking worldwide teleconference, how would you make sure that all of the participants were included in the conversation? How would you begin the conference and establish rules to make sure that each person had an equal opportunity to be heard and understood? Write up a plan for the conference and a list of guidelines you'd ask the participants to observe.

6.11. Do communication technologies save time for workplace communicators? In groups of three or four, consider one communication tool and one medium, listing the advantages and disadvantages of each. Is the time a communicator invests in a communication technology worthwhile?

READING

6.12. In "The Impact of New Technologies on Technical Communication," professor and technical communicator Henrietta N. Shirk discusses the ways in which new communication technologies force communicators to reconsider their work and the ways they create and deliver it.

From *The Impact of New Technologies on Technical Communication*

HENRIETTA N. SHIRK

ELECTRONIC PENS TO ASSIST CREATIVITY

During the past decade, the work of technical communicators has been most greatly influenced by the widespread use of computers in the document creation, design, and publication processes. As "electronic pens," computers enable technical communicators to write, edit, and create graphics and page layouts more quickly, and to publish documents with professional appearance from the convenience of their own desktops. In addition to the cost benefits of greater efficiency and improved packaging techniques, these technologies engage their users in the creative space of a computer environment. And this environment, in turn, affects the creativity inherent in technical communication.

Changing Previous Mindsets

The mindset about technical communication that is evolving because of this group of electronic pen technologies is different from the one that technical communicators had in their old paper-based environment. Although computers may indeed be only "dumb machines," they are used and managed by the "smart machines" of human minds. Many technical communicators have made the mistake of assuming that computer technologies will do their professional work for them, or, minimally, do their work better than they are able to do it. They have failed to recognize that technologies are just tools; as tools they do not automatically confer talent, although they can be used with talent. In short, technical communicators have not been serious enough in addressing the challenges and consequent changes that these technologies command.

Such challenges and changes involve responding appropriately to the overabundance of electronic options for the creation of communication products. The wide variety of fonts, styles, type sizes, and page layout formats makes graphical expertise a necessity for technical communicators. Prospective employers now assume that technical communicators know about more than writing. This demand requires learning about skills such as visual design and page layout. Even if technical communicators do not know how to draw, they must learn something about effective visual communication and, at least, how to recognize it. For those who are verbally rather than visually oriented, this may be a daunting task, requiring creativity on many levels.

VIEWING COMPUTERS AS PARTNERS

A more comprehensive view of the pen metaphor as a category for computer technologies relates to the creative environment in which technical communicators participate when they use these technologies to assist in the communication process. To produce communication products with an electronic pen means not only to engage in the greater efficiency of revisions, but also to enter fully into an environment in which the computer becomes a partner as well as a tool. This partnership can enhance creativity, if technical communicators are willing to take risks with these technologies and not only undergo the effort in learning them but also to have the desire to experiment with them. Many technical communicators have gotten caught in word processing software and have never explored any of the other software options available to assist them in communicating more effectively. . . .

A significant electronic pen that many technical communicators are now wielding is that of the creation of on-line communications—information intended to be presented solely via the computer rather than on paper. Those who are creating on-line publications know that there are no easy rules for doing so. Research in this area of communication has been limited, and technical communicators have tended simply to repeat what they already know about paper-based communication. Because computer screens are not the same as pieces of paper, new ways of presenting information must be explored. Technical communicators have to learn about software design and software psychology, and then take from these fields whatever will help them communicate effectively in the on-line medium. The challenges are great, and, in a sense, everyone is a beginner in relation to on-line communication, especially that which enables linkages.

QUESTIONS FOR DISCUSSION

a. In what ways are word processors more than pens? What do they require of their users?
b. How do word processors encourage writers to understand and act on principles of information design?
c. What options do word processors provide for writers as they draft and revise? How do these options save writers time?
d. What are some of the options that Shirk suggests writers neglect to learn about their word-processing programs? How can these programs help writers in composing, using information design, and preparing graphics?
e. In what ways do word-processing systems intimidate new users? What ideas from "The Impact of New Technologies on Technical Communication" would you use to explain the advantages and limitations of word processors to writers who have not used them before?
f. What are some of the new mindsets that communication technologies require of communicators?
g. In what ways can communication technologies become communication partners? What computer-based tools other than the word processor can contribute to effective communication?

PART III

The Technical Writing Process

When authors develop a writing project, they typically use a flexible process to plan, organize, draft, edit, and revise. In the workplace, technical communicators also use this process, but with some important differences. First, parts of a workplace communication project—such as the subject, the audience, or the medium—may be defined for the author. In addition, technical authors have many communication choices, including written language, graphics, and information design. For technical communicators, the creativity of the writing process lies in developing imaginative ways to make communication clear, helpful, and easy for readers to use, and then making sure that the finished project lives up to its potential. Second, workplace communication is often developed collaboratively. The technical writing process must be adapted for groups when teams make decisions about a project.

The chapters in **Part III The Technical Writing Process** describe the writing process and provide practical approaches to help you make decisions about planning, drafting, editing, and revising. These decisions are important whether you work individually or collaboratively. **Chapter 7 The Technical Writing Process: Planning and Drafting** introduces the Project Plan Sheet, a set of easily adapted questions to help define a project's audience and goals. Chapter 7 also provides practical advice about organizing a project and developing a working draft. **Chapter 8 The Technical Writing Process: Reviewing and Revising** explains ways to assure that each completed project reflects the goals identified through the Project Plan Sheet. Chapter 8 suggests checks for:

- completeness and accuracy
- easy-to-follow organization
- clear style
- effective graphics and design
- consistency of usage and writing conventions

Experienced workplace communicators typically save time by spending it on planning. They produce effective projects by revising carefully. Whether you work individually or as part of a team, the suggestions in Part III are designed to help you develop projects that meet goals and deadlines.

CHAPTER 7

The Technical Writing Process: Planning and Drafting

CHAPTER GOALS

This chapter:

- Suggests communication project planning strategies for individuals and groups
- Defines the Project Plan Sheet, showing its use in raising and answering important questions about a project
- Describes important principles for effective organization
- Explains and illustrates organizing patterns
- Shows the relationship between accurate headings and a project's parts
- Provides guidelines for effective drafting

INTRODUCTION

Getting started with writing is always a challenge. What is the subject? Who is the audience? How can the finished product best meet the requirements of the assignment? What plan can best help you to understand and organize the assignment?

Workplace communication projects are designed to meet practical needs, so your starting points will be answering questions about audience, rather than subject matter. Every part of your plan, including subject matter, organization, medium, information design, and graphics and other visuals, must serve your audience. Only when you understand who will use your information and how can you decide what and how much to present, the best way to arrange and express ideas, and the best use of information design and graphics and other visuals. From this plan you can write a draft.

This chapter provides strategies for defining your audience's needs and organizing your own work to meet them with a **project plan sheet**, a set of questions that can be adapted to any workplace communication project. Since many projects are team efforts, the chapter suggests planning strategies for teams as well as individuals. This chapter also discusses patterns of organization, provides guidelines for organizing the sequence of a project, and suggests ways of getting your plans on paper as you write a draft.

GETTING STARTED

As a workplace communicator, your responsibility is to plan so that the finished project meets the audience's needs. As you develop your plan, you must make the practical decisions that will allow you to complete the project successfully.

Time spent planning a communication project will save you time as you draft and revise. This planning aspect of communication is like the planning you do for any task you undertake. If you don't start by considering why you are doing something and how to do it efficiently, you will waste time and you may not be pleased with the results of your work. For example, you find that your garage is so full that you can't even open the garage door. You know that you need to clear out some space, so you set aside a full day to do the work. However, you hadn't really decided how much space inside the garage you need, if any of the items stored in the garage were worth keeping, or if you needed to find boxes, buy trash bags, or have a garage sale. Without careful advance planning, you will have wasted a whole day and made little progress.

The writing process operates in the same way. If you don't begin with a clearly defined purpose and a strategy to help you meet it, you will find it difficult to know what to do or where to start. However, if you plan your projects by first answering the larger **who** and **why** questions to define your audience and purpose, finding answers to the practical questions—**what**, **when**, and **how**—becomes much easier.

Suppose that in April the chair of the New Student Orientation Committee asks your class to develop a brochure to help new students and campus visitors learn their way around campus and locate college services. However, the committee chair hasn't explained why a brochure would be helpful. When you ask, she says that although new students all have a complete college catalog, it is a large book, too big to use for quick reference on campus. While the campus map helps new students and visitors locate buildings, it doesn't explain the services available at different locations. Now you have a clearer idea of the problem: people who are new to campus need to learn quickly what is available and where.

THE PROJECT PLAN SHEET: DEFINING THE PROBLEM

To define exactly what information you will need to develop and to determine the communication tools and strategies you will use, you need more answers. The Project Plan Sheet is a flexible guide—a set of questions you can use to guide your thinking about a communication problem. By responding to these questions, you'll first define your project's audience and purpose. Your answers—if they are thorough, complete, and well researched—will lead you to informed, practical decisions about tasks and working deadlines.

The Project Plan Sheet starts by asking you to answer the most important questions:

- Who is your audience?
- Why does your audience need information?
- What specific kinds of information does your audience need?

The answers to these questions allow you to define your audience and your project in order to support the audience's needs. Only when you understand your audience and its needs can you successfully define your project's purpose. A generic project plan sheet is shown on page 149 and on the back endpaper of this textbook.

This generic form can be adapted to suit each communication project. Note how it has been slightly altered in the completed Project Plan Sheet for the campus brochure that follows.

PROJECT PLAN SHEET

FOR SMITHVILLE COLLEGE CAMPUS GUIDE

Audience

- Who will read the Guide?

  New students, visitors to campus

- How will readers use the Guide?

  People will use the brochure to locate campus services both when they are physically on campus and during orientation presentations.

  The Guide will serve both students and visitors, but both audiences will be unfamiliar with the campus. Whether they use the Guide when they are walking around campus or listening to a presentation, they will want a quick, easy-to-use reference.

- How will your audience guide your communication choices?

  Since the catalog provides too much information and the campus map provides too little information, the Guide needs to provide the best of both in a brief, easy-to-hold brochure that is easy to read and follow.

Purpose

- What is the purpose of the Guide?

  The Guide will welcome those new to campus by making it easy for them to find what they need without getting lost or wasting time. It will also help them learn their way around.

- What need will the Guide meet? What problem can it help to solve?

  The Guide will allow visitors and new students to locate what they need on campus quickly and easily. In doing this, it will also make users feel at home and welcome.

- What title will most clearly reflect the Guide's purpose?

  The Guide should be called "Campus Guide" since it will help people find their way to what they want.

Subject

- What is the Campus Guide's subject matter?

  Services available on campus and the buildings where these services are located

- How technical should your discussion of the subject matter be?

  Very nontechnical! The Guide should make no assumptions about what people who are new to campus know.

- Do you have sufficient information to complete the subject? If not, what sources or people can help you to locate additional materials?

  The catalog and campus map are available and current. These will be the best-written sources.

  The chair of the New Student Orientation Committee, committee members who conduct campus orientations, and volunteers from campus information booths will be able to tell me some of the questions and problems people who are new to campus have.

Author

- Will the project be a collaborative or an individual effort?

  A team of five classmates will be working together to develop the Campus Guide.

- If the project is collaborative, what are the responsibilities of each team member?

  One team member has conducted orientations for the New Student Orientation Committee, so she will be able to collect information. Another team member is a graphic artist who can adapt the map and design pages, and the other team members can plan, draft, and edit the Campus Guide.

- How can the project's developer(s) evaluate the success of the completed Campus Guide?

  We can ask visitors to campus to review a draft of the Campus Guide to make sure that it answers all of their questions and is as easy to use and understand as possible.

Project Design and Specifications

- Are there models for organization or format for the Campus Guide?

  The city has self-guided walking tour brochures that could be helpful models.

- In what medium will the completed project be presented?

  The finished Campus Guide will need to be a printed brochure. The team will prepare text, information design, and graphics.

- Are there special features the completed Campus Guide should have?

  Yes, the Guide will need a map, clearly written text, an attractive design (perhaps featuring the college logo), and lists of buildings and services, with a brief description of each one keyed to the map.

- Will the Guide require graphics and other visuals? If so, what kinds and for what purposes?

  Yes, graphics will appear in the design of the map and the key to different campus buildings and services for ease of location.

- What information design features can help the project's audience?

  The Guide will need to be large enough so that the map and the lists will be easy to read and small enough so that it can fit into a backpack or purse. A sheet of $8\frac{1}{2}$ by 14 inch paper folded into five $8\frac{1}{2}$ by $2\frac{3}{4}$ inch sections would be easy to handle and inexpensive to duplicate.

Due Date

- What is the deadline for the completed project?

  The New Student Orientation Committee would like to begin using the Campus Guide for sessions in August.

- How long will the project take to plan, research, draft, revise, and complete?

  Six weeks is a reasonable time frame for the project, including asking for reviews from visitors and the New Student Orientation Committee. The committee can then duplicate and distribute the Guide.

- What is the timeline for different stages of the project?

One week: *Working with the chair of the New Student Orientation Committee, develop a plan for the brochure. Make any revisions to the plan. Assign tasks to team members.*

Two weeks: *Plan graphics and other visuals and design information.*

Contact and interview at least two volunteers from campus information booths.

Perform preliminary research from the catalog and campus map.

Develop an outline.

As a team, review the design plan and the outline.

One week: *Draft the map and the text and sketch the map.*

Review these drafts as a team. Check the draft with the chair of the New Student Orientation Committee.

One week: *Revise and edit the map and the text.*

Incorporate the text and the brochure design.

Proofread the finished brochure for correctness, completeness, and consistency.

One week: *Ask visitors and committee members for comments.*

Incorporate any suggestions.

Review the Guide one last time for correctness and consistency.

As you can see, the Project Plan Sheet allowed the developers of the Campus Guide to make many decisions, from the title to subject matter, organization, graphics and other visuals, language, and schedule. If you use a project plan sheet to answer the important **who**, **what**, and **why** questions that define your project's audience and purpose, you will find it easier to decide **how** and **when** to get started with the practical details.

While a project plan sheet helps members of communication teams to develop ideas together, it is equally helpful for individual authors as they define the audiences and purposes of their projects. If you complete a project plan sheet for each communication project, you will find that decisions about organizing and drafting will be easier to make. A project plan sheet adapted for the various types of workplace communication appears in each of the chapters in Parts IV and V of this text.

REMINDER

Communicators need to start by understanding each project's audience and that audience's needs. These define the communication problem that the project can help to solve.

ORGANIZING THE PROJECT

Once you have completed a project plan sheet, you need to develop a strategy for presenting the information. One strategy is using an organization plan sheet. For examples, see Chapter 13 Reports, Chapter 15 Correspondence, and Chapter 16 Oral Communication. A project plan sheet and an organization plan sheet—each is to be adapted for the particular communication project—appear on the back endpapers of this book. Another strategy for presenting information is a guide which prompts you to select and organize material, depending on your audience and purpose. For examples, see Chapter 9 Process Explanation, Chapter 10 Description, and Chapter 11 Definition. You need to determine the sequence of the information being presented and decide which patterns will allow you to explain ideas most clearly.

An organizing plan is a map of the parts of a writing project. When you develop an organizing plan, you place the sections of your discussion in the order in which they will appear as you write your rough draft. Careful planning forces you to review your subject, becoming more familiar with what you want to say and how you want to say it. Reviewing your completed Project Plan Sheet and your notes will help you to make your organization plan successful.

If you are developing your project as part of a team, all of the team members need to discuss the Project Plan Sheet and determine organization. It is especially important for the team members responsible for graphics and other visuals and information design to take part, since organizing plans must indicate the location of each visual and explain any information design features.

The example below illustrates an organizing plan for the Smithville College Campus Guide project discussed earlier in this chapter. The organizing plan shows comments from the team member who will develop graphics and manage information design. Other team members have added comments about types of development and completeness of information.

The team member developing the map must work closely with other team members to assure that the map corresponds to the information and sequence

The organizing plan classifies campus facilities by type. Is the list complete?

We must provide the campus information number so that our readers can get more information if they need it! Where should it appear in the Guide?

Each type of facility needs a simple and distinctive symbol that will also appear on the campus map. Will the map be big enough to let us use these symbols?

Most of the Guide's information will appear in the lists and the map. The introduction needs to be especially clear so that readers will know how to use the map and the lists.

Sections III and IV need brief introductions to explain their purpose and arrangement. That will make these parts easier to use.

Smithville College Campus Guide

I. Introduction

II. Map of the Campus

III. Campus Locations by Name
(Alphabetical listing of names that appear on the campus map go here.)

IV. Campus Services
- Library
- Classroom Buildings
- Administrative Buildings
- Store
- Theaters and Auditoriums
- Athletic Facilities
- Student Recreational Areas
- Information Booths

of the organizing plan. Likewise, the team member handling information design must assure that headings, page numbering, and the size of the Guide will be easy for readers to see and use.

REMINDER

Audience needs (as defined by the Project Plan Sheet) guide effective project organization.

The organizing plan should reflect the subject, the sequence, and the audience's needs. It is a working document, one you can easily revise. As you develop your organizing plan, consider how your readers can best follow your ideas. Will they read the entire project? Will they look up parts for reference? Your organizing plan should meet their needs.

INTRODUCTION, BODY, AND CLOSING

To be clear and usable for its audience, every piece of planned communication needs three important parts: a beginning, a middle, and an end. The names sound simple, but planning and developing these parts and understanding their importance for readers can be difficult.

Beginning: Introducing the Project

The beginning, or introduction, of a communication project briefly defines a project's subject, purpose, and parts so that the audience will have some idea of what is to follow. Experienced technical communicators sometimes become so involved with their subject matter that they forget that the project's purpose and organization are not necessarily obvious to an audience.

Helpful introductions briefly provide answers to the following questions.

- What is this project about?
- How technical or detailed is the information?
- What is the project's purpose?
- How is the project organized?

Every motivated reader or listener deserves clear answers to these questions.

An introduction to a procedure for telephone registration at Smithville College appears below. The procedure must teach students who are new to telephone registration how to complete the process successfully. It must also help students who have registered by telephone previously but who need to review the procedure or look up individual steps for quick reference. The introduction must explain what the procedure covers and how it is organized so that all student readers can quickly understand and use the information they need to register. Since the procedure's secondary purpose is to make students feel comfortable about registering by telephone, the introduction addresses them directly.

How to Register by Telephone
for Classes at Smithville College

We're glad you have decided to register for classes at Smithville College. These instructions will help you quickly and easily sign up for classes from any location with a touch tone telephone.

To make your registration faster, we suggest that you use the current Smithville Course Schedule (available at all campus locations) to make a list of the classes you'd like to take. A schedule planner appears on page two of this booklet. The planner asks you to list the information you will need during your telephone registration.

Once you have completed the schedule planner, you are ready to begin. A brief overview of the steps for the procedure appears on page three. We suggest that you read this overview to prepare for the step-by-step procedure, which begins on page four. You can follow the steps as you dial our toll-free information number and begin your telephone registration. We have left space on each page for your notes. When you complete the steps, which takes most people four minutes or less, you will have selected your classes, learned your schedule and the amount of your fees, and selected a method of payment.

If you need help or have any difficulty with telephone registration, please call our toll-free Smithville College registration hot line at 1–666–153–2079. You can use this number to access recorded information 24 hours a day. During registration, our information operators are available to answer your questions seven days a week from 9:00 a.m. until 9:00 p.m.

We want your registration to be easy. Please let us know how we can help.

Body: Developing the Parts of a Document

Imagine a textbook that provided no chapter breaks, no paragraphs, and no headings. No matter how well written, wonderfully organized, or valuable the subject matter might be, such a text would require so much concentration that it would be unusable for reading, reviewing, or reference. The middle, or body, of a document is helpful to readers only when it presents information in chunks that are easy for readers to follow, skim, or look up selectively. In other words, it is important for you to divide the body of any discussion into sections: larger ones, such as chapters, and smaller ones, such as paragraphs and lists. In technical communication, headings that accurately describe the content of each section make the parts of a document even easier to follow or to read selectively.

As you organize your project's text, give each major division a clear descriptive title or heading—one that you are sure your reader can understand or that states or answers a question your reader might ask. Creating this outline of headings will make it easy for you to develop the sections of text in your rough draft. A clear heading structure also guides your readers in following your ideas or locating parts of your discussion. Below are headings that can help readers understand the parts of the procedure for telephone registration.

1. Introduction to Telephone Registration
2. Preparing to Register: The Registration Planner
3. Overview of Telephone Registration
4. Step-by-Step Instructions for Registering
   a. Entering Your Personal Identification
   b. Selecting Classes
   c. Arranging for Payment
   d. Concluding Your Registration

These headings identify each part of the discussion as clearly as possible for readers.

By listing headings, you name the major parts of your text. By listing them in an order you have chosen, you determine the sequence in which you will present the information and the relative importance of each part.

Ending: Giving Closure

Every project needs a brief ending to give the audience closure. Your ending may be a **summary,** a statement that reminds readers of the key points of your discussion, or a **conclusion**, which first summarizes the key points of a discussion and then goes on to comment. Some projects need only a brief comment. For example, the procedure for telephone registration at Smithville College will simply give the student reader the last step of the procedure: how to leave the system. Whether your discussion ends with a summary, a conclusion, or a brief comment depends on the information you present and your purpose in presenting it.

Here are two examples to illustrate the difference between a summary and a conclusion. The Smithville College Campus Guide could end with a brief summary to remind readers of the guide's key features, advise readers of other sources of information, and express the hope that the guide has been helpful. A report that shows the need for a campus guide, however, might conclude with a summary of visitor and new student problems and complaints and a recommendation for better written information to help newcomers to campus.

A summary, conclusion, or closing statement need not be long to be effective, but, like the introduction, it must address the audience's needs.

METHODS OF DEVELOPMENT: PATTERNS OF ORGANIZATION

To allow readers to understand and follow new ideas, you must use logical patterns, or methods of development, as you develop paragraphs, sections of text, and whole documents. The patterns you select must clearly develop the subject matter and fulfill the purpose of your project. Typical methods of development that can be used to organize complete documents or their selected parts are discussed below. Each method of development is followed by a brief example.

Cause and Effect

Cause and effect explains why an event or condition takes place or will take place. This organizing pattern is effective only if you can provide convincing evidence to demonstrate that the cause leads to the effect. In the case of cigarette smoking, ample and reliable research links smoking with the incidence of cancer. Personal opinion, unsupported evidence, or a failure to link cause and effect will not lead to a logical or convincing conclusion.

The following example explains that aspirin (cause) has many advantages, as well as some disadvantages (effects).

> Since its introduction in Germany in 1853, aspirin (the acetyl derivative of salicylic acid) has proved to be a miracle drug. Aspirin banishes headache, reduces fever, and eases pain. It is not without its dangers, however. Taken in excess, aspirin will cause stomach upset, bleeding ulcer, and even death. Nonetheless, aspirin is responsible for more good than harm. It may even prevent heart attacks. According to current research, it is possible that aspirin contains anticoagulant properties that prevent thrombi, the blood clots that cause heart attacks by clogging the coronary arteries.

Narrative (Time Sequence)

Narrative arranges events in chronological order from first to last. Narrative is used, for instance, to explain the sequence of events or the steps in a procedure.

The following paragraph uses narrative to show the cycle of a refrigerant in a cooling system.

> A compression system refrigerates by changing a refrigerant from a liquid to a gas and back to a liquid again. The cycle of a refrigerant in a cooling system is very simple.

1. First the refrigerant passes through the compressor where it is changed from a low-pressure liquid to a hot, high-pressure vapor.
2. It then passes through the condenser where it is changed to a hot liquid.
3. The hot liquid is pushed out of the condenser by the compressor into the metering device, usually a capillary tube.
4. This capillary changes the liquid to a vapor and spreads it through the evaporator where it picks up heat from external air.
5. In the evaporator it becomes a low-pressure liquid, and this liquid feeds back into the compressor.

An example of a compression-type refrigeration unit is an electric refrigerator.

Spatial Description

Spatial description explains an object or a place by presenting an orderly view of the physical arrangement of parts or features. You must determine a clear way to present your subject—from top to bottom, side to side, or according to its dimensions or shape, for example. The subject of the description will determine whether you discuss the object in terms of its surroundings or its own features. For example, a description of the location of a building lot will explain the lot's location relative to familiar landmarks, such as buildings and streets. A description of a piece of equipment, such as a microscope, will typically explain the physical relationship between the parts—in this case, barrel, lens, and base.

In the following paragraph, which explains a shop layout, the details are described in spatial order.

> The interior of the maintenance shop at Hopkins Engineering is efficiently organized. At the left of the entrance is a set of lathes, turret lathes, and engine lathes. The left back corner of the shop houses a tool room where mechanics can check out tools and materials. To the right of the tool room is a cutting saw. The center of the shop provides an open area for grinding machines, including surface grinders, tool-cutter grinders, cylindrical grinders, and centerless grinders. Near the center back of the shop is the furnace area used in heat treating. To the right of the entrance are vertical and horizontal milling machines and multispindle and radial drill presses.

General to Particular Order

In general to particular order, the discussion focuses on information that explains, describes, or substantiates the general opening statement. This method of development is helpful in paragraphs that open with a general thesis sentence. Each particular should support and define the general thesis. The general to particular order is helpful for busy readers who need to understand the largest issues and who may or may not need all of the supporting information. It helps readers understand the main point, which is presented at the beginning of the discussion.

The following paragraph illustrates general to particular order.

> Lures for bass fishing should be selected according to the time of day you plan to fish. For early morning and late afternoon fishing, use top-water

baits that make a bubbling or splashing sound. For midday fishing, select deep-running lures and plastic worms. For night fishing, try any bait that makes a lot of noise and has a lot of action. The reason for selecting different baits for different times of day is that bass feed on top of shallow water in the early morning, late afternoon, and at night. The only way to attract their attention is to use a noisy, top-water lure. During the middle of the day, however, bass swim down to cooler water. At this time, a deep-running lure is best.

Particular to General Order

Particular to general order starts with evidence and specifics and carefully builds to a conclusion. Discussions that use specific to general order place the main point at the end, allowing audiences to follow and understand the supporting points. This method of development is especially effective for strong conclusions, first summarizing the main points of the preceding discussion, then providing the general recommendation that follows from these points. Particular to general order places your most important ideas where your audience can best remember and understand them: at the end, supported by evidence.

The following paragraph illustrates particular to general order.

The human body requires energy for growth and activity, energy which is provided by the calories in food. The weight of an adult is determined in general by the balance of food intake with the expenditure of energy in activity and growth. When energy intake equals energy outgo, weight will stay the same. Similarly, weight is reduced when the body receives fewer food calories than it uses for energy. Therefore, weight can be controlled by regulating the amount of food eaten, the extent of physical activity, or both.

Classification (Order by Grouping)

Classification sorts related specifics into larger groups, or classes. A great deal of scientific and technical communication involves grouping related types, so this kind of grouping is important as a method of development. Classification is a useful organizing pattern for many kinds of practical communication. In a letter of application, for example, items could be sorted into one group on the basis of work experience and into another group on the basis of education. Classification means grouping together items that have common characteristics.

The following paragraph classifies batteries as primary or secondary types.

Batteries are broadly classified as primary or secondary types. Primary batteries generate power by irreversible chemical reaction. This requires that the battery parts consumed during discharge be replaced, or that the batteries themselves be replaced after discharge is complete. Secondary batteries, on the other hand, involve reversible chemical reactions, which means that their reacting material can be restored to its original "charged" state when a reverse, or charging, current is applied. In general, primary batteries provide higher energy, density, and specific power than do secondary batteries.

Partition (Order by Dividing)

Partition arranges ideas by dividing a singular subject into mutually exclusive categories and then describing each category separately. The following discussion uses partition to show how a lamp socket can be divided into groups of outer and inner parts. The groups of parts are mutually exclusive, and the division is complete. Every part of the lamp socket is accounted for; no part is left out.

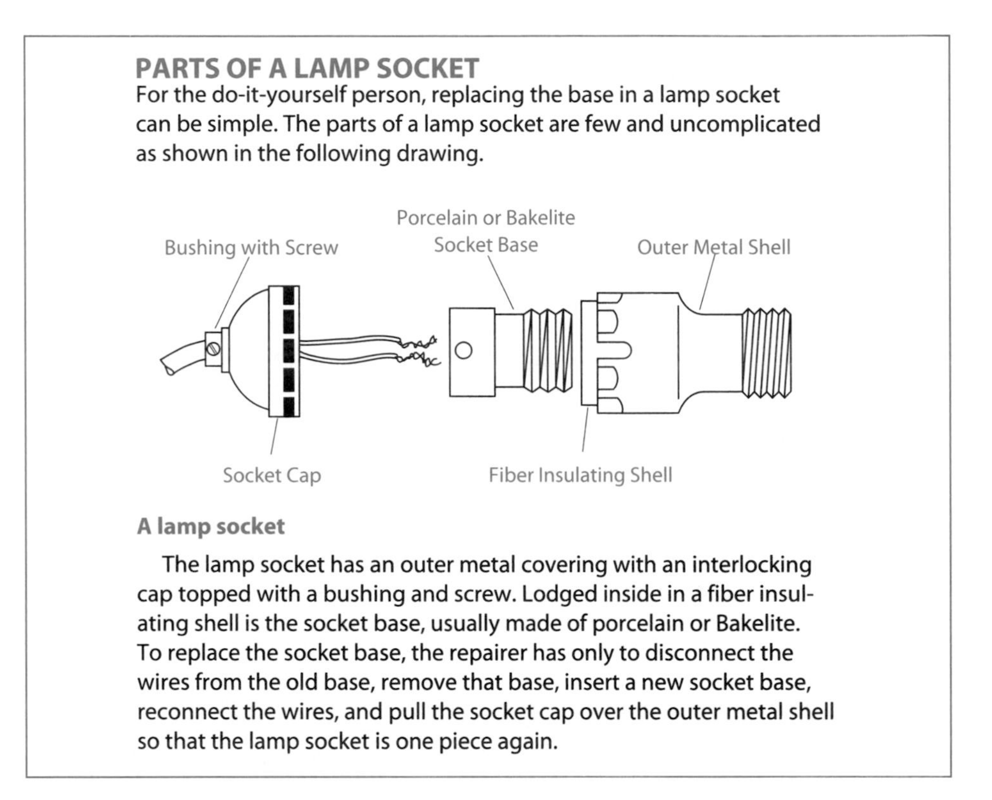

PARTS OF A LAMP SOCKET

For the do-it-yourself person, replacing the base in a lamp socket can be simple. The parts of a lamp socket are few and uncomplicated as shown in the following drawing.

A lamp socket

The lamp socket has an outer metal covering with an interlocking cap topped with a bushing and screw. Lodged inside in a fiber insulating shell is the socket base, usually made of porcelain or Bakelite. To replace the socket base, the repairer has only to disconnect the wires from the old base, remove that base, insert a new socket base, reconnect the wires, and pull the socket cap over the outer metal shell so that the lamp socket is one piece again.

REMINDER

The best methods of development for a project as a whole and for its parts are compatible with the project's subject matter and the audience's needs.

WRITING THE ROUGH DRAFT

Whether you are developing a project on your own or in a group, you will typically write several drafts. When you write a preliminary draft, you are acting on all of the plans you have made earlier in the project. As you continue to draft, you are relying on your knowledge of your audience, your subject, language, and writing conventions. To be successful, you need to concentrate on your ideas and on your organizing plan, writing quickly and deliberately.

At every point in the planning and organizing process, a project should remain open to revision. Successful drafting means acting on careful plans, not starting over with new ones.

A rough draft is just that: rough. Remember that early drafts are preliminary to the completed project. Expect to make mistakes, but do not worry about correcting them at this point. Correct errors during the revision process.

Even if you are well prepared, you may find it difficult to write the first draft. You may find it helpful to work on short, rather than long, sections of a draft. If you concentrate on effective communication and what you know, you'll find drafting easier. Remember that the early drafts are steps toward a completed project. The editing and revision process follow.

When you complete the drafting process put the project aside for a time. When you return to the draft to edit and revise, you will see what you have written from a different perspective. When you complete all of the sections of your draft, all of your planning will have paid off: your ideas will be down on paper. Chapter 8 Revising and Reviewing will show you how to rework your draft to make sure that it best reflects your plan, your subject matter, and your clearest writing—and meets the needs of your audience.

GENERAL PRINCIPLES for Planning and Drafting

- Effective communicators keep the audience's point of view, concerns, and needs in mind as they plan.
- In determining what subject matter a project should include, communicators plan to locate the best and most current information, but no more than the audience needs.
- Choices about medium, organization, format, and graphics and other visuals must be appropriate to the context in which the audience will use the project.
- Whether a project is developed individually or by a team, communicators must define the parts of the project and establish realistic deadlines for completing each part.
- Each team member must be actively involved in the planning process. Each member must understand the plan before the team begins collaboratively organizing the project.
- Responsible team members maintain regular contact with one another about progress, problems, and possible changes for the plan.
- Writers prepare intellectually for a drafting session by reviewing the Project Plan Sheet, the working organizing plan, and any notes that may be helpful in supporting ideas.

CHAPTER SUMMARY

PLANNING AND DRAFTING

During every stage of the planning process, the best advice for all technical communicators—whether they work in teams or as individual authors—is to keep the audience and the audience's needs in mind. The Project Plan Sheet helps to define a project's audience and purpose. It also answers practical questions about a project's medium, category, visual communication, information design, and deadlines. Organizing by dividing a project into its parts, determining an effective sequence, and developing an organizing plan moves the communication project further toward drafting. In drafting, however, an author must act on all the steps that have come before, relying on the Project Plan Sheet, research, the organizing plan, and, above all, an awareness of audience needs. The rough draft is rough, but it provides the important material that communicators will then shape into effective completed projects.

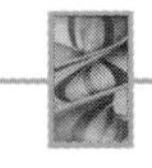

ACTIVITIES

INDIVIDUAL AND COLLABORATIVE ACTIVITIES

7.1. For a writing assignment in one of your classes, complete a project plan sheet (see page 149). Then in groups of three or four, review one another's plans. Discuss how each plan could help in analyzing the assignment.

7.2. Consider the ways in which you typically draft a writing assignment. What works for you and what doesn't? List ways in which you can plan your writing to make your drafting faster and more efficient.

7.3. Read the introduction to one of your textbooks. Had you read the introduction before? What information does it provide? As the audience for the text, did you find that the introduction prepared you to learn from the book? Can you suggest any changes? Write a brief summary of your findings, suggesting ways to make introductions helpful to readers.

7.4. The rough draft of the introduction to the Smithville College Campus Guide appears below. Compare the draft to the completed Project Plan Sheet which appears on pages 136–138. In what ways does the draft successfully reflect the team's plans?

!!!!WORKING DRAFT!!!!

Introduction to the Campus Guide

The purpose of this brochure is to direct newcomers to the SC Campus to the facilities and locations and functions they need without any problems. The guide is intended to allow users to locate particular buildings as well as services that each building or location offers. Emergency and information SC telephone listings also appear at the end. A map is included to direct newcomers to specific locations and the easiest way to get there. All facilities on the map are identified by locator icons and by name. Additional brochure information consists of a list of campus sites (alphabetized for convenient use). A second alphabetized listing provides SC services (with location).

PROJECT PLAN SHEET

Audience

- Who will read the communication project or hear it as a presentation?
- How will readers use the project?
- How will your audience guide your communication choices?

Purpose

- What is the purpose of the communication project?
- What need will the project meet? What problem can it help to solve?

Subject

- What is the communication project's subject matter?
- How technical should the discussion of the subject matter be?
- Do you have sufficient information to complete the project? If not, what sources or people can help you to locate the additional information?
- What title can clearly identify the project's subject and purpose?

Author

- Will the project be a collaborative or an individual effort?
- If the project is collaborative, what are the responsibilities of each team member?
- How can the developer(s) evaluate the success of the completed project?

Project Design and Specifications

- Are there models for organization or format for the communication project?
- In what medium will the completed project be presented?
- Are there special features the completed project should have?
- Will the project require graphics and other visuals? If so, what kinds and for what purpose?
- What information design features can help the project's audience?

Due Date

- What is the final deadline for the completed project?
- How long will the project take to plan, research, draft, revise, and complete?
- What is the timeline for different stages of the project?

CHAPTER 8

The Technical Writing Process: Reviewing and Revising

CHAPTER GOALS

This chapter:

- Defines a *global review*
- Explains the role of the Project Plan Sheet in a global review
- Provides checks for conducting a global review
- Provides strategies for doing a *consistency review*
- Shows the roles of team members in reviewing a collaborative project
- Describes the importance of peer review for individual authors
- Shows how authors act on reviews to revise a project
- Describes strategies for testing a revised project for success

INTRODUCTION

Without practical strategies for reviewing and revising, you may find a rough draft difficult to face. There seems to be so much left to do and to fix. Is the draft complete? How can you identify and correct mistakes? Are the proposed information design and visuals helpful? And, above all, does the draft reflect the project's intended audience, purpose, and plan?

This chapter suggests answers to all of these questions—and more. It provides practical strategies for reviewing your own draft and for reviewing the work of others. Because it can be difficult for you to recognize the successes and shortcomings of your own draft, the chapter emphasizes the importance of peer review. In this process, an individual or a team helps an author by making constructive comments about a draft. Last, the chapter shows ways to act on reviews by revising drafts to assure that revised projects are successful in meeting audience needs. See also Reviewing a Communication Project, front endpapers of the textbook.

WHAT IS REVIEWING?

Reviewing is the process technical communicators use to check a draft for success in meeting its intended purpose. A constructive review recognizes a draft's strengths and clearly suggests any needed changes. You can review your own draft, or you can request that others review it for you. However, if you ask others to help, it is important for them to understand the project's intended purpose before they begin their review; be sure to supply a copy of the completed Project Plan Sheet.

Reviews must address and coordinate every element of a draft, large and small. To address elements that reflect a project's purpose, it is helpful to begin with a **global review.** The global review examines the large, or global, issues such as subject matter, information design, visuals, and language—all relative to the project's purpose and the audience's needs. A second review, the **consistency review,** considers details of language, visual design, and format such as page numbering, spelling, capitalization, and punctuation. Although the details addressed by a consistency review may seem less significant than the larger issues of a project's audience and purpose, they are still important. Without internal consistency, a completed project will be less helpful to its intended audience. After all, errors can distract, confuse, and even offend readers. The findings of a global review and of a consistency review help to direct **revision,** a process of adapting, polishing, and coordinating all parts of a project.

REVIEWING FOR GLOBAL ISSUES

A global review is a process that allows authors and reviewers to determine which elements of a draft meet a project's purpose and which need changes. Experienced communicators find that time spent on project planning helps them review a draft's strengths and weaknesses more efficiently. The completed Project Plan Sheet serves as an outline that authors and reviewers can use to check a draft for success as they conduct a global review.

In the global review process, reviewers consider the project's proposed audience, purpose, subject matter, order, information design, visuals, and language. The groups of global checks that follow describe issues to consider for each of these areas and provide questions for conducting a thorough global review, whether you are reviewing your own draft or someone else's.

Global Checks: Audience and Subject Matter

As you review the draft, first consider the project's audience and subject matter. Has the draft met the audience's needs? Will the audience read, skim, or follow steps one by one? Does the draft provide accurate and complete information? To find answers, start your review by addressing the following questions.

Review Questions: Audience

- Does the draft reflect the Project Plan Sheet's definition of audience needs? How might the draft be improved?
- Does the title of the draft accurately reflect the project's purpose?
- Do writing style, information design, and visuals help the project's audience? Do any of these elements need to be changed?

- Do the order and design of the draft reflect the context in which the audience will use the completed project?

Review Questions: Subject Matter

- Is the information complete? What, if anything, is missing?
- Does the draft provide all of the information the audience needs at an appropriate level of detail?
- Is the information correct and accurate?

Reviewing a draft for subject matter may suggest that changes are needed. Does the draft need to include more background information? Is there too much detail? Making suggestions based on the answers to these questions helps authors make decisions as they revise.

Global reviewing is also a time to consider the effectiveness of a project's original goals and project plan. At this point, you may want to slightly adapt a project's purpose. Reviewing gives you a chance to adjust, rework, and polish a draft, not just to "fix" or "correct" it, but this is certainly not the time to start all over! A draft, no matter how rough, will always provide constructive reviewers with material for revision. Whether you are a global reviewer or acting on reviewer comments, always keep the project's audience and purpose in mind.

To illustrate a global review for audience, purpose, and subject matter, below are reviewer comments on the draft of an introduction to an explanation of how to sharpen a ruling pen for a beginner.

Reviewer's Comments

Could the title be shorter and clearer so that a beginner can quickly grasp the subject?

Audience and Completeness: A beginner will need more information about the pen in the introduction. What is the pen used for, exactly? What are "nibs"? More information needed here—for beginners, anyway!

It would be helpful for the opening to reflect the purpose and importance of the procedure. A clearer discussion of why it's important to sharpen the pen would be helpful. Maybe you could address the reader directly as "you"?

The sequence is a little hard to follow. Maybe you could start by explaining what the pen does in drafting, why it needs to be sharp, and what the steps in sharpening are. The equipment could come last.

It would be easier to list the equipment under a heading. Is it necessary to define a "crocus cloth"? What kind of sharpening stone is needed?

Keeping the Ruling Pen Ready for Use

To sharpen a ruling pen (which is used regularly in drafting), it is important for the user to make sure that the right equipment a sharpening stone and a crocus cloth—is available. It is important to keep the pen sharp so that it will be possible to draw well because the nibs tend to wear down after use.

The procedure is really easy, involving only a few simple steps.

Below are the author's response and a revised draft showing how the author acted on the reviewer's comments.

The answers to global questions about a project's audience and subject matter will shape the review of other elements that support audience needs: order, information design, graphics, style, and tone.

Author's Response

Good idea about the title! The new one is much clearer for beginning drafters.

I was assuming that my reader already knew what a nib pen was—and that isn't necessarily correct. Now I begin with a clear definition. I also needed to explain why the procedure is important. Beginners don't automatically know this information.

I needed to define "nibs," and I needed to be much clearer about the kind of sharpening stone to use. However, a "crocus cloth" is easy to find and buy, and it's optional, after all. It's easier for my readers if I discuss this cloth later on in the procedure.

Headings will certainly help me break up text and sequence ideas.

The needed materials should appear in a bulleted list so that they will be easy to see.

How to Sharpen a Ruling Pen

The ruling pen is a piece of drawing equipment used to ink straight lines and noncircular curves with a T square and other drawing guides. The shape of the nibs, which are tweezerlike blades at the tip of the pen, is important because the nibs are the part of the pen that puts ink on paper. To hold ink in an ink space, the nibs must be rounded (elliptical). Since the nibs wear down and lose their shape after extended use, it is important to keep the ruling pen sharp to assure good, clean, and consistently inked lines.

Sharpening a ruling pen is easy if you have the right materials and follow the steps carefully.

Materials

- 3- or 4-inch sharpening stone (preferably an Arkansas knife piece that has been soaked in oil for several days)
- A crocus cloth to polish the nibs (optional)

Global Checks: Order

Order is the pattern that reflects a logical and easy-to-follow sequence in developing ideas. The effectiveness of a sequence depends on the ways in which the audience will use the material. For example, users of a pocket guide to automobile parts will want to skip quickly to the sections they need, while readers of an e-mail describing a new payroll procedure will want to read the entire message, perhaps rereading some sections. In both examples, information design decisions must enhance organizational ones.

In reviewing the draft of an entire document, it is important to consider how the sequence of topics serves the audience's needs. Some review questions that address order appear below.

Review Questions: Order of the Project as a Whole

- Is the draft's sequence consistent with the project's medium?
- Is the project's organizing pattern easy for the audience to recognize and follow?
- Does the planned information design complement the draft's sequence?

Responses to these questions may determine how much change and what kinds of changes you should make to clarify the project's patterns and sequence for the audience.

Audiences often need to navigate or selectively locate parts of a project as much as they need to recognize the project's organizing pattern. The following questions address elements that can make a project easy to use.

Review Questions: Connecting the Project's Parts

- Does the project's introduction forecast, or announce, the subject, purpose, and sequence?
- Is the text chunked, or divided, into sections?
- Are headings accurately worded to describe the sections they introduce?
- Does the closing reflect the project's purpose?

Responses to these questions will indicate any changes needed to help readers follow written information and locate parts of text easily.

The sequence of ideas in the text must be easy for readers to recognize and follow. The following questions address the relationships between ideas at the section and paragraph levels.

Review Questions: Unifying the Project's Text

- Does each section of the project begin with a forecasting statement?
- Does each section include sufficient information?
- Is it easy to follow the sequence of ideas?
- Is it easy to understand the relationships between sections and paragraphs?

Responses to these questions can highlight any changes needed to strengthen the draft's coherence, or logical flow, for readers.

To illustrate a global review for order, below are reviewer comments about one section of the draft of instructions for sharpening a nib pen.

Reviewer's Comments

The last sentence needs to go first to make the steps easy to follow.

The warning should go with the steps that describe holding the pen and stone, not here in the overview.

I usually pick up the stone first. Isn't that an easier sequence for the beginner to recognize?

The sentence listing the steps is long, but it is also easy to follow. The "polishing the nibs" step needs to appear where it belongs in the list.

The optional step is polishing the nibs.

Warning: This procedure can be hazardous if you do not hold the pen and stone correctly. Close the nibs, hold the ruling pen in the left hand, hold the stone in the right hand, round the nibs, sharpen the nibs, and test the pen. These are the seven basic steps.

Below are the author's responses and a revised draft showing how the author acted on the reviewer's comments.

Author's Response

Good point about beginning the overview with an introduction! A heading would help identify what this short section offers readers.

I'll keep the long sentence listing the steps as it is, adding the optional step in sequence. Keeping the order of steps in sequence seems much easier to follow than breaking the steps into shorter sentences or into a list.

Yes, it's easier (and safer) to pick up the stone before picking up the pen.

Overview of the Procedure

Sharpening a ruling pen involves following seven steps: close the nibs, hold the ruling pen in the left hand, hold the stone in the right hand, round the nibs, sharpen the nibs, polish the nibs (optional), and test the pen.

Global Checks: Information Design, Visuals, and Language

Information design, visuals, and language all complement a project's purpose, but in different ways. For this reason, it is helpful to conduct separate reviews for each element.

A draft should show the project's plan for information design, including headings, page design, and organizing features. This plan must be in keeping with the way the reader will use the completed project. The following questions can help reviewers evaluate the information design plan.

Review Questions: Information Design

- Does the project's table of contents or abstract help the reader understand the project's purpose and uses?
- Does the medium planned for the project best meet the audience's needs in context?
- Is the planned design attractive and easy to see, read, and skim?
- Do headings guide the audience to different parts of the project?

Although drafts will probably include only rough sketches of graphics and other visuals, it is important to provide them early so that reviewers can consider their role in enhancing the project's purpose. The following questions can help reviewers evaluate a project's planned visuals.

Review Questions: Graphics and Other Visuals

- Do visuals support the draft's information and purpose?
- Do visuals include a suitable level of detail and are they appropriate to the subject matter?
- Are visuals placed helpfully and discussed in the text?
- Is each visual easy for the audience to understand and recognize?
- Is each visual clearly labeled with an accurate caption?

Responses to these questions will help you in refining and clarifying a project's use of graphics and other visuals.

No matter how technical a project's subject may be, the project's language must be as clear and concise as possible. After all, no reader should have to struggle with language to understand ideas. A project's tone, the attitude suggested by its use of language, should be suitable to its purpose and audience. The following questions are useful in evaluating a project's language.

Review Questions: Language

- Will the audience understand technical terms? What terms need defining?
- What tone, or attitude, does the draft's wording convey? Is this tone appropriate for the project's audience, subject matter, and purpose?
- Are any sentences needlessly long and difficult to follow?
- Are any passages made up of short, choppy sentences?

Responses to these questions will suggest changes to make the project's use of language appropriate for its audience.

Summarizing and Responding to a Global Review

Global checks for purpose, order, information design, visuals, and language determine the important changes communicators need to make in a draft to assure that the completed project meets the audience's needs. To determine the kinds of changes needed, it is helpful to compile the results of global reviews into a short summary. To make such a summary, you can classify reviewer comments into groups that correspond to major areas of the Project Plan Sheet or to the major categories of review questions. Developing such a summary helps you understand the kinds of comments you have received and makes it easier for you to respond as you determine the changes you wish to make in your draft.

The summary below sorts reviewer comments about one section of a draft explaining the procedure for sharpening a nib pen into three categories.

Reviewer's Comments

Language
This passage is very formal and hard to follow because it uses so much passive voice. It would be easier and less formal to use active voice, calling the reader "you." Active voice would also make sentences shorter.

Information Design
The heading "Steps" needs to be larger and easier to distinguish from the text.

These steps are hard to follow because they are written as one long paragraph. Readers would find numbered steps easier to read and use one at a time.

The warning needs to be separate and visually different from the steps so that it will be easy for users to see in time.

Visuals
Readers need to see how to hold the pen, particularly since this step is closely related to safety. A simple line drawing should appear after the caution.

Steps.
The nibs must be closed. This is accomplished when the adjustment screw, located above the nibs, is turned until the nibs touch. The nibs can then be sharpened to exactly the same length and the same shape. Next, the sharpening stone must be held in the left hand in a usable position. With the stone lying across the palm of the left hand, it must be grasped by the thumb and fingers so that there is the best possible control and grip of the stone. The ruling pen is then held in the left hand. The pen is picked up and held between the thumb and index finger of the left hand as if it were a drawing crayon. The other fingers should rest lightly along the length of the pen. Also, it is important to hold the sharpening stone and the ruling pen correctly or they may slip, injuring the hand and ruining the pen.

Below are the author's responses to each category of comment. The changes the author has made to the draft also appear below.

Author's Response

Language
Good point about passive voice! It makes language more difficult and tone needlessly formal. I've also tried to leave out needless repetition. The technical level of language at this point seems suitable for beginners, so I have not made any changes in technical terms (which I have already defined.)

Information Design
By separating steps on the page and numbering them in order, I have made it much easier for readers to read and act on one step at a time. Also, by highlighting the warning and placing it with the information it comments on, I may help readers avoid injury or damage.

Visuals
I have sketched a simple line drawing of a right hand holding a pen and placed it after the warning. Readers will easily see it, especially since I refer to it in step 3.

Steps

1. Close the nibs by turning the adjustment screw (located above the nibs) until the nibs touch. Then it is easy to sharpen them to exactly the same length and shape.
2. Hold the sharpening stone in a secure and usable position. Lay the stone across the palm of your left hand, grasping it firmly with the thumb and fingers. This grip allows you the best control of the stone.
3. Hold the ruling pen in your left hand between thumb and index finger as you would hold a drawing crayon. Rest your fingers lightly along the length of the pen. (See Figure 1.)

WARNING: *Hold the sharpening stone and the ruling pen correctly and firmly. If they slip, you may injure your hand and ruin the pen.*

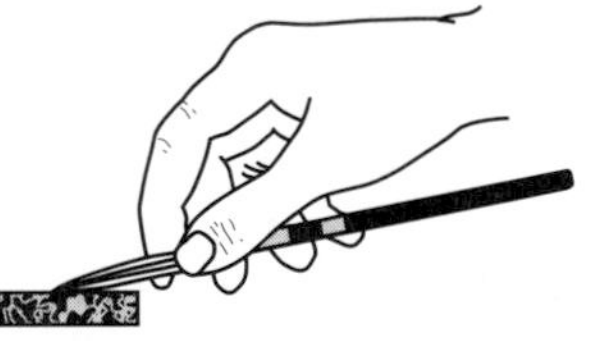

Figure 1. Holding the pen for sharpening.

REVIEWING FOR CONSISTENCY

A consistency review is a process in which authors or reviewers locate inconsistencies, make comments, and suggest corrections. Inconsistencies in written language, such as spelling and punctuation errors, are the focus of such a review. A consistency review also addresses the use of visual language. Misleading headings, misnumbered pages, and visuals listed out of sequence are as distracting as careless errors in writing.

Consistency Checks: Information Design, Visuals, and Language

As you review a draft for consistency, you must identify and correct problems. Your role in a consistency review is to make sure that all of a project's details are compatible with the larger decisions you have made about audience and purpose. Every element of information design, visuals, and language must be consistent to help the audience read, use, and understand the finished project.

It can be difficult to see the details of your own draft, so peer reviewers are especially helpful in a consistency review. They can locate, correct, and address small problems that you may have missed. If you are a reviewer, make your comments and corrections in the same spirit that you would like to receive them: carefully, tactfully, and clearly.

Since information design, visuals, and language require different kinds of review, it is practical to consider these issues one at a time. Some questions for checking consistency appear below.

Review Questions: Information Design

- Do headings consistently mark different sections of text?
- Are pages numbered consecutively? Do page numbers appear in the same place on each page?
- If the project includes documentation, is the documentation consistent in style and placement?

Review Questions: Graphics and Other Visuals

- Is each visual numbered and captioned?
- Is each visual placed with the text it explains?
- Does the text discuss each visual appropriately?

Review Questions: Language

- Does use of technical terms conform to the global decisions about style?
- Is word choice consistent in tone?
- Is spelling correct and consistent throughout the draft?
- Is the text free from grammatical errors?
- If passive voice is used, is it appropriate?
- Is the text punctuated correctly and consistently?

Spell and grammar checks included with many word processing programs can identify inconsistencies. A spell check, however, makes no distinction between "in," "inn," and "ion," since all three are correctly spelled words. But which one did the author intend? The spell check cannot choose. Likewise, grammar checks sometimes suggest grammatical corrections in a sentence that has no error. Programs such as spell and grammar checks can make suggestions, as can reviewers, but the author must ultimately decide which choices are appropriate for a document's purpose, audience, and meaning.

Writing Comments and Making Corrections

Careful, readable comments and notes on the draft are an important part of the review process. They form the basis for decisions about changes. They also specify places in the draft where corrections and changes should appear. Corrections can be as minor as indicating spelling errors, and comments can be as global as suggestions about the sequence of a discussion. The following example of marked text shows how thorough but tactful reviewer comments can help authors in making corrections.

Description of accoiunt Types

A pety cash fund is is a specific amount of money set aside for the purpose of making quick purchase fore items ore services that can not be purchase with the State MasterCard. The the majority of those types of purchasing situation exists for certain services (not goods). And when the vendor are not incorporated (businesses that are solely or partnership owned). If used to it's potential, and according to these guidelines a petty petty cash fund can significantly reduce the number off purchase Orders generated, for small purchases. And free up staff time to concentrate on more immediate tasks at hand.

As the example above shows, the reviewer has suggested corrections and asked questions about points that the author needs to clarify. Authors welcome such comments. However, comments should be tactful and thorough.

Acting on a Consistency Review

The results of a consistency review require a different response from the response to a global review. In a consistency review, the reviewer makes comments and suggestions directly on the page. After a consistency review, an author makes the final decisions about what is correct and appropriate. The author reads each page, considering the comments, making corrections, and looking for any further inconsistencies.

REMINDER

Reviewing can be a constructive process for strengthening and polishing a draft when authors and reviewers share the same view of a draft's audience and purpose.

STRATEGIES FOR PEER REVIEW

Peer reviewers who are new to your draft can often see its merits and needs more clearly than you can. Any review must begin with the clear understanding of your project's goals as stated in the Project Plan Sheet. Before beginning their review, peer reviewers or team members must discuss and agree on the project's goals with you, the draft's author. After all, team members and peer reviewers should be helpful partners, not negative adversaries.

In collaborative projects, it is helpful to assign team members to review parts of the project they have not drafted. The team member working on graphics and other visuals, for example, might review text, while the team member planning information design might review the use of visuals. After the reviews are completed, the team can then meet to take a look at all of the comments and questions about the draft, agree on changes, and assign revision tasks.

STRATEGIES FOR REVIEWING YOUR OWN DRAFT

If you must review your own draft, it is helpful and efficient to conduct a series of combined global reviews. For example, in one reading, review audience and document design; in a second reading, address subject matter and language. While the global review process sounds easy to manage, it's not. All of the review elements are closely connected. For example, you may recognize a need for additional information at the same moment as you notice a well defined term and a spelling error. By structuring reviews in stages, you can better see how a draft meets a project's purpose and what remains to be done.

If you must review your own draft for consistency, here are some helpful strategies:

- Try to allow time between review stages. When you return to the draft, you will find it easier to see and address inconsistencies.
- Read your draft aloud, or ask someone to read it aloud to you, as you carefully follow your own written version. You can catch and correct the errors and unclear wording that you hear.
- If you use a word processor, print out sections of text for review. The change in medium from screen to page can allow you to see small inconsistencies in the text and the page design more clearly.
- Review sections of text out of their sequence in the draft in order to see what is on the page, not what you expect to find there.

These strategies can help you keep annoying inconsistencies from slipping into the completed project.

WHAT IS REVISION?

Revision is the process of incorporating the suggestions raised in global reviews and acting on the corrections and comments made in consistency reviews. Careful and complete revision shapes the draft into its finished form.

It is easiest to act on reviewer suggestions after both global and consistency reviews are complete. Then you can summarize global comments about larger issues, such as audience and purpose, and make appropriate changes in sequence, information design, visuals, and language. It is important to make changes thoroughly and consistently, acting on comments and working according to plan. A working draft of the Smithville College Campus Guide (see the completed Project Plan Sheet in Chapter 7, pages 136–138) and global reviewer comments appear on page 162.

!!!!WORKING DRAFT!!!!

Introduction to the Campus Guide

The purpose of this brochure is to direct newcomers to the SC Campus to the facilities and locations and functions they need without any problems. The guide is intended to allow users to locate particular buildings as well as services that each building or location offers. Emergency and information SC telephone listings also appear at the end. A map is included to direct newcomers to specific locations and the easiest way to get there. All facilities on the map are identified by locator icons and by name. Additional brochure information consists of a list of campus sites (alphabetized for convenient use). A second alphabetized listing provides SC services (with location).

Reviewer's Comments

Audience
If the introduction to the Campus Guide is supposed to be helpful and welcoming, it needs to be more reader-friendly. We can better support the Guide's purpose by addressing the reader directly. Also, we can't assume that newcomers to campus will know the kinds of facilities available at Smithville College. Maybe examples would help?

Subject Matter
The introduction should give an overview of the purpose and content of the Guide, so subject matter is not as important an issue as it will be later on in the Guide.

Order
If the introduction should explain the purpose and sequence of the Guide, perhaps it should also discuss the parts of the Guide in the order in which they appear? The parts of this paragraph seem out of sequence.

Information Design
This short section of text seems awfully condensed on the page. Could it be broken into shorter chunks? Each chunk of text could discuss a different section of the Guide. Also, shouldn't this section be captioned "introduction"? The self-guided tour booklets of Smithville Township have similar headings for each section, which make it easy for readers to find and use parts of the booklets quickly.

Language
Here's where lots of changes can help the text meet readers' needs. By addressing readers as "you" and beginning with a short welcome, the Guide can establish a friendly tone. Also, sentences need to be shorter than the first one in the draft to make style easier to read. More active voice (rather than passive voice) will also create a clear and direct style. Perhaps it would be less intimidating to call the college hotline an "information number," not an "emergency number"? After all, the 24-hour phone number *does* provide information service. SC is our campus abbreviation for Smithville College, but visitors won't necessarily know that. It might be a good idea to close the introduction in the same way as it opened, by addressing the reader directly.

As these comments show, the reviewer evaluated the draft and made constructive comments relevant to the project's purpose and audience, which will help the author revise to meet the Campus Guide's goals.

As revision proceeds, it is helpful to ask for comments about changes from a peer reviewer at regular intervals. The reviewer's responses can assure that the revisions reflect the project's purpose. A final consistency review at the end of the entire revision process is important, so take special care in performing it or ask a peer reviewer for help. It is useful to compare the draft to the finished version of a project to evaluate changes. Compare the working draft for the Campus Guide on page 162 to the final version shown below.

FINAL VERSION

Introduction

Welcome to Smithville College! This Campus Guide is designed to help you find your way around the campus quickly and easily. You can use it to find a specific building or to locate the on-campus services you need.

The Campus Guide map identifies buildings (such as libraries) and facilities (such as parking lots and playing fields). As a key to the map's locations, the Campus Guide provides two lists, one an alphabetical listing of buildings and facilities and the other an alphabetical listing of campus services.

If you need help, our 24-hour campus information service will gladly answer any questions you may have. The information service telephone number appears on every page of the guide.

We hope that you will find the Campus Guide helpful in making you feel at home on the Smithville campus.

As the above example shows, revision does far more than correct written text. Revision involves completing and integrating all of a project's communication elements. In collaborative projects, the team member developing information design and the team member developing graphics and other visuals need to work closely with the author. Together they can review the project in its final form. Whether prepared by an individual or a team, the completed project must demonstrate a clear and consistent relationship between visual and written language.

TESTING FOR SUCCESS

When is a project finished? How can an author or a team be sure that a completed project serves its audience as it should? A project's timeline should provide for planning and review to define the project's purpose. The timeline also needs to provide time to make sure that the completed project meets the needs of its audience.

To evaluate the success of a revised project, ask reviewers who are unfamiliar with the project to read and use it as the project's intended audience will do. For example, in the case of the Smithville College Campus Guide, the communication team could ask visitors who are new to the campus to use the guide to locate a bookstore. A series of follow-up questions could ask the visitor whether the guide was helpful, accurate, easy to handle, and welcoming in tone. The responses to these questions could be useful in revising the guide, in developing other materials, and in helping the team improve their performance on future projects.

Since communication is a continuing process, audience response to revised projects is always helpful. It allows communicators to improve their skills and to remain sensitive to the needs of others.

GENERAL PRINCIPLES for Reviewing and Revising

- Reviewers should begin work by studying the draft's Project Plan Sheet. Only then can they understand the draft's merits and flaws.
- Wise communicators invite the participation of careful but tactful peer reviewers who can share a project's plan and review the draft for global issues as well as for consistency.
- Consistency reviews require concentration. Communicators who review their own drafts can use strategies to review their own writing objectively as they check for inconsistencies and errors.
- Members of peer reviewer teams should begin their work with a shared understanding of a draft's goal.
- Peer reviewers should comment on all of the elements of a project, including text, information design, and graphics and other visuals.
- Reviewing is time-consuming; therefore, team members should share the work and agree on specific duties.
- Revision means considering and acting on all comments, large and small, to allow a document to meet the readers' needs.
- Revision involves coordination among *all* of a project's elements, verbal and visual.
- Wise communicators can improve projects by seeking comments from peer reviewers at regular intervals during revision.
- Testing the success of a completed project is a practical way for communicators to learn from their work and remain sensitive to the needs of their audiences.

CHAPTER SUMMARY

REVIEWING AND REVISING

Effective communicators know the value of time spent reviewing and revising drafts. They also welcome the comments of peer reviewers. After all, no author has all of the insights into an audience's needs, and no rough draft is more than a beginning. To help authors, peer reviewers must understand a project's purpose. They must also be supportive in their comments and suggestions, noting the parts of the draft that are consistent with the project's purpose and the parts that need change or correction. For authors and reviewers, meeting audience needs must be a shared goal. In a revised project, all of the elements—verbal and visual—must work together to meet that goal.

ACTIVITIES

INDIVIDUAL AND COLLABORATIVE ACTIVITIES

8.1. What elements have you found most difficult to recognize and change in your own drafts? Are these global or consistency issues? Make a checklist to use in reviewing and revising your own work.

8.2. In groups of three or four, discuss your review and revision process. What do you typically spend the most time on when you review and revise your drafts? Make a list of strategies for using time efficiently.

8.3. Bring to class a sample document from a business or a campus organization. In groups of three or four, define the communication's audience and purpose. Then conduct a global review, commenting on subject matter, order, document design, and visuals. Does the communication meet its goal for its audience? Why or why not? Suggest changes for revision.

8.4. Below is a draft of an introduction to petty cash, developed for use by employees of a state agency. The draft is designed to help new employees who are unfamiliar with policy, procedures, and language. It is also designed for employees who need to review procedures for using petty cash. Since agency funds are involved, it is important that employees use petty cash correctly. Employees have complained that earlier versions of the procedure have been too hard to read and intimidating.

In a group of four or five, conduct a global review of the draft. What changes can you suggest to better meet the needs of readers? What corrections and notes would be helpful to the author in a consistency review?

!!!!WORKING DRAFT!!!!

WHAT IS PETTY CASH & HOW CAN I USE IT?

Description of Account Types

A petty cash fund is a specific amount of money set aside for the purpose of making quick purchases for items or services that cannot be purchased with the State MasterCard. The majority of these types of purchasing situations exist for certain services (not goods) and when the vendor is not incorporated (businesses that are solely or partnership owned). If used to its potential and according to these guidelines, a petty cash fund can significantly reduce the number of IFS purchase orders generated for small purchases and free up staff time to concentrate on the more immediate tasks at hand.

Review the following list to determine whether your needs are for a Petty Cash fund account or for one of the other available types of accounts. Petty Cash accounts are authorized by the Office of the Comptroller of Public Accounts and are initiated through the Financial Management section and maintenance of general account information is kept in the Accounts Payable section and subject to the TPW procedures and Comptroller guidelines (APS 010). Other accounts can be obtained by contacting the listed branch sections noted for each type of account.

Petty Cash accounts Used for small purchases and reimbursed with purchase vouchers; account balance remains constant (cash plus outstanding vouchers). Established through the Financial Management section, maintained by the Accounts Payable section.

Change Fund accounts Used only for issuing change for currency; account balance remains constant. Established through the Financial Management section and maintained by the Revenue section.

State Treasury accounts Used as temporary holding accounts for transferring funds to the Comptroller-State Treasury. The only transactions that can be posted to these accounts are deposits; no checks or withdrawals are allowed. Established through the Financial Management section and maintained by the Revenue section.

Escrow accounts	Used for controlling funds held in escrow or in trust, pending fulfillment of specific conditions or agreements; includes principal and interest. Established through and reconciled by the Financial Management section.
Evidence account	Used for the purchase of evidence and/or information and surveillance solely by Law Enforcement.
Operation Game Thief (OGT) account	Used only to maintain the OGT fund, promote the operation of the OGT fund and pay authorized rewards and death benefits. Information prepared by Law Enforcement and quarterly reports prepared from statements by the Financial Management section.

Used by permission of Frances Stiles.

8.5. On pages 168–170 are the completed Project Plan Sheet sections on audience, purpose, subject, and overall design specifications and the revised version of the Smithville College Campus Guide. In groups of three or four, discuss the ways in which the finished Guide meets the needs of its intended audience.

8.6. In groups of three or four, evaluate the effectiveness of the completed Smithville College Campus Guide. Analyze information design, visuals, and language.

PROJECT PLAN SHEET

SMITHVILLE COLLEGE CAMPUS GUIDE

Audience

- Who will read the Guide?

  New students, visitors to campus
- How will readers use the Guide?

  People will use the Guide to locate campus services both when they are physically on campus and during orientation presentations.
- Will more than one audience use the project? If so, how can the writing project meet the needs of all of its audiences?

  The Guide will serve both students and visitors, but both audiences will be unfamiliar with the campus. Whether they use the Guide when they are walking around campus or listening to a presentation, they will want a quick, easy-to-use reference.
- How will your audience guide your communication choices?

  Since the catalog provides too much information and the campus map provides too little information, the Guide needs to provide the best of both in a brief, easy-to-hold Guide that is easy to read and follow.

Purpose

- What is the purpose of the Guide?

  The Guide will welcome those new to campus by making it easy for them to find what they need without getting lost or wasting time. It will also help them learn their way around.
- What need will the Guide meet? What problem can it help to solve?

  The Guide will allow visitors and new students to locate what they need on campus quickly and easily. In doing this, it will also make users feel at home and welcome.
- What title will most clearly reflect the Guide's purpose?

  The Guide should be called "Campus Guide" since it will help people find their way to what they want.

Subject

- What is the Campus Guide's subject matter?

  Services available on campus and the buildings where these services are located
- How technical should your discussion of the subject matter be?

  Very nontechnical! The Guide should make no assumptions about what people who are new to campus know.

- Do you have sufficient information to complete the Guide? If not, what sources or people can help you to locate the best research materials?

  No. The catalog and campus map are available and current. These will be the best-written sources.

  The chair of the New Student Orientation Committee, committee members who conduct campus orientations, and volunteers from campus information booths will be able to tell me some of the questions and problems people who are new to campus have.

Project Design and Specifications

- Are there models for organization or format for the Campus Guide?

  The city has self-guided walking-tour guides that could be helpful models.

- In what medium will the completed project be presented?

  The finished Campus Guide will need to be a printed brochure. The team will prepare text, information design, and graphics.

- Are there special features the completed Campus Guide should have?

  Yes, the Guide will need a map, clearly written text, an attractive design (perhaps featuring the college logo), and lists of buildings and services, with a brief description of each one keyed to the map.

- Will the Guide require graphics and other visuals? If so, what kinds and for what purposes?

  Yes, graphics will appear in the design of the map and the key to different campus buildings and services for ease of location.

- What information design features can help the project's audience?

  The Guide will need to be large enough so that the map and the lists will be easy to read and small enough so that it can fit into a backpack or purse. A sheet of $8\frac{1}{2}$ by 14 inch paper folded into five $8\frac{1}{2}$ by $2\frac{3}{4}$ inch sections would be easy to handle and inexpensive to duplicate.

SMITHVILLE COLLEGE

Welcome to Smithville College! This campus Guide is designed to help you find your way around the campus quickly and easily. You can use it to find a specific building or to locate the on-campus services you need.

The Campus Guide map identifies buildings (such as libraries) and facilities (such as parking lots and playing fields). As a key to the map's locations, the Campus Guide provides two lists, one an alphabetical listing of buildings and facilities and the other an alphabetical listing of campus services.

If you need additional help, our 24-hour campus information service will gladly answer any questions you may have. The information service telephone number appears on every panel of the guide.

We hope that you will find the Campus Guide helpful in making you feel at home on the Smithville campus.

Need further information?
Call 341/284-3791

1

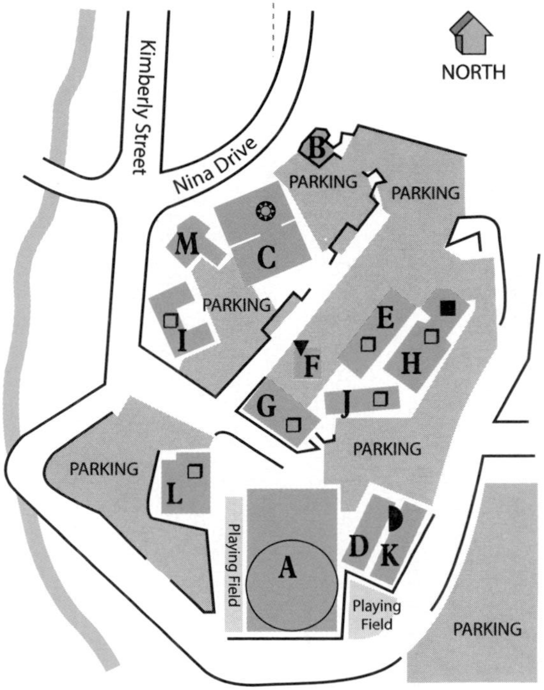

SMITHVILLE COLLEGE

Need further information?
Call 341/284-3791

2

Need further information?
Call 341/284-3791

3

Buildings and Facilities

Below is an alphabetical listing of Smithville College buildings and facilities by name. Each building is identified by a letter, which allows you to locate the building on the Smithville Campus map. Parking facilities and playing fields are identified on the map.

A Armstrong Stadium & Gymnasium
B Campus Police Center
C Fonte Administration Building
D Freeman Dormitory (Women)
E Hill Math Studies Building
F Information Kiosk
G Merriam Technical Studies Building
H Murray English Building
I Powers Business Building
J Renau Humanities Building
K Renfro Dormitory (Men)
L Stuart Fine Arts Building
M Zimmer Memorial Library

Need further information?
Call 341/284-3791

4

Services

Below is a list of Smithville College services, each identified with a symbol. This symbol allows you to use the map to locate types of services.

Classrooms ☐
College Press ■
Dormitories ◗
Freeman (Women) D
Renfro (Men) K
Information ▼
Student Services ✪
Business Office
Counseling
Financial Aid
Testing Center
Veterans Affairs

Need further information?
Call 341/284-3791

5

PART IV

Situations and Strategies for Technical Communication

What kinds of documents do technical communicators typically prepare? The chapters in **Part IV: Situations and Strategies for Technical Communication** discuss the categories of documents that professionals use to present workplace information. Each chapter describes a different category, describing its conventions and explaining appropriate contexts and audiences. In addition, each chapter provides a project plan sheet and procedures to help you prepare such documents. To help you consider the effective and ethical use of each category, all of the chapters in this section include a reading and a case study.

Chapter 9 Procedure: Explanation Showing How describes one of the most useful categories. Explaining can take the form of instructions, showing a reader the steps necessary to perform a task, or process, describing how a process takes place. For example, instructions might explain how to tell the age of a tree. A process explanation might explain how an additional circle forms each year in the life of a tree.

Chapter 10 Description: Using Details discusses a category of communication that explains the details of an item, an object, a place, a mechanism, or a situation. This chapter explains the difference between a representative model and a specific model.

Chapter 11 Definition: Explaining What Something Is covers a category of communication that helps readers understand the meaning of terms. Sometimes a term is an unfamiliar word for a familiar idea, and sometimes the term is a completely new idea for a reader, requiring extensive discussion. Chapter 11 provides guidelines and strategies for defining terms effectively.

Chapter 12 Summaries: Getting to the Heart of the Matter discusses a category that calls for critical reading, logical planning, and clear writing. The summary, or brief synopsis, is a useful way to introduce workplace audiences to new ideas.

Chapter 13 Reports: Shaping Information describes the common elements of a category of communication that can be adapted for different audiences, purposes, and situations. In each case, the report provides factual information to readers in useful ways. Chapter 13 demonstrates the report conventions and practices that can make any type of report easy to use.

Chapter 14 Proposals: Using Facts to Make a Case explains a category of communication that allows businesses to earn customers. Authors of proposals win approval by presenting information usefully adapted to the needs of readers. When proposing to undertake a job, a communicator persuades by describing how and when a project will be undertaken, who will work on it, how long the work will take, and how much the project will cost. Chapter 14 explains the conventions and ethics for writing effective proposals.

Chapter 15 Correspondence: Sending and Responding to Messages discusses a diverse and important category of workplace communication: memos and letters. Correspondence focuses on an addressed audience and purpose. With the development of electronic delivery systems such as e-mail and fax, messages travel quickly around the world. Chapter 15 explains conventions and strategies to make correspondence clear and reader-centered.

The content of the chapters in Part IV reflects current workplace practices. As a workplace problem solver, you will adapt these categories to your needs.

CHAPTER 9

Procedure: Explanation Showing How

CHAPTER GOALS

This chapter:

- Defines *procedure explanations*
- Discusses audiences and uses for procedure explanations
- Outlines the author's responsibilities in procedure explanations
- Identifies the role of graphics and other visuals in procedure explanations
- Emphasizes the importance of information design in procedure explanations
- Distinguishes between procedures that explain what to do (instructions) and procedures that explain what is done or what happens (process)
- Explains how to plan and construct procedures for instructions and processes

INTRODUCTION

All aspects of life are affected by procedures—from brushing your teeth to filling out income tax forms, from playing soccer to tuning an electric guitar, from understanding how a diamond is cut and polished to how a human heart is transplanted. Being able to explain and follow procedures is essential wherever you are—at home, in a classroom, on vacation, in a sports arena, and at work. You may think that explaining a procedure is simple. When you consider audience and purpose and other factors affecting the explanation, however, you will find that explaining a procedure requires critical thinking, planning, analysis, effective information design, and perhaps even research.

The following pages discuss audiences and uses for procedure explanations, the author's responsibilities, graphics and other visuals, and information design. Suggestions for planning and composing procedure explanations, for instructions and processes, are also included. The purpose of this chapter is to help you learn how to prepare clear, accurate, and complete procedure explanations.

WHAT IS A PROCEDURE EXPLANATION?

A procedure explanation is a description of a sequence of actions expressed as steps or stages. The way in which you explain the procedure depends on audience and purpose. The purpose may be to explain a sequence of actions that an audience follows to perform a particular operation. You explain **what to do** and expect the audience to act, to carry out or perform the sequence of steps. For example, your purpose might be to explain how to assemble an entertainment center, how to troubleshoot problems with a VCR, how to set up a new computer and printer, or how to build a multilevel deck. Such procedure explanations are typically referred to as instructions.

Or your purpose may be to explain a method, operation, or sequence of events so that the audience will comprehend the concept of what is done or what happens. For example, you might explain how sound is transmitted, how bricks are made, or how a tornado forms. No individual could possibly perform some processes, for example, those that occur in nature. Audiences can, however, understand what happens as these natural processes occur; for instance, they can understand how sound is transmitted or how a tornado develops. Further, they can understand how bricks are made, but few will ever make a brick. Such procedure explanations are typically referred to as processes.

The difference in purpose for procedure explanations is reflected in the choices of subjects and verbs. The following chart compares these choices.

Procedure explanation for understanding **what to do (instructions)**	Procedure explanation for understanding **what is done (process)**	Procedure explanation for understanding **what happens (process)**
(You) Clean the wound and then place a dressing over it.	After the wound has been cleaned, a dressing is placed over it.	After the nurse cleans the wound, he places a dressing over it.
1. Imperative mood (orders or commands) 2. Active voice (subject does the action)—"Clean" and "place" 3. Second person (person spoken to is subject); subject is understood to be *you.*	1. Indicative mood (statement of fact) 2. Passive voice (subject is acted on)—"has been cleaned," "is placed" 3. Third person (thing spoken about is subject)—"wound," "dressing"	1. Indicative mood (statement of fact) 2. Active voice (subject acts)—"cleans," "places" 3. Third person (person acting; describes what person does)—"nurse cleans," "he places"

To explain a procedure, you must thoroughly understand the material.

Audience and purpose also determine whether the procedure explanation is written or oral.

AUDIENCES AND USES FOR PROCEDURE EXPLANATIONS

Explaining procedures in a way that is clear, easy to follow, and understandable requires thinking and careful planning. Two of the first considerations are (1) Who is the audience? and (2) What is the purpose of the explanation? Knowing your audience and your purpose helps you to make decisions about content, language choices and sentence structure, and information design. An explanation of how to operate the latest model x-ray machine would differ for an experienced x-ray technician and for a student just being introduced to x-ray equipment. Instructions on how to freeze corn would differ for a food specialist at General Mills, for a homemaker who has frozen other vegetables but not corn, and for a seventh-grade student beginning a nutrition course. An explanation of how a poisonous spider bite affects a young child would differ for parents of elementary school children and for licensed practical nurses.

It Could Be Worse!

Interstate 55 Exit Ramps To County Line Road Project

"What's happening?"
Beginning Monday, October 28, 1996, the Mississippi Department of Transportation (MDOT), in conjunction with the Mississippi Concrete Industries Association (MCIA), will begin replacing the asphalt pavement on the exit ramps to County Line Road from Interstate 55 north and southbound with a new, high-tech concrete. MCIA is a nonprofit trade association of the Mississippi concrete industry.

"Is this work really necessary?"
Yes! The existing pavement is worn and must be replaced. The fact that advanced technology is being used is an added bonus. After the new pavement is down, both MDOT and MCIA will continue to monitor the area for up to three years in order to gather invaluable data for use on similar projects.

"How long will it take?"
The entire project is expected to take just *one week*, depending on Mother Nature, of course.

"Will the Interstate 55 exits to County Line Road remain open?"
Yes! At no time will the ramps be completely closed. However, there will be lane closures within the project, so if you choose to go through that area, expect traffic to be moving slowly and *please be patient.*

"Are there other roads in the area that will be closed?"
Yes. Both the east I-55 frontage road from Ridgewood Court to County Line Road will be closed as well as the ramp from Highway 51 to County Line Road. Detour signs will be in place to direct traffic.

"What's so special about this new concrete?"
The concrete, which has never before been used in Mississippi, is expected to last longer and cost less. It's ultra-thin, quick drying, and can be driven on within 18 hours after application. Within the concrete are specially designed plastic fibers, which make it stronger and less likely to break apart.

Construction Necessary to Make High-Tech Improvements to County Line Road.

CONSTRUCTION MAPS

Source: Mississippi Department of Transportation.

Audiences exhibit a variety of characteristics. After you identify your audience and purpose, try to determine the background, experience, specialized knowledge, or skills the audience may have. Audiences with some knowledge and experience will understand specialized language. Audiences with little knowledge and experience will need simple, direct language. If you must use specialized terms, define each term. (See also Chapter 2 Defining Workplace Readers: Who Are They?)

An example of procedure explanation appears above and continues on the next page. Distributed by the Mississippi Department of Transportation (MDOT), the purpose is to alert drivers who use Interstate 55 exit ramps to County Line Road in Jackson, MS, to the construction process.

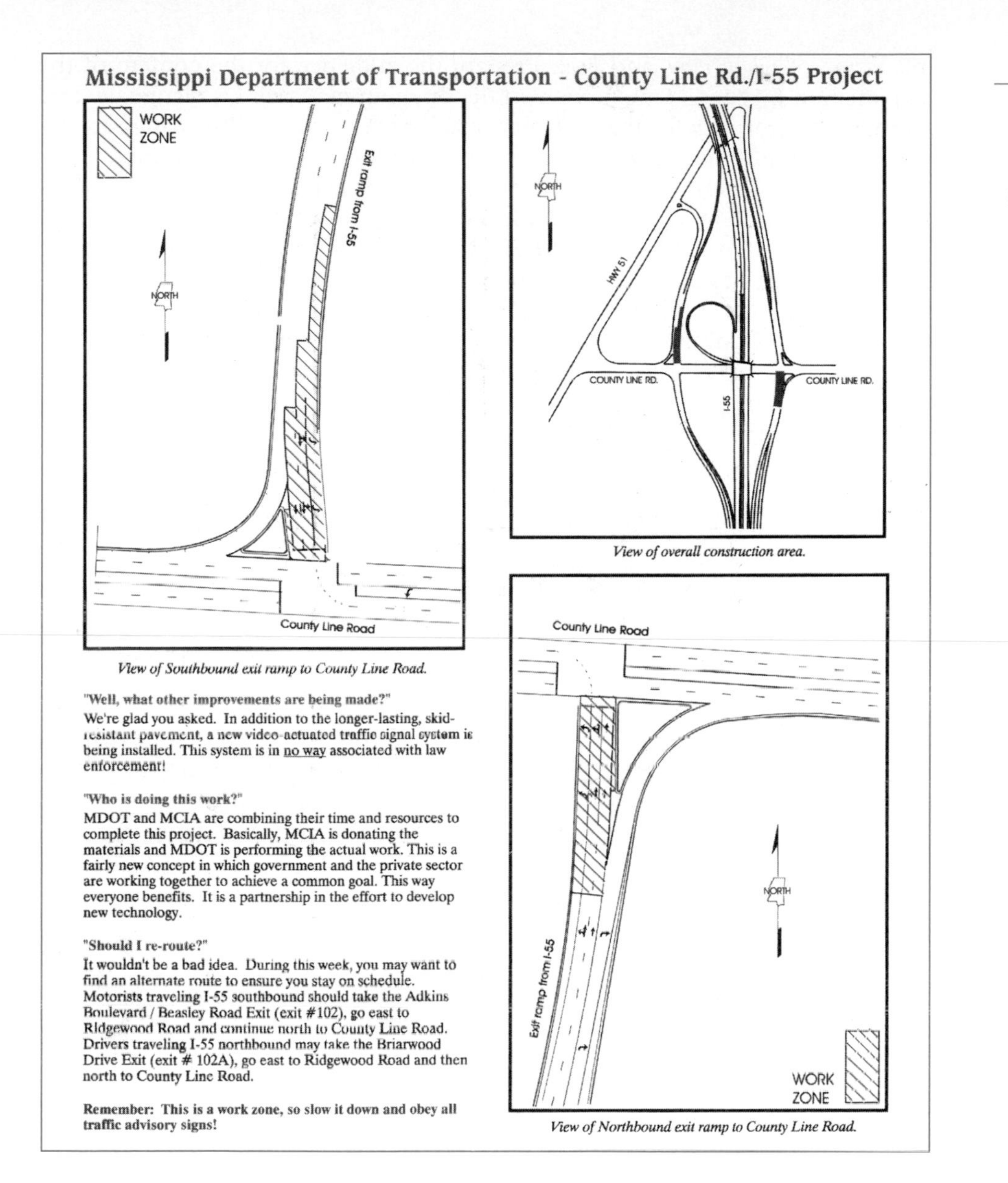

Mississippi Department of Transportation - County Line Rd./I-55 Project

View of Southbound exit ramp to County Line Road.

View of overall construction area.

"Well, what other improvements are being made?"
We're glad you asked. In addition to the longer-lasting, skid-resistant pavement, a new video-actuated traffic signal system is being installed. This system is in no way associated with law enforcement!

"Who is doing this work?"
MDOT and MCIA are combining their time and resources to complete this project. Basically, MCIA is donating the materials and MDOT is performing the actual work. This is a fairly new concept in which government and the private sector are working together to achieve a common goal. This way everyone benefits. It is a partnership in the effort to develop new technology.

"Should I re-route?"
It wouldn't be a bad idea. During this week, you may want to find an alternate route to ensure you stay on schedule. Motorists traveling I-55 southbound should take the Adkins Boulevard / Beasley Road Exit (exit #102), go east to Ridgewood Road and continue north to County Line Road. Drivers traveling I-55 northbound may take the Briarwood Drive Exit (exit # 102A), go east to Ridgewood Road and then north to County Line Road.

Remember: This is a work zone, so slow it down and obey all traffic advisory signs!

View of Northbound exit ramp to County Line Road.

By using questions and answers to present both instructions and processes, the writers and designers meet their readers' needs. The language is simple and direct, the graphics are clear and easy to understand, and color [omitted in the reprinted example] calls attention to critical areas. The overall tone suggests MDOT regrets the inconvenience but wants drivers to understand that access will be improved when construction is completed.

REMINDER

Always be aware of audience characteristics and audience needs. Use this awareness as you make decisions about procedure explanations.

THE AUTHOR'S RESPONSIBILITIES

As author of a procedure explanation, you bear specific responsibilities. First, you owe the audience a clear, accurate, complete, impartial, and unbiased explanation of *what*, *why*, and *how*. Prepare the audience for the content of the explanation by giving an overview in the opening material. Use language and graphics appropriate for the audience and for the purpose. For written explanations, use information design to enhance reading and understanding.

Clarity and completeness are essential in explaining a procedure for audience understanding. A clear, complete explanation of what happens when a person is on a respirator is critical for medical personnel who will assist such patients. An explanation of how natural phenomena interact to create hurricanes is essential for meteorologists. When writing such explanations, you must include all information needed for clear understanding.

You also bear significant responsibility for the safety and well-being of the audience. This responsibility manifests itself in a variety of ways. For example, many individuals do not read through a procedure explanation before carrying out individual steps. They typically read and act alternately, or they read and act simultaneously. Therefore, you must supply information when the audience needs it and use information design to help readers easily find their place in the procedure.

Identify any dangers—warnings, cautions, or precautions—**before** the audience could become involved in these dangers. A user is not helped by the warning "Be sure to disengage the blade before removing the work piece" at the end of a set of instructions if she is already missing a fingertip. If a patient with severe stomach pain is following directions to prepare for a medical procedure and Step 2 directs her to drink magnesium citrate, she does not want to read in Step 10, "Do not drink magnesium citrate if you are currently pregnant or are currently experiencing severe stomach pain." If a procedure involves hazardous steps or materials, identify these hazards **before** your reader has opportunity to carry out a hazardous step or use hazardous material. For example, instructions on how to build a deck should tell the reader at the beginning to check with telephone, gas, electric, and cable companies before digging in the yard where lines may be buried.

Call attention to any warnings, cautions, or precautions by highlighting information about dangers. Use techniques such as uppercase letters, color, rules, boxes, boldface, and font size to highlight important information.

Also keep in mind that audiences may read and follow procedure explanations in unpleasant and distracting environments. Readers may be hot, sweaty, and tired. They may be under extreme pressure. Clear, easily followed explanations decrease audience anxiety and the likelihood of error.

Be careful to supply precise measurements and timing, when needed. For example, write "Breaker points in use more than 10,000 miles need cleaning or replacing," not "Breaker points in use more than a few thousand miles need attention." Obviously the nonprecise phrase "more than a few thousand miles" could be interpreted differently by different readers. To ensure that readers interpret data as you intend, use precise measurements and timing.

Finally, if you write instructions that accompany an appliance or other item—how to assemble a Sauder computer desk, how to operate a Toro lawnmower, how to install a Whirlpool electric cooktop—be aware that these instructions may determine how a buyer feels about a product and the company that makes it.

GRAPHICS AND OTHER VISUALS FOR PROCEDURE EXPLANATIONS

Include appropriate graphics and other visuals to enhance and clarify procedure explanations—to make them clear, easy to follow, and understandable for an audience. Graphics and other visuals such as drawings, photographs, and diagrams are helpful in written procedure explanations. Real objects, models, demonstrations, slides, and projected materials enhance oral procedure explanations.

Do not include graphics and other visuals, however, unless they benefit your audience. Evaluate your audience's needs and the purpose of the explanation. Ask yourself, "Would graphics or other visuals make the explanation clearer and easier to understand?" A good rule of thumb is that more experienced audiences may need fewer but more complex graphics and other visuals; inexperienced audiences may need more graphics, but simpler ones.

Graphics and other visuals allow the audience to *see* what a writer or speaker is explaining. Note the graphics (line drawings) in the procedure explanation for performing the Heimlich Maneuver on page 180. The drawings are as important as the verbal explanation in conveying what a person should do to help the victim.

Consider the use of graphics and visuals in the examples of instructions on how to measure for a garage door, page 184, and how to use water and ice dispensers on a Whirlpool refrigerator, page 185. Look at graphics in the process explanation of how the heart works, page 196, and how a black and white advertisement is produced, pages 202–205. Also, look at the suggested objects to be shown during an oral presentation on how Alexander Fleming discovered penicillin, pages 486–487. Showing the objects adds visual interest to the spoken word. For a detailed discussion of graphics and other visuals, see Chapter 5.

INFORMATION DESIGN

In procedure explanations, information design is of prime importance. An overall design planned to provide ready access to needed information, clearly presented in a convenient format, is essential for audience understanding and use. Arrange material on the page to achieve maximum ease in reading and comprehension. For example, simply dividing the material into chunks of related information, using a heading to identify each chunk, and leaving white space around each chunk significantly increases ease of reading and understanding. Review the examples on pages 181–182. In each example, techniques such as line drawings, numbered steps, white space, double columns,

Example of Graphics Complementing Verbal Explanation

THE HEIMLICH MANEUVER

<u>FOOD-CHOKING</u>
<u>What to Look For.</u> Victim cannot speak or breathe; turns blue; collapses.

<u>To perform the Heimlich Maneuver when the victim is standing or sitting</u>

1. Stand behind the victim and wrap your arms around his waist.
2. Place the thumb side of your fist against the victim's abdomen, slightly above the navel and below the rib cage.
3. Grasp your fist with the other hand and press your fist into the victim's abdomen with a quick upward thrust. Repeat as often as necessary.
4. If the victim is sitting, stand behind the victim's chair and perform the maneuver in the same manner.
5. After the food is dislodged, have the victim seen by a doctor.

Position of hands

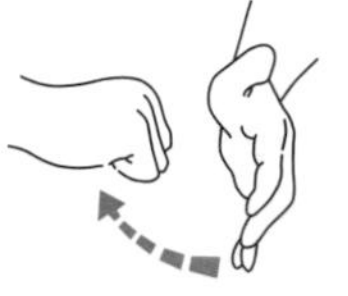

Direction of hand and fist movement

<u>To perform the Heimlich Maneuver when the victim has collapsed and cannot be lifted</u>

1. Lay the victim on his back.
2. Face the victim and kneel astride his hips.
3. With one hand on top of the other, place the heel of your bottom hand on the abdomen slightly above the navel and below the rib cage.
4. Press into the victim's abdomen with a quick upward thrust. Repeat as often as necessary.
5. Should the victim vomit, quickly place him on his side and wipe out his mouth to prevent aspiration (drawing of vomit into the throat).
6. After the food is dislodged, have the victim seen by a doctor.

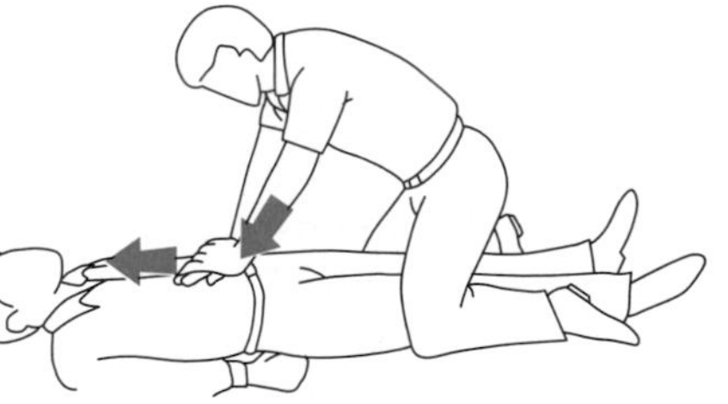

Position of hands

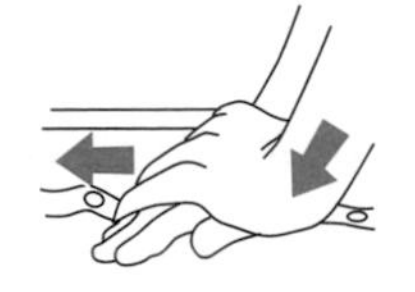

Direction of hand movement

NOTE: If you start to choke when alone and help is not available, an attempt should be made to self-administer this maneuver.

Source: New York City Department of Health

headings, and boldface contribute to an overall appealing and effective design for the audience and purpose.

Consider two versions of the opening page of an explanation on how to determine the resistance of a carbon resistor, a version with minimal information design (page 181) and a version with significant information design (page 182). Both versions give the same information. The positive page design, however, creates a document that is easier to read and comprehend:

- The material is divided into chunks of related material.
- The page has plenty of *white space;* the material is uncluttered and is placed on the page so that the eye easily and quickly sees the relationships of chunks of material.
- The *headings* provide key words for each section.
- Emphasis is achieved in several ways:

  uppercase letters for the headings

  the *box* to set off material

  the *closed bullets* for a list in which sequence is not important

  the *numbers* for the major steps (where sequence is very important)

  the *letters of the alphabet* (for substeps in a sequence)

For a detailed discussion of how to design effective, readable documents, see Chapter 4 Information Design.

Version 1 Page 1 of a Procedure Explanation with Minimum Information Design

HOW TO DETERMINE THE RESISTANCE OF A CARBON RESISTOR

An essential component of most electronic circuits is the carbon resistor. The carbon resistor opposes current flow, thereby providing a usable load into which a circuit may operate. For this load to be effective, the value of the carbon resistor must have the correct resistance. A carbon resistor looks like a firecracker with a fuse on both ends. Resistors normally come in sizes of $\frac{1}{8}$ watt, $\frac{1}{4}$ watt, $\frac{1}{2}$ watt, 1 watt, and 2 watts. The $\frac{1}{8}$ watt resistor is about $\frac{1}{4}$ inch long and $\frac{1}{16}$ inch in diameter, and the 2 watt resistor is about 1 inch long and $\frac{5}{16}$ inch in diameter. To facilitate reading the value of a resistor, the electronic technician will need to know the color code or have access to the color code: black = 0, brown = 1, red = 2, orange = 3, yellow = 4, green = 5, blue = 6, violet = 7, gray = 8, white = 9; for tolerances, silver = ±10% and gold = ±5%. The Greek letter Ω (omega) is the universal symbol for *ohms,* the unit in which resistance is measured. Each resistor has four colored bands; the first three indicate the amount of resistance, and the fourth indicates the amount of tolerance.

The first step in determining the resistance is to look at Band 1. Determine its color, note the digit value (0 to 9) of the color from the color code, and record the digit.

Version 2 Page 1 of a Procedure Explanation with Significant Information Design

Uppercase letters, in color

HOW TO DETERMINE THE RESISTANCE OF A CARBON RESISTOR

Ample margins

An essential component of most electronic circuits is the carbon resistor. It opposes current flow, thereby providing a usable load into which a circuit may operate. For this load to be effective, the value of the carbon resistor must have the correct resistance.

A carbon resistor looks like a firecracker with a fuse on both ends. Resistors normally come in these sizes:

List format
Closed bullets

- $\frac{1}{8}$ watt (about $\frac{1}{4}$ in. long, $\frac{1}{16}$ in. diameter)
- $\frac{1}{4}$ watt
- $\frac{1}{2}$ watt
- 1 watt
- 2 watts (about 1 in. long, $\frac{5}{16}$ in. diameter)

Headings, underlined

Color code

To facilitate reading the value of a resistor, the electronics technician will need to know the color code or have access to a code chart, as shown in Table 1.

Table

TABLE 1 Color Code for Carbon Resistors[a]

Color	Value	Color	Value
black	0 Ω	green	5 Ω
brown	1	blue	6
red	2	violet	7
orange	3	gray	8
yellow	4	white	9

[a]Tolerances: silver = ±10%
gold = ±5%

Box

> NOTE: The Greek letter Ω (omega) is the universal symbol for *ohms,* the unit in which resistance is measured.

Each resistor has four colored bands; the first three indicate the amount of resistance, and the fourth indicates the amount of tolerance.

Steps In Determining the Resistance

Numbers for major steps and for substeps

1. Look at Band 1.
   1.1 Determine its color.
   1.2 Note the digit value (0 to 9) of the color from the color code chart.

PROCEDURE EXPLANATIONS OF HOW TO DO SOMETHING (INSTRUCTIONS)

A procedure explanation telling an audience how to do something is typically referred to as instructions. All aspects of life, personal and professional, are affected by instructions—how to assemble a Barbie dollhouse, how to brush and floss teeth, how to write an effective letter, how to use a computer, how to work collaboratively, how to buy stocks and bonds, how to fly an F-16. To survive in this world, we must be able to give and follow instructions.

Clear, accurate, complete instructions save people time, help them do a job faster and more satisfactorily, and allow them to get better service from a product. The ability to give and follow instructions is essential for any person on the job. Certainly, in order to advance to supervisory positions, employees must be able to give intelligent, specific, accurate instructions; and they must be able to follow the instructions of others.

In this section, you will learn how to analyze a situation requiring instructions and prepare clear, accurate, and complete instructions that will meet audience needs and serve your purpose.

Adapting Instructions to Audience and Purpose

The audience and purpose for instructions determine the kind and extent of details included and the manner in which they are presented. Therefore, you must know who will be reading or hearing the instructions and why.

Consider the panel of instructions on how to measure for a garage door, shown on page 184. The panel is one part of a pamphlet entitled "How to Choose the Perfect Garage Door" from the Clopay Building Products Company. The pamphlet guides a lay audience (anyone shopping for a garage door) in selecting and installing the door.

The next example is a page of instructions from a Whirlpool side-by-side refrigerator/freezer operator's manual on page 185. This page primarily gives instructions for using the water and ice dispensers. It also gives a brief explanation of how the ice dispenser works. The simple language, sentence structure, and graphics indicate an awareness of the audience's needs. Also, ease of use is enhanced by the two-column format, lists marked with squares and numbers, headings, a readable font, uppercase letters, and boldface type. These information design features allow the audience to follow the procedures with ease.

The final example, on handwashing (page 186), is designed for a specialized audience. For a lay audience, such instructions would likely include three or four steps. For a nurse, however, the instructions are more complex. They include ten steps, with a rationale for each step.

> **REMINDER**
>
> Remember that your audience is probably not as familiar with the subject as you are. Therefore, tell your audience enough detail so that they can follow the instructions successfully.

Planning and Organizing Instructions

Planning instructions requires decisions about topics such as audience, purpose, subject matter, author, design, and deadlines. This textbook guides you in making these decisions with a communication project plan sheet that is

How To Measure for Your New Garage Door

Please take the following measurements (refer to drawing and directions below) to determine the proper garage door size:

[The author and designer of this panel were keenly aware of their audience. Notice the graphic that supports the verbal explanation by illustrating exactly how to measure the opening. Also notice the lines provided for the record of measurements.]

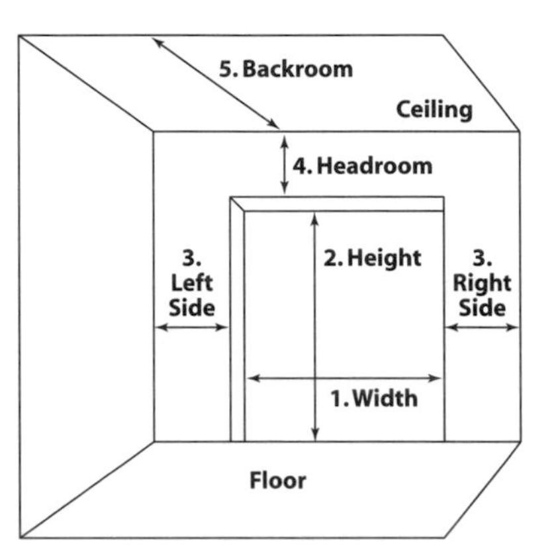

Step #1
Measure the width of your garage door opening at the widest point—this is the distance (in feet) between the jambs on left and right side of opening.

1. ____________________

Step #2
Measure the height of your garage door opening at the highest point—this is the distance (in feet) between the inside "finished" floor and jamb header.

2. ____________________

Step #3
Measure the width of the area labeled "left side" and "right side." $3\frac{3}{4}$" is required on each side for installation of the vertical track for standard extension springs, $5\frac{1}{2}$" for the EZ-Set Spring™.

3. (left) ____________________ 3. (right) ____________________

Step #4
Measure area labeled "headroom" distance (in inches) between top of door opening ("jamb header") and "ceiling." Headroom must be clear of obstructions for the full backroom required (See Step 5).*

4. ____________________

Our doors require 10" headroom for standard extension spring or EZ-Set Spring™. Special kits are available from Clopay that reduce the minimum headroom requirements to $4\frac{1}{2}$."

Step #5
Measure area labeled "backroom"—distance is measured from the back of the garage door into the garage. Door height plus 18" is required.

5. ____________________

*Additional headroom is required for attachment of an automatic garage door opener (headroom plus 2 to $3\frac{1}{2}$"). Check with your opener manufacturer for headroom requirements.

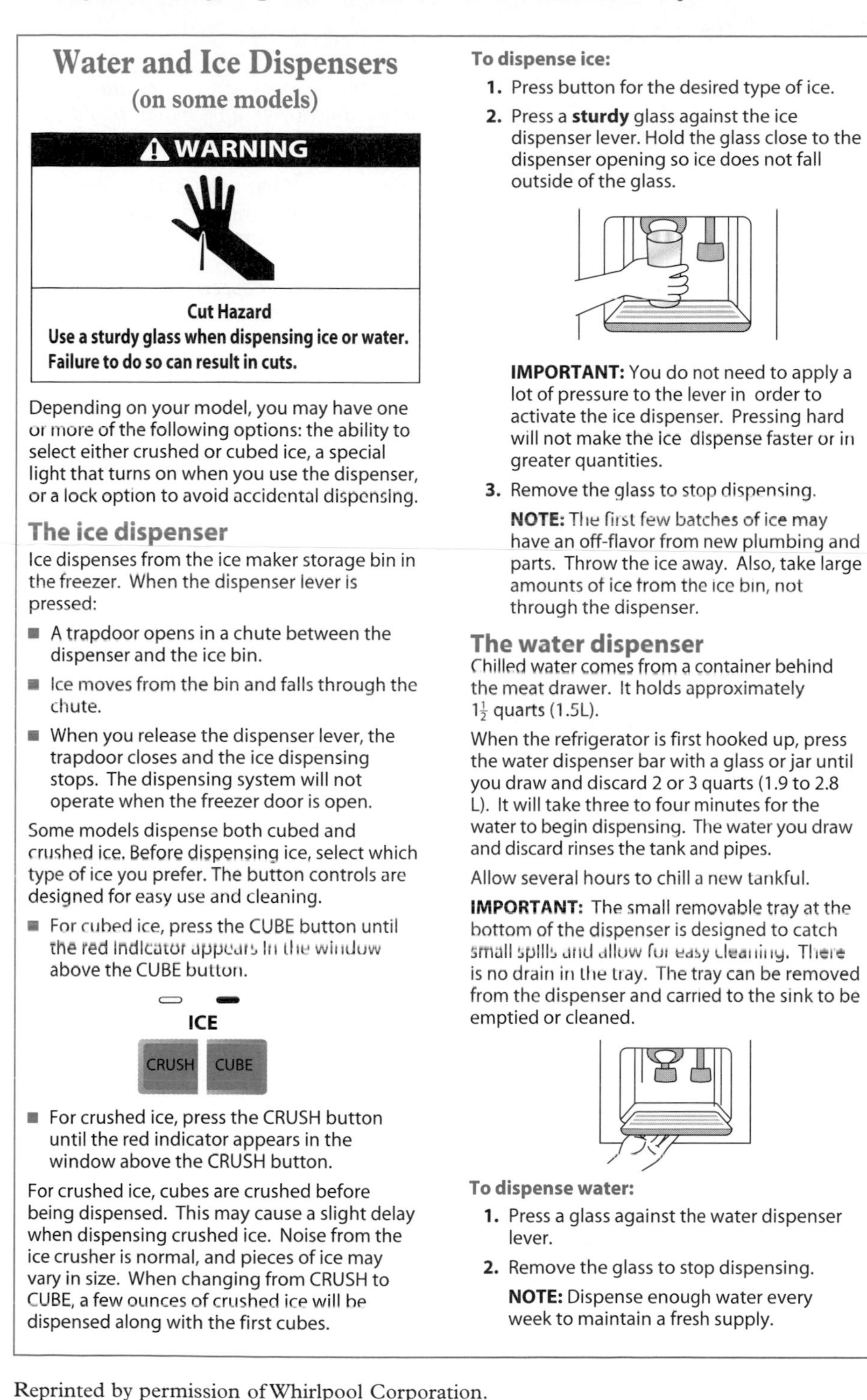

Water and Ice Dispensers

(on some models)

⚠ WARNING

Cut Hazard

Use a sturdy glass when dispensing ice or water.

Failure to do so can result in cuts.

Depending on your model, you may have one or more of the following options: the ability to select either crushed or cubed ice, a special light that turns on when you use the dispenser, or a lock option to avoid accidental dispensing.

The ice dispenser

Ice dispenses from the ice maker storage bin in the freezer. When the dispenser lever is pressed:

- A trapdoor opens in a chute between the dispenser and the ice bin.
- Ice moves from the bin and falls through the chute.
- When you release the dispenser lever, the trapdoor closes and the ice dispensing stops. The dispensing system will not operate when the freezer door is open.

Some models dispense both cubed and crushed ice. Before dispensing ice, select which type of ice you prefer. The button controls are designed for easy use and cleaning.

- For cubed ice, press the CUBE button until the red indicator appears in the window above the CUBE button.

ICE

CRUSH CUBE

- For crushed ice, press the CRUSH button until the red indicator appears in the window above the CRUSH button.

For crushed ice, cubes are crushed before being dispensed. This may cause a slight delay when dispensing crushed ice. Noise from the ice crusher is normal, and pieces of ice may vary in size. When changing from CRUSH to CUBE, a few ounces of crushed ice will be dispensed along with the first cubes.

To dispense ice:

1. Press button for the desired type of ice.
2. Press a **sturdy** glass against the ice dispenser lever. Hold the glass close to the dispenser opening so ice does not fall outside of the glass.

   IMPORTANT: You do not need to apply a lot of pressure to the lever in order to activate the ice dispenser. Pressing hard will not make the ice dispense faster or in greater quantities.
3. Remove the glass to stop dispensing.

   NOTE: The first few batches of ice may have an off-flavor from new plumbing and parts. Throw the ice away. Also, take large amounts of ice from the ice bin, not through the dispenser.

The water dispenser

Chilled water comes from a container behind the meat drawer. It holds approximately $1\frac{1}{2}$ quarts (1.5L).

When the refrigerator is first hooked up, press the water dispenser bar with a glass or jar until you draw and discard 2 or 3 quarts (1.9 to 2.8 L). It will take three to four minutes for the water to begin dispensing. The water you draw and discard rinses the tank and pipes.

Allow several hours to chill a new tankful.

IMPORTANT: The small removable tray at the bottom of the dispenser is designed to catch small spills and allow for easy cleaning. There is no drain in the tray. The tray can be removed from the dispenser and carried to the sink to be emptied or cleaned.

To dispense water:

1. Press a glass against the water dispenser lever.
2. Remove the glass to stop dispensing.

   NOTE: Dispense enough water every week to maintain a fresh supply.

Reprinted by permission of Whirlpool Corporation.

Example 3 Adapting Instructions to Audience and Purpose

Handwashing

Purpose

1. Reduce the numbers of resistant and transient bacteria from the hands.
2. Prevent transfer of microorganisms from the hospital environment to the client and from the client to hospital personnel.

Assessment

- Inspect hands for breaks or cuts in skin or cuticles.
- Identify appropriate times for hand-washing before and after client contact.
- Identify need to repeat handwashing if hands become contaminated during a procedure.

Equipment

Warm, running water

Soap. Most hospitals supply liquid soaps, containing a germicidal agent, in dispensers at each sink.

Paper towels.

Procedure

1. Remove all rings except a plain wedding band. Push watch 4 to 5 inches above wrist.
   Rationale: Microorganisms lodge in the irregular surfaces of jewelry.
2. File nails short. Refrain from wearing nail polish or artificial fingernails.
   Rationale: Microorganisms harbor under long nails, which are hard to clean. Microorganisms may hide in cracked nail polish crevices or along adhesive edges of paste-on nails.
3. Turn on the water and adjust the temperature to warm. Do not splash water or lean against the wet sink.
   Note: Faucets may be controlled by your hands or may be operated by knee levers or foot pedals.
   Rationale: Warm water removes less protective oils from the skin than hot water and reduces chapping of hands from frequent handwashing. Microorganisms need moisture to thrive. Avoid water splashing and sink contact on clothing to prevent contamination of uniform.
4. Hold hands lower than elbows and thoroughly wet hands and lower arms under running water.
   Rationale: Hands are more contaminated than lower arms: water should flow from least to most contaminated areas.
5. Apply soap. If bar soap is used, rinse bar before lathering and rinse bar again before returning it to the dish.
   Rationale: The number of surface bacteria is to be reduced on the soap bar.
6. Rub palms, wrists, and back of hands firmly with circular movements. Interlace fingers and thumbs, moving hands back and forth. Continue using plenty of lather and friction for 15 to 30 seconds on each hand.
   Note: Timing of scrub may vary depending on purpose of wash.
   Rationale: Mechanically loosens and removes dirt and microorganisms on all hand surfaces.
7. Clean under fingernails using fingernails of other hand and additional soap. Use orangewood stick if available.
   Rationale: Microorganisms are frequently harbored under nails.
8. Rinse hands and wrists thoroughly with hands held lower than forearms.
   Rationale: Washes away microorganisms and dirt and prevents recontamination of clean skin surfaces.
9. Dry hands and arms thoroughly with paper towel, wiping from fingertips toward forearm. Discard in proper receptacle.
   Rationale: Drying hands prevents chapping and cracking of skin. Dry from cleanest area (fingertips) toward least clean to reduce chances of contamination.
10. Turn off faucets using clean, dry paper towels.
    Rationale: Prevents transfer of microorganisms from faucet to hands.

Home Care Modifications

- Visiting nurse should bring to client's home bactericidal soap in a plastic container and paper towels.
- If running water is unavailable, disposable cloths or alcohol may be used as an alternative to hand-washing. Both these agents are drying to the hands if used often.

Reprinted by permission from Ruth F. Craven and Constance J. Hirnle, *Fundamentals of Nursing* (Philadelphia: Lippincott, 1996) 525–526.

adaptable to any communication project. A sample project plan sheet for instructions appears at the end of the chapter. A completed project plan sheet adapted to instructions appears on pages 189–190.

Instructions are designed to allow the audience to carry out a sequence of steps in a prescribed order. Therefore, instructions are typically presented as a list of steps, often with arabic numbers (1, 2, 3, etc.) or words (first, second, third, etc.) indicating the order of the major steps. Each step is written as a command, in the imperative mood. Often the *you* is implicit.

EXAMPLE

1. Place the cursor on the word File.
2. Depress the left button on the mouse and hold the button in that position.
3. Gently drag the cursor down to the Save As line.

In the example above, each step to follow to save a file is numbered to show sequence, and each step is stated as a command with *you* understood, not stated.

Tell your audience not only *what* to do but also *how* to do each activity (step), if there is any chance that they might not know how. You may add *why* the activity is necessary.

EXAMPLE

Remove the cap (tells *what*). Use the fingers and thumb to turn the cap counterclockwise until it can be removed (tells *how*). You can then insert the spout of the gasoline can to fill the tank with gasoline (tells *why*).

Use good judgment in selecting details. For example, if you are explaining how to pump gas, you do not need to say that a vehicle and a gas tank are required.

Generally, avoid recipe or telegraphic writing. Include *a*, *an*, and *the*, unless space is limited. Define any unfamiliar terms. For nonsequential lists (such as a list of needed materials or equipment) and for indicating emphasis, use symbols such as the open or closed bullet ○ ● or box □ ■, the dash —, an asterisk *, or various other symbols available with computer programs.

Several conditions determine the length of instructions: the complexity of the operation, the knowledge level of the audience, and the purpose of the instructions. In explaining how to carry out a simple operation such as taking a temperature or installing a new printer cartridge, the entire presentation may be very brief. Explaining how to design and set up a Web page is longer and more complex.

Below is a guide for organizing instructions. Remember, however, that ultimately you must select and organize material for the particular audience and purpose.

- Introduce the instructions.
    1. State the operation to be explained.
    2. If applicable, give the purpose and significance of the instructions and indicate who uses them, when, where, and why.

3. If applicable, relate the instructions to a larger whole.
4. Indicate any needed preparations, skills, equipment, or materials.
5. Give a brief overview of the operation.

- Develop the body of the instructions.

6. List each step one by one, and develop each step fully with sufficient detail about what to do and how to do it. Also tell why a step is necessary, if applicable.
7. Include every step necessary to complete the operation.
8. If the operation is complex, subdivide each major step.
9. Emphasize particularly important points and caution the reader where mistakes are most likely to be made and where injury can occur.
10. Plan the information design using elements such as headings, graphics, space, and visual markers to identify sections and to enhance reading and understanding.

REMINDER

Time spent in careful planning and organizing will reward you with an easier, less frustrating writing experience.

- Close the instructions, as appropriate.

11. End the instructions with the explanation of the final step.
12. You may summarize the main steps.
13. Add a comment on the significance of the operation.
14. Mention other methods by which the operation is performed.

Student Example: Instructions

The student example explains instructions, which follow the guide for organizing instructions (shown above), for building a street-side mailbox support. Included is the Project Plan Sheet adapted for the instructions. The writer's purpose is to explain to a lay reader how to build the support. The writer assumes that the audience has some knowledge of basic woodworking tools such as a saw, hammer, screwdriver, square, and posthole digger. Following the instructions, the reader should be able to construct a support.

PROJECT PLAN SHEET

FOR INSTRUCTIONS

Audience

- Who will read or hear a presentation of the instructions?

  A lay audience (homeowners and do-it-yourselfers)
- How will readers use the instructions?

  To learn how to construct a street-side mailbox support to add an attractive yet functional decoration in front of their houses
- How will your audience guide your communication choices?

  Some readers will have considerable knowledge of simple carpentry; others will have little, if any, knowledge. Therefore, I will need to use simple language and graphics and other visuals. I will need to emphasize the sequence of activities.

Purpose

- What is the purpose of the instructions?

  The instructions will guide the reader in constructing a street-side mailbox support.
- What need will the instructions meet? What problem can they help to solve?

  The instructions will guide readers in constructing a street-side mailbox support to make delivery and retrieval of mail easier. Also, the support will add an attractive asset to a house.

Subject

- What is the instructions' subject matter?

  How to construct a street-side mailbox support
- How technical should your discussion of the subject matter be?

  Since the instructions are intended for a lay audience, the language should be non-technical.
- Do you have sufficient information to complete instructions? If not, what sources or people can help you to locate additional information?

  Yes. I have constructed such a support in front of my house, so basically, I will record the steps I followed. I will need to measure the lengths of the different pieces of the lumber to be cut.

- What title can clearly identify the instructions' subject and purpose?
  How to Construct a Street-Side Mailbox Support

Author

- Will the instructions be a collaborative or an individual effort?
  Individual effort
- How can the developer(s) evaluate the success of the finished instructions?
  Maybe I can ask someone who does a lot of carpentry work to look over my finished draft.

Project Design and Specifications

- Are there models for organization or format for the instructions?
  I have copies of brochures, manuals, and other sources of instructions.
- In what medium will the completed instructions be presented?
  Written
- Are there special features the completed instructions should have?
  The completed instructions should be especially clear and easy to understand. Complex sentence structure and technical terms will confuse my audience.
- Will the instructions require graphics? If so, what kinds and for what purpose?
  Simple line drawings to show shape of the brace piece and the pieces of lumber as they are joined together and a photograph of a finished support with mailbox attached
- What information design features can help the instructions' audience?
  Chunking information, headings, and graphics

Due Date

- What is the final deadline for the completed instructions?
  The deadline for the instructions is November 10.
- How long will the instructions take to plan, research, draft, revise, and complete?
  I will spend approximately three weeks on planning and drafting the instructions.
- What is the timeline for different stages of the instructions?
  I will spend two weeks in planning, drafting, and revising. This will give me time to put the instructions aside for a while and then reread. I need to ask someone to review the content, keeping in mind that carpenters are busy.

FINAL DRAFT

CONSTRUCTING A STREET-SIDE MAILBOX SUPPORT

MATERIALS

- 10 ft. of 4 x 4 treated lumber
- 60 lb. bag of concrete mix
- eight 16d nails
- paint, if desired

TOOLS

- claw hammer
- flatblade screwdriver
- square
- saw (hand or electric)
- measuring tape
- posthole digger
- wood chisel
- pencil
- level

STEPS

Measure and Cut the Lumber

1. Using a saw (hand or electric), cut the 10 ft. piece of 4 x 4 treated lumber into three pieces, measured as follows:
    - one 6 ft. length
    - one 34 in. length
    - one 14 in. length
2. On the 6 ft. piece of lumber, use the measuring tape to measure 14 inches from one end. With the lead pencil, mark the distance with a dot or a dash.
3. Place the square against the side of the piece so that the blade aligns with the mark. With the lead pencil, draw a heavy line along the edge of the blade. This line will be used as a guide to make a saw cut in the piece.
4. With the saw, make a cut halfway through the piece.
5. Measure 3 1/2 inches from the cut and mark the distance with the lead pencil.
6. Place the square against the side of the piece so that the blade aligns with the mark and draw a heavy line along the edge of the blade to mark the second saw cut.
7. Make a second cut halfway through the piece.
8. With the wood chisel, chisel out the wood between the two saw cuts to create a notch.

9. On the 34-inch piece, measure 9 inches from one end.
10. Repeat steps 3–8.
11. On each end of the 14-inch piece, use the square and the pencil to mark a 45° angle line. One side of the piece remains 14 inches long. See Figure 1.

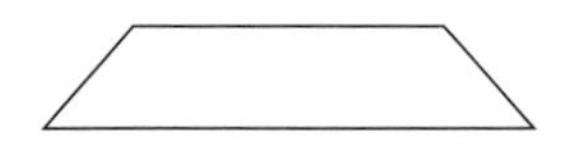

Figure 1. The brace piece

12. Cut along the lines on both ends. This piece will be used to brace the other two pieces of lumber.

Put the Pieces Together

1. Align the notches on the long piece and the short piece so that the notches face. Press the two pieces so that they fit together. You may need to tap the two pieces lightly with the hammer to make the pieces fit together.
2. Into each side where the two pieces fit together, hammer two16d nails (4 nails).
3. With the hammer, nail the 45°-angle piece to support the 34-inch piece to the 6-foot piece, using two nails on each end. See Figure 2.

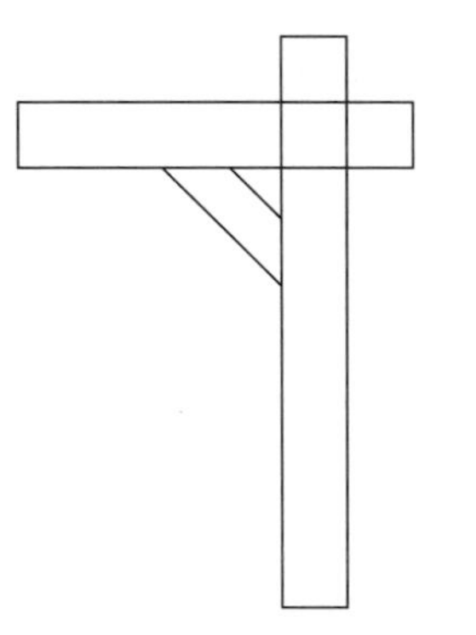

Figure 2. The three pieces nailed together

Place the Support in the Hole

1. With the posthole digger, dig a hole two feet deep and eight inches in diameter.
2. Place the end of the support opposite the brace piece into the hole. Approximately 2 feet of the piece will be in the hole. The longest end of the cross piece should face the street.
3. Place the level on one side of the post, checking to see that the post is vertical (plumb). Repeat this step on an adjacent side of the post.

4. Fill the hole with concrete mix, packing it with the sole of your shoe. With the level, continue to check that the post is vertical.
5. Thoroughly dampen the concrete mix with water to set the concrete mix.

COMMENTS

To attach a mailbox to the crossbar, follow the instructions that come with the mailbox. Once the mailbox is attached, you have a variety of options. It is useful to paint or attach metal, plastic, or ceramic numbers to identify your house number for postal delivery service personnel and emergency personnel. You can paint the mailbox support and the mailbox any color or you can leave the natural lumber. Since the lumber is treated, it will last a long time. You can also plant flowers or foliage around the base of the mailbox. Consider plants that will not grow too tall or too bushy and cover the opening to the mailbox. And if you are having a party, you can tie balloons to the mailbox support to direct your guests. See Figure 3.

Figure 3. A support with mailbox

GENERAL PRINCIPLES for Instructions

- **Knowledge of the subject matter is essential.** Consult knowledgeable people, textbooks, reference works in the library, and other sources to gain further understanding and information about the subject.
- **Audience determines what information is to be presented and how.** Instructions should be presented so that they neither talk down to the audience nor overestimate the audience's knowledge or skills.
- **Effective instructions are accurate and complete**. The information must be correct. Instructions should cover the subject adequately, with no step or essential information omitted.
- **Graphics and other visuals help to clarify instructions**. Instructions can often be clearer with the inclusion of graphics such as maps, diagrams, graphs, pictures, drawings, slides, demonstrations, and real objects.
- **Instructions require careful information design**. An integral part of effective instructions is consideration of how material is placed on the page and of how the document as a whole is presented.
- **Conciseness and directness contribute to effective communication.** An explanation that is stated in the simplest language with the fewest words is usually the clearest. Terms included that may be unfamiliar to the audience or familiar terms used with specialized meanings should be explained.
- **Instructions that can be followed have no unexplained gaps in the procedure or vagueness about what to do next.** Well-stated instructions do not require the audience to make inferences, to make decisions, or to ask, "What does this mean?" or "What do I do next?"

PROCEDURE EXPLANATIONS OF HOW SOMETHING IS DONE OR HOW IT HAPPENS (PROCESS)

A process explanation of how something is done or how something happens helps an audience gain understanding of a sequence of events or an operation. The audience is unlikely to perform the operation. For example, a service manager may explain to you what happened to cause your car's *Check Engine* light to come on and what a technician did to solve the problem. You understand what happened, but you will probably not attempt to solve the problem yourself if the light comes on again.

The following chart gives examples of processes carried out by people, machines, and nature.

Obviously, individuals cannot carry out processes carried out by machines

Processes carried out by people	Processes carried out by machines	Processes carried out by nature
How steel is made from iron	How a mainspring clock works	How sound waves are transmitted
How glass is made	How a copier works	How rust is formed
How a computer programmer designs a new program	How a gasoline engine operates	How food is digested
		How mastitis is spread

and by nature, but they can understand what happens or what is done during the process.

Adapting Process Explanations to Audience and Purpose

Process explanation requires consideration of audience and purpose. It also requires making decisions about information design to meet audience needs. Questions to be answered include: Who is my audience? What is my purpose for writing? What do I want the audience to know and understand when they finish reading? How can I arrange information on pages to help my audience use the material as easily as possible? Following are several examples to illustrate how material can be adapted to audience and purpose.

The first example, on page 196, a description of how the heart works, is directed to a lay audience. Note the use of drawings as well as simplified language in describing the process.

Example 2 on page 197, "Your Car's Engine," is excerpted from a pamphlet, *What You Should Know About Motor Oil,* produced by the Quaker State Corporation. While this pamphlet is directed to a lay audience, the writer assumes that the reader has some knowledge of automobile engine terminology, words such as *spark plug*, *electrode*, *connecting rod*, and *drive shaft*. The writer does define *cylinder* and describes valves as looking like "oversized golf tees." The designer of the pamphlet also uses chunking and bold headings to guide the reader.

A specialized audience, as the term implies, has at least an interest in a particular subject and probably has background, either from reading or from actual experience, to understand an explanation of a process related to that subject. For example, a person whose hobby is working on cars and reading about them would understand a relatively technical description of a fuel injection system. The specialized reader may even have a high level of knowledge and skill. For example, sanitation engineers, fire chiefs, inhalation therapists, research analysts, programmers, or machinists would understand technical explanations related to their fields of specialization.

Example 3, the boxed article (page 198) from *Code One*, April 1999, explains how a jet engine produces power. Obviously, the article is written for a specialized audience. Nevertheless, it is written in a clear, concise style and includes a supportive visual of the jet engine (visual omitted here).

Example 1 Process Explanation Adapted to Audience and Purpose

How the Heart Works

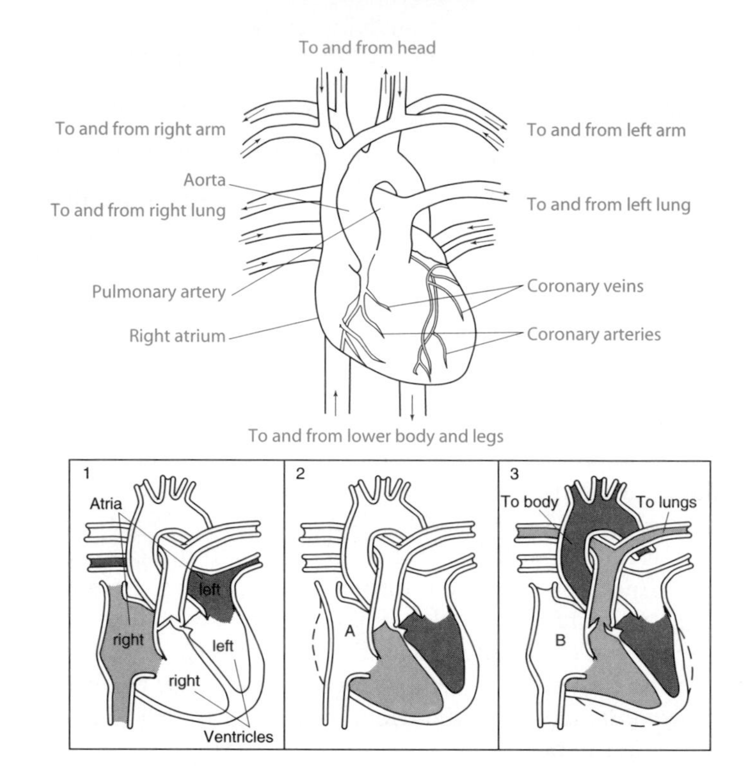

1. During the heart's relaxed stage (diastole), oxygen-depleted blood from body flows into right atrium, oxygenated blood from lungs into left. 2. Natural pacemaker, or sinoatrial node (A), fires electrical impulse and atria contract. Valves open and blood fills ventricles. 3. In pumping stage (systole), the electrical signal, relayed through atrioventricular node (B), causes ventricles to contract, forcing oxygen-poor blood to lungs, oxygen-rich blood to body.

If you have any doubt about your audience's level of knowledge, clarify any part of the explanation that you think the reader might have difficulty understanding. Design information to make the audience's use of the material as easy and productive as possible.

Planning and Organizing Process Explanations

Planning a process explanation requires that you make certain decisions. These decisions include audience, purpose, subject matter, author, design, and timeline. A project plan sheet, adapted for process explanations and completed for a sample explanation, is shown on pages 200–201. Use it as a guide for planning process explanations, making changes as dictated by your audience and your purpose.

In organizing a process explanation, you may arrange the steps or stages in a numbered list or in paragraphs. Remember: The purpose of the explanation is to make the reader understand *what is done* or *what happens*. One way

Your Car's Engine

The modern engine is of French and German origins. The theory is French. But its practical development is German. In 1862, Beau de Rochas conceived the theory for the ingenious four-stroke principle. In 1876, Germany's Dr. Nikolaus Otto turned it into reality. And gasoline and motor oil, which work hand in hand to power, lubricate and cool engines, became necessities.

Since then, there have been thousands of improvements, particularly in the areas of design, metallurgy, and support systems. But, in most cases, the de Rochas four-stroke principle is still the basic idea behind it.

How It Works. In a four-stroke engine, the strokes occur at very high speed inside a hollow piece of steel called a cylinder. In most conventional car engines, there are three holes at the top of the cylinder. One is sealed by a spark plug, which has electrodes that reach down into the hollow chamber. The other two holes are opened by valves that look like oversized golf tees. One is the intake valve and the other the exhaust valve. (Some of today's more advanced engines have two intake and exhaust valves.)

Inside the cylinder, a close fitting piston moves up and down four times in every power cycle. Most engines have four, six, or eight cylinders.

Stroke one. The intake valves open and a mist of fuel, which is a combination of gasoline and air, is sucked or injected into the vacuum created as the piston moves down.

Stroke two. Both valves are closed. The piston moves up again, compressing the fuel mixture, until the piston reaches its highest point.

Stroke three. Both valves are closed. A spark is created at the tip of the spark plug electrodes, which ignites the fuel, and the resulting explosion drives the piston down. This creates the power that drives the car, through a connecting rod that connects to the piston and ultimately to your car's driveshaft.

Stroke four. The exhaust valves open. The piston moves up once more, and the spent fuel is forced out of the cylinder chamber and away through the exhaust system.

to explain what is done or what happens is to use the third-person, present tense, active or passive voice.

EXAMPLES

- An officer (third person) fingerprints (present tense, active voice) a suspect by. . . .
- Glass (third person) is made (present tense, passive voice) by. . . .

Or, you can use *-ing* verb forms to explain what happens.

EXAMPLES

- Peeling (*-ing* verb form) the thin layers. . . .
- The final stage is removing (*-ing* verb form) the seeds. . . .

Example 3 Process Explanation Adapted to Audience and Purpose

JSF119-611 Primer

A jet engine produces power by compressing air, adding fuel, and sustaining a continuous combustion. The hot gas of combustion expands rapidly out of the back end of the engine, which produces forward thrust. The primary parts of a military jet engine are the compressor, combustor, turbine, and augmentor. The compressor forms the front part of the engine. The first three sets of blades of the JSF119-611 form the three stages of the low-pressure compressor (also called a fan). The next six sets of blades form the six stages of the high-pressure compressor. These nine compressor stages draw air in, pressurize it, and deliver it to the combustor, where it is mixed with fuel and ignited. A portion of the compressed air bypasses the combustor as well. This bypass air is used to cool hot portions of the engine and to provide airflow for the augmentor. The Lockheed Martin STOVL variant uses bypass air to power the roll ducts in the wings.

Extremely hot and rapidly expanding gases produced in the combustor enter the turbine section, which consists of three stages of alternating stationary and rotating blades. The first stage of the turbine (the high-pressure turbine) is connected by a shaft to the high-pressure compressor in the front of the engine. The back two stages of the turbine (the low-pressure turbine) are connected by another independent shaft to the first three stages of the compressor (the fan). The turbine stages essentially absorb enough energy from the hot expanding gases to keep the compressor stages rotating at an optimum speed.

The hot gases exit the turbine and enter the augmentor (also called the afterburner), where they are joined with bypass air. When the engine is in the afterburner, additional fuel is injected into the augmentor. This secondary combustion produces a significant amount of additional thrust.

The JSF119-611 shares a common engine core with the F119 that powers the F-22 Raptor. The core consists of the high-pressure compressor, the combustor, and the high-pressure turbine.

Reprinted by permission from Eric Hehs, editor, *Code One* magazine.

Remember to be consistent with person, tense, and voice in a single presentation. For example, if you select third-person, present tense, active voice for the major steps or stages, use it for all major steps or stages throughout the explanation. Do not needlessly shift to another person, tense, or voice. (See pages 671–672, 674.)

The length of a presentation is determined by the complexity of the process, the knowledge of the audience, and the purpose of the presentation. In a simple process, such as how a cat laps milk or how a stapler works, the steps may be developed adequately in a single paragraph, with perhaps a minimum of two or three sentences for each step. In a more complex process, the steps may be listed and numbered, and the explanation of each step or stage may require a paragraph or more. The closing is usually brief.

Below is a guide for organizing process explanations. Remember, however, that ultimately you must select and organize material for the particular audience and purpose.

- Introduce the process.
  1. State the process to be explained and identify or define it.
  2. If applicable, give the purpose and significance of the process.
  3. Briefly list the main steps or stages of the process, preferably in one sentence.
- Develop the steps (or stages).
  4. Use the list of the main steps given in the introduction to organize this section.
  5. Take up each step in turn, developing it fully with sufficient detail.
  6. Subdivide major steps as needed.
  7. Insert headings for at least the main steps.
  8. Use graphics whenever they will help to clarify, explain, or emphasize.
- Close the process explanation.
  9. If the purpose is simply to inform the audience about the specific procedure, the closing may be the completion of the last step, a summary, a comment on the significance of the process, or a mention of other methods for performing the process.
  10. If the presentation serves a specific purpose, such as evaluation of economy or practicality, the closing may be a recommendation.

Student Example: Process Explanation

The student-written process explanation describes how a black and white ad for a college newspaper is produced. Following the adapted Project Plan Sheet shown on pages 200–201 and the guide for organizing process explanation shown above, the student completed the final draft shown on pages 202–205.

PROJECT PLAN SHEET

FOR A PROCESS EXPLANATION

Audience

- Who will read the process explanation or hear it as a presentation?

  Students beginning a major in graphic design
- How will readers use the process?

  To understand the basic stages of producing a black and white advertisement for promoting Hinds Community College
- How will the audience guide communication choices?

  I will need to use simple language and graphics. I will need to emphasize the sequence of events in the process.

Purpose

- What is the purpose of the process explanation?

  The explanation will show how a black and white ad is produced.
- What need will the explanation meet? What problem can it help to solve?

  The explanation will introduce students beginning a major in graphic design to the procedure for producing black and white advertisements.

Subject

- What is the process explanation's subject matter?

  How a black and white ad to promote Hinds Community College is produced
- How technical should your discussion of the subject matter be?

  Since the explanation is intended for a beginning graphic design major, the language should be accurate but not highly technical. Definitions may be needed.
- Do you have sufficient information to complete the explanation? If not, what sources or people can help you to locate additional information?

  I have a basic understanding of the process, but I will need to read more information about it and perhaps talk to someone more familiar with the process. I will interview the editor of the college newspaper, a layout artist, and a printer.
- What title can clearly identify the explanation's subject and purpose?

  How a Black and White Ad to Promote Hinds Community College Is Produced

Author

- Will the explanation be a collaborative or an individual effort?

  Individual effort

- How can the developer(s) evaluate the success of the finished explanation?

  Maybe I can ask an editor, a layout artist, or a printer to look over my finished draft.

Project Design and Specifications

- Are there models for organization or format for the explanation?

  Reference books for graphic design examples from Hinds Community College publications

- In what medium will the completed explanation be presented?

  Written

- Are there special features the completed explanation should have?

  The completed explanation should be especially clear and easy to understand. Complex sentence structure and technical terms will alienate my audience.

- Will the explanation require graphics? If so, what kinds and for what purpose?

  Camera-ready copy sample, line negative sample, simulation of a masking sheet to understand the steps in the process

- What information design features can help the explanation's audience?

  Chunking information, headings, and graphics

Due Date

- What is the final deadline for the completed explanation?

  I will submit the explanation for publication in the college newspaper on October 5.

- How long will the explanation take to plan, research, draft, revise, and complete?

  Approximately one and a half weeks

- What is the timeline for different stages of the explanation?

  Research will take three days. I also need to allow at least three days for my final reviewer and another three days to make any changes and final revisions. I need to allow time to gather visuals and to practice delivering.

FINAL DRAFT

Title reflects the purpose of the article.

How a Black and White Ad to Promote Hinds Community College Is Produced

[Marginal notes have been added to show organizational plan.]

Background information explains how the need for the ad evolved.

A listing of the main stages in the process, which also provides headings for the article

Hinds Community College uses a variety of advertisements to promote interest in the college. Recently, when the college put into service an 800 number to call for information about the college, the Institutional Advancement Office and the Public Relations Office wanted to advertise the availability of this service. One plan was to announce the service through a black and white ad in *The Hindsonian,* the Hinds Community College newspaper. Once this decision was made, college personnel produced the ad through a process that included discussing ideas, producing camera-ready copy, preparing the copy for printing, and printing the ad in the newspaper.

Introduction and development of first stage in the process; discusses content of the ad, text, and visual

Discussing Ideas

First, Institutional Advancement and Public Relations personnel met to discuss ideas for an ad. What should the ad convey? What effect should the ad have? What information should the ad include? After many ideas were discussed, the decision was made to use a simple advertisement with the message that at Hinds Community College a student can get the desired education for the future and can learn about the options available by calling the 800 number. The theme chosen was "Your future's on the line" (a play on the word "line" since the ad would also advertise the 1-800 telephone line). Other text selected was "Call 1-800 HINDS CC." The ad would also include the college logo (an H superimposed over a C) and the college motto, "The College for All People."

Introduction and explanation of second stage; explains the layout artist's experimentation in designing the ad and submission of a suggestion to appropriate individuals for approval

Producing Camera-Ready Copy

This information then went to the layout artist. Using a computer, she experimented with different fonts and type sizes and the placement of text on the page to achieve the best balance and effect. As she experimented, she decided to use a visual of a rising sun with a telephone in the center to suggest that people need to make the call *now*. She printed a copy to show what the final ad might look like. Using this copy, she worked further with the placement of the different elements on the page—text and graphics—to achieve the best possible effects. She proofread the text and used a computer spell check to be sure all words were spelled correctly.

Satisfied with the placement of the elements on the page, the layout artist printed a copy of the planned advertisement and sent it to Institutional Advancement and Public Relations personnel. After their approval, the layout artist printed a final copy of the advertisement for the printer. This copy is called a mechanical or camera-ready copy. See Figure 1.

Visual showing a mechanical of camera-ready copy

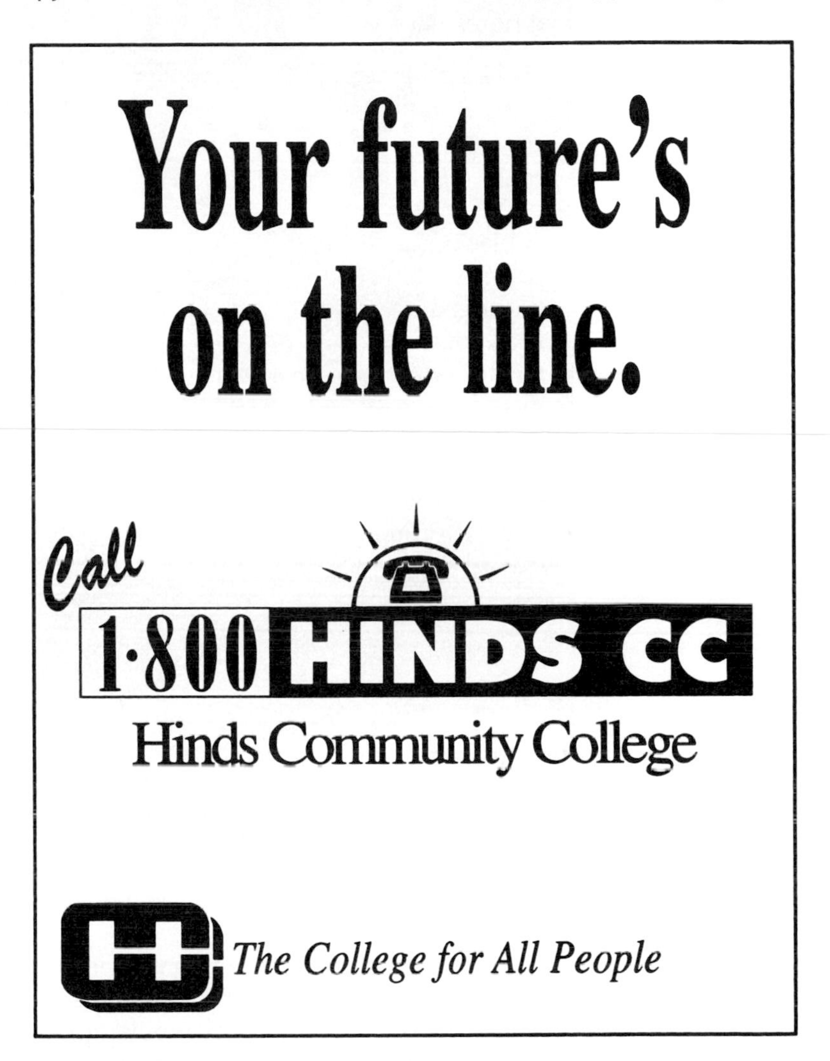

Figure 1. A mechanical or camera-ready copy for the ad

Introduction and development of third stage; explains how the printer prepared copy for printing in the newspaper

Preparing the Copy for Printing

The printer used the mechanical and a process camera to shoot a line negative showing all details of the advertisement. See Figure 2.

Visual showing a line negative of the ad

Figure 2. A line negative of the ad

The printer then stripped the negative into a masking sheet or goldenrod flat (named for the color of the paper) to position the copy so that when the plate was made, the copy would appear in the desired location on the printed page. The masking sheet, available in many sizes, has markings to guide the printer in placing the material. Also masked to this sheet is all other copy that will appear on the same page with the ad. Figure 3 is an example of a masking sheet, without the goldenrod color.

Many newspapers and magazines use computers for layout , making masking sheets unnecessary.

Introduction and development of fourth stage; explains the printing method for producing the ad page in the newspaper.

Printing the Ad

The printer made a plate—a thin, flexible metal sheet from which impressions are transferred during a printing operation.

Visual showing a masking sheet.

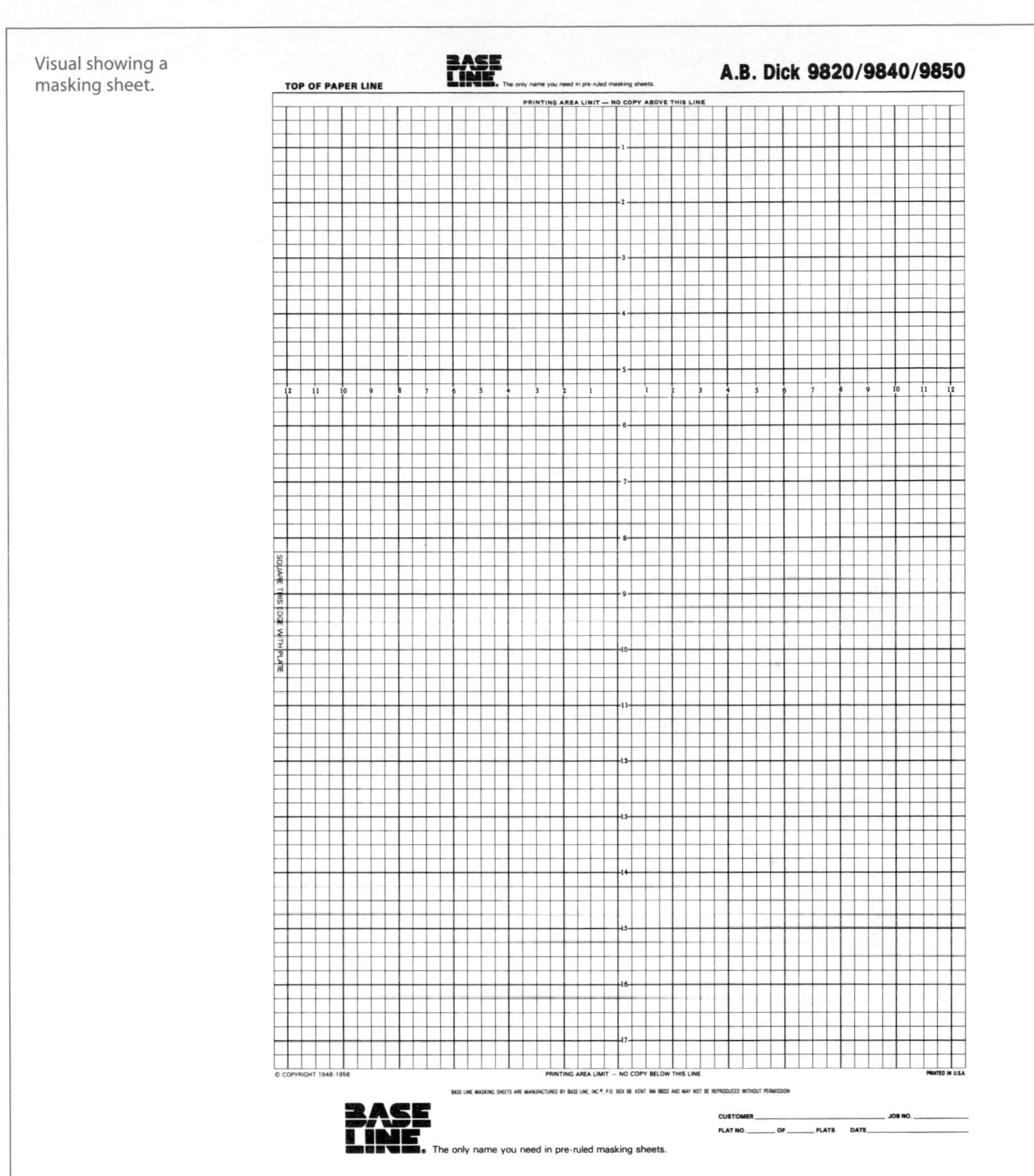

Figure 3. An example of a masking sheet (does not show goldenrod color)

The printer placed the plate on a cylinder. This cylinder transfers the image to a rubber blanket cylinder which prints the image on paper. This process is called offset printing. Once the ad page and other pages of *The Hindsonian* were printed and collated, the campus newspaper was distributed to various campus locations.

GENERAL PRINCIPLES for Explaining a Process

- **The purpose of process explanation is audience understanding, not audience action.**
- **Audience influences the kind and extent of details included and the manner in which they are presented.** The audience's knowledge and understanding of the subject determine the level of language.
- **Audiences may be grouped into two broad categories: lay audiences and specialized audiences.** Lay audiences are a mixture, a cross section of people. Specialized audiences have mutually shared professional knowledge.
- **Accuracy and completeness are essential**. The process explanation should be accurate in its information, and it should adequately cover all necessary aspects of the process.
- **Graphics and other visuals can enhance a process explanation**. Graphics and other visuals—such as flowcharts, diagrams, drawings, real objects, and demonstrations—can clarify and emphasize a verbal explanation.

CHAPTER SUMMARY

PROCEDURE

Procedure explanations use sequential steps or stages to meet an audience's need to understand what to do to carry out a procedure or to understand what happens or is done during a procedure. Such procedures are typically referred to as instructions (which explain what to do) and processes (which explain what happens or what is done). Audience and purpose help you make decisions about content, language choices, sentence structure, and information design. A major consideration for authors of procedure explanations is making audiences aware of cautions, dangers, and hazards, at the appropriate place in the explanation and through attention-getting techniques.

You must be aware of the audience's need for graphics and other visuals to make explanations clear and easy to understand. Decisions about information design are also critical in choosing format and providing easy access to needed information.

Planning and organizing instructions and process explanations start with analysis of audience, purpose, subject matter, author, design, and deadlines. A project plan sheet offers suggestions for this analysis. A guide for organizing instructions and process explanations helps you make decisions about content and arrangement of information.

ACTIVITIES

INDIVIDUAL AND COLLABORATIVE ACTIVITIES

9.1. Make a list of three people to interview about their jobs. After interviewing each of the three, make a list of examples showing how they use procedure explanations on the job. From this experience, what can you speculate about the importance of procedure explanations in your own future work? Discuss your findings as a class. What can you speculate about the importance of procedure explanations in the workplace?

9.2. Make a collection of procedure explanations. These are readily available at hardware stores, toy stores, building materials stores, and so on. Working in groups of two or three, analyze the collection. Ask such questions as:
   a. Is the writer/designer aware of audience and purpose? How do you know?
   b. Is the content sufficient for audience understanding? Use?
   c. Describe graphics and information design. Evaluate the effectiveness of each.

9.3. Find instructions that a manufacturer included with a product.
   a. Evaluate in a paragraph the information design of the instructions.
   b. Evaluate in a paragraph the clarity and completeness of the instructions by applying the General Principles for Instructions, page 194.

   Attach a copy of the instructions to your evaluation.

9.4. Assume that you are a manager or supervisor with a new employee. Explain to the employee in writing how to carry out some simple operation.

For Activity 9.5

a. Review the pages on planning and organizing instructions, pages 187–188.
b. Adapt and complete a Project Plan Sheet for Instructions, pages 214–215.
c. Following the guide for organizing instructions, pages 187–188, plan the content and organization of your instructions.
d. Using decisions from the Project Plan Sheet and the content and organization guide for instructions, write a preliminary draft.
e. Revise the draft, incorporating design decisions.
f. Prepare a final draft.

9.5. Choose a topic from the following list or one from your own experience for an assignment on instructions. You may need to limit the selected topic before you begin to plan the instructions. For example, if you choose "change oil in a car," you will need to identify the make and model of the car. Write the instructions for a lay audience.

How to:

Take a patient's blood pressure
Sharpen a drill bit
Produce a business letter—individualized to several people—on a word processor
Cut a mat for a picture
Operate a piece of heavy-duty equipment
Administer an intramuscular injection
Open a checking account
Change a tire on a hill
Change oil in a car
Hang wallpaper
Set out a shrub
Sterilize an instrument
Make a tack weld
Operate an office machine
Prepare a laboratory specimen for shipment
Develop black and white film
Fingerprint a suspect
Download a file from the Web

For Activities 9.6–9.7

a. Review the pages on planning and organizing process explanation, pages 194–199.
b. Adapt and complete a Project Plan Sheet for a Process Explanation, page 216.
c. Following the guide for organizing a process explanation, page 199, plan the content and organization.
d. Using decisions from the Project Plan Sheet and the content and organization guide for process explanations, write a preliminary draft.
e. Revise the draft, incorporating design decisions.
f. Prepare a final draft.

9.6. Identify some event or activity that takes place at your college (students' notification of grades, library's getting a book ready to place on the shelf, disciplinary process), at your place of employment (how a new employee is hired, how payroll is prepared and distributed), or on your job. Explain the process to show a new student or employee the stages from beginning to completion of the event or activity. Then prepare the explanation for inclusion in a student or personnel manual.

9.7. Choose one of the following processes or another similar topic of interest to you. Write an explanation of the process for a lay audience.

How

Ceramic tile, bricks, tires, sugar (or any other material) is (are) made
A person becomes a "star"
A site is chosen for a business or industry
Flood prevention helps to eliminate soil erosion
A buyer selects merchandise for a retail outlet
A college cafeteria dietitian plans meals
A telephone answering service works
A space satellite works
A photocopier (or similar machine) reproduces copies
A computer copies a disk
An automatic icemaker works
A mechanical cotton picker (or any other piece of mechanical farm or industrial machinery) operates
A lathe (or any other motorized piece of equipment in a shop) works
An autoclave works
A printing press works
An electronic calculator displays numbers
Aging affects one of the senses (such as seeing or hearing)
Infants change during the first year of life
A tadpole becomes a frog
Freshwater fish spawn
Sound is transmitted
An amoeba reproduces

9.8. As directed by your instructor, adapt one of the procedure explanations you prepared in the activities above for oral presentation. Ask each member of the audience to evaluate your speech by filling in an Evaluation of Oral Presentations (see page 508).

9.9. The following procedure explanation is excerpted from *The Shores of the Cosmic Ocean* by Carl Sagan. Read the selection. In small groups or as a class, respond to the questions for discussion.

READING

From *The Shores of the Cosmic Ocean*

CARL SAGAN

ANCIENT DISCOVERY THAT THE EARTH IS ROUND

The discovery that the Earth is a *little* world was made, as so many important human discoveries were, in the ancient Near East, in a time some humans called the third century B.C., in the greatest metropolis of the age, the Egyptian city of Alexandria. Here there lived a man named Eratosthenes. One of his envious contemporaries called him "Beta," the second letter of the Greek alphabet, because, he said, Eratosthenes was second best in the world in everything. But it seems clear that in almost everything Eratosthenes was "Alpha." He was an astronomer, historian, geographer, philosopher, poet, theater critic and mathematician. The titles of the books he wrote range from *Astronomy* to *On Freedom from Pain.* He was also the director of the great library of Alexandria, where one day he read in a papyrus book that in the southern frontier outpost of Syene, near the first cataract of the Nile, at noon on June 21 vertical sticks cast no shadows. On the summer solstice, the longest day of the year, as the hours crept toward midday, the shadows of temple columns grew shorter. At noon, they were gone. The reflection of the Sun could then be seen in the water at the bottom of a deep well. The Sun was directly overhead.

It was an observation that someone else might easily have ignored. Sticks, shadows, reflections in wells, the position of the Sun—of what possible importance could such simple everyday matters be? But Eratosthenes was a scientist, and his musings on these commonplaces changed the world; in a way, they made the world. Eratosthenes had the presence of mind to do an experiment, actually to observe whether in Alexandria vertical sticks cast shadows near noon on June 21. And, he discovered, sticks do.

Eratosthenes asked himself how, at the same moment, a stick in Syene could cast no shadow and a stick in Alexandria, far to the north, could cast a pronounced shadow. Consider a map of ancient Egypt with two vertical sticks of equal length, one stuck in Alexandria, the other in Syene. Suppose that, at a certain moment, each stick casts no shadow at all. This is perfectly easy to understand—provided the Earth is flat. The Sun would then be directly overhead. If the two sticks cast shadows of equal length, that also would make sense on a flat Earth: the Sun's rays would then be inclined at the same angle to the two sticks. But how could it be that at the same instant there was no shadow at Syene and a substantial shadow at Alexandria?

The only possible answer, he saw, was that the surface of the Earth is curved. Not only that: the greater the curvature, the greater the difference in the shadow lengths. The Sun is so far away that its rays are parallel when they reach the Earth. Sticks placed at different angles to the Sun's rays cast shadows of different lengths. For the observed difference in the shadow lengths, the distance between Alexandria and Syene had to be about seven degrees along the surface of the Earth; that is, if you imagine the sticks extending down to the center of the Earth, they would there intersect at an angle of seven degrees. Seven degrees is something like one-fiftieth of three hundred and sixty degrees, the full circumference of the Earth. Eratosthenes knew that the distance between Alexandria and Syene was approximately 800 kilometers, because he hired a man to pace it out. Eight hundred kilometers times 50 is 40,000 kilometers: so that must be the circumference of the Earth.*

This is the right answer. Eratosthenes' only tools were sticks, eyes, feet, and brains, plus a taste for experiment. With them he deduced the circumference of the Earth with an error of only a few percent, a remarkable achievement for 2,200 years ago. He was the first person accurately to measure the size of a planet.

The Mediterranean world at that time was famous for seafaring. Alexandria was the greatest seaport on the planet. Once you knew the Earth to be a sphere of modest diameter, would you not be tempted to make voyages of exploration, to seek out undiscovered lands, perhaps even to attempt to sail around the planet? Four hundred years before Eratosthenes, Africa had been circumnavigated by

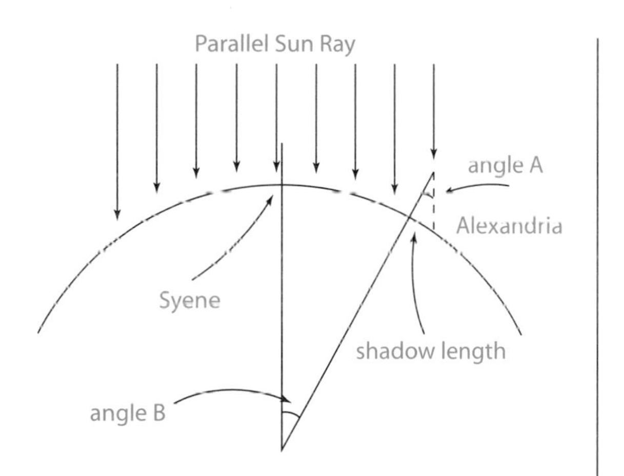

From the shadow length in Alexandria, the angle A can be measured. But from simple geometry ("if two parallel straight lines are transected by a third line, the alternate interior angles are equal"), angle B equals angle A. So by measuring the shadow length in Alexandria, Eratosthenes concluded that Syene was A = B = 7° away on the circumference of the Earth.

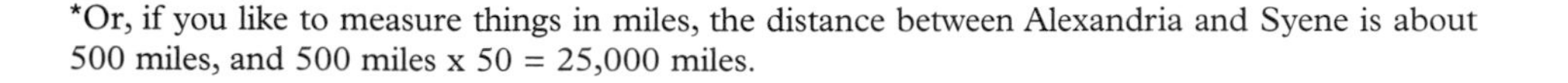
*Or, if you like to measure things in miles, the distance between Alexandria and Syene is about 500 miles, and 500 miles x 50 = 25,000 miles.

a Phoenician fleet in the employ of the Egyptian Pharaoh Necho. They set sail, probably in frail open boats, from the Red Sea, turned down the east coast of Africa up into the Atlantic, returning through the Mediterranean. This epic journey took three years, about as long as a modern Voyager spacecraft takes to fly from Earth to Saturn.

After Eratosthenes' discovery, many great voyages were attempted by brave and venturesome sailors. Their ships were tiny. They had only rudimentary navigational instruments. They used dead reckoning and followed coastlines as far as they could. In an unknown ocean they could determine their latitude, but not their longitude, by observing night after night, the position of the constellations with respect to the horizon. The familiar constellations must have been reassuring in the midst of an unexplored ocean. The stars are the friends of explorers, then with seagoing ships on Earth and now with spacefaring ships in the sky. After Eratosthenes, some may have tried, but not until the time of Magellan did anyone succeed in circumnavigating the Earth. What tales of daring and adventure must earlier have been recounted as sailors and navigators, practical men of the world, gambled their lives on the mathematics of a scientist from Alexandria?

QUESTIONS FOR DISCUSSION

1. What judgment can you make about the intended readers?
2. What comments can you make about the writing to illustrate that the author took into account the intended readers? Language choices? Sentence length? Graphic design?
3. Explain how Eratosthenes deduced that the world is curved, not flat. How did he figure the circumference of the Earth?
4. Analyze the value of the graphic.

CASE STUDY

9.10. In a class discussion, small group discussion, or a short written assignment, respond to the following case study by identifying the ways in which the case illustrates the nature of procedure explanations.

Software? Tough Luck!

Latesha Saunders, a training manager, has purchased a sophisticated new database program for Harvey Fields, her administrative assistant. Since Harvey keeps track of all of the training programs, trainers, and enrolled employees, he had urged Latesha to purchase the program, which would allow the entire department to quickly update and organize its activities. However, the software has been on Harvey's desk for seven weeks, and the system still isn't in place. Latesha decides to ask why.

"Harvey," she says. "What happened to that new database program that seemed so promising? You told me that you could hardly wait to get everything updated after all of the research we did to get the right product. Have you been too busy to get started?"

"Oh, no," Harvey replies. "That's not the problem. The software has every feature we need. It's just too hard to use!"

"What's this? You're the company software guru, Harvey! What's so hard about it?"

"Take a look at it yourself, Latesha. And look at the user's manual, too. It's. . . . well, see for yourself."

"Hmmmm. This book must weigh four pounds," mutters Latesha. "It must be complete, but it's got awfully small print. What's wrong with the pages? They're numbered in groups, not in sequence. That sure makes it hard to look up commands. Here. Let me try."

"Go for it, Latesha. Give it your best shot."

"Okay . . . let's see. I guess I'll try to copy the description of our August 15 session on word processing into the template for training sessions you've set up. Okay. First I copy from the old text. Now the manual says to. . . . wait a minute! What should I be opening here? There seems to be a difference between what I see on screen and what the manual says will happen!"

"Just wait."

"Hey! What is this? I just erased the whole file! And. . . . oh, blast! The warning in the manual came too late for me to see it. Now what are we going to do, Harvey?"

"Well," Harvey suggests, "we could try the free help line, but it's been busy every time I tried. And I tried day after day for over a week. There isn't a very clear program overview or a usable table of contents in the manual, and the online help doesn't seem to work. The warnings come too late, and the writing is so unclear that I'm not sure who is supposed to do what! Lots of passive voice and technobabble."

"That settles it," Latesha snaps. "We have a 90-day warranty on this hunk of junk. It's going back to the vendor! Let's look for a piece of software with fewer options and better instructions."

QUESTIONS FOR DISCUSSION

a. Is Harvey lazy because he refuses to spend the time to read and reread the manual and to explore the software system on his own?
b. What is the new software's biggest problem for its users?
c. What features should computer manuals include to help users read? Learn? Look up information for review?
d. Harvey's new software may be the best on the market, but what does poor information do to the product's reputation?
e. What problems make the software manual unusable? Accuracy? Organization? Warnings? Unclear writing? How would you suggest revising it?
f. What is the software company's responsibility to its customers? Is information part of that responsibility?

PROJECT PLAN SHEET

FOR INSTRUCTIONS

Audience

- Who will read or hear a presentation of the instructions?
- How will readers use the instructions?
- How will your audience guide your communication choices?

Purpose

- What is the purpose of the instructions?
- What need will the instructions meet? What problem can they help to solve?

Subject

- What is the instructions' subject matter?
- How technical should your discussion of the subject matter be?
- Do you have sufficient information to complete the instructions? If not, what sources or people can help you to locate additional information?
- What title can clearly identify the instructions' subject and purpose?

Author

- Will the instructions be a collaborative or an individual effort?
- How can the developer(s) evaluate the success of the finished instructions?

Project Design and Specifications

- Are there models for organization or format for the instructions?
- In what medium will the completed instructions be presented?
- Are there special features the completed instructions should have?
- Will the instructions require graphics and other visuals? If so, what kinds and for what purpose?
- What information design features can help the instructions' audience?

Due Date

- What is the final deadline for the completed instructions?
- How long will the instructions take to plan, research, draft, revise, and complete?
- What is the timeline for different stages of the instructions?

PROJECT PLAN SHEET

FOR A PROCESS EXPLANATION

Audience

- Who will read the process explanation or hear it as a presentation?
- How will readers use the process explanation?
- How will the audience guide communication choices?

Purpose

- What is the purpose of the process explanation?
- What need will the explanation meet? What problem can it help to solve?

Subject

- What is the process explanation's subject matter?
- How technical should your discussion of the subject matter be?
- Do you have sufficient information to complete the explanation? If not, what sources or people can help you to locate additional information?
- What title can clearly identify the explanation's subject and purpose?

Author

- Will the explanation be a collaborative or an individual effort?
- How can the developer(s) evaluate the success of the finished explanation?

Project Design and Specifications

- Are there models for organization or format for the explanation?
- In what medium will the completed explanation be presented?
- Are there special features the completed explanation should have?
- Will the explanation require graphics and other visuals? If so, what kinds and for what purpose?
- What information design features can help the explanation's audience?

Due Date

- What is the final deadline for the completed explanation?
- How long will the explanation take to plan, research, draft, revise, and complete?
- What is the timeline for different stages of the explanation?

CHAPTER 10

Description: Using Details

CHAPTER GOALS

This chapter:

- Defines *description*
- Discusses the role of audience and purpose in description
- Distinguishes between description of a representative model and a specific model
- Analyzes the author's responsibilities in description
- Emphasizes the importance of accurate and precise terminology in description
- Evaluates graphics and other visuals and information design in description
- Emphasizes description of mechanisms
- Explains how to plan and construct descriptions of a representative model and a specific model of a mechanism

INTRODUCTION

Communication frequently relies on description to convey information. Depending on purpose and the needs of your audience, description may be the focus of a presentation or only a part of it. The focus of bid specifications, for example, is describing exact features or characteristics required for a piece of equipment. An owner's manual for a piece of equipment may first describe the dangers inherent in operating the equipment and then continue with other information, such as parts, steps for assembly, steps for operating, procedure for troubleshooting problems, and other information. Whether it is the sole focus or part of a whole, effective description relies on accurate and complete details, and often on supplementary graphics, to meet audience needs.

This chapter discusses the audiences and purposes for description, the distinction between description of a representative model and a specific model, your responsibility as an author, the need for accurate and appropriate terminology, use of graphics and information design, and mechanism descriptions.

WHAT IS DESCRIPTION?

Description is written or oral communication that creates a mental picture. Through visual language, a communicator can describe objects (such as ballet shoes, the Rodrigues fruit bat, specifications for construction of a bridge), places (such as a computer laboratory, the habitat of an endangered species, a personal care home for patients with Alzheimer's disease), mechanisms (such as a jet engine, a hay baler, an incubator), or other things (such as architects' plans for converting a Victorian house into a bed and breakfast inn, fire codes for residence halls on college campuses). In describing an object, place, mechanism, or situation, think of yourself as an observer describing what you see, selecting details and organizing them to meet the needs of a specific audience. Technical description appears in a wide variety of technical communications but specifically in manuals, proposals, marketing and advertising materials, product specifications standards, reports, and training materials. Description may be the purpose of an entire document, or it may be a segment of a longer document. Any object, place, mechanism, or situation can be described from multiple perspectives, so you must identify your purpose and determine audience needs in order to choose a perspective.

AUDIENCES AND USES FOR DESCRIPTION

Once you decide that description is needed, you must answer several questions before you select details. Who is the audience? How will the audience use the description? To identify something? To evaluate something? To assemble an object? To repair it? To manufacture it? To understand it in more detail? To operate it? You select details to provide the audience with necessary information.

Consider the student-written description of pointe shoes on page 220. The audience is parents of young ballerinas who are just starting to dance *en pointe*. The description tells parents about the construction of the shoe so that they can make a wise purchase. Ballet teachers can also use the description as a handout for parents to educate them about the construction and purpose of the shoe.

Read the description on page 221 on electrical wall outlets. The description appears in a brochure, *Switches and Outlets,* prepared for Habitat for Humanity. Available at a materials outlet operated by Habitat for Humanity, the brochure serves do-it-yourself clients who want to install wall outlets.

DESCRIPTIONS OF REPRESENTATIVE MODELS AND SPECIFIC MODELS

Whether you are describing an object, place, mechanism, or situation, you can describe it as a representative model or as a specific model, depending on the audience and purpose. For example, you can describe an ideal configuration for a computer laboratory or you can describe the configuration of the computer laboratory at XYZ Corporation. The description of a representative model emphasizes characteristics of a group or class of objects, places, mechanisms, or situations. The description of a specific model emphasizes particular characteristics, aspects, qualities, or features, as well as what sets one model apart from other models.

Consider the descriptions of glass used in automobile windshields. The first below describes laminated safety glass for a segment on *glass* in *World Book Encyclopedia.* The description is obviously written for a lay audience and the purpose is to describe laminated safety glass as a class of glass to help the audience understand its characteristics.

> Laminated Safety Glass is a "sandwich" made by combining alternate layers of plastic material and flat glass. The outside layer of glass may break when struck by a flying object, but the plastic layer is elastic and stretches. This holds

Example 1 Description Adapted to Audience and Purpose

Construction of the Pointe Shoe

Pointe shoes, or toe shoes, are constructed to support the arch of the foot as the ballerina dances on the tip of her toes. The shoe has three main parts: the box, the shank, and the fabric upper.

THE BOX

The **box** of the shoe is constructed out of layers of leather that have been glued together. The layers are molded into the toe portion of the shoe: rounded at the top and flat on the bottom and side. The box can be very square or tapered according to the dancer's needs. Beginner shoes have a very stiff box. More experienced dancers prefer a more pliable box.

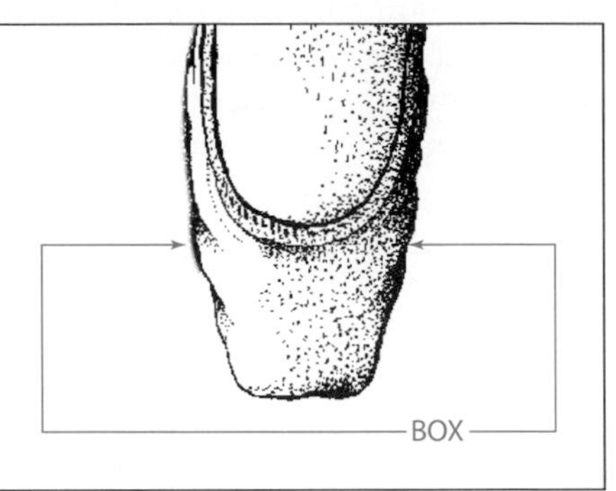

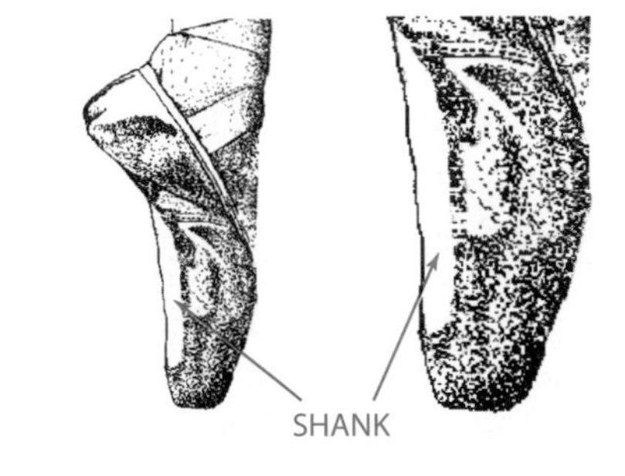

THE SHANK

The **shank** is a stiff piece of leather that forms the main support of the shoe. It is sewn to the bottom of the shoe and reaches from the ball of the foot to the end of the heel. Beginning dancers need the strength of a hard shank to support the arch while dancing. More experienced dancers need more flexibility.

THE FABRIC UPPER

The **fabric upper** part of the shoe is usually made of satin fabric. It is attached to a cotton lining and pulled around the box, then fastened down by the application of the shank. The function of the upper is primarily aesthetic—shoes used in class are pink, while the fabric can be dyed to match costumes for performance. The binding around the edge of the shoe houses a thin cotton string. This string is used to pull the fabric around the foot so that the top of the shoe is snug, allowing the shoe to move without slipping during exercises or performance. Ribbons are usually included with the shoes, but each dancer must determine the correct placement of the ribbons and sew them on herself.

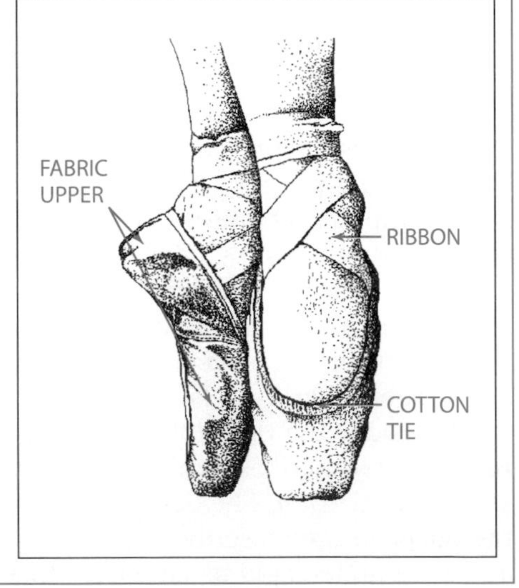

ABOUT OUTLETS (RECEPTACLES)

A receptacle is frequently called a wall receptacle. It is the point of electrical service into which you insert the plug of a lamp, appliance, clock, or other electricity-using equipment. There are several varieties of receptacles. Some are designed for outdoor use, some to handle the heavy-duty requirements of major appliances, some are integrated into light fixtures, and some are combined with switches. The most common home receptacle is the duplex receptacle that is rated at 15 or 20 amperes and 120 volts **(Fig. 5)**. A duplex receptacle has two outlets and accommodates two pieces of electrical equipment.

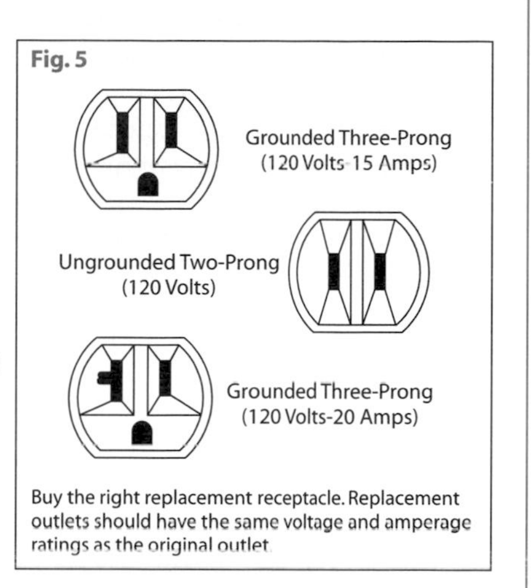

Buy the right replacement receptacle. Replacement outlets should have the same voltage and amperage ratings as the original outlet.

Although many homes have two-hole outlets, three-hole grounding outlets are required in all new houses and should be used for all replacements where a ground is available. Such outlets include one hot wire, one neutral wire, and one ground wire. Most people recognize the hole for the grounding prong, but many do not realize that the other two slots are different sizes. The shorter slot is connected to the hot wire and the longer slot to the neutral wire. This distinction is important with the increased use of electronic equipment in the home. Many of the plugs for this type of equipment are polarized, meaning that one prong is wider than the other because internal switches and other components must connect to the current in proper sequence.

Courtesy of Creative Homeowner Press

the broken pieces of glass together and keeps them from flying in all directions. Laminated glass is used where broken glass might cause serious injuries, as in automobile windshields.

Excerpted from THE WORLD BOOK ENCYCLOPEDIA. Copyright © 1999 World Book, Inc. By permission of the publisher. www.worldbook.com

Compare the preceding description of laminated safety glass with the following description of genuine Volvo windshields made of laminated safety glass on page 222. This description from a brochure on *Genuine Volvo Windshields* is also written for a lay audience, but the purpose here is to persuade readers that the laminated safety glass used in Volvo windshields is superior to ordinary windshield glass. Notice the inclusion of a cutaway view of the windshield glass to emphasize the layers.

Job descriptions are another example of representative versus specific models. A job description can describe what is typically required of a construction manager (i.e., education; knowledge of concrete, steel, wood, and stone con-

A Persuasive Description of Windshield Glass

THROUGH THICK AND THIN

Genuine Volvo windshields have an overall thickness much greater than that found in ordinary windshields. Each is made from two strong layers of glass, of different thicknesses, which are reinforced with an invisible plastic laminate. This makes them stronger and more resistant to chipping and shattering. Damage caused by flying stones, for example. It also minimizes the risk of splintering glass in the passenger compartment during a collision.

Outer Layer
2.5 mm thick

Plastic Laminate
0.8 mm thick

Inner Layer
2.1 mm thick

struction; understanding of electrical instrumentation, heating and cooling systems, roofing systems, landscaping, site preparation, and building foundations; awareness of environmental regulations) or it can describe what a construction manager with a particular company does. A quick search of the Internet shows literally thousands of specific job descriptions. The example on page 223 was located online through a search for available government job openings.

Another frequently used description of a specific model is bid specifications. They name a product or service and describe the unique characteristics, qualities, or features the product or service must have to meet a need. The audience for bid specifications is a manufacturer, business, or individual who can supply a product or service. The purpose of the bid specifications is to locate and buy the product or service.

While there is no standard form for recording or presenting them, bid specifications must clearly and precisely identify the required qualities. The manufacturer, business, or person who can supply the product or service at a stated price must know *exactly* what is expected.

The sample bid specifications shown on pages 224–225 are for feed to be used at the BullTest Station, a part of the Hinds Community College agriculture program. These pages clearly set forth what the college expects in the feed required.

Audience and purpose affect decisions on everything related to description, including content, language, complexity, organization, type of presentation, length, format, page layout, and information design. For instance, descriptions of a stereo speaker differ in emphasis and in detail, depending on whether they are written so a music buff can construct a similar speaker, so a prospective buyer can compare the speaker to a similar one, or so the general public can understand what a stereo speaker is. A description of a laptop computer for a lay audience who wants to know what a laptop computer is, what

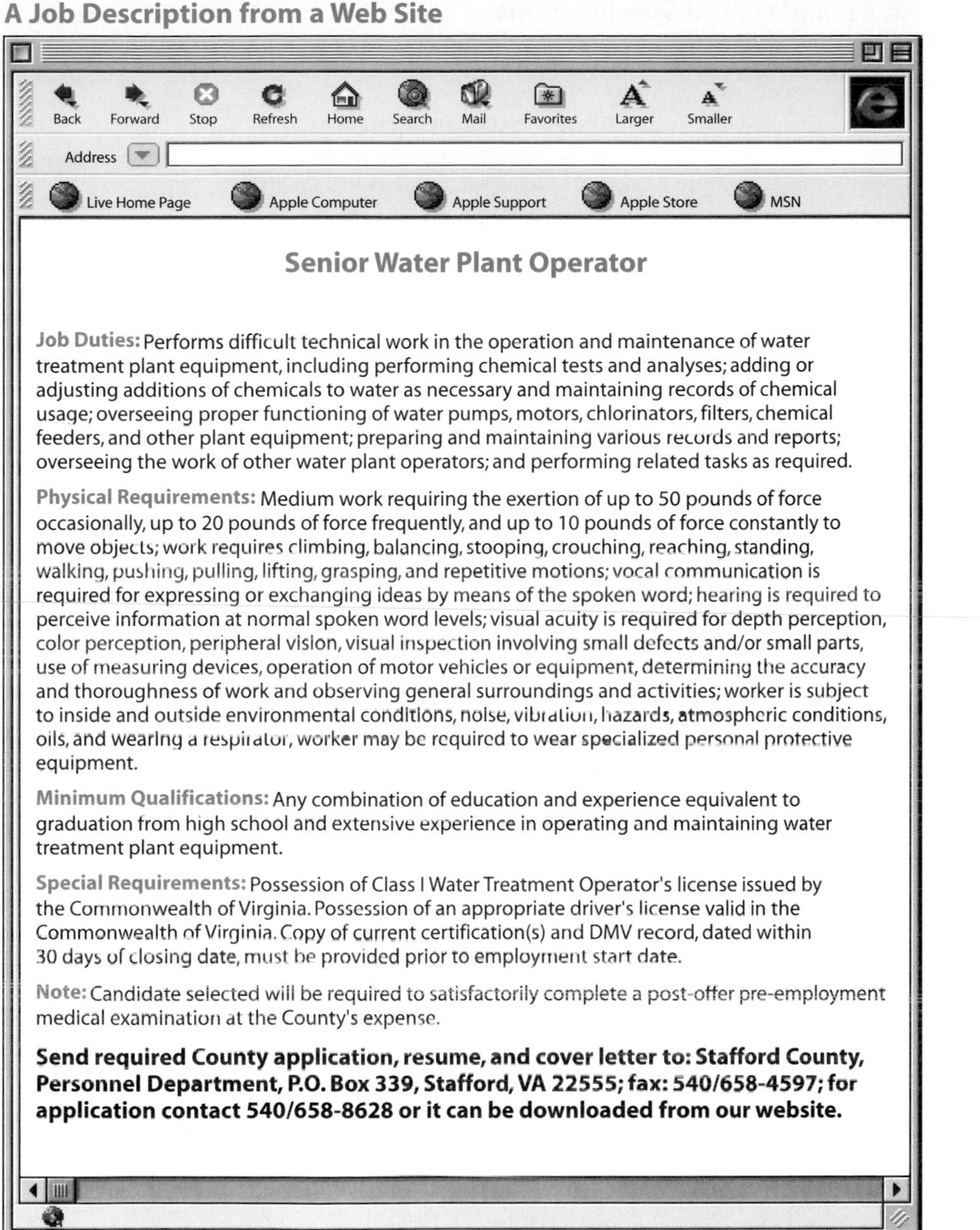

Senior Water Plant Operator

Job Duties: Performs difficult technical work in the operation and maintenance of water treatment plant equipment, including performing chemical tests and analyses; adding or adjusting additions of chemicals to water as necessary and maintaining records of chemical usage; overseeing proper functioning of water pumps, motors, chlorinators, filters, chemical feeders, and other plant equipment; preparing and maintaining various records and reports; overseeing the work of other water plant operators; and performing related tasks as required.

Physical Requirements: Medium work requiring the exertion of up to 50 pounds of force occasionally, up to 20 pounds of force frequently, and up to 10 pounds of force constantly to move objects; work requires climbing, balancing, stooping, crouching, reaching, standing, walking, pushing, pulling, lifting, grasping, and repetitive motions; vocal communication is required for expressing or exchanging ideas by means of the spoken word; hearing is required to perceive information at normal spoken word levels; visual acuity is required for depth perception, color perception, peripheral vision, visual inspection involving small defects and/or small parts, use of measuring devices, operation of motor vehicles or equipment, determining the accuracy and thoroughness of work and observing general surroundings and activities; worker is subject to inside and outside environmental conditions, noise, vibration, hazards, atmospheric conditions, oils, and wearing a respirator; worker may be required to wear specialized personal protective equipment.

Minimum Qualifications: Any combination of education and experience equivalent to graduation from high school and extensive experience in operating and maintaining water treatment plant equipment.

Special Requirements: Possession of Class I Water Treatment Operator's license issued by the Commonwealth of Virginia. Possession of an appropriate driver's license valid in the Commonwealth of Virginia. Copy of current certification(s) and DMV record, dated within 30 days of closing date, must be provided prior to employment start date.

Note: Candidate selected will be required to satisfactorily complete a post-offer pre-employment medical examination at the County's expense.

Send required County application, resume, and cover letter to: Stafford County, Personnel Department, P.O. Box 339, Stafford, VA 22555; fax: 540/658-4597; for application contact 540/658-8628 or it can be downloaded from our website.

Source: "Employment Opportunities." County of Stafford, Virginia. <http://www.co.stafford.va.us/personnel> 20 January 2000.

it is used for, and how it works differs in emphasis and detail from a description in a sale catalog offering laptop computer accessories. The primary readers of accessories descriptions are likely to be professionals and experts who already know what a laptop computer is, what it is used for, and how it works. Such professionals and experts read the catalog descriptions to find a laptop computer with certain features or characteristics to meet their needs.

An Example of Bid Specifications

HINDS COMMUNITY COLLEGE
PURCHASING DEPARTMENT
ADMINISTRATION BUILDING-ROOM 201
RAYMOND, MS 39154

The Hinds Community College District is now receiving sealed bids in the District Purchasing Office, Raymond Campus, Raymond, Mississippi, on the items listed on the attached specifications sheets. All bids must be received in the district Purchasing Office by 2:00 p.m. on the date given below. The Hinds Community College District reserves the right to reject any and all bids, and to waive any and all informalities.

BID #: 2374

BID OPENING DATE: JULY 20, 1999

ALL ITEMS SHIPPED FOB: RAYMOND

ADDRESS ALL CORRESPONDENCE TO:

Hinds Community College District
Purchasing Department
Administration Department
Administration Building—Room 201
Raymond, MS 39154
Bid #: 2374

INSTRUCTION TO BIDDERS

All prices shall be quoted FOB to the location indicated above. State and Federal taxes are not applicable due to the exemption status of the College. All bids must be submitted on the attached forms. Any vendor submitting a price on an alternate item shall be prepared to provide a sample of that alternate. All alternate bids should be clearly identified on the attached bid form. Failure to do so will result in bid rejection. If you have any questions concerning the attached specifications, please contact the District Purchasing Office at (601) 857-3368.

DATE: ________________

We propose to furnish the following items at prices opposite each, delivered and installed to the address above.

Firm: ________________

Address: ________________

____

By: ________________

Title: ________________

Phone: ________________

BID NO.: 2374
OPENING DATE: JULY 20, 1999
BID NAME: BULLTEST FEED-RAYMOND

GENERAL INFORMATION TO BIDDERS

1. All items to be priced with delivery included.
2. The owner reserves the right to define equals, to reject any and all bids, to adjust quantities, and to waive any and all informalities.
3. All bids shall be good for a period of 6 months after bid opening date.
4. All bids shall be extended from unit price to total price or marked no bid.
5. All bids shall be dated and signed in ink.
6. Price of feed could vary from base bid. Prices will be based on the Chicago Board of Trade-plus delivery on whole corn and cottonseed meal.
7. Size of pellet 3/32".
8. All loads will be approximately 12 tons.
9. Alpha toxin shall be less than 25 p.p.m.
10. Please address all bid inquires to: K.P. Lewis 601-857-3368

BID NO.: 2374
OPENING DATE: JULY 20, 1999
BID NAME: BULLTEST FEED-RAYMOND

HINDS COMMUNITY COLLEGE SPECIAL MIXED FEED

Ingredient	%	Lbs./Fed
Cracked corn	49.25	985#
*Soybean Meal	16.50	330#
Cottonseed Hulls	27.50	550#
Molasses	5.00	100#
Feedgrade Limestone	1.20	24#
Trace-Mineralized Salt	.50	10#
Selenium Premix	.05	1#

Approximate As-Fed Analysis	%
Dry Matter	88.90
Crude Protein	12.30
T.D.N.	65.00
Crude Fiber	14.60
Crude Fat	2.50
Calcium	.50
Phosphorus	.35
Potassium	.75
Selenium	.18 ppm

1. Add 20 grams of Bovatec and 4.0 grams of EDDI per ton of ration.
2. Add a Vitamin A-D-E premix to provide the following minimum concentrations per ton: 5 million I.U. Vitamin A; 500,000 I.U. Vitamin D; 20,000 I.U. Vitamin E.
3. Add Probios at the rate of 1,340 grams per ton.
4. Add Sodium Bicarbonate at the rate of 25# per ton.

* Cottonseed Meal can be substituted for Soybean Meal.

Bid Price per ton for September 1999 is: ____________________

THE AUTHOR'S RESPONSIBILITIIES

In description, as in all other types of communication, you are responsible for providing accurate, complete information that meets the needs of an audience. Typically, an audience expects answers to questions such as: What is it? What does it do? What does it look like? Who uses it? When? For what? What is it made of? How does it work? Omitting needed information or misleading the audience through unclear or inaccurate information can lead to serious consequences—a lawsuit, financial loss, physical injury, even death.

Also, based on audience need and purpose, you must decide how technical a description should be. Look at the following two descriptions of the Rodrigues Fruit Bat. The first description, titled *A Personal Portrait of the Rodrigues Fruit Bat*, was written for readers who are interested in understanding the bats and may be potential donors to bat research. Thus, the selection is written to present technical information in an appealing, understandable way for a lay audience. The second description, a Web page from the Philadelphia Zoo, communicates observable details about the Rodrigues Fruit Bat.

Most technical description should be as objective and impartial as is humanly possible. You may include evaluations such as "durable," "safe," "easy to use," and so on. Be sure, however, that such evaluations are verifiable. Your audience will more readily accept such evaluations if you supply details to support them. For example, if you describe a pair of needle nose pliers as durable, you might explain that they are made of heat-treated steel.

You bear responsibility for clearly indicating cautions, warnings, dangers, or hazards. The operator's safety manual for chain saw owners, published by Husqvarna Sweden, for example, consists of twenty-three pages of description and illustrations to ensure safe use of the chain saws. The excerpt on page 229, describing the felling of a tree, is taken from that manual.

You are ethically responsible for accurate and clear descriptions that meet your audience needs, including all information essential to their understanding and safety.

ACCURATE AND APPROPRIATE TERMINOLOGY

Accurate and appropriate terminology is essential in description. Although it is easier to use broad terms such as *thing*, *good*, *large*, *narrow*, *tall*, and so forth, such terms obviously have different meanings for different audiences. To someone who lives in a small city, a tall building may be one with ten floors; to someone who lives in a large city, it may be one with thirty floors.

In description, you help your audience visualize the appearance or the composition of an object, place, mechanism, or situation. Audiences visualize in five basic ways—by size, shape, color, texture, and position. The examples on page 230 show you some of the language choices you can make to help your audience visualize the subject.

A Personal Portrait of the Rodrigues Fruit Bat

BY KIM WHITMAN

"PROBABLY THE RAREST BAT IN THE WORLD" is how the Rodrigues fruit bat was described in the early 1970s. Today, thanks to conservation efforts on Rodrigues Island and at captive breeding centers around the world, that dubious distinction no longer applies; instead, I've dubbed this flying fox "probably the cutest bat in the world."

Admittedly, I'm biased, having spent more than eight years contemplating the various attributes of this species. I've observed Rodrigues fruit bats "up close and personal" in captivity, as well as framed against the verdant leaves of a Rodriguan forest.

Flying foxes are an attractive group in general, but Rodrigues fruit bats go beyond the typical flying fox. To begin with, they're adorably fuzzy. Their chocolate brown fur is accented with splashes of pure gold, which in some individuals extend from the shoulders all the way down the back. In the glow of a strong tropical sun, these highlights are even more striking—prompting Rodriguans to give their endemic chiroptera the nickname "golden bat." But what really sets the Rodrigues fruit bat apart is its face. Its foreshortened muzzle, liquid black eyes, and ever-mobile triangular ears make it look like a small black chow dog with wings. Okay, so maybe that's not everyone's idea of the perfect face, but it works for the bats.

Weighing in at less than a pound with a two-and-a-half-foot wingspan, Rodrigues fruit bats are medium-sized as flying foxes go. But don't be fooled by their diminutive stature and beguiling appearance: when caught in a situation they don't like, these bats have an attitude. Their teeth, designed for grasping and holding fruit, also serve extremely well for grasping and holding the fingers of bat biologists. Fortunately, a good pair of leather gloves is enough to diminish the bite to a painful pinch.

These bats also possess a powerful pair of lungs, capable of producing a startling scream. Their favorite occasions for demonstrating this ability appear to be during breeding and while being retrieved from mist nets. Observations have shown that the volume, pitch, and duration of the bat's vocalization is directly proportional to its proximity to your ear—and your level of fatigue.

No description of Rodrigues fruit bats would be complete without a reference to their aroma. Whether housed in an exhibit or hanging 100 feet up in a tree, these bats carry with them a distinctive, ripe odor capable of eliciting comments from even the most nasally challenged. Produced in dermal scent glands and rubbed onto nearby branches, leaves, and other bats, fruit bat musk is a potent scent believed to aid in identifying each other in social interactions.

With their relatively short, broad wings, shaped for maneuverability, Rodrigues fruit bats are not the strongest of fliers. I've watched them fly furiously against the wind for several minutes without making any headway, only to become exhausted and be swept backwards. Unfortunately, Rodrigues Island sits directly in the middle of the cyclone belt. The destructive winds of this area's annual cyclones can knock bats out of trees and sweep them out to sea. Only adequate forest cover can protect the bats from Mother Nature's ire. Rodriguans are doing their best to help: reforestation programs and protection of existing forests are increasing the habitat available to the bats. With all of these good efforts and a little luck, the golden bats will continue to flap ponderously over Rodrigues for many years to come.

The article on page 227 is reprinted from BATS magazine with permission from Bat Conservation International (BCI). For more information about bats, BATS magazine, or membership in BCI, please visit the BCI Web site at <www.batcon.org> or write or call: Bat Conservation International, P.O. Box 162603, Austin, Texas 78716, 512-327-9721. Basic membership, which includes a one-year subscription to BATS magazine, is $30.

Objective Description

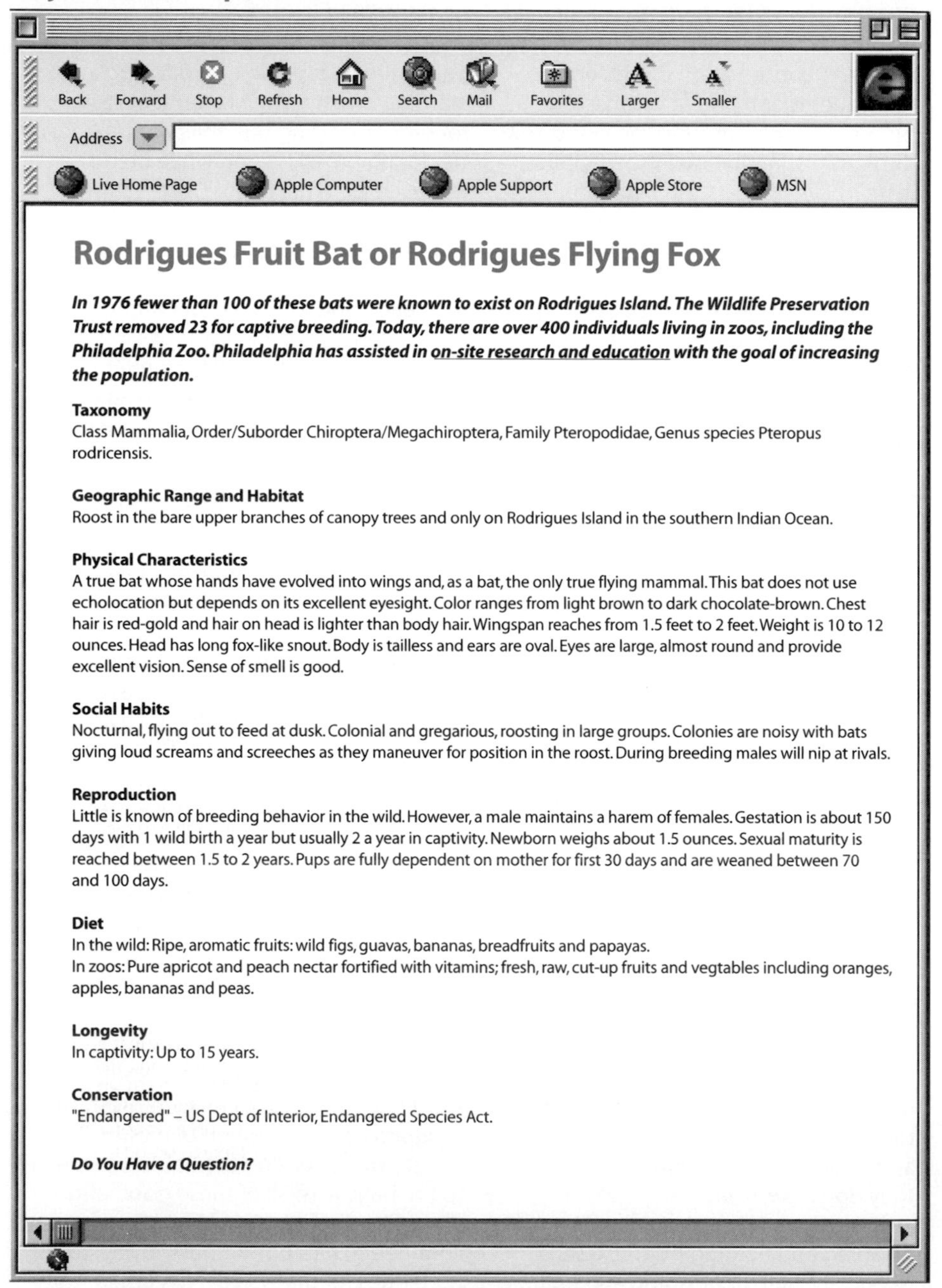

Rodrigues Fruit Bat or Rodrigues Flying Fox

In 1976 fewer than 100 of these bats were known to exist on Rodrigues Island. The Wildlife Preservation Trust removed 23 for captive breeding. Today, there are over 400 individuals living in zoos, including the Philadelphia Zoo. Philadelphia has assisted in on-site research and education with the goal of increasing the population.

Taxonomy
Class Mammalia, Order/Suborder Chiroptera/Megachiroptera, Family Pteropodidae, Genus species Pteropus rodricensis.

Geographic Range and Habitat
Roost in the bare upper branches of canopy trees and only on Rodrigues Island in the southern Indian Ocean.

Physical Characteristics
A true bat whose hands have evolved into wings and, as a bat, the only true flying mammal. This bat does not use echolocation but depends on its excellent eyesight. Color ranges from light brown to dark chocolate-brown. Chest hair is red-gold and hair on head is lighter than body hair. Wingspan reaches from 1.5 feet to 2 feet. Weight is 10 to 12 ounces. Head has long fox-like snout. Body is tailless and ears are oval. Eyes are large, almost round and provide excellent vision. Sense of smell is good.

Social Habits
Nocturnal, flying out to feed at dusk. Colonial and gregarious, roosting in large groups. Colonies are noisy with bats giving loud screams and screeches as they maneuver for position in the roost. During breeding males will nip at rivals.

Reproduction
Little is known of breeding behavior in the wild. However, a male maintains a harem of females. Gestation is about 150 days with 1 wild birth a year but usually 2 a year in captivity. Newborn weighs about 1.5 ounces. Sexual maturity is reached between 1.5 to 2 years. Pups are fully dependent on mother for first 30 days and are weaned between 70 and 100 days.

Diet
In the wild: Ripe, aromatic fruits: wild figs, guavas, bananas, breadfruits and papayas.
In zoos: Pure apricot and peach nectar fortified with vitamins; fresh, raw, cut-up fruits and vegtables including oranges, apples, bananas and peas.

Longevity
In captivity: Up to 15 years.

Conservation
"Endangered" – US Dept of Interior, Endangered Species Act.

Do You Have a Question?

Used by permission of R. N. Sloane, Docent, The Philadelphia Zoo.

FELLING

Felling is more than cutting down a tree. You must also bring it down as near to an intended place as possible without damaging the tree or anything else.

Before felling a tree, carefully consider all conditions which may affect the intended direction of fall, including:

Inclination of the tree.
Shape of the crown.
Snow load on the crown.
Wind conditions.
Obstacles within tree range: e.g., other trees, power lines, roads, buildings, etc.

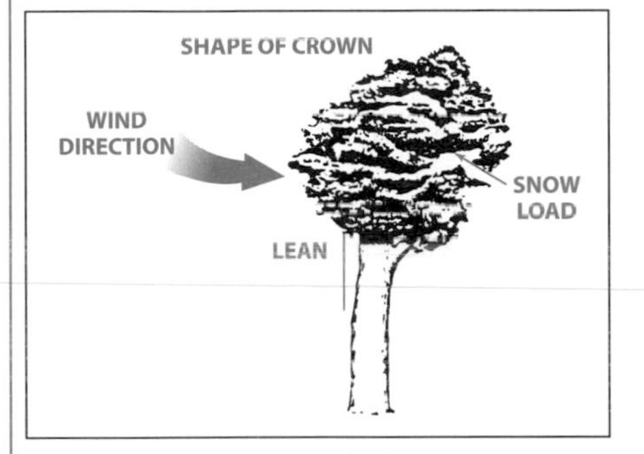

⚠ **WARNING**
Always observe the general condition of the tree. Look for decay and rot in the trunk which will make it more likely to snap and start to fall before you expect it. Look for dry branches, which may break and hit you when you are working.

Always keep animals and people at least twice the tree length away while felling.

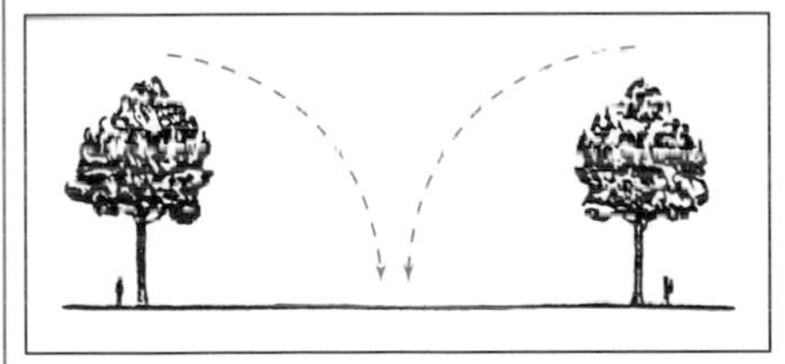

Clear away shrubs and branches from around the tree.

Prepare a path of retreat diagonally away from the felling direction.

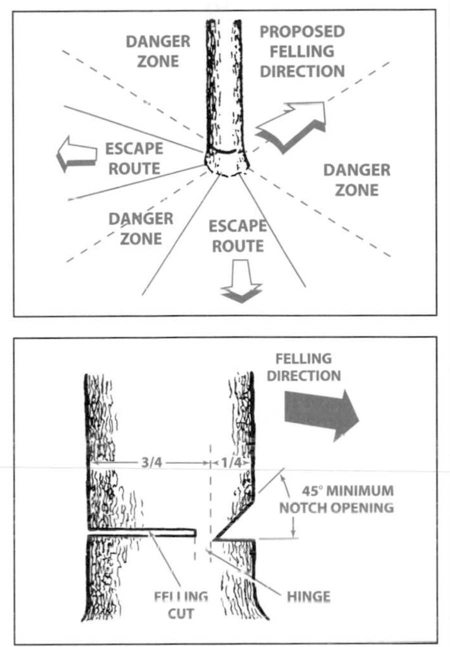

BASIC RULES FOR FELLING TREES

Normally the felling consists of two main cutting operations—notching and making the felling cut. (see figure above).

Start making the upper notch cut on the side of the tree facing the felling direction. Look through the kerf as you saw the lower cut so you do not saw too deep into the trunk.

The notch should be deep enough to create a hinge of sufficient width and strength. The notch opening should be wide enough to direct the fall of the tree as long as possible.

Saw the felling cut from the other side of the tree between one and two inches (3–5 cm) above the edge of the notch.

Never saw completely through the trunk. Always leave a hinge.

The hinge guides the tree. If the trunk is completely cut through, you lose control over the felling direction.

Insert a wedge or a felling lever in the cut well before the tree becomes unstable and starts to move. This will prevent the guidebar from binding in the felling cut if you have misjudged the falling direction. Make sure no people have come into the range of the falling tree before you push it over.

EXAMPLES OF LANGUAGE CHOICES FOR VISUALIZATION

Size	Shape	Color	Texture	Position
physical	round	green	grainy	east
dimensions (i.e., 5″ by 3″)	square	yellow	coarse	west
	triangular	red	nubby	in front
	conical	blue	pebbly	behind
	cylindrical	bright	grooved	to the left side
	spherical	pastel		above
	hexagonal	neon		below

Another effective way to describe size, shape, color, texture, and position is to compare the unfamiliar to the familiar. For example, you can make comparisons using phrases such as A-framed, star shaped, finger tight. You can also use comparison to suggest size. For example, regular guests on a once-popular television game show typically included the question, "Is it bigger than a bread box?" as they attempted to identify objects.

Describing situations and concepts can be especially difficult. Using analogy in such descriptions often helps to clarify meaning. For example, the U.S. Census Bureau, in describing the rapid growth of the world's population, used the following analogies: The world population is increasing by 78 million people each year. That is the equivalent of adding a city nearly the size of San Francisco every three days, or the combined populations of France, Greece, and Sweden every year. Another way the Bureau puts this growth into perspective is by stating that every 20 minutes the world adds another 3,500 human lives.

In description, always be precise. Write "The chart shows manufactured goods representing 44 percent of the goods produced in Manitoba," not "The chart shows manufactured goods representing a large percentage of goods produced in Manitoba." Write "Dividers may vary in size from 2 to 12 inches," not "Dividers may vary in size." Give exact statistics, measurements, dimensions, weights, and other identifying details.

In general, avoid overly complicated technical terms or jargon, depending on your audience and purpose. A builder discussing with a client the color for exterior house trim above the brick walls and below the roof would likely use the phrase *color for the exterior wall trim*. The same builder, discussing trendy exterior house trim colors with other builders, would use the phrase *color for the fascia and soffit*.

Be careful to refer to sizes by the standard method of measurement. The size of an electric drill, for instance, is referred to by the maximum diameter of the chuck (such as ¼-inch or ½-inch), not by the length of the barrel or the dimensions of the handle. The height of a horse is measured in hands (a hand equals 4 inches). Bicycle sizes are based on the wheel diameter.

If you are not certain about precise words, consult available sources, such as dictionaries; general encyclopedias; textbooks; specialized dictionaries, handbooks, and encyclopedias; knowledgeable people; advertisements; mail order catalogs, and instruction manuals.

GRAPHICS AND OTHER VISUALS IN DESCRIPTION

Graphics and other visuals can vividly enhance description. Often, a graphic can illustrate details that are difficult to explain in words. Pictorial illustrations, such as photographs, drawings, and diagrams, help audiences visualize objects, places, mechanisms, or situations. Photographs can provide a clear external or internal view of a subject; they can show size, color, and texture. Drawings, ranging from simple, freehand sketches in pencil, pen, crayon, or brush, to engineers' or architects' minutely detailed computer designs can do the same. Diagrams can show the parts, operation, and assembly of an object or a system. Drawings and diagrams can show exterior views, the relationships between parts, the functions of parts as they work together, and cross-sections or cutaway views. Exploded views show parts disassembled but arranged in sequence of assembly, which can be helpful in describing mechanisms.

Graphics do not have to be complicated or sophisticated to be effective. Even a rough sketch may be helpful. Almost everyone has access to computers that offer an array of graphic capabilities. From simple line drawing programs to highly complex graphics programs, the means to create effective visuals are available. For a detailed discussion of how to put graphics, and graphics programs, to work for you, see Chapter 5 Visuals and Chapter 6 Technology.

MECHANISM DESCRIPTION

The description of a mechanism is one of the most common kinds of technical description. People constantly cope with mechanisms: hair dryers, computers, traffic lights, coffee makers, compact disc players, cash register terminals, forklifts, shopping carts, stethoscopes, pagers, fax machines. Users need to understand these mechanisms: what they do, what they look like, what parts they have, how these parts work together, and how to use them.

As an employee, you might describe mechanisms when writing specifications for product specialists, a memorandum for a repair person, or a purchase request for a supervisor, or when demonstrating a new piece of equipment to a potential customer. Users need descriptions when learning how to use new equipment.

A mechanism can be described as a representative model or as a specific model, depending on the audience's needs and your purpose. Descriptions of a representative model focus on characteristics of a group or class of mechanisms. Examples are a word processor, a camera, a ball point pen, a power supply, or a calculator. Descriptions of a specific model, conversely, focus on particular characteristics of an identified model, style, or brand of mechanism. Examples are a Polaroid 600 Business Edition Instamatic Camera, a

Parker classic Black Matte Ballpoint Pen No. 67839, or an HP Pavilion 8250 computer.

Describing a Representative Model of a Mechanism

A description of a representative model identifies and explains aspects usually or typically associated with the mechanism. These aspects may include what the mechanism can do or can be used for; what it looks like; what its parts (components) are, what each part looks like, what each part does, and how these parts interact to perform the function of the mechanism.

Frames of Reference in Describing a Representative Model

Logically, there are three frames of reference (points of view or organizing principles) from which to describe a representative model of a mechanism: its function, its physical characteristics, and its parts.

Function. A mechanism performs a particular function or task. A wedging board, for example, eliminates air bubbles in clay. A fire hydrant provides a conduit of water for fire fighting. A hypodermic needle and syringe make possible the injection of medications into the body. A microscope makes a very small object, such as a microorganism, appear larger so that it is clearly visible to the human eye. An automobile serves as a means of transportation. A kidney separates water and the waste products of metabolism from the blood. An elevator transports people, goods, and equipment vertically between floors.

The key element in a description of a representative model is an explanation of its function. What is it used for? This question raises other questions: When is it used? By whom? How? All of these questions must be answered for the description to be adequate.

Physical Characteristics. A description of a representative model points out the physical characteristics of the mechanism. The purpose is to help the audience see, or visualize, the object, to give an overall impression of its appearance. Physical characteristics such as size, shape, weight, material, finish, color, and texture are helpful. Comparison with a familiar object is also useful. Sometimes photographs or drawings of the mechanism and its parts are especially helpful.

Parts. A third frame of reference in describing a representative model of a mechanism is its parts, or components. The mechanism is divided into its parts, the purpose of each part is given, and the way that the parts fit together is explained. For example, as a percussion instrument, a drum is described as having two major parts: a hollow shell or cylinder and a drumhead stretched over one or both ends. When the drumhead is beaten with the hands or with an implement such as a stick or wire brush, sound is produced. In the part-by-part description, the hollow shell or cylinder and the drumhead are each described in turn. In essence, each of these parts becomes a new mechanism and is subsequently described according to function, physical characteristics, and parts.

Planning and Organizing a Description of a Representative Model

Planning a description of a representative model of a mechanism begins with answering a series of questions to identify audience, purpose, subject matter (including graphics and other visuals), format and medium, and deadlines. A Project Plan Sheet for Description, shown on page 261, will help you to determine specific questions. This project plan sheet can be adapted for a specific project such as describing a representative model of a mechanism.

Having answered the questions suggested on the Project Plan Sheet and any others you may add, you can begin to gather material for the description. Once you have sufficient material to begin, think about which method of organization is most appropriate. Typical methods include chronological, spatial, and functional. The very names of the methods suggest the organizational pattern. Chronological organization focuses on the order in which a mechanism is assembled or put together. Spatial organization focuses on the relationship of parts, such as left to right, top to bottom, or inside to outside. Functional organization focuses on how a mechanism works. You may use these patterns in combination. For example, you may first show how a mechanism is assembled (chronological organization) and then show how it works (functional organization). Below is a guide for arranging information in a description of a representative model of a mechanism. Remember, however, that ultimately you must select and organize material, depending on your audience and purpose.

- Introduce the mechanism (an overview).
  1. Identify or define the mechanism and indicate why the description is important, if appropriate.
  2. Explain the function, use, or purpose of the mechanism.
  3. If applicable, state who uses the mechanism, when, where, and why.
  4. If the mechanism is part of a larger whole, show the relationship between the part and the whole.
  5. List the points (frames of reference) to be described. These could be stated as a list of items in a series or as a bulleted or numbered list. Considering your audience and purpose will help you to decide which arrangement of points is appropriate.
- Describe the mechanism part by part (component by component).
  6. Describe the physical characteristics of the mechanism, including, as applicable, size, shape, weight, material, color, texture, and so on.
  7. Identify the parts of the mechanism by
     a. listing the parts in the order in which they will be described.
     b. describing each part separately by telling its function and its physical characteristics.
  8. Explain the relationship of each part to the other parts.
  9. Subdivide the part into its components and give their functions and physical characteristics, as appropriate.

- Close the description.

10. Show how the individual parts work together.
11. Mention variations of the mechanism, such as optional features, other types, and other sizes; or comment on the importance or significance of the mechanism, as applicable.

Student Example: Description of a Representative Model of a Mechanism

Following is a student-written example of a description of a representative model of a mechanism. First the student adapted a Project Plan Sheet for Description and answered the plan sheet questions as a way of thinking about the description. Then the student reviewed the guide for organizing a description of a representative model. The Project Plan Sheet is shown on pages 235–236. After planning, writing, and revising, the student prepared the final draft which appears on pages 237–239.

PROJECT PLAN SHEET

FOR A DESCRIPTION OF A REPRESENTATIVE MODEL OF A MECHANISM

Audience

- Who will read the description or hear it as a presentation?
  Individuals who will use a sprayer, a lay audience
- How will readers use the project?
  To gain basic knowledge of manual sprayers
- How will your audience guide your communication choices?
  The description will require nontechnical language and simple explanation.

Purpose

- What is the purpose of the description?
  To familiarize the audience with manual sprayers
- What need will the description meet? What problem can it help to solve?
  The audience can use the description to gain an understanding of manual sprayers and make a decision about using such a sprayer.

Subject

- What is the description's subject matter?
  Manual sprayers
- How technical should the discussion of the subject matter be?
  Nontechnical because of the audience
- Do you have sufficient information to complete the project? If not, what sources or people can help you to locate additional information?
  No. I will read several sources such as brochures and gardening books that discuss manual sprayers. I will also talk to individuals who are familiar with manual sprayers.
- What title can clearly identify the description's subject and purpose?
  A General Description of a Manual Sprayer

Author

- Will the description be collaborative or an individual effort?
  Individual effort

- How can the developer(s) evaluate the success of the finished description?

  I will ask someone at a garden center to read the finished description.

Project Design and Specifications

- Are there models for organization or format for the description?

  Yes. Descriptions of representative models are available in encyclopedias and other reference works, in various published literature at home and garden centers, and in textbooks.

- In what medium will the finished description be presented?

  Written

- Are there special features the finished description should have?

  No

- Will the description require graphics and other visuals? If so, what kinds and for what purpose?

  A cutaway view of a sprayer to show its parts and thus give the audience an understanding of how the sprayer works

- What information design features can help the description's audience?

  Headings and chunked material, white space to mark sections

Due Date

- What is the final deadline for the completed description?

  The description is due in three weeks.

- How long will the description take to plan, research, draft, revise, and complete?

  I should be able to accomplish these tasks through drafting in two weeks. This will allow a week to revise and complete the description.

- What is the timeline for different stages of the description?

  Plan and research – one week

  Draft – one week

  Revise and complete – one week

FINAL DRAFT
A General Description of a Manual Sprayer

Title reflects purpose of the document.

[Marginal notes have been added to show organization.]

Sprayer defined, including function and uses

A manual sprayer is a device used by homeowners, horticulturists, landscape technicians, and others to spray chemicals such as herbicides and pesticides. While a sprayer may have some variations, most look basically the same, have the same parts, and work on the same principle.

Forecasting sentence listing topics to be explained

Caution to alert readers to a common danger in using a manual sprayer

CAUTION Before you use any sprayer, *always* read the manufacturer's owner's manual carefully. A sprayer is operated with liquid under pressure. If you do not observe caution and follow instructions for operating and cleaning, you can be seriously injured.

Physical Characteristics

Physical characteristics: appearance

A typical sprayer looks like a plastic jar with an open handle (shaped somewhat like the handle on a briefcase) on top and a hose attached to one side. The jar holds the liquid chemical. The handle is manually pushed up and down to build pressure in the tank and force the liquid to eject through the hose onto a surface.

Purpose of jar

Basic operating principle

Parts

Parts listed to show order of explanation

A sprayer typically has four major sections: the tank, the handle, the pump, and the hose and nozzle. See the following cutaway view of a Roundup Herbicide Sprayer, a typical manual sprayer.

Graphic showing parts

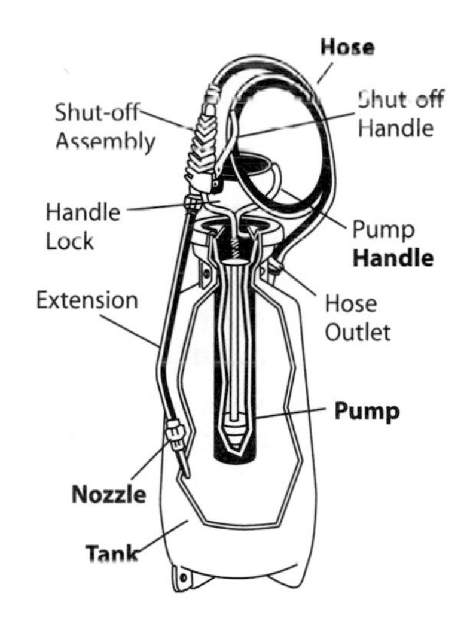

Source of visual: *Owner Manual for a Roundup Herbicide Sprayer.* Published by the Fountainhead Group, Inc.

Each major section described by physical characteristics and function

The Tank

The tank is the jarlike container that holds the liquid to be sprayed. It is most always made of hard plastic. The opening in the top center is shaped like a shallow bowl with a hole in the center.

To one side of the neck of the tank is a small tubelike opening molded as a part of the tank. The hose attaches over this opening. The size of the tank varies from a half-gallon to three gallons. Lines on the tank mark the liquid levels. For example, on a two-gallon tank, lines mark 1 gallon, 1½ gallons, and 2 gallons. Lines also mark 4 liters, 6 liters, and MAXIMUM FILL. Anything larger than two gallons would be too heavy to carry when the tank is filled with liquid.

The Handle and Handle Lock

The handle is made of a metal rod with a plastic hand grip for ease of use. Shaped like an open rectangle, the handle is easy to grasp. The ¼-inch diameter metal rod comes together below the hand grip. One end of the metal rod angles outward approximately ½ inch. This end can be pushed into the lid and turned so that the handle is secured to the pump assembly. The other end of the metal rod fits through a washer and a spring and into the pump shaft. This piece allows the up-and-down movement of the handle to build pressure.

The Pump Assembly

The pump assembly looks something like a large medicine dropper. A tube, made of hard plastic, has an air pump inside with a lid on the top. Beneath the lid, the tube is threaded to allow it to fit into a threaded center opening of the tank. A rubber seal at the top of the tube threads compresses and maintains an airtight seal. The lid fits over the edges of the bowl-shaped top. At the end of the exterior of the tube is a circular piece of flexible rubber that fits tightly over the end opposite the handle. It seals the tube to prevent liquid from entering the pump assembly and maintains pressure in the tank.

The Hose Assembly

The hose assembly includes a hose outlet, a hose, a shut-off assembly, and a nozzle. The hose outlet is molded as a part of the tank. The flexible plastic hose fits over the opening and is secured to the tank with a metal clamp. Typically, the hose is approximately three to four feet long. Attached to the end of the hose with a metal hose clamp is a shut-off assembly. It operates with a handle lever that, when compressed, allows a flow of liquid. A hard plastic wand fits into the end of the shut-off assembly. The wand has a slight bend four or five inches from its end to allow a user to direct the flow of liquid. At the end of the wand is a threaded opening where a nozzle can be screwed onto the wand. Different nozzles can be attached for different spray patterns:

- a foaming nozzle for spot spraying, edging, and killing brush

- a fan nozzle for preparing large areas for planting
- an adjustable cone nozzle for spot spraying, edging, killing brush, and preparing beds for planting

An Overview of How the Sprayer Operates

CAUTION Before you operate any sprayer, read the operator's manual that comes with the sprayer. Operating a sprayer is more complex than it appears.

Following is a general overview of the operating procedure.

Sprayer in operation

Use the manufacturer's directions to prepare the tank mix. Check the level of the tank mix to be sure that it does not exceed "Maximum Fill." Secure the pump assembly to the tank by turning the handle to the right until the assembly is seated. Release the handle by grasping it firmly and turning it $\frac{1}{4}$ turn counterclockwise and pulling it up from the tank. Then move the handle up and down until the handle responds stiffly to hand pressure. Push the handle against the lid and turn it $\frac{1}{4}$ turn to the right to lock it. You can then carry the tank by the handle.

The tank now has sufficient air pressure to operate the sprayer. Hold the tank in one hand and the wand on the hose assembly in the other hand. Depress the shut-off lever and spray as desired.

Describing a Specific Model of a Mechanism

In contrast to the description of a representative model, the description of a specific model emphasizes *particular* characteristics, aspects, qualities, or features. The description of a specific model focuses on what sets this model apart from other models.

Description of a specific model of a mechanism is used in bid specifications, advertisements, or any description for an audience that wants to know particular features of a specific model of a mechanism. Descriptions are found in general merchandise catalogs such as J.C. Penney, Sears, or Speigel. They are also found in specialized catalogs offering computer hardware and software, automotive parts and accessories, electronic equipment and components, parts for appliances—the list goes on and on.

Advertisements often incorporate description to highlight a particular product or model. The following description advertises a Nikon Coolpix 700 digital camera.

▲ Nikon Coolpix 700

Nikon Coolpix 700

The Nikon Coolpix 700 digital camera captures 2.11 megapixel high-definition images 1600 x 1200ppi (UXGA) resolution with the 1/2" CCD for photographic image quality. This lightweight (9.5 oz) point-and-shoot camera features fully automatic focus, flash, 256 element matrix metering, and white balance for perfect pictures every time. With the 8MB removable CompactFlash™ memory card, the Coolpix 700 enables you to store up to 128 of your favorite images! Select one of the two shooting resolutions for a variety of image sizes, which you can view through the 1.8" LCD display. Plus, you can save your pictures in uncompressed TIFF or 3 JPEG modes for different storage options.

Reprinted by permission of Nikon

A final example of a description of a specific mechanism, from the World Wide Web, is a description of Toro Wheel Horse 300 Series Tractors. Clearly described features, specifications, and options provide a potential customer with details with which to make a decision about purchasing the tractor.

Planning and Organizing a Description of a Specific Model

Planning a description of a specific mechanism requires you to make decisions about audience, purpose, subject matter, design, and time. A Project Plan Sheet for a Description is shown on page 261. This plan sheet, adapted for a description of a specific model of a mechanism and completed by a student for a writing assignment, is shown on pages 243–245. You can adapt the

Back Forward Stop Refresh Home Search Mail Favorites Larger Smaller

Address

Live Home Page Apple Computer Apple Support Apple Store MSN

Toro® Wheel Horse® Classics – modern tractors with a generation of proven performance.

FEATURES

KeyChoice™ Reverse Operation System To improve the safety of our products without diminishing their performance, we have developed a new safety system for all of our tractors. The KeyChoice™ system allows the operator to mow or operate attachments while in reverse if he or she feels that it is safe to do so (i.e. there is absolute certainty that no children will be in the area).
Dial-A-Height Change cutting height quickly and easily with our exclusive Dial-A-Height control.
Engine Toro 300 Series Tractors feature 12.5 or 14 HP Kohler® Command engines.
Front Axle Solid cast-iron front axles out-perform and out-last stamped steel and do it under a range of working conditions other tractors can't even approach.
Mowing Decks Toro 300 Series Tractors offer a choice of four rugged high performance mowing decks. Both models accept Recycler® mulching kits.
Tires Wheel Horse 300 Series tractors ride on wide, large diameter turf tires designed to reduce tracking especially in turns.

Toro Wheel Horse 300 Series Tractors Specifications		
	Model 312–8	**Model 314–8**
ENGINE		
Make	Kohler® Command	
Horsepower	12.5	14
Construction	Cast-iron sleeve, single cylinder OHV	
Certification	EPA and CARB certified	
TRANSAXLE		
Make	Uni-Drive®	
Type	8-speed	
Ground speed fwd.	.5–5.5 mph	
Ground speed rev.	.7–2.6 mph	
EQUIPMENT		
Hourmeter	Yes	
Gauges	N/A	Yes
Headlights	Sealed Beam Headlights	
Battery	210 CCA	
Steering Wheel	15" Comfort Grip	
Seat	High	
Tires	15 x 6.00 front 23 x 8.50 rear	16 x 6.50 front 23 x 9.50 rear
Hubcaps	Optional	
Attachment Lift System	Manual	

Used by permission from The Toro Company

Project Plan Sheet for a Description, making changes as dictated by your audience and purpose.

The length of a description of a specific model is determined by the audience and purpose. A bid specification describing required features on an AO Series 10 Microscope would be brief. A description of a microscope in a beginning science laboratory manual would be longer and more detailed.

Below is a guide for organizing a specific mechanism description. Remember, however, that ultimately you must select and organize material, depending on your audience and purpose.

- Identify the mechanism (an overview).
  1. Give the brand name and/or model number.
  2. State the function or purpose of the mechanism, who uses it, when, why, and where, as applicable.
  3. Introduce the distinguishing features or characteristics of the mechanism.
- Describe the distinguishing features or characteristics.
  4. Identify each feature or characteristic by selecting information about the mechanism to set it apart from other similar mechanisms. Consider such features as size, shape, weight, material, overall appearance, available colors, cost, warranty.
  5. Describe each feature or characteristic in detail.
  6. Use headings and visuals, when appropriate, to clarify meaning.
- Close the description.
  7. You may end the description after the discussion of the last feature or characteristic.
  8. You may summarize the features or characteristics.
  9. You may mention other models, different colors, other prices, or different designs.

Student Example: Description of a Specific Model of a Mechanism

The following student-written description of a specific model of a mechanism was prepared as an oral presentation. The example includes a completed project plan sheet.

The description focuses on the features that make the Winsome Professional Quiet 1800 blow dryer desirable to use. Unusual aspects are emphasized. The function, physical characteristics, and parts of this specific blow dryer are very similar to those of any other blow dryer. Treatment of these aspects, therefore, is kept to a minimum.

PROJECT PLAN SHEET

FOR DESCRIPTION OF A SPECIFIC MODEL OF A MECHANISM

Audience

- Who will read the description or hear it as a presentation?

  Sales personnel; individuals who use a hair dryer often, particularly around wet clothing and wet surfaces

- How will readers use the description?

  Potential buyers will learn about a model of blow dryer that has several safety and performance features.

  Individuals who use a hair dryer frequently and around wet clothing and wet surfaces. The details included will emphasize safety features, the point of the presentation.

- How will your audience guide your communication choices?

  The audience will need simple language presented in an informal style.

Purpose

- What is the purpose of the description?

  The presentation will explain the four major safety and performance features of the blow dryer.

- What need will the description meet? What problem can it help to solve?

  Information for sales personnel. The description will supply information about a hair dryer that may be useful to individuals in specific situations: professional hairdressers who use a dryer often and around wet surfaces and materials and swimmers and scuba divers who need a model of hair dryer that is safe to use often and around water. Also, the fact that the dryer is quiet may be of importance if the dryer is used around other individuals, as in a hair salon.

Subject

- What is the description's subject matter?

  Description of the four major safety and performance features of the blow dryer

- How technical should the discussion of the subject matter be?

  Since the description is intended for a lay audience, it should be nontechnical.

- Do you have sufficient information to complete the project? If not, what sources or people can help you to locate additional information?

  No. I will need to acquire further information. I will look at the "Use and Care" pamphlet that came with the hair dryer and I will talk with Shunya Fenton, a Winsome Products distributor.

- What title can clearly identify the description's subject and purpose?

  Description of the Winsome Professional Quiet 1800 Blow Dryer

Author

- Will the description be a collaborative or an individual effort?

  Individual effort

- How can the developer(s) evaluate the success of the finished description?

  Maybe I can ask Shunya Fenton, a Winsome Products distributor, to look over my finished draft.

Project Design and Specifications

- Are there models for organization or format for the description?

  Yes. I can listen to speeches and review other descriptions of mechanisms.

- In what medium will the finished description be presented?

  Oral presentation

- Are there special features the finished description should have?

  No.

- Will the description require graphics and other visuals? If so, what kinds and for what purpose?

  The oral presentation will be supported by screens:

  Screen to show title of presentation

  Screen to show list of safety and performance features

  Screen to show first safety and performance feature

  Screen to show second safety and performance feature

  Screen to show third safety and performance feature

  Screen to show fourth safety and performance feature

  Screen to show four switch settings

  I will also use a Winsome Professional Quiet 1800 Blow Dryer to show features.

- What information design features can help the description's audience?

  Screen content should be keyed in a clearly readable font and in a large enough point size to be easily read. Individual screen should not be cluttered.

Date Due

- What is the final deadline for the completed description?

  March 1

- How long will the description take to plan, research, draft, revise, and complete?

  Two weeks total. Planning and research will take approximately one week. I also need to allow one week for my final review and to make any changes and final revisions. I need to allow time to prepare visuals and to practice delivering the talk.

- What is the timeline for different stages of the description?

  I'll plan an outline, locate sources, and take notes during the first week. I'll spend one week on drafting and preparing PowerPoint slides, revising, and rehearsing.

FINAL DRAFT

A Description of the Winsome Professional Quiet 1800 Blow Dryer

Screen shows title. Title clearly reflects the content of the presentation. Identification of mechanism

[Marginal notes have been added to show the organizational plan and suggest screens for projection. Visual aids are shown in bold.]

Emphasis on safety and performance features

Designed for use by hair care professionals, the Winsome Professional Quiet 1800 is an especially safe, handheld blow dryer. For home use it may seem a bit expensive at $59.95, but I opted to purchase it rather than one of the $9 to $15 models because of the safety and performance features.

Hold up a Winsome Professional Quiet 1800 blow dryer. Personal requirements for a hair dryer

I am an avid swimmer and a scuba diver. As you can see, I have long hair. Like professionals in hair care salons, I use a hair dryer quite often—sometimes three or four times a day. I need a hair dryer with variable settings. If I'm off to class, I need a quick, thorough dry—HIGH HEAT. For personal comfort at home, I like the LOW COOL or LOW HEAT setting, depending on outside temperatures when I come in from swimming or diving.

Control sentence for the description

Note the four distinguishing safety and performance features of the Winsome 1800 blow dryer.

Screen shows list of four safety and performance features.

- Built-in GFCI (ground fault circuit interrupter) for safe use near water
- Two built-in safety devices to prevent overheating
- 1800-watt extra quiet motor with dual ventilation
- Heavy duty switches (two speeds, four heat settings, two cool settings)

Screen shows first safety and performance feature. Hold up dryer and point to the feature. Description of first feature

The most outstanding safety feature of the Winsome Professional Quiet 1800 dryer is the GFCI. GFCI means ground fault circuit interrupter. It is a sensing device that immediately stops electricity flow when a small electric current leak occurs.

This GFCI feature could save your life. Why? Electricity and water are a deadly mix. If a typical hair dryer falls in water while it's plugged in, the electric shock could kill—even if the switch is "off." The Winsome 1800 has a built-in GFCI. When this hair dryer comes in contact with water, the GFCI automatically shuts off the dryer. This feature runs up the price of the dryer, but I'm willing to pay the extra dollars for the protection.

Screen shows second safety and performance feature. Hold up dryer and point to the feature.

The second distinguishing safety feature of the Winsome 1800 is the two built-in safety devices to prevent overheating. The wiring in the dryer contains two electronic chips for overheating protection. One electronic chip is activated if there is a surge in electric current. Such a surge can occur if the electricity supplier is experiencing difficulty. For instance, the lights may blink on and off, suddenly become dim, or go off for a couple of minutes and then come back on. In such cases, the power surge chip in the hair dryer prevents a small internal "explosion" that could overheat the dryer and cause it to burn out. When a hair dryer burns out, typically there is no smoke or unusual odor—the hair dryer just doesn't work anymore.

Description of second feature, details about two electronic chips as protectors against power surges and overheating

Another electronic chip in the wiring of the hair dryer activates when overheating could occur. If the High Heat (HI H) setting runs for more than sixty seconds, the dryer automatically reverts to Low Heat (Lo H). After another sixty seconds, the dryer cycles to OFF. Let's say I'm trying to quickly dry my hair and the phone rings. I go to the bedside phone and forget to turn off the hair dryer. Ten minutes later, I'm dressed and leave for class. When I return two hours later, I remember that I left the hair dryer on. Thank goodness it automatically switched itself off after two minutes.

Example to illustrate the value of the power source cut-off feature

Screen shows third safety and performance feature. Hold up dryer and point to the feature.
Description of third feature

The third safety feature of the Winsome 1800 is the 1800-watt extra quiet, heavy-duty motor with dual ventilation. An 1800-watt hair dryer puts out a lot of wattage. That's the same as holding eighteen 100-watt bulbs a few inches from your head. Quick, intense heat! That quick heat could burn out an ordinary hair dryer. The Winsome 1800, however, with its unusually rugged construction, has a high power fan for maximum velocity.

In addition, the dryer has dual ventilation, that is, air vents on both sides of the dryer. The dual ventilation system is needed to accommodate the high-power fan and the heat created by 1800 watts of power.

Screen shows fourth safety and performance feature.
Description of fourth feature

The last of the four safety and performance features I would like to emphasize is the heavy-duty switches. The dryer has two speeds, four heat settings, and two cool settings.

The next screen gives a close-up view of the four switches.

Screen shows four switch settings. Hold up dryer and point to each setting.

The Four Switches

Switch 1	Warm	Switch 3	High
	Hot	Switch 4	On
	Cool		Off
Switch 2	Low		

The fact that the switches are heavy duty helps to ensure many years of trouble-free use in changing rapidly from, say, High Hot to Low Warm or Low Cool.

In the "Use and Care" pamphlet, the manufacturer recommends that the dryer be switched to Low Cool for a few seconds before switching the dryer to Off. This permits the dryer motor to cool at a slower rate, prolonging the life of the dryer.

Screen lists the four safety and performance features.
Summary of the safety and performance features

In summary, the characteristics that distinguish the Winsome professional Quiet 1800 blow dryer are its safety and performance features:

- GFCI
- Overheating prevention
- Motor with dual ventilation
- Heavy-duty switches

In closing, let me remind you that most other hair dryers do not have these safety and performance features, particularly the ground fault circuit interrupter (GFCI). Be careful about using a typical hair dryer around water.

Closing reemphasizes speaker's need for safety and performance features

As I mentioned at the beginning, I use a hair dryer a lot—sometimes several times a day. The extra cost of this model is a good investment. For me, the safety and performance features are a must.

GENERAL PRINCIPLES for Description

- Audience and purpose affect all decisions about description. The audience and the purpose must be clearly defined. These two basic considerations determine the extent and the kind of details included and the manner in which they are given.
- Description may be the main purpose of a communication project or it may be part of a longer project.
- Accurate and precise terminology is essential. Dictionaries, encyclopedias, textbooks, knowledgeable people, instruction manuals, and merchandise catalogs can supply appropriate terminology.
- Graphics and other visuals can effectively enhance description. Typical visuals include pictorial illustrations, such as photographs, drawings, and diagrams.
- A description of a representative model gives an overall view of what the mechanism looks like, what it can do or can be used for, and what its components are.
- The three frames of reference in a description of a representative model are function, physical characteristics, and parts. Although these three frames of reference are closely related, ordinarily they should be presented separately and in logical order.
- A specific model description emphasizes particular characteristics or aspects of a mechanism identifiable by brand name, model number, and so on.

CHAPTER SUMMARY

DESCRIPTION

Description, whether written or oral, creates a mental picture. Through visual language, a communicator can describe objects, places, mechanisms, or situations to meet the needs of an audience.

The audience and the purpose affect decisions about everything related to description including content, language, complexity, organization, type of presentation, length, format, page layout, and information design. Whether it focuses on a representative model or a specific model, effective description relies on accuracy, completeness of details, and often on graphics and other visuals to meet the audience's needs. The communicator bears tremendous responsibility for the accuracy, completeness, and usability of a description. Omitting needed information or misleading the audience through unclear, inaccurate, or carelessly arranged information can have serious consequences—a lawsuit, financial loss, physical injury, even death.

Description is used frequently to explain mechanisms:

- how they look
- what they do
- what they are made of
- what their components are
- how the components interact to allow the mechanisms to function

A description of a representative model of a mechanism would supply such general information.

Description of a specific model focuses on specific features or characteristics of a particular model or brand.

ACTIVITIES

INDIVIDUAL AND COLLABORATIVE ACTIVITIES

10.1 Working individually or in groups of three or four, sort the sentences below into three general groups under the headings: Function, Physical Characteristics, and Parts. Write the letters of the sentences under the appropriate headings.

a. The jaws are made of cast iron and have removable faces of hardened tool steel.
b. The machinist uses a small stationary holding device called the machinist's bench vise to grip the work securely when performing bench operations.
c. For a firmer grip on heavy work, serrated faces are usually inserted.
d. In addition to the typical bench vise described here, there are many other varieties and sizes.
e. It is essential for holding work pieces when filing, sawing, and clipping.
f. A vise consists of a fixed jaw, a movable jaw, a screw, a nut fastened in the fixed jaw, and the handle by which the screw is turned to position the movable jaw.
g. To protect soft metal or finished surfaces from dents and scratches, false lining jaws are often set over the regular jaws.
h. This holding device is about the size of a small grinding wheel and is fastened to the workbench in a similar manner.
i. These lining jaws can be made from paper, leather, wood, brass, copper, or lead.
j. A smooth face is inserted to prevent marring the surface of certain work pieces.

After sorting the sentences into three groups, arrange the sentences in a logical order to form a paragraph. Start by placing sentence b as the first sentence of the paragraph, sentence f as the fourth sentence, and sentence d as the tenth sentence. Arrange the other sentences to form a paragraph. You may find writing out the paragraph helpful.

When you have completed the paragraph, answer these questions individually, in a small group, or as a class.

1. What clues in individual sentences helped you decide how to order the sentences?
2. If you worked in groups, did you agree on the order? Why? What persuaded the group to accept one order of sentences over another?

10.2 Read the following short description on Broccoli Raab from the Co-op Food Stores, a publication of the Hanover Consumer Cooperative Society, Inc., Hanover and Lebanon, New Hampshire. Working alone or in groups of three or four, identify the descriptive techniques used in the selection and cite examples.

> Broccoli raab—also called broccoli rabe, *cima di rape*, or *rapini*—is the tender shoot of the wintered-over turnip known as *Brassica rapa* or *B. campestris.* With its little broccoli-shaped florets, the vegetable looks like a bunch of turnip greens on their way to becoming broccoli.
>
> Both the leaves and florets are cooked; the stalks may be cut away if tough or left on if young and tender. Broccoli raab has a slightly bitter, peppery, pungent taste and is a favorite of Italians who saute, steam, or braise it. *Orecchiette con Broccoli di Rabe* is a classic pasta dish in several regions of Italy.
>
> As a member of the Brassica family, broccoli raab is related to broccoli, cauliflower, cabbage, kale, and turnips. When purchasing, look for small, tender stems and crisp deep green leaves without a hint of yellow. Florets may or may not be present, and most should be closed. Raab tastes best when it is six to eight inches long. Older, longer shoots will be thick and tough at the base.
>
> Broccoli raab is low in fat and calories, and high in fiber, potassium, and calcium as well as vitamins A and C.

10.3. Bring to class two examples of technical description. Working alone or in groups of three or four, evaluate the descriptions. Consider the following questions.

a. Who is the intended audience?
b. Are the details adequate for the audiences? Are they appropriate?
c. Is the information design effective? Is it easy to follow?
d. Are graphics or other visuals included? Are they useful? Explain.
e. Is the writing subjective or objective? Explain.

Choose one effective example of technical description from each group. Show it to the class and explain the characteristics that make it effective.

10.4 Choose a mechanism that you have used and know well. Design an advertisement for the mechanism to appear in an appropriate publication or a sales brochure for a lay audience. Adapt the Project Plan

Sheet on page 261 to help you plan the description and adapt the guide for organizing a description on page 242.

10.5 Identify a job in which you are interested. Find out what is required to carry out the job. Look at sources such as the *Occupational Outlook Handbook,* the *Encyclopedia of Careers,* or the Internet, and interview a person who is employed in a similar job. Then write a description for an entry-level job.

10.6. Divide into teams of three or four. As a team, identify a product or a service that would improve the classroom, computer laboratory, office, drama group, or some similar place or group. Then assign each team member an aspect of the product or service to research and identify characteristics appropriate for the desired use. Once all data has been gathered, write a bid specification for the product or service.

For Activities 10.7–10.8

a. Review the pages on planning and organizing a description of a representative model of a mechanism, pages 233–234.
b. Adapt and complete the Project Plan Sheet for Description, page 261.
c. Following the guide for organizing a description of a representative model of a mechanism, pages 233–234, plan the content and organization.
d. Using decisions from the Project Plan Sheet and the content and organization guide for description of a representative model of a mechanism, write a preliminary draft.
e. Revise the draft, incorporating design decisions.
f. Prepare a final draft.

10.7. Choose a mechanism from the following list or another one from your own experience. Plan a description of the mechanism as a representative model for a lay audience who must use the mechanism.

a. A disk drive, monitor, keyboard, or other computer component
b. A drill press, power saw, sander, other similar tool
c. A lie detector, pair of handcuffs, service revolver, or other law enforcement tool
d. A vaccinating needle, autoclave, stethoscope, incubator, or other health-related mechanism
e. A compact disc player, modem, bag phone, or other electronic mechanism
f. A flashlight, wristwatch, camera, calculator, stapler, or other common mechanism
g. A fax machine, telephone answering machine, photocopier, or other office machine
h. A toaster, hair dryer, blender, coffee maker, or other small appliance

10.8. Select a mechanism that you know well, one you may have used or know well from other experience. Describe the mechanism for the following audiences and purposes. As a class, analyze each audience and purpose; discuss the details that each audience would need in order to carry out the responsibility described. Use the Project Plan Sheet on page 261 as a guide in analyzing and planning each description. Individually, write the three descriptions.
   a. Describe the mechanism so that an employee can locate it and put an inventory number on it.
   b. Describe the mechanism for a person who is unfamiliar with it but must use it.
   c. Describe the mechanism for a technician who must repair or replace a broken part.

For Activity 10.9

a. Review the pages on planning and organizing a description of a specific model of a mechanism, pages 240, 242.
b. Adapt and complete a Project Plan Sheet for Description, page 261.
c. Following the guide for organizing a description of a specific model of a mechanism, page 242, plan the content and organization.
d. Using decisions from the Project Plan Sheet and the content and organization guide for description of a specific model of a mechanism, write a preliminary draft.
e. Revise the draft, incorporating design decisions.
f. Prepare a final draft.

10.9. Choose a mechanism from the list in Exercise 10.7. Identify a specific brand or model of the mechanism. Plan a description of the mechanism as a specific model for a lay audience that is considering purchasing the mechanism.

10.10. As directed by your instructor, adapt one of the mechanism descriptions you prepared in the activities above for oral presentations. Ask each member of the audience to evaluate your speech by filling in an Evaluation of Oral Presentations (see page 508) or one like it.

READING

10.11 The following excerpt describes Anna Pigeon's entry into a section of Lechuguilla, a cave she has entered with two other national park rangers to rescue a friend, Frieda. The excerpt describes the Wormhole.

Excerpt from *Blind Descent*

Nevada Barr

"The Wormhole," Iverson said. Clicking on his headlamp, he ran the beam along the traverse to where it was anchored on the far side. Below the jug handle securing the line was an irregularity in the stone about the size and shape of an inverted Chianti bottle. The opening was flush with the wall: no ledge, lip, or handholds; no nooks or crannies to brace boots in.

"You're kidding," Anna said hopefully.

"I'll admit it looks tricky," Iverson said. "You want to go first, Holden?"

"And rob you of the glory? No indeed."

Iverson peeled off his pack and began strapping on ascenders.

The one time Anna had been on a tyrolean traverse it had been a simple horizontal move over a river valley under kind blues skies with the music of frogs to keep her company. They'd not used ascenders, just strung the traverse line through a trolley on the web gear and scuttled across with much the same movements used when shinnying along a rope. Because of the steep tilt of this traverse—close to forty-five degrees—ascenders were needed.

Mechanical ascenders were a relatively simple invention that had revolutionized climbing. A one-way locking cam device about the size of a pack of cigarettes and shaped like a tetrahedron was strapped to the right boot above the instep. An identical device was attached to the left foot but on a tether that, when pulled out to full length, reached the climber's knee. This ascender was tied to a thin bungee cord and hooked over the shoulder. Once this awkward arrangement was complete, the rope to be climbed was hooked through both ascenders and a roller on the climber's chest harness. Thus married to the rope, it was a not-so-simple matter of walking, as up an invisible ladder. Raise the right foot; up comes the Gibbs ascender. Put weight on the right foot: cam locks down on the line. The foot is firm in its stirrup, and the body is propelled upward. This movement tightens the bungee, which in turn pulls the second ascender up along the rope. When the left foot steps down, the cam locks and another "stair" is provided.

Anna had used Gibbs ascenders enough that she was proficient, but she always enjoyed watching a master. The ascenders, so arranged, were called a rope-walker system. On the right climber that appeared literally true. Anna had seen men walk as efficiently up two hundred feet of rope as if they had walked up carpeted stairs in their living room.

"Croll," Iverson said as he rigged a third ascender into his seat harness. "Ever used one?"

Anna shook her head.

"Like falling off a log," he assured her.

Rotten analogy, Anna thought, but she didn't say anything.

"Get the packs ready," Iverson said to Holden. "Once I get settled, I'll bring them across with a haul line."

"We'll use the haul line, too," Holden told Anna. "Ever so much more civilized."

As Iverson rigged himself to the traverse, Anna watched with a keen interest. One lesson, then the test. It crossed her mind how much better a student she would have been if in school the options had been learn or die.

Crouching on a thumb of rock as big around as a plate, eighteen inches of it thrust over the chasm, Oscar attached his safety, clipped his Croll into the rope, wrestled his two foot ascenders onto the line, then pushed till his torso and buttocks hung like a side of beef a hundred twenty feet above God knew what.

Anna kept her headlamp trained where he would find it useful, kept the light steady and out of his eyes. It was all she could do. There was no room for a second pair of hands to help him.

Holden was occupied with the business of tying the packs into trouble-free bundles that could easily be hauled across. Along with personal gear were medical supplies requested by Dr. McCarty, among them oxygen. In the case of a head wound it might be the only thing that could keep Frieda's brain tissue from permanent damage.

Iverson finished and snaked an elbow over the line so he could hold his head up. "Check my gear?" he said to Holden.

"Right-o."

Having been checked out by a fresh pair of eyes, Oscar began rope-walking over the rift. Suspended at four points along the line, his body was nearly horizontal, spine toward the center of the earth. As always in a traverse, there was an element of sag. It made the operation of the ascenders inefficient, and progress was measured in inches. As the line began to angle, rising steeply toward the Wormhole, the slack was taken up. After a few strangled kicks to get the cams to lock down, Oscar climbed like a pro, covering the last thirty feet in as many seconds. The rig was a work of art; not so much slack that he was head-down at the bottom, but enough so that he was standing upright when he reached the hole.

In the scattering light, Anna could see his skinny arms weaving some sort of magic near the top of the traverse, where his shoulders blocked her view.

"That's the tricky part alluded to earlier," Holden said. Anna was startled at the nearness of his voice. She could feel his breath against her cheek. In their absorption, they'd knelt shoulder to shoulder on the edge of their limestone block like a couple of White Rock fairies in a troglydyte nightmare.

"What's he doing?" The words gusted out, and Anna realized she'd been holding her breath.

"There's no place to be, and the Wormhole's so tight it's sort of pay-as-you-go. He's changing ropes. Hooking onto the rebelay. He'll let himself loose a little bit at a time as he gets that part of himself into the hole. The hard bit is feeling for the Gibbs down at your feet. Sort of like tying your shoelaces when you're halfway down a python's throat."

Iverson squirmed a moment more. Anna and Holden so entranced they forgot to talk. Then it was as if the rock swallowed his head. It was gone, and only the body remained, twitching with remembered life. The shoulders were next, melting seemlessly into the cliff, the pathetic stick legs kicking in short convulsive movements. Hips vanished, and Anna and Holden's lights played over a pair of size-thirteen boots flipping feebly. The image was ludicrous, but Anna didn't feel like laughing. Hands reappeared, only the hands, gloved and bulky. With no assistance from a corporeal body, they fussed and fluttered and danced like a cartoon drawn by an artist on acid.

"Removing the foot ascenders," Holden said. "Last link."

Abruptly, the hands were sucked back into the limestone. The boots followed. All that remained to indicate that any life-form had ever existed was the gentle swaying of the rope.

"He's good," Holden said admiringly. "The man is good. He could thread himself through all four stomachs of a cow and never even give her the hiccups."

Anna eased back from the edge of the rock and rearranged legs grown stiff from too long without movement. "Me next," she said, and was pleased at how normal her voice sounded.

"Not yet. Oscar will holler when he's ready. There's no room to turn around in the Wormhole. He's got to crawl on up a little ways. There's a chimney. He'll go up, spin himself about, then come back down so he can get the gear and pull you in. One of you guys will do the same for me.

"The creepy-crawly part of the hole's fifty feet or so," Holden answered Anna's unasked question. "Then it opens right up. Big stuff like we've been doing."

Anna nodded. She'd not wanted to ask, but the Wormhole was eating away at her hard-won control. Fifty feet; she measured it out in her mind. Frieda's backyard in Mesa Verde was fenced to keep her dog, Taco, from wandering. The posts were eight feet apart. Anna had helped replace four of them at the beginning of the summer. Fifty feet: six postholes. She could swim that far underwater. Sprint that far in high heels and panty hose. Fifty feet. Nothing to it.

"All set." The words reverberated across the chasm, and Anna and Tillman trained their lights on the orifice that had swallowed Oscar. A disembodied hand waved to them from the solid fortress that was their destination.

"Now you next," Holden said. "Have fun." He smiled and she smiled back, glad to know him, glad to have him with her. Holden Tillman's smile was better than a bottle of Xanax.

Anna was always careful with climbing gear. This time her attention to detail was obsessive. The contortionist movements required to hook ascenders with half her body bobbing over a crevice she couldn't see the bottom of required that each movement he thought through before it was attempted. She was glad the ranger from Rocky Mountains who had taught her to climb had insisted she practice everything one-handed, and by touch, not sight. Tillman trained his lamp on her hands and watched.

"Looks good," he said when she had finished. "See you on the other side."

Anna nodded and pulled herself hand-over-hand along the rope till she could begin to kick in with her rope-walkers. For a brief eternity she floundered like a fish in a net. Because the Gibbs wouldn't lock regardless of how she angled her feet in an attempt to get the cams to catch, she was muscling her way along, using the strength of her arms and shoulders. Bad form, and something she couldn't maintain for any length of time. Exertion made the line sway. Coupled with the wild dancing of shadows every time she moved her head, it was dizzying. Adrenaline, already high, rose to poisonous levels.

Closing her eyes, she forced herself to relax, let the rope and metal take the weight. After a steadying breath, she began again her crippled walk, this time with more success. The rope inclined, tension increased, the cams locked and unlocked fluidly. Much to her amazement, she found she was enjoying herself. Deep in the bedrock of the Southwest, and she was probably as high as she'd ever climbed. She laughed aloud and hoped Holden wouldn't mistake it for hysteria.

The ascent was over too soon. Head against the wall, feet over the rift, she dead-ended. By craning her neck she could look back far enough to see the opening she was supposed to get into. It was nothing short of miraculous that Iverson had managed it without aid. She felt utterly helpless.

"Oscar?"

"Right here." An ungloved hand, looking sublimely human, came out several inches above her face and the fingers waggled cheerfully. "Put your arms over your head like you're diving."

It took a minor act of will, but Anna got both hands off the traverse line and did as she was instructed. Fingers locked around her wrists, and she was drawn into the Wormhole. Stone brushed against her left shoulder and her face was no more than two inches from the rock above. The dragging pulled her helmet over her eyes. Newly blind, her first sense that the environment had changed was the warm sweet smell of sweat mixed liberally with cotton and dust. An unseen hand tipped her hat back so she could see. Rock had been replaced by a maroon tee shirt with pink lettering so close to her face she couldn't read it. She was less than an inch below Iverson's chest. The space was closing in. Her lungs squeezed and her throat constricted. An image of fighting like a maddened cat, clawing back out to fall into the abyss, ripped through her mind.

Now's not a good time, she warned herself. "What's next?" she asked, needing to move.

"Is your rear end in?"

"Feels like it."

"Okay. There's space to your right. You're going to have to stretch over, reach down, and let loose your foot ascenders."

Anna bent to the right at her waist. She tried to bring her knee up, but cracked it painfully against the rock. Second try and she could feel the ascender on her boot. After a moment of fumbling she pulled the quick-release pin securing the cam and plucked the line out. "Got it," she said.

"Left is harder, but you're short and flexible," Iverson said encouragingly.

Anna grunted and pretzeled farther over in the slot that sandwiched her in. The left proved easier. Small blessings. She'd take what she could get. "I'm loose."

"Get your body as straight as you can," Iverson told her. She squirmed her bones into a line. Every movement was dogged by difficulty. Clothing and gear caught and dragged. Her hard hat pulled, choking then blinding her. Fear built. She began to count in her head to drown out the distant buzz of panic, a sound like a swarm of angry bees.

"Okay. Good. Grab my knees and pull yourself the rest of the way in."

Anna worked her arms back over her head and felt the warmth of Oscar's thighs. Hooking her fingers behind his knees, she dragged herself toward them.

"Watch your light!"

She stopped just before she rammed his groin with the apparatus. "Caving is not conducive to human dignity," she grumbled.

"Nope. Just no long and meaningful relationships," Oscar said, and Anna laughed to let him know she appreciated the effort. "You're home free. Take off your helmet and push it ahead. Keep going till you come out in a place that looks like it's made of cheese. Wait for us there."

Inching along like a worm, Anna oozed between Oscar's legs. The passage opened enough so she could roll over onto her stomach, then closed down again so tight she had to turn her head sideways to make any progress. The bees whined, the noise threatening sanity.

She scratched ahead with her toes. Tugged forward with fingertips and stomach muscles. Humped along like a caterpillar, ignoring the rake of stone knives down her shoulder blades. Each foot achieved was a goal to be celebrated. When she thought she couldn't take any more she was granted a reprieve. The Wormhole bored out of the wall into a large room, and she spewed gratefully to the floor.

a. Identify descriptive words or phrases that you find effective. Explain why you think they are effective.
b. How does the author use description to let the reader know Anna fears her descent into the Wormhole?
c. Is jargon used in the description? If so, is it appropriate or inappropriate? Explain your choice.
d. Should fiction writers be expected to accurately describe places, objects, mechanisms, and situations? Why or why not?
e. How does the author use simile and metaphor to describe? Cite examples from the excerpt.
f. Who is the audience for this description? Does the author consider her audience in the description? Explain your answers by referring to the excerpt.

CASE STUDY

10.12. In a class discussion, small group discussion, or a short written assignment, respond to the following case study by identifying the ways in which the case illustrates the nature of description.

Get the Picture?

Claude Snopes and his fellow sales manager, Darlene Wong, are worried as they discuss the sales figures for their clerks at the Video Bargain Mart, a large store that sells high quality audio, video, and computer equipment at discount rates.

"What's going on here?" Claude complains. "I was impressed with the price and the features of that first version of the Micron's new Scanomatic Video Camera. But now there's a problem. Since they put out that new version, sales are down a whopping 45 percent since the last report."

"Well, Claude, I had a feeling you'd be wondering about that," says Darlene. "We sure can't blame our staff. Sales for all products are up 17 percent, and sales for video cameras are up 11 percent, too. Their morale is high, and they are all committed to helping customers get the best buy. I talked to a group of them at lunch yesterday, and I think I know what's wrong."

"Let's hear it."

"Micron is a new company, as you know, and it spends a great deal of effort on design and new features for each product. However, Micron is not very careful about updating sales information. The user manual for that first Scanomatic was great. We had some terrific product descriptions, too. The sales staff all had technical specifications, and Micron sent us product overviews for customers. It was easy for customers to see what they were buying and for our clerks to answer any technical questions."

"So what's the problem, Darlene?"

"Micron got carried away with improvements for the new Scanomatic. It looks complicated, it has lots more features, and somehow we never received product information for customers. Our sales clerks have the technical specifications, but they're so complicated that customers get bored and frustrated listening to clerks trying to explain."

"I don't suppose the sales staff has tried to get the information from Micron. Doesn't the company have a toll-free number and Web page?" grumbles Claude.

"Oh, sure," Darlene replies. "They have tried to locate what they need, but Micron just hasn't compiled it. Sure, they've updated the user instructions, but we need good, clear product descriptions for the customers. Our clerks are uneasy about selling an expensive camera, even one as good as the Scanomatic, if Micron isn't forthcoming about technical information. Remember our sales motto? 'If we can't help you use it, we won't sell it.'"

"I certainly do remember that motto. I wrote it. Score one for our sales staff! Maybe we should propose dropping the Micron line. We can't afford to carry products that don't sell themselves with good information and simple design. What do you think, Darlene?"

QUESTIONS FOR DISCUSSION

a. What responsibility does Micron have to its customers? Who are its customers, the stores that carry Micron products or the people who purchase and use them?
b. Does the Video Bargain Mart have a responsibility to prepare new product materials? Why or why not?
c. Is Micron's problem the complicated design of the new camera? Is it the lack of information to explain it? Is it both?
d. What kinds of information should a product description provide for sales clerks? For video camera customers? How technical should such information be?
e. In what ways does a product description reflect the quality of a company's products?
f. How can information design best help nontechnical customers who want to understand a product's features quickly and easily? How should this information be arranged on the page?
g. What do you recommend that Claude and Darlene do to substantiate the Video Bargain Mart's sales motto?

PROJECT PLAN SHEET
FOR A DESCRIPTION

Audience

- Who will read the description or hear it as a presentation?
- How will readers use the description?
- How will your audience guide your communication choices?

Purpose

- What is the purpose of the description?
- What need will the description meet? What problem can it help to solve?

Subject

- What is the description's subject matter?
- How technical should the discussion of the subject matter be?
- Do you have sufficient information to complete the project? If not, what sources or people can help you to locate additional information?
- What title can clearly identify the description's subject and purpose?

Author

- Will the description be collaborative or an individual effort?
- If the project is collaborative, what are the responsibilities of each team member?
- How can the developer(s) evaluate the success of the finished description?

Project Design and Specifications

- Are there models for organization or format for the description?
- In what medium will the finished description be presented?
- Are there special features the completed description should have?
- Will the description require graphics and other visuals? If so, what kinds and for what purpose?
- What information design features can help the description's audience?

Due Date

- What is the final deadline for the completed description?
- How long will the description take to plan, research, draft, revise, and complete?
- What is the timeline for different stages of the description?

CHAPTER 11

Definition: Explaining What Something Is

CHAPTER GOALS

This chapter:

- Defines *definition*
- Explores when to define a term
- Relates definition to audience and purpose
- Discusses the role of graphics and other visuals in definition
- Lists and illustrates typical methods for defining terms
- Demonstrates three steps for arriving at a formal definition
- Explains extended definition
- Outlines steps for organizing an extended definition

INTRODUCTION

Many of you have had the experience of hearing or reading terms you did not understand—in a medical facility, a classroom, a lawyer's office, or in a book, a letter, or some other document. As a careful speaker and writer, you should consider your audience and define any terms with which they might not be familiar. Definition of terms is required when you think that your audience may not understand a term's meaning or may not understand the meaning of the term as you have used it.

The following pages introduce you to the major considerations in writing definitions: what a definition is, when to define a term, where to place a definition, adapting definition to audience and purpose, the writer's responsibilities in defining, and incorporating graphics and other visuals in definition. The text also explores three approaches to definition: informal, formal, and extended.

WHAT IS DEFINITION?

Definition gives an understandable meaning to something new or unfamiliar. It explains what something is—an object (a Phillips screwdriver), a process (saving information to a hard drive), or a concept (Ohm's law).

Definition may be brief or extended, depending on audience and purpose. A brief definition reflects the essence or primary characteristic of a term. The essence of a refrigerator, for instance, is that it preserves food by keeping it at a constant, cold temperature.

An extended definition includes information beyond the essence or primary characteristic of a term. For example, the brief definition of refrigerator could be extended to include the information that this is a large home appliance, it may be self-defrosting, and it usually comes in combination with a freezer. A brief definition of *tornado* is "a severe storm with winds whirling at up to 400 miles per hour, occurring primarily in the central section of the United States, usually in the spring." To extend this definition, the writer could include information on how a tornado forms, what features or characteristics a tornado has, and areas where tornadoes most frequently occur.

Definition may be the main purpose of a communication; more often, however, a definition is an important part of a longer communication. A writer or speaker defines any terms the audience needs to comprehend the entire message. For example, the article on page 198, whose main purpose is to explain how a jet engine produces power, includes brief definitions of two terms, *compressor* and *augmentor.* An article on pages 287–289 defines *executor.* After an opening formal definition, the article provides information on who can be an executor, the responsibilities of an executor, and the costs incurred by an executor.

Definition helps an audience understand your intended meaning. Because many words have multiple meanings, you must often define a term to assure mutual understanding. Consider, for example, the word *pitch.* Certainly, the word *pitch* is familiar to most people. However, when used in certain contexts, the word has many possible meanings.

> A **drafter** writing about *pitch* for other drafters can be sure that her readers know that the term refers to the slope of a roof, expressed by the ratio of its height to its span.
>
> An **aeronautical technician** automatically associates *pitch* with the distance advanced by a propeller in one revolution.
>
> A **geologist** thinks of *pitch* as the dip of a stratum or vein.
>
> A **machinist** considers *pitch* the distance between corresponding points on two adjacent gear teeth or the distance between corresponding points on two adjacent threads of a screw, measured along the axis.
>
> A **musician** knows *pitch* as the quality of a tone or sound determined by the frequency or vibration of the sound waves reaching the ear.
>
> A **construction worker** sees *pitch* as a black, sticky substance formed in the distillation of such substances as coal tar, wood tar, or petroleum, and used for waterproofing, roofing, and paving.
>
> The average **lay reader** may think of *pitch* as a verb meaning "toss" or "throw," as in "*Pitch* the ball." Others may use the word in such informal expressions as "It's *pitch* dark" or "I see marketing has developed a new *pitch*."

A communicator using the word *pitch* with a specific intended meaning would likely include a definition. Suppose, for example, an architect is discussing a design for a house with a new client. When referring to the *pitch* of the roof, the architect may need to define the term *pitch* and explain how pitch relates to the design of the house, to the appearance of the roof, and to construction costs.

It is easy to see how misunderstanding occurs. Defining terms assures mutual understanding, a key factor in technical communication between you and your audience.

WHEN TO DEFINE A TERM

A good rule of thumb is to define a term whenever you believe that the audience may not understand the term in the same way that you do. The following suggestions can help determine whether a definition is needed.

1. Define a term that is unfamiliar to the audience.

   Example An electronics technician, writing a repair report for a typical customer, would likely need to explain the meaning of his use of *zener diode* or *signal-to-noise ratio*.

2. Define a term that has multiple meanings for the audience.

   Example Consider the possible meanings of words such as *economy*, *produce*, *package*, *range*, *cap*, *outlet*, and so on.

3. Define a term that has a meaning other than the one the audience associates with it.

   Example A teenager tells his grandfather that he received a CD for a gift. The teenager is referring to a compact disk, but the grandfather knows CD as designating a certificate of deposit.

4. Define a term to show your meaning in a specific presentation (sometimes referred to as stipulative definition because you "stipulate" an intended definition out of several possible ones).

   Example Form (the shape of landscape features) in a landscape design can be represented by such features as flowers, trees, and shrubs.

REMINDER

If there is any possibility that your audience may misunderstand your intended meaning, define the term.

WHERE TO PLACE A DEFINITION

Once you decide that you need to define certain terms in a document, you must then decide where to place the definition. You have three choices:

- **In the text**. Placing definitions in the text is appropriate if the reader needs the information immediately or if the type of document (letter or memorandum, for example) would not likely include a glossary. Such definitions typically appear in parentheses, set off by dashes, or as separate sentences.
- **In footnotes**. If a definition in the text interrupts the flow of thought, place the definition in a footnote at the bottom of a page. This way the definition is readily available for the reader who needs it or easily ignored by the reader who does not.
- **In a glossary**. A glossary is a list of selected terms (words or phrases) briefly defined. The glossary defines terms that are used in a document

or that the reader would find helpful in understanding it. Typically, the glossary appears at the end of a document, although a short list of selected terms may appear at the beginning. The terms in a glossary are listed in alphabetical order for ease of use. To help readers, indicate in the text which words are included in the glossary. You might place "see glossary" in parentheses after a term or place an asterisk after the term with direction to the reader that an asterisk marks words that appear in the glossary.

ADAPTING DEFINITION TO AUDIENCE AND PURPOSE

In business and industry, communicators often define terms because they realize that their audiences may not understand their meanings. For example, a dentist who tells a patient that she has gingivitis might define the problem as inflamed and swollen gums. If a physician diagnoses a patient with mitral incompetence, the patient would need to know that he has a heart valve problem that is allowing blood to leak from a lower heart chamber back into an upper heart chamber. As a communicator, you must know when to define a term and you must define it for a particular purpose that is appropriate for the audience and the situation.

Substantial amounts of writing cover technical subjects for lay audiences. Think of the appliances, tools, and equipment the average person buys. Some kind of written material accompanies each item:

- directions for assembling a Ping Pong table
- directions for the safe operation of a circular saw
- instructions on troubleshooting problems with a computer
- instructions for constructing a multilevel deck
- instructions for installing an oven unit

As more lay readers learn and apply more technical information, the need to define terms grows and the challenge of defining increases.

Consider the following definitions of the Internet. Although they are based on the same term, each definition is designed for a different audience and a different purpose or use.

EXAMPLE 1

Audience and Purpose	general definition designed for a lay audience to give a basic understanding of what the Internet is
Use	introductory overview

Definition The Internet is a vast network of thousands of telecommunications lines that connect network computers. Online services, such as AOL, allow users to retrieve information, send electronic mail, establish private links between sites, and connect sellers and customers.

EXAMPLE 2

Audience and Purpose readers who want to learn to use the Internet for library research

Use part of an article or brochure to introduce users to Internet research resources

Definition A bibliographic network is a reference tool that gives easy access to the catalogs of other libraries. As a telecommunications network, this tool is an informal linkage made up of libraries—research, government, and academic. The catalog of each library is the main data. Once users get online, they can search the catalog of any of the networked libraries. Quickly and easily users can locate any needed book, periodical, or other online document. Other network resources include online databases such as full-text periodicals, books, and government documents; statistics; business and travel information; and indexes. Software, pictures, and sounds can also be located and downloaded.

EXAMPLE 3

Audience and Purpose readers who want to connect to and understand available network services

Use part of promotional information describing the services of an Internet provider

Definition The Internet is the world's largest electronic information exchange, with three potential connection paths.

1. Computer connection to a LAN (Local Area Network) whose server is an Internet host
2. Dial access to an Internet host using SLIP (Serial Line Internet Protocol) or PPP (Point-to-Point Protocol)
3. Dial access into an online service

The LAN connection gives the user access to everything the Internet has to offer: e-mail, the World Wide Web, news, and more. The SLIP/PPP connection also gives access to everything the Internet has to offer, but at slower speeds. The online service connection gives access to the Internet services of a particular online service. Some services offer only e-mail; others offer e-mail, Web browsers, newsgroups, and more.

In the next example of definition, note the writer's consideration of audience. The definition of *relay* appeared in the mid-February 1999 issue of *Farm Journal.* The definition was included in "Demystify Electric Gadgets, Gizmos," an article explaining to farmers the language they need to understand how their equipment works, how to diagnose problems, and how to do repairs. With an understanding of the basic principle of a relay, farmers can learn how to test, diagnose, and replace a relay. The writer assumes that farmers have a basic understanding of electrical current and magnetism.

> A relay is a two-part switch that uses a small electrical current to turn a larger electrical current on and off. A low amperage current enters one side of a relay and energizes a tiny electromagnet. The resulting magnetism closes contacts to allow a high amperage electrical current to flow through the relay and power lights, motors, and other components with high current draw.
>
>

Following the definition on relays, the writer explains the use of relays in specific farm equipment.

Below are three examples of definition for lay audiences. The first example is from a U.S. government publication on the metric system. In a section titled "Questions and Answers," the first question is "What is the metric system?" The answer is a brief definition for a lay audience.

> The metric system is a decimal-based system of measurement units. Units for a given quantity, such as length or mass, are related by factors of 10. Calculations involve the simple process of moving the decimal point to the right or to the left. This modern system is called *Le systeme International d'Unites* or the International System of Units (abbreviated SI).

The second example includes definitions excerpted from student reports on health records.

- **Your medication record** is a list of medicines that have been prescribed or given to you. This form often lists any medication allergies you may have.
- **Consultation** is an opinion about your condition given by a physician other than your primary care physician. Sometimes a consultation is performed because your physician would like the advice and counsel of another physician. At other times, a consultation occurs when you request a second opinion.
- **Imaging and x-ray reports** are documents describing x-ray results, mammograms, ultrasounds, or scans. The actual films are usually maintained in the radiology or imaging department.

The final example is from a panel on safety information included in a manual for installing and operating a garage door opener. The writer simply and concisely defines the terms *danger*, *warning*, and *caution* to emphasize the reader's need for safety precautions.

SAFETY INFORMATION

OVERVIEW OF POTENTIAL HAZARDS

Overhead Doors are large, heavy objects that move with the help of springs under high tension and electric motors. Since moving objects, springs under tension, and electric motors can cause injuries, your safety and the safety of others depend on your reading the information in this manual. If you have questions or do not understand the information presented, call your nearest service representative.

In this Section and those that follow, the words **Danger, Warning,** and **Caution** are used to emphasize important safety information. The word:

- ! **DANGER** means that severe injury or death will result from failure to follow instructions.
- ! **WARNING** means that severe injury or death can result from failure to follow instructions.
- ! **CAUTION** means that property damage or injury can result from failure to follow instructions.

The word **NOTE** is used to indicate important steps to be followed or important considerations.

POTENTIAL HAZARD	*EFFECT*	*PREVENTION*
MOVING DOOR	**WARNING:** Can Cause Serious Injury or Death	Keep people clear of opening while Door is moving. **Do Not** allow children to play with the Door operator. **Do Not** operate a Door that jams or one that has a broken spring.
ELECTRICAL SHOCK	**WARNING:** Can Cause Serious Injury or Death	Turn off power before removing operator cover. When replacing cover, make sure wires are not pinched or near moving parts. Operator must be properly grounded.
HIGH SPRING TENSION	**WARNING:** Can Cause Serious Injury or Death	**Do Not** try to remove, repair or adjust springs or anything to which Door spring parts are fastened, such as wood blocks, steel brackets, cables, or other like items. Repairs and adjustments must be made by a trained service person using proper tools and instructions.

Used by permission from the Genie Company

Each of the preceding examples demonstrates an awareness of audience and purpose and an adaptation of the definition to audience needs.

THE AUTHOR'S RESPONSIBILITIES

In any writing situation, you are responsible for defining terms a reader needs for clear understanding. In most situations, this means you should give definitions that inform and avoid bias and opinions that may influence your reader. As a rule, a definition should not include your personal opinions. For example, if you define *metric system* as "a group of units used to make any kind of measurement, such as length, temperature, time, or weight," you have given an objective definition. If, however, you define *metric system* as "a crazy system of measurement that confuses everyone who tries to use it to measure length, temperature, time, or weight," you have given an opinionated, biased definition.

GRAPHICS AND OTHER VISUALS IN DEFINITION

Visuals can be helpful in defining. Photographs and drawings can show what a physical object looks like, illustrate its parts, and explain how it works. For example, consider the parts of a laser shown in the graphic below.

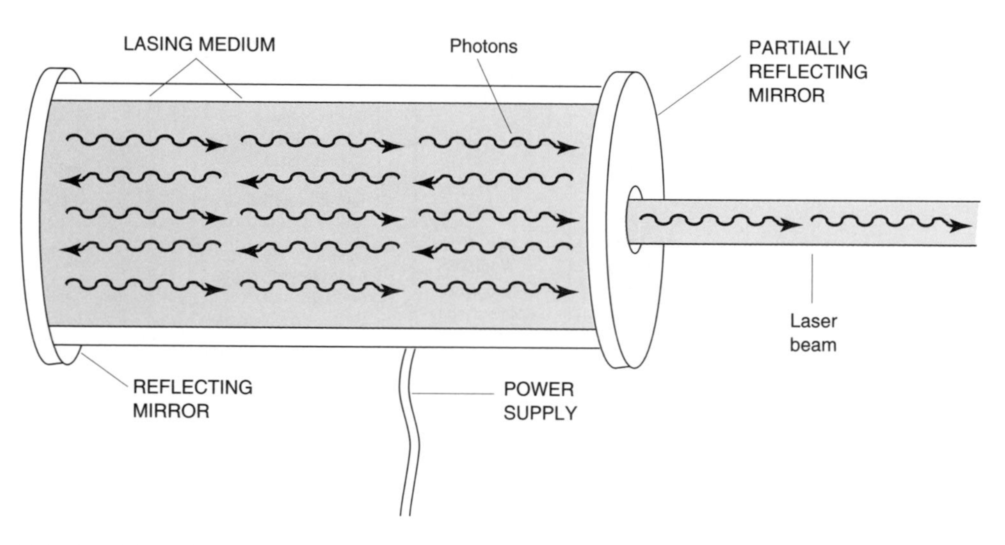

This simplified diagram of a common laser shows its three major parts: a lasing medium (solid, liquid or gas), two mirrors (one at either end), and a power supply.

The power supply provides the energy to charge the atoms of the lasing medium, so that they emit photons. The photons bounce back and forth between the mirrors and with each pass, the energy is amplified. Some of it escapes through the partially reflecting mirror at one end as a laser beam.

To define *laser*, the writer explains how a laser works; then she adds an illustration to reinforce and enhance the verbal explanation.

Graphics can also illustrate concepts. Consider, for example, the following definition of *horsepower*. Note how each of the three distinguishing characteristics (raising 33,000 pounds, distance of 1 foot, in 1 minute) is visually illustrated below.

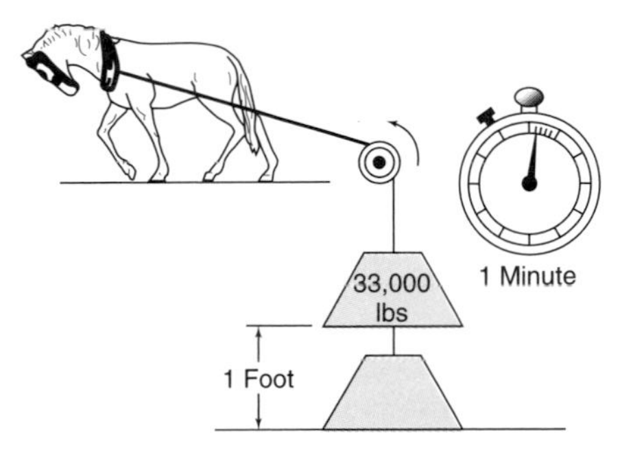

One horsepower is the rate of doing work equivalent to raising 33,000 pounds a distance of 1 foot in 1 minute.

REMINDER

Use graphics to amplify definition. The graphic typically does not take the place of a verbal definition, but enhances and supports it.

THREE APPROACHES TO DEFINITION: INFORMAL, FORMAL, AND EXTENDED DEFINITION

A definition can be informal, formal, or extended, depending on the audience's needs and the definition's purpose.

Informal Definition

An informal definition uses a familiar word or phrase to explain an unfamiliar word or phrase.

> **EXAMPLE**
> A myocardial infarction is a heart attack.

An informal definition can be incorporated into the text in a variety of ways. One way is illustrated above. Three other choices are illustrated below.

- Definition in parentheses
  The optimum (most favorable) cutting speed of cast iron is 100 feet per minute.
- Definition following a dash
  Consider, for instance, lexicographers—those who write dictionaries.
- Definition following "that is"
  The city acquired the property by eminent domain, that is, the legal right of a government to take private property for public use.

Formal Definition

A well-established, three-step method for giving a formal definition includes stating

1. The term—the word to be defined
2. The class—the group or category of similar items
3. The distinguishing characteristics—essential qualities that set the term apart from all other terms of the same class

Term

The term is the word you wish or need to define.

Class

The class identifies a group or category of similar terms. For example, a chair is a piece of furniture; a stethoscope is a medical instrument.

When determining class, be as specific as possible. Placing a screwdriver in the class of small hand tools is more specific than placing it in a class such as pieces of hardware. The more specific the class, the simpler it is to give the distinguishing characteristics, that is, the qualities that separate the term being defined from all other terms in the class. If, for instance, in a definition of *trowel*, the class is given simply as a device, you would have difficulty placing the trowel with a group of similar items. Such items as staple guns, rockets, tractors, phonographs, keys, pencil sharpeners, and washing machines are also devices. If, however, the class for *trowel* is "a handheld, flat-bladed implement," then the distinguishing characteristics "having an offset handle and used to smooth plaster and mortar" can be added to complete a formal definition that is clear and meaningful.

Distinguishing Characteristics

The distinguishing characteristics make a definition accurate and complete. They identify the essential qualities that set the term apart from all other terms of the same class.

A *chair* (term) is a piece of furniture (class). But there are other pieces of furniture that are not chairs; thus, the chair as a piece of furniture must be distinguished from all other pieces of furniture. The characteristics "that has a frame, usually made of wood or metal, forming a seat, legs, and backrest and is used for one person to sit in" differentiate the chair from other items in the class.

To write a definition, you must first have an understanding of the *essence* of the term being defined. Essential to the nature of a brick is that it is made out of baked clay. The color, methods of firing, size, shape, cost, and so on are peripheral information. Essential to the nature, the essence, of a refrigerator is that it preserves food by keeping it at a constant, cold temperature. Information that this large home appliance may be self-defrosting and usually comes in combination with a freezer has no place, ordinarily, in a formal definition. Peripheral information would be included, however, in an extended definition (see page 276).

Planning and Stating Formal Definitions

A formal definition covers in a prescribed order—term, class, distinguishing characteristics—only one meaning of a word. In planning formal definitions, it is helpful to set up three columns to identify the term, class, and distinguishing characteristics. Once you have planned the content of the formal definition, you can restate it in sentence form. See the following examples.

EXAMPLE 1

Term	Class	Distinguishing Characteristics
program	set of organized instructions	composed of words and symbols and used to direct the performance of a computer

Content restated A program is a set of organized instructions composed of words and symbols and used to direct the performance of a computer.

EXAMPLE 2

Term	Class	Distinguishing Characteristics
rivet	permanent metal fastener	shaped like a cylinder with a head on one end; when placed in position, the opposite head is formed by impact

Content restated A rivet is a permanent metal fastener that is shaped like a cylinder and has a head on one end. When placed in position, the opposite head Is formed by Impact.

EXAMPLE 3

Term	Class	Distinguishing Characteristics
photojournalist	photographer	works with newsworthy events, people, and places; works for newspapers, magazines, television; must possess skill in using cameras and in composing pictures

Content restated A photojournalist is a photographer who works with newsworthy events, people, and places. This person works for newspapers, magazines, and television and must possess skill in using cameras and in composing pictures.

EXAMPLE 4

Term	Class	Distinguishing Characteristics
scrolling	movement on a computer screen	vertically moving from one line to another, up or down, continuously through the text, or horizontally one-quarter inch at a time or continuously; allows user to move from page to page within an open document

Content restated Scrolling is a movement on a computer screen vertically from one line to another, up or down, continuously through the text; or horizontally one-quarter inch at a time or continuously. Scrolling allows the user to move from page to page within an open document.

Depending on audience and purpose, formal definition may require a paragraph or several paragraphs. The following paragraph defines *adolescence*.

> Adolescence is the period of physical and emotional transition when a person discards childish ways and prepares for the duties and responsibilities of adulthood. During this period the adolescent is neither child nor adult, although having the characteristics of both. "To grow to maturity," the meaning of the derivative Latin word *adolescere*, describes the state of the adolescent. This period of transition is customarily divided into three phases: preadolescence, or puberty, the approximately two-year period of sexual maturation; early adolescence, extending from the time of sexual maturity to the age of 16½; and late adolescence, extending from the age of 16½ to 21.

Problems to Avoid in Formal Definition

In writing a formal definition, avoid the following problems.

- Including a vague class
- Omitting the class

EXAMPLES
Scissors are things that cut. (vague class)
Scissors are used for cutting. (omitted class)

The above definitions of *scissors* equate scissors with razors, saws, drills, cookie cutters, plows, knives, sharp tongues, and everything else that cuts in any way.

REVISED DEFINITION
Scissors are a two-bladed cutting implement held in one hand. The pivoted blades are pressed against opposing edges to perform a cutting operation.

- Omitting distinguishing characteristics

EXAMPLE
Slicing is something you ought not to do in golf. (vague class; no distinguishing characteristics)

In this definition, the class is not specific, and the definition provides no distinguishing characteristics. The phrase "ought not to do in golf" is meaningless because it is negative and refers to golfers, not to slicing.

REVISED DEFINITION
Slicing is the golf stroke that causes the ball to veer to the right.

- Including a form of the word being defined in the class or distinguishing characteristics (sometimes referred to as circular definition)

EXAMPLE
Oxidization is the process of oxidizing. (term and distinguishing characteristics are not differentiated)

The key term *oxidize* is still unexplained. A reader who knows what oxidizing means has a pretty good idea of what oxidization means; however, if a reader doesn't know either word, the definition is only confusing. This kind of definition, which uses a form of the term as either the class or the distinguishing characteristics, is known as *circular* because it sends the reader in circles, never reaching the point: what the word means.

REVISED DEFINITION
Oxidization is the process of combining oxygen with a substance, thereby reducing the substance's strength.

- Using *is when* or *is where* (also called "faulty predication")

EXAMPLE
A stadium is where games are played. (no class; vague distinguishing characteristics)

This definition includes no class and no specific distinguishing characteristic because the *is where* or *is when* structure does not denote a class. Besides, only certain types of games—sports events—are typically associated with a stadium.

REVISED DEFINITION
A stadium is a large, usually unroofed building where spectator activities, primarily sports events, are held.

Extended Definition

An extended definition goes further than the essence, or primary characteristic, of a term. The central focus in an extended definition is on stating what something is—an object, a concept, or a process—by giving a full, detailed explanation suitable to the audience's needs.

An extended definition may include information such as

- synonyms
- origin of the term or item
- data concerning discovery and development
- analysis of parts
- physical description
- mention of necessary conditions, materials, or equipment
- description of function
- explanation of uses
- instructions for operating or using
- examples and illustrations
- comparisons and contrasts
- different styles, sizes, and methods
- data concerning manufacture and sale
- causes
- principles underlying a concept

The length of an extended definition depends on the audience, the purpose, and the complexity of the term you are defining. What you include will depend on the nature of the term and on the audience's needs.

Planning and Organizing an Extended Definition

As you compose an extended definition, a logical way to begin is by asking questions that lead you to decisions about key areas such as audience, purpose, the term to be defined, the author, the format, and the timeline for the project. Look at the questions posed on the Project Plan Sheet for an Extended Definition, pages 291–292. These questions are guides to preliminary decision making. You may think of additional questions you need to ask as you make decisions about the project.

Once you make decisions about the key areas listed above, gather information about the term being defined. Considering the audience and purpose, identify information you already know and information you need to find; for example, the origin of the term and its history, a graphic to interpret the meaning, terminology for accurate and clear description, and other information to meet the audience's needs. See also pages 513–525, on research strategies.

After you have gathered information, evaluate it. Is it appropriate for the audience and purpose? Do you have enough information to begin organizing and drafting the extended definition? If you are not sure the information is consistent with your plan, continue your research. When you know your information is appropriate and sufficient, begin developing your organization plan. If you find a gap in your information, do additional research to fill it. You may also decide to strengthen your plan with more information.

As you organize, think of an extended definition as two chunks, one smaller and one larger. The first chunk identifies the term, and the second chunk provides the specific information that the reader needs to understand the term fully. A formal closing is optional, although extended definitions often end with a brief comment or summarizing statement.

Below is a guide for organizing an extended definition. Remember, however, that ultimately you must select and organize material for the particular audience and purpose.

- Introduce the extended definition. (Chunk 1)
  1. State the term to be defined.
  2. Give a formal definition.
  3. Indicate the reason for giving a more detailed definition.
  4. State the kinds of additional information to be given.
- Develop the text of the extended definition. (Chunk 2)
  5. Select additional information as applicable for audience and purpose.
  6. Organize the information. Include whatever details are needed to give the audience an adequate understanding of the term.
  7. Include appropriate graphics and other visuals to enhance understanding of the term.
- Briefly comment on or summarize the definition. (optional)

Student Example: Extended Definition

Below is a student-written extended definition of *actuary*, in which readers will learn about a career they may be considering. The Project Plan Sheet, adapted for an extended definition, is included. In addition, marginal notes show how the writer followed the organizational plan discussed above.

PROJECT PLAN SHEET

FOR AN EXTENDED DEFINITION

Audience

- Who will read the definition or hear it as a presentation?

  Students and workers changing careers; career advisors

- How will readers use the definition?

  Readers will use the definition to learn about a career in order to consider it, making a serious life choice. Career advisors outside the field will use the definition, but their needs and attitudes about the subject will be similar to those of the primary audience.

- How will your audience guide your communication choices?

  Language should be simple; terms must be defined.

Purpose

- What is the purpose of the definition?

  The definition will give a complete view of a career, including its history, values, and professional requirements.

- What need will the definition meet? What problem can it help to solve?

  The definition will help informed readers make a serious career decision.

Subject

- What is the definition's subject matter?

  The term to be defined is "actuary."

- How technical should the discussion of the subject matter be?

  Since the definition is intended for readers new to actuarial roles, it should be nontechnical.

- Do you have sufficient information to complete the definition? If not, what sources or people can help you to locate additional information?

  No. Sources must be current in order to help readers make a sound career decision. I will need to read resource material and talk to someone familiar with this career.

  Many excellent and current sources will be available as software or on the Web. If needed, a local professional organization for insurance agents can provide suggestions for research or completeness of coverage.

- What title can clearly identify the definition's subject and purpose?

  Actuary: A Career Overview

Author

- Will the definition be a collaborative or an individual effort?

  Individual effort

- How can the developer(s) evaluate the success of the completed definition?

  Maybe I can ask a local actuary or career advisor to look over my finished draft.

Project Design and Specifications

- Are there models for organization or format for the definition?

  Since my definition will be part of a series of business career brochures in print, I can look at earlier brochures to find models of style, format, and organization.

- In what medium will the completed definition be presented?

  Written

- Will the definition require graphics and other visuals? If so, what kinds and for what purpose?

  Except for a timeline to support my history section, no visuals will be needed. The career ideas are conceptual, not physical.

- What information design features can help the definition's audience?

  Headings will make the information easier to follow.

Due Date

- What is the final deadline for the completed definition?

  No later than July 25.

- How long will the definition take to plan, research, draft, revise, and complete?

  Two weeks.

- What is the timeline for different stages of the definition?

  I'll plan an outline during one day. I'll spend three days locating sources and taking notes. I'll spend two days on drafting and one day on revision, allowing my reviewer two days to read the definition and contact me with suggestions. I will have two days to prepare the final draft.

FINAL DRAFT
ACTUARY: A CAREER OVERVIEW

Title reflects definition's purpose.

[Marginal notes have been added to show organization. Note: Documentation is American Psychological Association (APA) style.

Chunk 1: Identification of term and formal definitions of *actuarial science* and *actuary*

Actuarial science is a field that evolved from the insurance industry. The purpose of the field is to calculate risks, and the individual who calculates risks is an actuary. An actuary is a person "who applies a mathematical mind to solving long-term financial problems, primarily in connection with insurance, pensions, and investment" (Nation, 1998). An actuary is also a statistician who is knowledgeable about future risks, develops ways to invest, and makes monetarial decisions pertaining to rates or premiums (Choices, 1998).

Forecasting statement indicating topics for additional information

This brochure offers a brief discussion of the history of actuarial science, areas of focus, requirements to enter and advance in the field, and wages and working conditions.

A Brief History

Beginning of insurance

Chunk 2: Discussion of history, areas of focus, requirements to enter and advance, and wages and working conditions First topic developed: history

The insurance field, from which actuarial science evolved, began in early times and developed during the seventeenth to the twentieth centuries. Insurance dates back to A.D. 230 when Domitius Ulpianus made a table of annuity values. This table was used until 1814 (Warren, 1996). An early form of life insurance was the burial benefits of the Greek and Roman religious societies. These societies had no actuarial calculation, however; and a fixed amount was paid by everyone.

Early annuity system

One of the early annuity systems, the Tontine Annuity System, arranged people in groups according to age. Every person in the group paid a fixed amount of money. At the end of the year, interest was calculated and divided among the group. The last living person in the group was given the interest and the principle. The person who arranged the group held a position similar to a modern day actuary.

Blaise Pascal's theory of probability and law of large numbers; influence on actuarial science

A significant contribution to actuarial science occurred around 1645 when Blaise Pascal derived the theory of probability and the law of large numbers. In 1671, John Dewitt applied Pascal's theory of probability to annuities. In 1693, Edmund Halley then combined birth and death records with the theory of probability and created the first modern life table (*Microsoft Encarta*, 1998).

List of events occurring in England that advanced the field

On his Web site, Harold Warren lists several events in England that advanced the field of actuarial science.

- 1725 Abraham Demoivre developed the derivative theory based on DX = Constant.

- 1750–1752 Thomas Simpson introduced whole life insurance with level annual payments.
- c. 1770 Condorcet related mathematics to social goals, bringing about a need for actuaries.
- 1806 M. Duvillard constructed life tables for males and females. (Warren, 1996)

First fire and first life insurance companies in the United States

In 1752, Benjamin Franklin helped found the first mutual fire insurance company in the American Colonies, and in 1759, the first life insurance company. Both of these companies, the Philadelphia Contributorship for the Insurance of Houses from Loss by Fire and the Presbyterian Ministers' Fund, still exist (Anderson, 1994).

Influence of computers on actuarial science

The invention and perfecting of computers in the twentieth century has changed the role of the actuary, creating many new applications for actuarial science. Also, computers enable actuaries to work faster, more accurately, more easily, and more efficiently.

Number of actuaries in United States in 1996 and areas of employment

Today actuaries are not a large segment of the working population. In 1996, in the United States, "about 15,944 individuals were employed as actuaries" (U.S. Department of Labor, 1998, p. 105). Of this number, "approximately half worked in the insurance industry, some were self-employed, and some taught in colleges and universities" ("Actuaries," 1997, p. 13).

Second topic developed: areas of focus

Three areas of focus: life, casualty, and consulting. Responsibilities within each area

Other possible areas of employment

Areas of Focus

Actuaries can focus on three major areas: life, casualty, or consulting ("Actuaries," 1997). Life actuaries calculate premium rates in life and health insurance and research ways to invest company assets (Nation, 1998). Life actuaries relate unexpected events such as death, sickness, or other catastrophes to the premium rate and the money that insurance companies pay in claims. Actuaries working in casualty specialize in property and liability insurance. Consulting actuaries may work for a consulting firm advising clients or they may formulate pension and welfare plans and employee benefits ("Actuaries," 1997). Actuaries may also work for a stock exchange or for the government (Nation, 1998).

Third topic developed: requirements to enter and advance

Desirable personal characteristics

Formal education

Requirements to Enter and Advance

Individuals who want to enter and excel in the actuarial science field must meet stringent requirements. They should have an "aptitude for mathematics, an inquisitive mind, motivation, and self-discipline" ("Actuaries," 1997, p. 14). They should complete a bachelor's degree in mathematics; essential courses include elementary and advanced algebra, differential and integral calculus, descriptive and analytical statistics, principles of mathematical statistics, probability, numerical analysis, and computer science. Courses in the arts and business are also helpful (Choices, 1998).

Since qualifying as an actuary is difficult, the pool of actuaries is limited; the demand for the services of actuaries, however, is growing (Nation, 1998).

Advancement through examinations administered by professional societies

To advance in the field, a person can take a series of examinations through the Society of Actuaries, the Casualty Actuaries Society, or the American Society of Pension Actuaries. The Society of Actuaries has a series of ten examinations. Successfully completing five of the ten examinations rates membership and associate status; successfully completing all ten rates fellow status. The Casualty Actuaries Society grants associate status to an individual who passes seven out of ten examinations. The Society of Pension Actuaries gives only seven examinations; passing two examinations rates membership; passing three actuarial examinations and two advanced consulting examinations rates fellowship status ("Actuaries," 1997). Examinations through these societies are given twice a year, in May and in November. Completion of a series of examinations takes five to ten years. Students should "complete the first two or three preliminary examinations while still in college because these examinations cover college course content" (U.S. Department of Labor, 1998, p. 105).

Other advancement possibilities

Advancing in the field also requires effective job performance, competence, and leadership. Some actuaries may advance to administrative or executive, even to chief executive officer or state insurance commissioner ("Actuaries," 1997).

Third topic developed: wages and working conditions

Wages and Working Conditions

Wages

Actuaries are typically well paid because the supply of qualified actuaries is limited and the job often requires great responsibility in handling large amounts of money (Nation, 1998). According to Choices 99, in 1996, a starting actuary with a bachelor's degree made about $37,600. New college graduates who had not passed any certification examinations made a slightly lower salary. Associate actuaries made about $78,600 and fellow actuaries made approximately $93,500. The *Occupational Outlook Handbook* verifies the good salary and further indicates that actuaries typically get other benefits such as vacation, sick leave, health and life insurance, and pension plans.

Other benefits

Working conditions

Most work is done in an office setting with a 40-hour work week the norm. However, consulting actuaries are often required to travel and put in overtime. Work can be stressful at times (U.S. Department of Labor, 1998).

References

Actuaries. (1997). In *The encyclopedia of careers and vocational guidance*. Chicago: Ferguson.

Anderson, D. (1994). Insurance. In *World book encyclopedia*.

Choices 99 [Computer software]. (1998). ISM Information Management Incorporated and ISM Information System Management Data base Corporation.

Microsoft Encarta. (1998). Redmond, WA: Microsoft, 1998.

Nation, R. (1998). Take this math job and love it [Web site]. Retrieved 18 August 1998. From the World Wide Web: http://www.miracosta.cc.ca.us/info/acad/Mathematics/math.htm

United States Department of Labor. (1998). *Occupational outlook handbook*. Washington, DC: U.S. Government Printing Office.

Warren, H. (1996). Introduction to actuarial science [Web site]. Retrieved 18 August 1998. From the World Wide Web: http://web.calstatela.edu/faculty/hwarren/a503/actuari0.htm

GENERAL PRINCIPLES for Definition

- Define a term if the audience does not know the term's meaning, if its meaning is different from what the audience expects, or if it has a special meaning within a given context.
- The crucial factor in giving a definition is understanding the essence of the term being defined. The communicator must understand the nature of the item or concept to define it accurately.
- The extent to which a term should be defined depends on the subject and goal. Factors determining the extent of a definition include the complexity of the term, the general knowledge and interest of the audience, and, primarily, the purpose for the definition.
- A formal definition has three parts: the term, the class, and the distinguishing characteristics. The term is the word to be defined. The class is the group or category of similar items in which the term can be placed. The distinguishing characteristics are the essential qualities that set the term apart from other terms in the same class.
- An extended definition gives information beyond stating the essence or the primary characteristics of a term. The extended definition includes such information as origin, development, analysis, physical description, and function, depending on the nature of the term and the needs of the audience.

CHAPTER SUMMARY

DEFINITION

Definition explains the meaning of an object, a process, or a concept. Because technical writing frequently includes terms or phrases the reader may not know, definition is essential. As a writer who accepts responsibility for and is concerned about readers' understanding, you must learn when to define. Answering four typical questions helps you to determine if definition is needed: (1) Is the term unfamiliar to the audience? (2) Does the term have multiple meanings? (3) Does the term have a meaning other than that which the audience expects? (4) Is the term used in a special way? As a concerned writer, you should learn how to adapt definition to audience and purpose, how to determine length, and when and where to include graphics to support verbal definition. Is a single-word definition or a short phrase sufficient? A sentence? A paragraph? Several paragraphs? A major question is where to place the definition: in the text, in a footnote, or in a glossary. Finally, you must choose an informal, formal, or extended definition as determined by the communication situation.

ACTIVITIES

INDIVIDUAL AND COLLABORATIVE ACTIVITIES

11.1. In class discussion or in small groups, suggest appropriate distinguishing characteristics to complete the following formal definitions.
   a. An orange is a citrus fruit. . . .
   b. A speedometer is a gauge. . . .
   c. A hammer is a hand tool. . . .
   d. Networking is an electronic. . . .
   e. Handcuffs are a restraining device. . . .
   f. An ambulance is a vehicle. . . .
   g. A printer is a data output device. . . .
   h. A library is a collection. . . .

11.2. In class discussion or small groups, analyze the following definitions, noting their degree of accuracy and usefulness. What pitfall does each definition illustrate? Suggest revisions to create adequate formal definitions for a lay audience.
   a. A compass is for drawing circles.
   b. A fire lane is where you should not park.
   c. A crime is a violation of the law.
   d. Sterilization is the process of sterilizing.

11.3. Write formal definitions of five of the following terms. Identify an audience and purpose for each definition you write.

technology	database	fee
computer language	depreciation	advertising
leader	terminal	disk
fax	memorandum	coagulation
e-mail	software	thermometer
antenna	online service	disinfectant
feedback	lineup	Internet appliance (IA)
market	arrest	
thermostat	multimedia	
cellular phone	graphics	
consumer	management	
herbicide	module	
network	flowchart	

11.4. Identify your major. Then choose five terms that an audience new to your technical field needs to understand. Assume that you are producing a newsletter to give basic information about your technical field. Design a glossary to include in the newsletter giving formal definitions of the five terms.

11.5. Choose from your major a single term that can be found in each reference work named below. Note how the term is defined in each reference work. In presenting each definition, include the title of each reference work.

- Give the meaning of the term as stated in a standard desk dictionary.
- Give the meaning as stated in a technical handbook or dictionary.
- Give the meaning as stated in the *McGraw-Hill Encyclopedia of Science and Technology* and its yearbooks, or a similar encyclopedia pertinent to your major.

a. Write a paragraph explaining the differences between the definitions in the reference works and the reasons for them.
b. Using the same term, write definitions as suggested below:
   - Write an informal definition of the term.
   - Write a formal definition of the term.
   - Define the term for a fifth-grade *Weekly Reader*.
   - Define the term for an instruction manual for workers.

For Activities 11.6–11.7

a. Review the pages for planning and organizing an extended definition, pages 276–277.
b. Adapt and complete the Project Plan Sheet for an Extended Definition, pages 291–292.
c. Following the guide for organizing an extended definition, page 277, plan the content and organization.
d. Using decisions from the Project Plan Sheet and the content and organization guide for an extended definition, write a preliminary draft.
e. Revise the draft, incorporating design decisions.
f. Prepare a final draft.

11.6. Select a term from the list in Exercise 11.3 or identify a term from your major or from a personal interest. Identify an audience and purpose. Then write an extended definition. You may need to consult sources. Cite and document each source. See pages 550–556.

11.7. Working in teams of three or four, identify a service or activity available to students on your campus. Plan an extended definition of the service or activity to use in marketing and other college promotional materials. As a team, analyze and complete a project plan sheet (pages 291–292). Assign responsibilities: researching information, interviewing appropriate college officials or students, identifying and preparing graphics. As a team, decide on writing responsibilities. Prepare a completed definition as a written document and as an oral presentation, with each team member assigned a section for writing and oral presentation.

11.8. As directed by your instructor, adapt the definitions you prepared in the activities above for oral presentation. Ask each member of the audience to evaluate your speech by filling in an Evaluation of Oral Presentations (see page 508).

READING

11.9. The following definition of *executor* appeared in a publication "Being an Executor" from Metropolitan Life Insurance Company. Read the selection. In small groups or as a class, respond to the questions for discussion.

Excerpt from *About . . . Being an Executor*

As executor, your duties include inventorying, appraising and distributing assets; paying taxes; and settling debts owed by the deceased. You are legally obligated to act in the interests of the deceased, following the wishes expressed in his or her will. If all this sounds a bit overwhelming, keep in mind that you can hire professional help—for example, an attorney to help with the probate process or an accountant to file taxes. This pamphlet provides an overview of the executor's role.

WHO CAN BE AN EXECUTOR?

Any U.S. citizen over the age of 18 who hasn't been convicted of a felony can be named the executor of a will. Some people choose a lawyer, accountant or financial consultant because of his or her expertise. Others choose to appoint a spouse, adult child, relative or friend, especially if the estate is small. Generally, a family member or friend expects little or no pay for settling the estate and is anxious to get things settled quickly and smoothly.

Being an executor can be a lot of work. You have to follow up on many details and may also be called upon to help defend the terms of the will against squabbling heirs or unwarranted claims by outside parties. You also need to be able to act quickly in order to preserve the value of the estate. For example, taxes must be filed in a timely manner to avoid penalties.

Because of the many responsibilities involved, it's wise to ask the person being named in a will if he or she is willing to serve as ex-

About . . . Being an Executor was produced by Metlife's Consumer Education Center and reviewed by the Division for Public Education of the American Bar Association and the Legal Services Corporation. To request a copy of this brochure, free of charge, please call 1-800-638-5433.

ecutor. If you've been named executor in someone's will but are unwilling or unable to serve, you need to file a *declination,* a document declining your designation as executor, with the court. The contingent executor named in the will then steps in. If no contingent executor is named, the court will appoint one.

RESPONSIBILITIES OF AN EXECUTOR

As executor, your first duty is to initiate *probate,* the formal process of proving the authenticity of the deceased person's will and confirming your assignment as executor. You'll need to file an application to appear before the probate court. The application form is available from the clerk of the probate court (found in the government listings of your local telephone directory). To help you perform your duties, you may want to consult an attorney. Attorney's fees are generally chargeable to the estate as expenses of administration. Next, you need to notify all parties named as beneficiaries that you have applied to the court to process the will.

When you appear in probate court you'll need copies of the will and death certificate. You should also be prepared to pay court costs, which are chargeable to the estate. The job of the probate court is to decide the validity of the will, generally a routine affair. However, this is also the time when parties may challenge or contest the will. A person who challenges a will, or part of a will, must file an objection with the court within a specified amount of time (check your state laws). Challenges to wills can be time-consuming and costly to the estate.

Once the will is determined to be valid by the probate court, you may begin to pay taxes and other claims against the estate and distribute assets to the beneficiaries. If the will is found to be invalid, the probate court will order that creditors and taxes be paid. Then the remainder of the estate will be distributed in accordance with state law.

Your last step is to finalize the estate by filing papers with the probate court. This usually involves providing the court with copies of notices to concerned parties, tax returns and bills paid. The executor must also provide evidence of distribution of the remaining assets, such as signed receipts from the beneficiaries. When the court recognizes the completion of the probate process, you are released from further responsibility as executor.

COSTS INCURRED BY THE EXECUTOR

Generally, an estate is responsible for paying the executor a fee. This fee may be specified in the will, or it may be determined by state regulation. The executor's fee may be waived. If the executor is an attorney, the law in most states prevents him or her from collecting both an executor's fee and an attorney's fee for legal advice on the estate.

Generally, an executor is entitled to be reimbursed from the proceeds of the estate for expenses incurred in settling the estate. For example, if you live in California and are named executor of an estate in New York, the estate is liable for your commuting costs.

QUESTIONS FOR DISCUSSION

a. What judgments can you make about the intended readers?
b. What comments can you make about the writing of the extended definition to show that the writer took the intended readers into account? That the writer took word choice into account? That the writer took sentence length into account?
c. Evaluate the headings, the selection of information included, and the organization of the material.
d. Find examples of different ways to define terms illustrated in the extended definition.

CASE STUDY

11.10. In a class discussion, small group discussion, or a short written assignment, respond to the following case study by identifying the ways in which the case illustrates the nature of definition.

In Terms of Pollution

The Environmental Regulatory Group of the State Water Commission has a big problem. Their field inspectors who inspect, report, and support groundwater cleanup activities are spending more and more time answering telephone calls from confused and annoyed property owners who do not understand the reports, letters, and enforcement notices they have been receiving from the Water Commission.

Sally Tijerina, a field inspector, has organized a Plain English Committee to help revise the report forms and letters that she and the other inspectors are required to use. However, she needs to get approval from the Water Commission administration and the commission's legal department in order to start work. Commissioner Wilkinson, Commission Attorney Claudie Oman, and Charles Davis, the commission's Public Relations Officer, have agreed to meet with Sally to discuss and solve the problem.

Commissioner Wilkinson is a busy man, and he knows how busy the field inspectors are. "Didn't we have those forms developed 15 years ago to save all of you people time?" he asks. "You need to spend your time monitoring cleanups, not wasting time on phone calls and letters."

Sally points out that over 50 percent of the time field inspectors spend in their offices consists of logged phone calls, and that the legislature's hiring freeze won't allow more staff members, adding that the language of the letters and forms are the problem. "We spend hours explaining the technical difference between a UST and an LUST, why the current landowner is responsible, what the IEPA has to do with it, and why a CAP precedes a CAR."

Charles Davis chuckles. "Well, we all know how you technical types write. Besides, all of this technical stuff is too confusing. Why not let us rewrite the forms in plain, nontechnical language? We'll make them simple—and friendly, too."

"Hold it a minute," says Claudie Oman. "Those are legal documents, and they need to be technical to address the issues in environmental law they discuss. Besides, Charlie, the field inspectors didn't write those forms. Our staff attorneys did. And the issues are serious. Some of those landowners could find themselves paying hundreds of thousands of dollars if they don't comply. And the law keeps getting more complicated all the time. You simply can't leave out technical information or ideas."

Sally agrees. "We want to help people solve their pollution problems quickly, legally, and as economically as possible. But we can't assume that landowners are engineers or that they understand all of the changes in the law. Some callers even accuse us of confusing them on purpose so that they can't comply with the law."

"All right! All right!" the Commissioner agrees. "We have a communication problem here. I suppose that we need to define our legal, technical, and Commission terms for the public we work with. But how are we going to do it? Where should the terms be defined? Can we just make up a glossary to send out with letters and reports? And who should do the defining?"

QUESTIONS FOR DISCUSSION

a. Who should define terms for polluters? Who best understands the questions they ask?
b. Is the commissioner right about working up a glossary of terms to include with other materials? Will including a glossary really cut down on telephone calls and complaints?
c. What role should the legal department play in revising the forms? If environmental law changes as quickly as Claudie Oman claims, how often should the forms be updated?
d. Is Charles Davis's claim that scientists and technicians are always poor communicators fair? How would you respond to his claim?
e. Why does the Water Commission need to make its letters and reports easy to understand? How can clear and well defined terms support the commission's goals?
f. Are revised form letters and reports enough to cause polluters to comply with the law?
g. What other kinds of helpful written materials can the Water Commission publish?

PROJECT PLAN SHEET

FOR AN EXTENDED DEFINITION

Audience

- Who will read the definition or hear it as a presentation?
- How will readers use the definition?
- How will your audience guide your communication choices?

Purpose

- What is the purpose of the definition?
- What need will the definition meet? What problem can it help to solve?

Subject

- What is the definition's subject matter?
- How technical should the discussion of the subject matter be?
- Do you have sufficient information to complete the definition? If not, what sources or people can help you to locate additional information?
- What title can clearly identify the definition's subject and purpose?

Author

- Will the definition be a collaborative or an individual effort?
- If the project is collaborative, what are the responsibilities of each team member?
- How can the developer(s) evaluate the success of the completed definition?

Project Design and Specifications

- Are there models for organization or format for the definition?
- In what medium will the completed definition be presented?
- Are there special features the completed definition should have?
- Will the definition require graphics and other visuals? If so, what kinds and for what purpose?
- What information design features can best help the definition's audience?

Due Date

- What is the final deadline for the completed definition?
- How long will the definition take to plan, research, draft, revise, and complete?
- What is the timeline for different stages of the definition?

CHAPTER 12

Summaries: Getting to the Heart of the Matter

CHAPTER GOALS

This chapter:

- Defines the summary as a form of workplace communication
- Explains the importance of summaries for workplace readers
- Describes the qualities of effective summaries
- Discusses and illustrates the three major types of summaries: descriptive, informative, and executive
- Provides instruction for planning and writing summaries

INTRODUCTION

Information is published so rapidly that it is time consuming for professionals to keep up with the latest developments in their fields. How can they find time to study so much new information? Summaries can certainly help. Summaries are brief overviews of the purpose and subject matter of longer documents, such as reports, proposals, and articles. By reading summaries, busy professionals can preview new material in order to determine what and how much to read in full. For this reason, well-written summaries are extremely important for workplace readers.

Writing such summaries is a challenge. To develop a useful summary, you must be a careful and critical reader of the material you summarize, organizing information logically and writing concisely. This chapter describes the three different types of summaries—descriptive, informative, and executive—explaining the audiences for and features of each type. The chapter also shows how to plan and write effective summaries.

WHAT IS A SUMMARY?

A summary, sometimes called an abstract (other synonyms include digest, précis, and synopsis), is a concise and factual overview of a longer document. Summaries are generally published with journal articles and often accompany proposals and long reports. By reading a summary, a workplace professional can easily determine which parts of which documents require careful study.

Whether you summarize a document you have written or a report or article by another author, you'll be helping readers preview, evaluate, and learn the material your summary describes.

QUALITIES OF A USEFUL SUMMARY

If a summary is to be helpful to readers, it must be short. In general, a summary should not exceed 15 percent of the length of the document it describes. After all, a summary nearly as long as the original document would not provide a reader with a timesaving brief overview. In addition to being short, a summary must also be easy for a reader to follow and understand. For this reason, while summaries must always be concise, they should not sacrifice clarity by leaving out articles ("a," "an," or "the") or by leaving out transitional elements ("however," "also," or "even so"). Such omissions would make a summary choppy and difficult to follow.

To give readers an accurate overview, a summary generally does not provide background or include any information not discussed in the original document. When you write a summary, your purpose is to present the main idea of the original document, not to discuss your own ideas.

REMINDER

To make your summaries useful to readers, keep them short.

Objective Summaries

In most cases, a summary that appears as part of a report presents an objective overview of the report's key points, focusing on subject matter and allowing readers to draw their own conclusions about the credibility, scope, style, and usefulness of the original document. Objective summaries give readers a brief informational preview of what is to follow.

The following summary presents an objective overview of an article from *Health*, a magazine on medical and fitness issues for nontechnical readers.

> Many people who rely on over-the-counter drugs are unaware of the risks of using a drug they do not need or of using it improperly. This article shows that over-the-counter drugs are powerful and can be harmful if not taken as directed. Specifically, the article discusses acetaminophen, antacids, antihistamines, aspirin, and decongestants, explaining when each drug should be used. The article also addresses side effects, the problems that call for a particular drug, and the indications that one over-the-counter drug should not be changed for another.
>
> Tennesen, M. (1999). "Before You Play Doctor." *Health*, 13(1), 100–103. (Reprinted by permission of Cheril Bible.)

This brief overview gives readers a clear and objective idea of the main points of "Before You Play Doctor."

Evaluative Summaries

To help readers keep up with new information, print and online publications called abstract services, publish summaries of new publications in virtually every field. Libraries, companies, and individuals can subscribe to these services, which can be delivered in print, on CD-ROM, or online. For example, *Computer and Control Abstracts* provides monthly summaries of international technical information. *Psychological Abstracts* provides print summaries of all literature in psychology and related disciplines. *AACN Nursing Scan in*

Critical Care provides print and online access to summaries of multidisciplinary articles related to critical care nursing. Each summary in *AACN Nursing Scan in Critical Care* provides commentary on the article's applications to nursing.

The following summary of "Before You Play Doctor" has been revised to include brief comments which, like the summaries in *AACN Nursing Scan in Critical Care,* evaluate the article's usefulness to health care professionals.

> Many people who rely on over-the-counter drugs are unaware of the risks of using a drug they do not need or of using it improperly. This clearly written article shows that over-the-counter drugs are powerful and can be harmful if not taken as directed. Specifically, the article discusses acetaminophen, antacids, antihistamines, aspirin, and decongestants, explaining when each drug should be used. The article also addresses side effects, the problems that call for a particular drug, and the indications that one over-the-counter drug should not be changed for another. This article provides a practical and nontechnical explanation that can be useful for patients who suffer from such chronic ailments as arthritis or allergies and for patients who self-medicate.
>
> Tennesen, M. (1999). "Before You Play Doctor." *Health*, 13(1), 100–103, with emendations.

As the above example shows, brief evaluative comments can be helpful to readers who use abstracts to preview new information. The needs of your audience will determine whether it is helpful and appropriate to include brief evaluative comments in a summary.

TYPES OF SUMMARIES

Whether objective or evaluative, different types of summaries meet different needs for different kinds of readers. The three types of summary—descriptive, informative, and executive—differ in purpose, the amount of detail, and the technical level of the information they provide.

Descriptive Summaries

A descriptive summary is a brief informational overview, generally only a few sentences long. The descriptive summary typically reflects only the main idea of the original. This simplicity gives readers a brief, helpful, and sometimes nontechnical understanding of the main points of the original document.

Two examples of descriptive abstracts from *Scientific American* appear below.

> **Unmasking Black Holes**
> *Jean-Pierre Lasota*
>
> Evidence for black holes was until recently all circumstantial. Distinguishing them at a distance from other highly compact, gravitationally massive bodies such as neutron stars is inherently problematic. Now astronomers have direct proof: energy is vanishing from volumes of space without a trace.

> **Killer Kangaroos and Other Murderous Marsupials**
> *Stephen Wroe*
>
> Australian mammals weren't always as cuddly as koalas. For tens of millions of years, the continent was home to ferocious marsupial wolves and lions, a pouched tiger and muscle-bound rat-kangaroos that terrorized smaller prey.
>
> Reprinted by permission of *Scientific American.*

Descriptive summaries like the examples above appear in the table of contents of each issue of *Scientific American,* briefly identifying the subject and main idea of each article. The difficulty in writing such a descriptive summary lies in explaining a long, often complex and technical discussion in only a few accurate sentences. However, these very short descriptive summaries allow the diverse readership of *Scientific American* to quickly review the articles in each issue and to learn about each article's main point.

The following example of a descriptive summary from *Technical Communication,* a journal for professional technical communicators, is very brief.

> "Going Online: Helping Technical Communicators Help Translators" Patricia Flint, Melanie Lord Van Slyke, Doreen Starke-Meyerring, and Aimee Thompson
>
> **Summary**
>
> - Explains why technical communicators should help translators and offers tips for creating "translation-friendly" documentation
> - Describes a research and design process for creating an online tutorial on writing and designing for translation
>
> Reprinted with permission from *Technical Communication,* the journal of the Society for Technical Communication, Arlington, VA, U.S.A.

This descriptive summary appears immediately before the article's text. The first sentence of the summary describes the article's purpose, and the second sentence indicates the article's main idea. This very brief summary allows technical communicators to quickly decide whether the article provides information that can meet their needs.

The following descriptive summary was written to accompany a student research report, "Martial Arts: Teaching the Whole Person." The short objective summary allows readers to preview the report's main ideas.

> This report provides a brief history of the Asian fighting arts. It discusses the importance of integrating the psychological and spiritual aspects of martial arts with the physical. The concepts of *ki* (inner strength) and of the martial arts philosophy as a way of life are also discussed. The report explains how martial arts teachers can show students ways to integrate all aspects of the martial arts in practice.
>
> Reprinted by permission of Kathy Judge, student.

REMINDER

A descriptive summary helps readers understand a document's main point, not the details.

Informative Summaries

In contrast to the descriptive summary, the informative summary provides a more inclusive view of an original document. The informative summary explains all of the original document's major headings, including key concepts, research methods, important statistical information, and findings, recommendations, or conclusions. Because the informative summary presents more information, it is generally longer than a descriptive one. The subjects of informative summaries are often research studies, written technically for professionals.

While a detailed discussion may call for several paragraphs of fairly technical information, the style of an informative summary should be as accessible as possible for readers. Quotations from the original are generally inappropriate, since summary authors can better condense information by paraphrasing the language of the original. Although the informative summary paraphrases and cites specifics from the document it describes, it does not need footnotes or endnotes, since it refers directly to the original article or report, and is often published or printed with it.

An example of an informative summary of a technical research study published in the *Journal of the American Dietetic Association* appears below.

Nutrient intakes and adequacy among an older population on the eastern shore of Maryland: The Salisbury Eye Evaluation

JAVIER CID-RUZAFA, MD, MPH; LAURA E. CAULFIELD, PhD; YOLANDA BARRON, MS; SHEILA K. WEST, PhD

ABSTRACT

Objective To describe the reported usual dietary intakes of the participants in the Salisbury Eye Evaluation (SEE) project and to estimate the prevalence of inadequate nutrient intakes using the probability approach.

Subjects/setting A representative sample of elderly residents (aged 65 to 85 years) of Salisbury, MD.

Design Cross-sectional survey, using a food frequency questionnaire to obtain nutrient intakes. We estimated energy and protein; percent of energy intake from carbohydrates, fat, and protein; as well as usual intakes of cholesterol, vitamin A, carotenoids, vitamin C, thiamin, riboflavin, vitamin B-6, vitamin E, niacin, iron, calcium, zinc, and folate. Estimates of prevalence of inadequate nutrient intakes were calculated using the probability approach among the 2,655 participants with complete nutrient intake information.

Statistical analyses performed The x^2 test for independence and analysis of variance. AP<.05 was considered significant in a 2-sided test.

Result On average, white participants of both genders reported higher mean energy and nutrient intakes than did black participants. Zinc had the highest estimated prevalences of inadequacy across all gender and race categories, followed by calcium, vitamin E, and vitamin B-6. Vitamin C, with estimated prevalences of inadequacy lower than 13%, and folate, with prevalences lower than 17%, had the lowest estimated prevalences of inadequacy across all gender, race, and age categories.

Conclusions In this population, there are race differences in estimated prevalences of inadequate nutrient intake. According to the current nutrient requirements for adults aged 65 to 85 years, many elderly persons have inadequate dietary intakes of key nutrients. *J Am Diet Assoc. 1999; 99:564–571.*

This example illustrates several important features of informative summaries. Since the article describes the results of an experiment, the summary is organized by headings that indicate the sections of discussion. Even though the text of the informative summary is technical, the authors are careful to show the relationships between ideas by carefully arranging sentences and using transitional phrases.

The following is an informative summary of "The Ecological Effects of Lead Shot: A Report for Duck Hunters," a student research report.

THE ECOLOGICAL EFFECTS OF LEAD SHOT: A REPORT FOR DUCK HUNTERS

This report discusses the long-range effect of lead birdshot on the marsh environment and on the ecosystem that supports aquatic game birds.

Discussion

In duck hunting, most lead birdshot, which is radioactive, lands in marshes and lakes. There the lead radiation degrades, contaminating plants and working through the food chain to become the food source for game birds. The reduction in numbers of game birds has been related to radiation (among other contaminants), leading to birth defects, high mortality, and fragile egg shells that shatter before hatching. While steel shot is manufactured in shells, hunters surveyed have been reluctant to use steel shot because they fear that this lighter shot will lessen the chance of hits. However, studies show that the scatter pattern of steel shot is only 12% wider than that of lead shot, allowing hunters an equally good opportunity of a hit as with lead shot.

Conclusions

- The use of lead shot by duck hunters is directly related to the decline of marsh hunting environments and the decreasing population of game birds.
- Steel shot provides as accurate a means of bird hunting as lead shot.
- Hunters must take an active role in protecting and preserving the environments in which they hunt.

Reprinted by permission of Michael Murrey, student.

This summary describes the report's main topic, discusses the main points, and lists conclusions. It provides a complete overview of the report, showing how the parts of the report's discussion are related and how the conclusions are derived from the discussion.

REMINDER

An informative summary gives readers the main idea, areas of discussion, and conclusions of an original document.

Executive Summaries

An executive summary, sometimes called a position paper or a white paper, provides a clear, readable, nontechnical overview of a publication, project, or proposal. This kind of summary is prepared for readers who may not have technical expertise but who need to understand a technical project's scope, subject, and important issues. Executive summaries often serve the needs of decision makers, such as legislators, review boards, and administrators, who need to remain up to date about the many activities of the organizations for which they are responsible.

While descriptive and informative summaries present brief previews of the subject matter of an original document for readers who are trying to determine which documents to read in full, executive summaries present information in such a way that nontechnical readers can understand the significance of the original document without reading it. For this reason, an executive summary should state key points concisely but comprehensively to provide readers with the background necessary to make decisions about policy, funding, or hiring. Reports designed for a wide range of readers often provide an executive summary as well as a descriptive summary in order to meet the widest possible range of audience needs. Following is the executive summary from a Texas Sunset Advisory Commission report on a review of the Texas Commission on Human Rights, pages 301 and 302.

The total length of the executive summary, which appears immediately after the table of contents, is 2½ pages of the report's total of 61 pages. This summary is designed to allow readers to quickly review the four findings, each of which is supported by discussion and recommendations. The report's discussion presents the same findings fully supported with data. The summary's page design, like its style, makes it easy for readers to follow and understand the main points of the Texas Commission on Human Rights report. Headings lead readers to key points, and numbered lists, a double-column format, and bold type allow readers to see and recognize the relationship between the findings and the supporting facts.

REMINDER

An executive summary should be clear, complete, and easy for nontechnical readers to follow and understand.

An example of a proposal with an executive summary in addition to a descriptive one appears in Chapter 14, pages 400–401.

WHAT TYPE OF SUMMARY SHOULD YOU WRITE?

Your decision whether to write a descriptive, an informative, or an executive summary depends on the content, length, and technical level of the information you need to summarize and on the kinds of audiences for whom you are writing.

For readers who need a brief preview of moderately technical information, a descriptive summary is a good choice. Its conciseness and generality give reviewers the kind of overview they need to determine whether they should read the original document.

For readers of lengthy, complex, and technical information, particularly original research or documents with a complicated organization, an informational summary is helpful. The careful organization, presentation of main points of discussion, and listing of findings and conclusions are useful for readers who need to understand a document's subject, scope, and conclusions

Executive Summary

The Texas Commission on Human Rights is responsible for enforcing state equal employment opportunity and fair housing laws that prohibit discrimination on the basis of such factors as race, sex, age, religion, national origin, and disability status. The Commission accomplishes its mission primarily by investigating and resolving employment and housing discrimination complaints as an alternative to litigation. The Commission also provides comprehensive training and technical assistance to state agencies and private businesses on the federal and state anti-discrimination laws.

To carry out its responsibilities, the agency had 46 employees and spent $2.6 million in fiscal year 1997. The agency is governed by a six-member Commission, appointed by the Governor, composed of one representative each from industry and labor and four public members.

The Sunset review focused on the Commission's ability to carry out its functions to reduce discrimination in the State of Texas. The issues in this report address improving the Commission's complaint resolution process, and strengthening its training and technical assistance efforts. Finally, staff focused on ensuring that the public has equal access to the agency's procedures and to legal remedies under the Texas Commission on Human Rights Act.

---

1. Enhance the Commission's Public Outreach and Investigator Training Efforts.

- Complaint resolution is the Commission's main activity to resolve citizens' employment and housing discrimination complaints as an alternative to litigation. An effective complaint resolution process requires that participants and investigators are knowledgeable about the process.
- The Commission does not have a toll-free telephone number or provide easy-to understand information in a readily-accessible format. This lack of outreach may cause confusion for the public and participants in the complaint resolution process.
- The agency does not provide a formally structured training program or a cohesive training manual to its investigators who perform a critical role in the complaint resolution process.

Recommendation

- **Require the Commission to make information more accessible to the public by establishing a toll-free telephone service and developing plain-language material about its complaint resolution process.**
- **Require all newly hired investigators to complete a formal training curriculum before conducting investigations and to complete an annual training update.**
- **Require the Commission to develop an investigation procedural manual to be updated biennially.**

2. Strengthen the Commission's Ability to Collect and Analyze Workforce Information and Its Technical Assistance and Training Responsibilities.

- The Commission relies heavily upon the authority of rider language in the General Appropriations Act to conduct its equal employment opportunity training and technical assistance activities. These activities include compiling statistics on the State's minority workforce composition and reviewing state agencies' personnel policy and procedural systems.
- Providing for Commission functions and policies in riders, rather than statute, may not serve the State's needs. Important training and technical provisions may not be considered independently through the legislative process, and unclear legal authority hinders the Commission's ability to provide guidance to state agencies and institutions of higher education. In addition, the Legislature has expressed an interest in placing riders into general law.

Recommendation

- **Require the Commission to conduct annual workforce analyses of state agencies and public institutions of higher education.**
- **Require the Commission to establish a technical assistance program on equal employment opportunity laws for state agencies and public institutions of higher education.**

- **Require the Commission to provide comprehensive equal employment opportunity training to all state agencies and public institutions of higher education.**
- **Require the Commission to collect and report statewide data on discriminatory activity in the state.**

3. Ensure Adequate Compensatory Relief for All Public Employees Who Suffer Employment Discrimination.

- The Texas Commission on Human Rights Act provides protection from employment discrimination to employees of all governmental entities, but does not provide all employees with equal access to compensatory relief. Employees of small governmental entities who cannot receive compensatory damages cannot be returned to the position they would have occupied had the discrimination not occurred.
- No rationale exists to prevent employees of a governmental entity from being able to seek compensation. Governmental employers receiving public funds paid by all citizens have an obligation not to discriminate against any citizen and should be subject to the same remedies as other public employers.

Recommendation

- **Specify that compensatory damages, already allowed under the Texas Commission on Human Rights Act, apply to all governmental entities, regardless of size.**

4. Continue the Texas Commission on Human Rights for 12 Years.

- Despite the enactment of anti-discrimination laws, employment and housing discrimination remains a problem in Texas. The Commission provides an alternative to litigation through its complaint resolution process, and provides training and technical assistance to prevent discrimination before it occurs.
- In fiscal year 1997, the agency resolved 1,258 employment complaints and 233 fair housing complaints. The Commission estimates that its complaint resolution process saved employers over $1 million by averting litigation and resulted in more than $1 million in benefits for individuals who filed complaints.
- Employers and housing providers who have used the agency's training and technical assistance have experienced a 5 percent reduction in discrimination complaints filed.
- Maintaining the Commission allows the State to administer anti-discrimination laws in Texas to be more responsive to state and local needs.

Recommendation

- **Continue the Texas Commission on Human Rights for 12 years.**

Fiscal Impact Summary

These recommendations, especially those regarding outreach, training, and codifying existing requirements, are intended to enable the Commission to better serve its functions within existing resources. Some recommendations, such as establishing and maintaining a toll-free telephone number, may have a slight fiscal impact. The recommendation to apply compensatory damages to all governmental entities may have a fiscal impact to the State, but the exact amount cannot be estimated. Finally, if the Legislature continues the Commission, as currently structured, the Commission's annual appropriation of approximately $2.6 million would continue to be required for operation of the agency.

REMINDER

Let the needs and knowledge level of your audience determine the kind of summary you write.

before they read it. The completeness of the informational summary makes it equally helpful for readers who may not read the original document but who need to know what the document discusses.

Nontechnical readers who may depend on a summary for complete information are best served by an executive summary. Readable, nontechnical style, conciseness, logical organization, and usable information design features give executive summary readers an overview of a document's ideas.

WRITING AND REVISING A SUMMARY

These steps outline the procedure for planning and writing an effective summary.

1. Read the original document once to understand the main idea. Then read it a second time for key ideas, noting subjects of discussion, internal headings, introduction, and conclusion.
2. Complete the Project Plan Sheet for a Summary. A project plan sheet will help you decide which kind of summary you need to write as you answer questions about your audience and your purpose.
3. Determine the kind of summary that will best help your readers:
   - A descriptive summary helps reviewers who need a brief informational preview of a document.
   - An informative summary gives readers of complex or technical information an inclusive view of a documents's headings, arguments, and conclusions.
   - An executive summary gives readers a complete, nontechnical discussion of a document, project, or issue.
4. List the key ideas of the original document. It is important to recognize the main points and to avoid being caught up in details, especially if you are the author of the original document.
5. Plan the sequence of the summary. Start with a forecasting sentence that states the main point of the information you wish to summarize. Then arrange the supporting points in the order in which they appear in the original document.
6. Draft the summary. Write quickly and steadily, concentrating on including *all* of the points you have listed.
7. Revise for order, clarity, and conciseness. As you revise, focus on the audience for your summary. Consider their level of knowledge in the language you use. Organize to make your summary easy to follow. If you plan to use a particular information design, it should support your subject matter and sequence of ideas.

Student Example: Descriptive Summary

Following is a student example of a summary of a comparison between two college electronics technology programs. The original article appears on pages 304–305. After quickly reading the article to understand its main ideas, the student completed a project plan sheet, which follows the article below, in order to make decisions about the summary. Then the student wrote the summary which appears on page 307.

A COMPARISON OF TWO ELECTRONICS TECHNOLOGY PROGRAMS IN CHARLOTTE: STEVENS TECHNICAL COLLEGE AND TINNIN COMMUNITY COLLEGE

A comparison of the electronics technology program at Stevens Technical College and at Tinnin Community College indicates that the Tinnin program has more to offer the student. This conclusion is based on a thorough examination of the catalog from each college and a visit to each college for a personal interview with the chair of the electronics technology department. This investigation shows that the Tinnin program costs less, has more lab equipment, and possibly has better instruction.

Cost (2000–2001)

The cost of tuition and books for each college is as follows:

Stevens Technical College

Tuition (18-month program)		$10,100.00
Books–included with tuition		0
	Total	$10,100.00

Tinnin Community College

Tuition (4-semester program at $750.00 per semester)		3,000.00
Books (approximately $250.00 per semester)		
4 semesters x $250.00 = $1,000.00		
resale value — 300.00		
$ 700.00		700.00
	Total	$3,700.00

The Stevens program costs $6,400.00 more than the Tinnin program. This large difference in cost is due primarily to Stevens being a private college and Tinnin being a public community college.

Lab Equipment (surveyed in 2000)

Each college has these major pieces of lab equipment:

Equipment	Stevens	Tinnin
Power supply		
Digital multimeter	15	50
Microprocessor trainer	12	40
Sine-square wave generator	4	12
Transistor curve tracer	6	25
Decade resistance/capacitance boxes	0	1
Digital trainer	0	40
Function generator	4	20
Dual-trace oscilloscope	2	30
Work stations	5	34
	4	12

The approximate retail value of the current Stevens equipment is $95,000. The total value of the current Tinnin equipment is well over $375,000.

The program at Tinnin Community College has more kinds and more pieces of each kind of lab equipment than does Stevens Technical College.

Instruction

The programs in both colleges differ in three aspects of instruction: qualifications of the instructors, daily class schedule, and required courses. Stevens Technical College has one electronics instructor. He has an Associate in Applied Science degree from Stevens, has taught three years (all at Stevens), and has worked two years in industry (at IBM). Tinnin Community College has three electronics instructors. Two of them have Bachelor of Science degrees and the other has a Master of Science degree. Each has taught in at least one other college, has taught at least five years, and has worked in industry at least three years. One instructor works in industry every other summer.

The daily class schedule is set up differently at each school. At Stevens, the students are in class from 8:00 A.M. until 12:00 noon and from 1:00 P.M. to 2:00 P.M. Monday through Thursday (there are no classes on Friday). For the electronics courses, the students have lecture from 8:00 to 10:30 and lab from 10:30 to 12:00. The afternoon hour is for a nonelectronics course. At Tinnin, the students have variable schedules from 8:00 A.M. to 3:00 P.M. five days a week, with electronics courses typically in two-hour blocks. The distribution of lecture and lab time is at the discretion of the instructor.

Both Stevens and Tinnin offer an Associate in Applied Science degree upon successful completion of their programs. Requirements for the associate degree differ slightly at the two schools. At Stevens, the required courses are College English I and II, Business Law I and II, Sociology, Math for Electronics, Digital Mathematics, Computer Mathematics, Basic Electronics, Fundamentals of Electricity, Digital Circuits, Semiconductors, and Transistor Circuits. At Tinnin the required courses are Technical Writing I and II, any course in social science, Physical Education, Industrial Psychology, Technical Mathematics I and II, Technical Physics I and II, Fundamentals of Drafting, Electricity for Electronics, Electron Devices and Circuits, Fundamentals of Fiber Optics, any four additional sophomore-level electronics courses, and one elective. Both programs offer the same amount of classroom instruction—18 months—but the Stevens program does not have a three-month summer break.

Thus, the instructors at Tinnin may be better qualified than the instructors at Stevens, the daily class schedule is more flexible at Tinnin, but the requirements for graduation are similar at both colleges.

Conclusions

A person planning to specialize in electronics technology should thoroughly explore the programs in prospective colleges. This report shows that two colleges (both in the same city) vary widely in their electronics programs. The electronics technology program at Tinnin Community College costs less, has more lab equipment, and possibly has better instruction than does the electronics technology program at Stevens Technical College.

PROJECT PLAN SHEET

FOR A SUMMARY

Audience

- Who will read the summary?

  High school students, parents, and high school counselors will be the primary audience.
- How will readers use the summary?

  Readers will use the summary to preview two college electronics programs.
- How will your audience guide your communication choices?

  My readers need to understand the main subject and the major areas of comparison. Since the article is very short, they need only the main ideas in order to decide whether it has information they can use.

Purpose

- What is the purpose of the summary?

  My summary should give readers a brief overview of the article's main idea.
- What need will the summary meet? What problem can it help to solve?

  My readers know what they want. My summary needs to help them decide whether this article has information to meet their needs.

Subject

- What is the summary's subject matter?

  I need to summarize "Comparison of Two Electronics Technology Programs in Charlotte: Stevens Technical College and Tinnin Community College."
- How technical should the summary be?

  Not technical or detailed.
- Do you have sufficient information to complete the summary?

  Yes, the article is all I need.
- What title can clearly identify the summary's subject and purpose?

  A Descriptive Summary of an Article Comparing Two Electonics Programs

Author

- Will the summary be a collaborative or an individual effort?

  Individual.
- How can the developer(s) evaluate the success of the completed summary?

  I'll ask a classmate to review the finished summary to make sure that it's clear and easy to follow.

Project Design and Specifications

- Are there special features the summary should have?

  No.
- What information design features can help the summary's audience?

  Since the summary will only be a few sentences long, no special design features will be needed.

Due Date

- What is the final deadline for the completed summary?

  I need to get this to our guidance office next Monday.
- How long will the summary take to plan, research, draft, revise, and complete?

  I can write this summary in three days, including time for review.
- What is the timeline for different stages of the project?

  I want to take an evening to read and study the article and to take notes. I'll take an hour to draft, and another hour to revise. It would be helpful if I could wait a day between drafting and revising so that my review will be complete. I'll take another day to get a classmate to review what I've written and to make any changes.

The final version of the summary appears below.

FINAL DRAFT: DESCRIPTIVE SUMMARY

A Descriptive Summary of an Article Comparing Two Electronics Programs

This article provides a comprehensive comparison between the electronics technology program of Stevens Technical College and that of Tinnin Community College to allow prospective students the opportunity to evaluate costs, lab equipment, and instruction. The report concludes that Tinnin Community College offers lower fees and a wider range of electronics technology equipment than Stevens College does.

Nontechnical readers and researchers can easily and quickly review this summary to determine whether the report provides information they need.

GENERAL PRINCIPLES for Summaries

- A summary must accurately reflect the subject matter it describes.
- The technical level of a summary depends on the subject matter of the original document and the readers' needs.
- Summaries should not run longer than 15 percent of the original document.
- Summary authors must read carefully and critically in order to be sure that they reflect a document's key points in their summaries.
- The type of summary an author writes depends on the needs and technical level of the readers.
- The descriptive summary provides a very brief overview of a document's main point.
- The informative summary presents a full view of a document's purpose, main points of discussion, and findings, recommendations, or conclusions.
- The executive summary presents a nontechnical overview of a document and is designed to inform the reader who may or may not read the original document but who needs to understand its main ideas.
- Summaries, no matter what type or how technical, should be as easy as possible to read, follow, and understand.

CHAPTER SUMMARY

SUMMARIES

Like other kinds of workplace communication, summaries must meet the needs of their audiences. Different types of summaries suit different readers. The descriptive summary provides a very brief overview of a document's main points, while an informative summary gives readers a more complete view of a document's discussion and conclusions. The executive summary provides a complete, nontechnical overview for the reader who may not read the original document in full.

Effective summaries allow readers to quickly grasp important ideas. Summaries are valuable for audiences when they are exact, clear, and orderly. To write such a summary, an author must understand the original document, distinguish between main issues and supporting ones, and be able to organize and adapt information for readers at different technical levels. Summary authors must read carefully and critically and write logically and concisely to communicate new ideas briefly and accurately to others.

ACTIVITIES

INDIVIDUAL AND COLLABORATIVE ACTIVITIES

12.1. In groups of three or four, make a list of the kinds of information you expect to need in your career. Where can you locate summaries of such information? In what ways could you use summaries in college classes? On the job?

12.2. In what ways does writing a summary deepen an author's understanding of a subject? Try writing summaries of class notes or highlights of a chapter from one of your textbooks. In a small group, discuss the uses of summary writing as a study skill.

12.3. Imagine that your city's Engineering Department has written a long and technical proposal for a new wastewater treatment plant to be located near a small, well-established neighborhood. Some of the readers who will want to understand the proposal are:

- Neighborhood residents
- City planners
- Engineers from other cities
- City council members
- Students of urban planning
- Members of city neighborhood organizations
- Reporters
- State legislators

What are the needs and the technical backgrounds of each group of readers? What kinds of summaries do you recommend that the city Engineering Department provide with the report?

For Activities 12.4–12.6

a. Review Writing and Revising a Summary on page 303.
b. Adapt and complete the Project Plan Sheet for a Summary, page 315.
c. Write a preliminary draft.
d. Using the Project Plan Sheet and Writing and Revising a Summary, analyze the draft.
e. Revise the draft, incorporating design decisions.
f. Prepare a final draft.

12.4. Revise the informative abstract of "Nutrient intakes and adequacy among an older population on the eastern shore of Maryland: The Salisbury Eye Evaluation" (page 298) as a descriptive abstract. What specifics can you summarize?
12.5. Write an executive summary of a front-page news article. What steps will you follow in preparing the summary?
12.6. Write a descriptive summary of "A Personal Portrait of the Rodrigues Fruit Bat," which appears in Chapter 10, page 227.

READING

12.7. In "How to Write a Useful Abstract," technical communicator Janis Ramey provides advice for the busy workplace author who needs to communicate new ideas to engineers.

How to Write a Useful Abstract

JANIS RAMEY

Engineers love formulas; programmers love flow charts; everyone likes easy-to-follow guides. This is an easy-to-follow guide for writing abstracts. There's nothing to it: It's as easy as 1-2-3!

1. Write a topic sentence.
2. Write two or three supporting sentences.
3. Tie everything together with logical order and good transition.

Voila!

With our ever-increasing need for quick access to information, useful abstracts are more important than ever. Think of those long articles you've seen pop up on the Internet. Wouldn't you appreciate it if the first few lines were a true abstract of the article so you'd know whether you should wait for the rest of it to load? How about when you look up a title in a library catalog? A useful abstract helps you decide whether to retrieve the publication, but a poor abstract leaves you wondering.

WRITE THE FIRST SENTENCE

After you've finished writing, sit back and think about the whole document. Think about its major idea. Think about its main conclusion or result. Think about its primary purpose. Think about what you can expect the reader to do with this document. Collect all this information in your mind and write a sentence. That is your topic sentence.

Notice, I said "Write a sentence." I didn't say "Write several sentences." This is the hard part. You need to write one sentence that covers the entire document—whether the document is a one-page letter or a thousand-page manual.

Look in Several Places for Inspiration

Look at the recommendations, conclusions, summaries, and results sections of your document. If you're abstracting a manual, look at the first tutorial for inspiration. These sections often are about the essence of the document. Don't pay much attention to the introduction section. It usually sets the stage, so to speak, so the main action can proceed.

Don't Depend on the Title

The title of the document may or may not help you write the topic sentence. Chances are the title is too vague to be much help. Parts of the title might serve as modifiers in your topic sentence, but you'll probably need to search beyond the title for inspiration.

Be Specific

The topic sentence must say specific things about the project. The tendency is to write something like "This report describes . . . [some general title]." What you need to write is something like "The results of this . . . [specific subject] . . . study show that . . . [actual result]."

FILL IN THE DETAILS

After you've settled on your topic sentence, write two or three supporting sentences. Each of these sentences must supply specific details about the ideas in the topic sentence. Think of the evidence in the document that supports the topic sentence. In your mind, answer those famous questions. Who? What? Where? When? Why? How? How much? You might give important statistics, results, conclusions, or recommendations that back up what the topic sentence says.

Try to limit yourself to two or three major supporting ideas. You might include some of the less important evidence as subordinate clauses and modifiers.

HOLD IT TOGETHER

Arrange the supporting sentences in a logical sequence. Add whatever transitions are needed to connect the supporting sentences to the topic sentence and to connect ideas within the sentences to each other. Rewrite the sentences, if necessary, to improve the connections.

THE RESULT

Now you have an abstract that is truly a digest of the material and will probably be useful to a teacher. And, surprisingly, this technique works for documents of any length. It also works for any kind of document—letters, reports, articles, materials, books, speeches, scripts, and just about anything else you have to write.

TWO EXAMPLES: ABSTRACTS FOR THIS ARTICLE

Overly Vague Abstract (or "Abstract Abstract")

This article describes how to write an abstract. The author relies on established rules of good composition. She also provides a number of helpful hints about writing abstracts. The article includes examples of both good and bad abstracts.

Useful Abstract

A useful abstract is a well-constructed paragraph with an informative topic sentence followed by two or three supporting sentences. The supporting sentences contain specific information about the topic. The sentences are arranged in a logical order and the ideas are connected with good transitions. Inspirations for the topic sentence should come from the recommendations, results, conclusions, tutorial, and summary sections of the document, rather than from the title or introduction.

SUGGESTIONS

- *Write the abstract after the document is written.* Abstracts written before the document is written are really previews or introductions, not true abstracts. If you are forced to write an abstract first, use the same technique: Think about the whole project, its purpose, etc., and then write a topic sentence. Keep in mind that you'll need to rewrite the abstract when the document is finished because it will no longer accurately reflect the contents of the document.
- *Before starting the abstract, make a random list of thoughts about the document.* Take several hours or days to develop the list. Group related items together and then prioritize the list, putting the most important group first. The first few groups on your list will probably be the core for your topic sentence. The rest of the groups will lead to supporting sentences.
- *If you find that you can't write a topic sentence, write the supporting sentences first.* The topic sentence may then become obvious. Caveat: The sentences must truly be supporting sentences—not topic sentences in themselves.

- *Write for an intelligent reader who's not necessarily up to speed in your subject area.* This technique is important because you never know who's going to pick up your abstract.
- *Choose words carefully.* Acronyms, abbreviations, and limited-use technical terms will leave many readers in a daze.
- *Define the scope of the project (that is, its limits) in the abstract.*
- *Include the name of the contracting agency or client.*
- *Read your abstract after several days have passed.* Delete all superfluous information.

Written by Janis Ramey and reprinted with permission from *Intercom,* the magazine of the Society for Technical Communication, Arlington, VA, U.S.A.

QUESTIONS FOR DISCUSSION

a. What kind of summary does Ramey's article describe? Is her "abstract" a descriptive, informative, or an executive summary?
b. Is Ramey's advice about critical reading useful and practical? How could you apply this advice to studying new information?
c. What is the difference between the two summaries of "How to Write a Useful Abstract"? What specific elements of writing tie ideas together in the "Useful abstract"? How does this example reflect the headings in the original article?
d. How do executive summaries differ from the kind of summary described in "How to Write a Useful Abstract"?

CASE STUDY

12.8. In a small group or in a short written assignment, discuss the ways in which the following case study illustrates the uses of an effective summary.

The Bottom Line

Becky Roylott, chairperson of the Benefits Committee at Inclusive Software, is worried. Two months ago her committee sent out a report about a new benefit, a generous rebate on college tuition for all company employees and for their dependents. Although Becky's committee worked hard to set up the new benefit, not a single employee has responded. At their monthly meeting, the committee tries to figure out why.

"We did our best. We really did. We sent that report out to every single company employee, all 932 of them," Becky says. "I know for a fact from Human Resources that dozens of Inclusive employees are taking college classes. They're so motivated! I sure wish they'd let us help. We budgeted $70,000 for tuition for this quarter, and we haven't had a single request so far. "

Jaime Lopez, who works in Engineering, asks, "Do you have a copy of that report?"

"Sure," replies Ralph Voss, Company Publications Manager. "Here it is. I wrote it myself. I wanted to make sure that it was easy to read and easy to use for reference. The report has a table of contents and an appendix with all of the application forms in it. It's everything that people need if they want to apply for the rebate."

"No question, Ralph," Jaime says, as he skims the nine-page report. "That report's easy to use, and clearly written, your usual good work. But it's still a long report! Our employees here at Inclusive are so busy that they probably don't have time to learn how to take advantage of the benefit."

"Ralph and I talked for a long time about just what the report needed to include," Becky says. "Since taxes and some new policy issues are involved, there was a lot to explain. We made everything as brief as we could without leaving out information that readers needed."

Ralph smiles. "Hey, wait a minute! What about a summary? We could write up a one-page summary to explain the main point of the report. That ought to help."

"That's a great idea," Jaime says. "We could even send the summary to our Marketing area to use in posters, newsletter articles, and even post on our in-house Web page. Then we'd *really* get some response. But what kind of summary should we write? Should it evaluate the report?"

"Let's look at the report's headings," Becky suggests. "Maybe we need more than one summary for more than one kind of reader. If we work on this together, we can have this problem solved in no time at all!"

QUESTIONS FOR DISCUSSION

a. Why has the Benefits Committee sent out the report? Is it this committee's responsibility to ensure that readers understand the benefit clearly?
b. Who are the audiences for a summary about the benefit? Are they the same as the audiences for the report?
c. Who are the different audiences for the report? How technical are these audiences? How much does each audience need to know? What kind of summary should be part of the report? What kind should be a handout? What kind should be the basis of text for posters, newsletter articles, and Web page announcements?
d. What additional media would you suggest that the Benefits Committee use to send out the summary?
e. For what audiences would an evaluative summary be most appropriate and most helpful?
f. In what ways does providing factual information about the new benefit support the purposes of the Benefits Committee? Of the employees? Of the company?
g. What suggestions do you have for the Benefits Committee?

PROJECT PLAN SHEET

FOR A SUMMARY

Audience

- Who will read the summary?
- How will readers use the summary?
- How will your audience guide your communication choices?

Purpose

- What is the purpose of the summary?
- What need will the summary meet? What problem can it help to solve?

Subject

- What is the summary's subject matter?
- How technical should the summary be?
- Do you have sufficient information to complete the summary? If not, what sources or people can help you to locate the additional information?
- What title can clearly identify the summary's subject and purpose?

Author

- Will the summary be a collaborative or an individual effort?
- If the summary is collaborative, what are the responsibilities of each team member?
- How can the developers evaluate the success of the completed summary?

Project Design and Specifications

- Are there models for organization or format for the summary?
- In what medium will the completed summary be presented?
- Are there special features the summary should have?
- What information design features can help the summary's audience?

Due Date

- What is the final deadline for the completed summary?
- How long will the summary take to plan, research, draft, revise, and complete?
- What is the timeline for different stages of the summary?

CHAPTER 13

Reports: Shaping Information

CHAPTER GOALS

This chapter:

- Defines reports as a type of communication
- Discusses workplace uses of reports
- Explains the responsibilities of report authors
- Distinguishes between formal and informal reports
- Distinguishes between special report formats and conventional report elements
- Describes typical parts of reports
- Discusses design options for reports
- Explains the use of graphics and other visuals in reports
- Describes three types of reports: observation, progress, and feasibility

INTRODUCTION

Reports are a firmly established means of communicating information in the workplace. Reports require extensive planning for the overall report project and for the organization and delivery of the report. This chapter is a guide for preparing reports. It demonstrates how to analyze the communication situation and the audience, incorporate design elements, and present the report to accomplish its purpose. The chapter also discusses various formats and models for reports and how to adapt them for the needs of each project.

REMINDER

A report must support the subject, the occasion, and the audience.

MULTIPURPOSE REPORTS

The frequency, the possible presentation media, and the flexibility of reports in the workplace attest to the rapid evolution and broadening usefulness of reports. Because of their flexibility, establishing specific classifications for reports is difficult. Effective report authors adapt the form to their purposes.

On page 318 is a report from an architect's firm to a client and a contracting firm, describing work in progress on a project. It illustrates the difficulty in classifying reports into neat categories. This sample could be classified as a periodic report, an observation report, or a progress report. It is a periodic report because such a report is prepared weekly to document work done. The report might also be described as an observation report because the writer observed on-site the activities and concerns identified in the report. It is also a progress report because it shows the progress of work being done to complete a project—a clubhouse.

Dean/Dale and **Dean**
architects a professional association field report

P.O. Box 4685
Jackson, MS
39216
1301 Mirror Lake Plaza
2829 Lakeland Drive
39208

(601) 939-7717

Project/Project No. CLUBHOUSE-HINDS COMMUNITY COLLEGE
RAYMOND, MISSISSIPPI
PROJECT NUMBER: 90054

Contact/Supt.: PHOENIX CONSTRUCTION COMPANY

Date 5 MARCH 2001 Time 7:45 AM Weather CLOUDY-COOL Approx. Temp. 60 F

Present at Site LARRY HOLDEN - BILL DICKEN

Remarks

WORK IN PROGRESS: Concrete

1. Contractor was in the process of pouring final exterior porch slab at north end of building when I arrived.

2. Testing lab personnel had been notified but had not shown up. Finally got to site approximately 8:30 a.m. Took cylinders from second truck — all appeared per specs.

3. I indicated to Contractor that I had a concern over water soaked spiral columns at bottom. Contractor indicated that he would add additional support to bottom of columns to strengthen this area.

END OF REPORT.

William A. Dicken, Jr.
WILLIAM A. DICKEN, JR.

WADjr : tms

cc: Troy Henderson
Claude Brown

Further demonstrating the broad scope of the term *report,* the following is a detailed plan and schedule (although the author uses the term *proposal*) for operating a concession stand at a football game.

**Proposal for Hinds Community College Honors
Football Concession Stand for
Thursday, September 13, 2001**

to: Kristi Sather-Smith
Director of Hinds Community College Honors Program

by Nathan Chisolm
Fundraising Committee Chair
and Committee Members
Mitzi Reed
Ginny Loomis

August 20, 2001

Proposal for Hinds Community College Honors
Football Concession Stand
Thursday, September 13, 2001

Purpose

The purpose of this proposal is to provide a detailed plan and tentative schedule in preparation for the Honors Program operating the football concession stands on Thursday, September 13, 2001.

Contacts

Dr. Barbara Blankenship, Dean of Students (x3232)

Dr. Blankenship is the initial contact. A date was requested via an interoffice memo during the middle of the Spring semester. Once the date was set and approved, the contact person was Judy Bufkin.

Judy Bufkin (x3374)

Mrs. Bufkin has the keys to the concession stand. Contact her to make sure that the stand will be unlocked the day of the game so that there is plenty of time to set up.

Kyle Mize (x3228)

Mr. Mize is the sponsor for the Marketing Organization on campus. He also runs the concession stand frequently. He can answer any questions that arise.

Lamar Currie (x3340)

Mr. Currie is in charge of the Meat Marketing Department on campus. He can supply the hamburger meat, hot dogs, grill, etc.

Michael (924-3688)

Michael is the manager of Papa John's in Clinton. He is able to sell us the pizzas for five dollars a piece. They should be bought on the Honors Program account, and he already has the tax exemption letter. Be sure he knows to cut the pizzas into sixths. Note—the telephone number is long distance from the Honors Center.

Greg Reese (878-2200)

Mr. Reese is the Division Manager of the Jackson Frito Lay Plant. He can help by donating chips or selling them at wholesale.

Dennis Parker (932-8502)

Mr. Parker is a contact at Earthgrains, a bread bakery. For charities, he can sell hamburger and hot dog buns that are a couple of days old for half price. Give him a call a few days before you will need the buns, and he will save them for you.

Concession Stand Layout

The present layout for the home side concession stand is depicted near the end of this report. The items needed that are not presently at the concession stand (i.e., tables, grill) are indicated in bold print. The visitors' side concession stand consists of a tent with 2 or 3 tables.

Materials Needed

The following is a list of materials that need to be purchased or donated.

Food

Drinks

item	quantity	notes
drinks - 3 liter	50 - Cokes 25 - Diet Cokes 25 Sprites	100 or more total
cups	1,000	hopefully donated
ice	6 cooler chests	facilities

Meats

hamburger meat	200 patties	Meat Merchandising
hot dogs	300	Meat Merchandising

Pizza

pizza	20 large: 15 pepperoni 5 cheese	cut into sixths: Papa John's

Bakery

hamburger buns	200	Earthgrains
hot dog buns	300	Earthgrains

Nachos

chips for nachos	3 boxes	
cheese for nachos	3 cans	need to be diluted
jalepeños	1 can	

Condiments

mayonnaise	2 boxes	in individual packets because mayonnaise will spoil in the heat
mustard	1 gallon	with push-type dispenser
ketchup	1 gallon	with push-type dispenser
relish	1 gallon	with 2 spoons
tomatoes	12	sliced; in covered tray
onions	6	sliced; in covered tray
lettuce	3 heads	pieces; in covered bowl
napkins	2,000	

Miscellaneous

item	quantity	notes
pickles	3 jars	2 on home side; 1 on visitors' side
brownies	4 dozen	donated by students; 2 in. x 2 in.; in sandwich bags
chips	50 large bags	to go on plate with hamburger; donated by Frito-Lay
pretzels/fritos	50–75 individual bags	Frito-Lay
cellular phone		for calling Papa John's for more pizzas
carpenter's apron	2	for the 2 cashiers
money bags	2	for excess money
candy	3 boxes	Snickers, Cracker Jacks, M&Ms; donated by Honors students

Grill and Utensils

item	quantity	notes
grill	1	Meat Merchandising
spatulas	2	for hamburgers
wire brush	1	for cleaning grill
aluminum pans	3	for hot dogs
glove and tongs	1	for turning hot dogs; from Meat Merchandising
aluminum foil	1 roll	heavy duty

Setup

item	quantity	notes
charcoal	6 bags @ 10 lb.	for grill
lighter fluid	1 bottle	for grill
matches	2 small boxes	for grill
paper towels	4 rolls	for wiping hands
disposable table cloths	10	plastic; speeds clean-up
ice chests	6	for ice
small coolers	2	for cooked hot dogs and hamburgers

Setup (continued)

item	quantity	notes
garbage cans	7–8	from facilities; designate one for plastic drink containers
fans	2–3	from facilities
tables	10	from facilities
crock pots	2	for nacho cheese
canopy	1	to keep items from stands from falling in grill

Serving

item	quantity	notes
ladle	1	for nacho cheese
serving spoons	2	for relish
tongs	9	from Valley Foods–cafeteria for hot dogs and patties
hot pads	5–10	from Valley Foods–cafeteria
plastic serving gloves	30 pair or more	from Valley Foods–cafeteria
sandwich bags	150	for pickles
small paper plates	300	for hamburgers
boxes	350	for hot dogs
boxes	200	for nachos
hand soap	1 bottle	have all workers wash their hands before working

Cleanup

item	quantity	notes
409	2 bottles	get students to donate
rags	2	for wiping off surfaces
broom	1	
garbage bags	20	
zip-lock bags	1 box	gallon size; for leftovers

Capital Needed

Money needed is as follows:

$125 - quarter rolls; $100 for home side, $25 for visitors' side

$ 75 - dollar bills; $50 for home side, $25 for visitors' side

Prices

Prices will be written on two posters for the home side and one for the visitors' side. Honors Program should be included somewhere on all of the signs. The signs should include the following:

Home Side	Visitors' Side
Drinks - $1.00	Drinks - $1.00
Hamburger w/Chips, pretzels, or fritos $3.00	Pickles - $0.75
Pizza - $1.50/slice	Candy - $0.50
Hot Dog - $1.50	Brownies - $0.50
Nachos - $1.50	
Pickles - $0.75	

Set-up
6 total
Tasks to be completed: 3:30 - 6:15

- fill coolers with ice
- pick up pizzas
- clean grill
- set up tables
- get charcoal ready
- set up visitors' side tent

Concession Stand Workers
21 total
6:00 - 10:00

Concession Stand Organization

There will be assigned positions in the concession stand. Workers should work in their assigned position during the night but should help others when they get behind. The positions are indicated below, and their locations are printed on the attached diagram of the concession stand:

2 - condiment table
2 - cashiers (one visitors' side, one home side)
2 - drink pourers
2 or 3 - filling ice
2 - on grill
4 - pizza, hamburger, hot dog, pickle servers
2 - nacho, pickles, candy, chips, drink servers
3 or 4 for visitors' side tent (before game through halftime)

Cleanup
All (except those who helped in setup)
To begin after 3rd quarter

Tasks to be completed:

- recycle all drink bottles
- wipe down tables
- sweep floors
- tie up trash bags
- clean out warmer
- put all utensils to be washed in a plastic bag and give to team leader
- put all remaining perishables in the gallon size zip-lock bags

Tentative Schedule

Friday, August 24—contact Steve Romano at Coca-Cola; check on Frito Lay donations; talk to Mr. Mize about quantities of hot dogs, hamburgers and pizza

Monday, August 27—announce plans during business meeting; pass sign-up sheet through each committee, a copy is attached to this proposal; ask committee chairs to dismiss workers 10 minutes early for a brief meeting with the fundraising committee; go over the minor details, and tell them about the Friday meeting

Tuesday, August 28—order hamburgers and hot dogs from Mr. Currie in Meat Merchandising; reserve the grill and order the 4 needed tables; ask Mr. Currie if the grill could be delivered by Wednesday afternoon; ask about the hot dog gloves; ask Judy Bufkin about the visitors' side facilities, order the 4 tables and the fans from her, and ask her about the ice; call Michael at Papa John's and check on making an account, and let him know about the 5 pizzas needed for Friday, and about the concession's pizza needs.

Wednesday, August 29—go over proposal with the Honors Council and ask for their help on September 5

Thursday, August 30—buy needed supplies at Sam's/Jitney Jungle, and store them at the HCCH Honors Center

Friday, August 31—team meeting; pizza and drink for dutch lunch; 12:00

Monday, September 3—Labor Day Holiday

Tuesday, September 4—check to see if concession stand is clean; if not clean it up; have Judy Bufkin inspect and sign off on it either way

Wednesday, September 5—remind the Forum class about the night's plans; get Mrs. Bufkin to unlock concession stand, and set up all non-perishable items; make sure that the grill and tables are in place; post signs with prices, one at each window; enlist student council help for setup

Tentative Schedule (continued)

Thursday, September 6—pick up meats from Meat Merchandising; get ice; workers need to arrive about 4:30; begin grilling burgers and hot dogs early so that there are enough to sell before the game—at least 50 each; prepare to start serving as early as 6:15; set up visitors' side tent:

- cover tables
- take over 2 chests of ice
- put up sign indicating what is available at other concession stand
- open before game through half-time; clean up after
- take one roll of paper towels
  (keep these supplies separate, in a worker's car on the visitors' side)

All workers need to wear their Honors Program T-shirts, those with long hair need to wear it up, and anyone who touches the food needs to wear gloves.

Concession Stand Sign-Up Sheet
Thursday, September 6

Name	Time
1.	4:30–8:00
2.	4:30–8:00
3.	4:30–8:00
4.	4:30–8:00
5.	4:30–8:00
6.	4:30–8:00
7.	6:00–10:00
8.	6:00–10:00
9.	6:00–10:00
10.	6:00–10:00
11.	6:00–10:00
12.	6:00–10:00
13.	6:00–10:00
14.	6:00–10:00
15.	6:00–10:00
16.	6:00–10:00
17.	6:00–10:00
18.	6:00–10:00
19.	6:00–10:00
20.	8:00–10:00
21.	8:00–10:00
22.	8:00–10:00
23.	8:00–10:00
24.	8:00–10:00
25.	8:00–10:00
26.	8:00–10:00

The concession stand needs at least 21 people at all times.

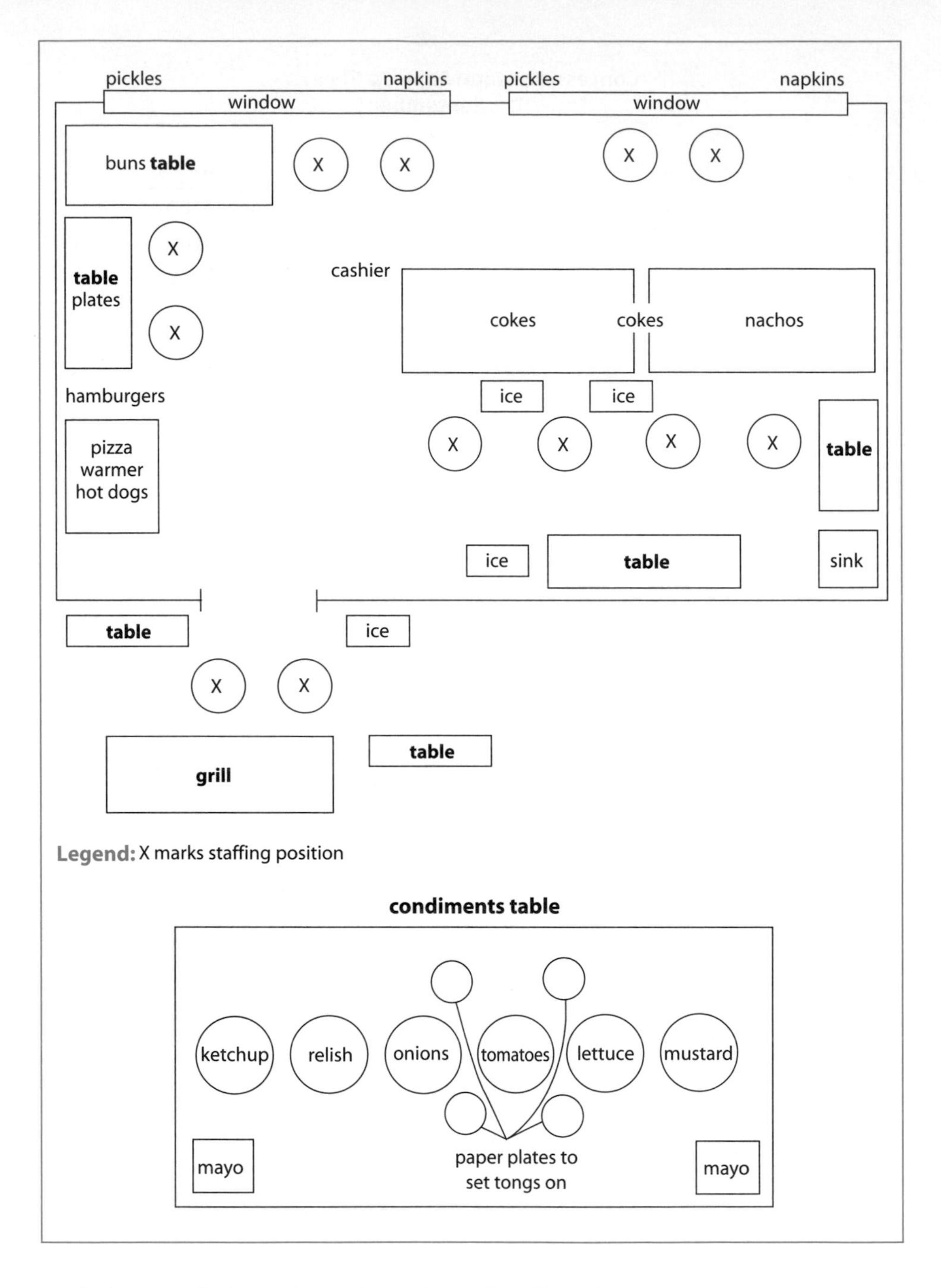

It is unrealistic to draw sharp boundary lines between types of reports or to try to cover all the situations and problems involved in preparing reports in this text. However, this chapter will help you to:

- become acquainted with the general nature of report preparation;
- develop self-confidence by learning basic principles of report preparation; and

- study, prepare, and present in writing and orally several types of reports that you are most likely to encounter in the workplace.

WHAT IS A REPORT?

A report can be defined as technical data, collected and analyzed, presented in an organized form. A report can also be defined as an objective, organized presentation of factual information that answers a request or supplies needed data. The report usually serves an immediate, practical purpose. The purpose may be to record particular data, as in an objective report or a trip report; or the purpose may be to present particular data as the basis for decision making. Generally, the report is requested or authorized by one person, is prepared for a particular audience, and is prepared sometimes by an individual but more often by a team.

Reports may be simple or complex, long or short, formal or informal, written or oral—all of which is determined mainly by the report's purpose and audience.

REMINDER

A report is an organized presentation of information that meets a practical need.

WORKPLACE USES OF REPORTS

The word *report* covers numerous communications that serve many purposes, both personally and professionally. You may be concerned about your monthly financial report (bank statement) or the job evaluation that you receive every six months. You may be polling union members about OSHA regulations, to support your committee's recommendations. You may be facing a due date for a budget proposal (with supporting data) for adding two new positions to your office management team.

In the workplace, reports are a vital part of communication. A detailed memorandum (memo) to the billing department, a weekly production report, a requisition for travel funds, a letter to the home office describing the status of bids for a construction project, a performance report on the new sorting machine, a sales report from the housewares department, a report on the availability of land for a housing project, a report to a customer on an estimate for automobile repairs are but a few examples of different kinds of reports and the many functions they serve.

QUALITIES OF REPORT CONTENT

Reports convey exact, useful information. That information, or content, should be presented with accuracy, clarity, conciseness, and objectivity.

Accuracy

A report must be accurate. If the information presented is factual, it should be verified by tests, research, documentation, authority, or other valid sources. Information that conveys opinions or probabilities should be labeled as such and accompanied by supporting evidence. Dishonesty and carelessness compromise the credibility of the report, its author, and its organization.

Clarity

A report must be clear. For the report to serve its purpose, the information must be understandable to the audience. The reader or listener should not have to ask: *What does this mean?* or What *is the author trying to say?* The author helps to ensure clarity by using exact language in easily understandable sentence patterns, by following conventional grammatical usage, and by organizing the material logically.

Conciseness

A report must be concise. Conciseness is saying much in a few words. Useful reports eliminate unnecessary wordiness yet provide complete information. Busy readers appreciate concise, timesaving reports that allow them to quickly understand the point. Examine the examples that follow.

Wordy: After all is said and done, it is my honest opinion that the company and all its employees will be better satisfied if the new plan for sick leave is adopted and put into practice.

Concise: The company should adopt the new sick leave plan.

Or consider the following report in memorandum format.

Memorandum Lacking Conciseness

To: J. Carraway
From: T. Jayroe
Subject: Ideas for revising "Ten Keys to Business Success" 3rd ed.
Date: 5 February 2002

It has come to my attention that it is time for us to consider revising the brochure "Ten Keys to Business Success" for a 4th edition. Would you be kind enough to take the time to answer some questions to give me some ideas about the new edition? Since you have been involved with the brochure since its first edition, I feel that your ideas are valuable.

First and foremost, do you feel that the 3rd edition has accomplished its purpose and should we therefore use the same basic content in the 4th edition? Second, what in your opinion are the changes we need to make to improve the brochure? Third and finally, what do you think we should do about distributing the brochures? Should we continue to distribute the 3rd edition, or should we wait for a new edition before distributing any additional brochures?

I need answers to these questions no later than 10 March.

Obviously the memo above is wordy. Also the layout of the content could be improved to make reading and comprehension easier. Consider the revised memo below.

Revised Memorandum

To: J. Carraway
From: T. Jayroe
Subject: Ideas for revising "Ten Keys to Business Success" 3rd ed.
Date: 5 February 2002

Please answer the following questions by 10 March.

1. Should the 4th edition brochure include the same basic content?
2. What changes can we make to improve the brochure?
3. Should we continue distributing the 3rd edition brochures in the meantime, or wait for the new edition?

Thank you.

Note that the revised memo, which is more concise and more pleasantly designed, conveys the same information as the original memo, but it can be read more quickly and understood more easily. Revise a report until it contains no more words than needed for accuracy, clarity, and correctness of expression. For further discussion of conciseness, see pages 36–39.

Objectivity

A report must be objective; that is, a report should present data fairly and without bias. Objectivity demands that logic rather than emotion determine both the content of the report and its presentation. Unless otherwise requested, the content should not reflect the personal bias and findings of the author. For instance, a report comparing new car warranties for six makes of automobiles should not be slanted toward the author's favorites.

Essential to objectivity is the use of the denotative meaning of words—the meaning that is the same, insofar as possible, to everyone. Denotative meanings of words are found in a dictionary; they are exact and objective. Denotation contrasts with connotation, which permits associated, emotive, or figurative overtones. The distinction between the single denotative meaning and the multiple connotative meanings of a word is illustrated by the following examples:

Word: war

Denotative Meaning: legally declared armed conflict between nations

Connotative Meaning: pacification, extermination, conquest

Word:	work
Denotative Meaning:	employment, job
Connotative Meaning:	paying bills, happiness, accomplishment, 9 to 5, satisfaction, alarm clock, fighting the traffic, income, new car, sweat, sitting at a desk

THE AUTHOR'S RESPONSIBILITIES

REMINDER

The author is responsible for complete, unbiased, and ethical reporting.

The author of a report is responsible for the content of a report and for its integrity. The author of a report is responsible for ensuring that information is complete and that the presentation of all information is free from bias. Omission of significant data, whether on purpose or through lack of thorough research, is unethical. A contractor, for instance, who misrepresents the warranty on a roof or lessens the agreed-upon percentage of wash gravel in a driveway is making an untruthful, and thus unethical, report to a client. The contractor has not provided full disclosure. Although readers of a report expect factual, objective information based on logical reasoning, they may be mislead by statements that play on sympathy or dramatic effect or that shift to emotional appeal. Such misleading reporting methods are unethical.

FORMATS FOR REPORTS

Most reports (including those presented orally) are put in writing to record the information for future reference and to ensure an accurate, efficient means of presenting the report when it is to go to people in different locations. A report may be given in **special formats:** on a printed form, as a memorandum, as a letter, or as an online template. The format may be prescribed by the person or agency requesting the report; it may be suggested by the nature of the report; or it may be left to the discretion of the author, who analyzes audience and purpose and then selects an appropriate format. Most reports, however, are presented using **conventional report elements.**

Special Formats

Printed Form

Printed forms are used for many routine reports, such as sales, purchase requests, production counts, medical examinations, census information, and delivery reports. Printed forms call for information to be reported in a prescribed, uniform manner with spaces left for responses. Information to be filled in on such a report, usually numbers or words and phrases, is expected to be a certain length.

Printed forms are especially timesaving for both the writer and the reader. The writer need not be concerned with structure and organization; the reader knows where specific information is given and need not worry about omission of essential items. However, printed forms lack flexibility: they apply to a limited number of situations. Further, they lack a personal touch that allows individuality. In making a report on a printed form, the primary considerations are accuracy, legibility, and conciseness.

Memorandum

Memorandums (memos) are used when the report is short and contains no visual materials. They are used primarily within a company. (For a detailed discussion of memorandums, see Chapter 15 Correspondence, pages 429–431.) Reports in memorandum formats are illustrated on pages 355–359 and 366–367.

Letter

As with the memorandum, the letter is often used for a short report (not more than several pages) that does not include visuals. The letter is almost always directed to someone *outside* the company.

The report letter should be as carefully planned and organized as any other piece of writing. The report letter follows conventional letter writing practices. (See Chapter 15 Correspondence.) A subject line is usually included, and the complimentary close is typically "Respectfully submitted." The report letter is often longer than other business letters and may have internal headings for better readability. The degree of formality varies, depending on the intended audience and purpose.

Some companies discourage the use of report letters to avoid the possible difficulty of report letters being filed with ordinary correspondence.

Online Template

Online templates can expedite the writing of some types of reports. By keying in words and phrases and responding to the various prompts, the report author can produce a readable report in a fraction of the time required to start from scratch.

Conventional Report Elements

Conventional report elements appear in both informal and formal reports. In actual practice, the terms *informal* and *formal* refer to report length and the degree to which the report is "dressed up" with such elements as a transmittal letter, title page, table of contents, and abstract.

Informal Reports

The informal report, usually one page to a few pages in length, is designed for circulation within an organization or for a named reader, and includes only the essential sections of the report proper: an introduction or statement of purpose, the information required for the report, and perhaps conclusions and recommendations. The informal report is far more common than the formal report. Examples of informal reports appear on pages 318, 319–328, 341–347, 355–359, 360–363, 366–367, and 560–562.

Formal Reports

The formal report has a stylized format evolving from the nature of the report and the readers' needs. The formal report is often long (from eight or ten to hundreds of pages), is usually designed for circulation outside an organization, and is intended for multiple readers. The formal report will probably not be read in its entirety by each person who examines it.

Following is a list of common elements of the formal report and a brief explanation of each element. The report author can combine, omit, or adapt these elements to accommodate the report's audience and purpose.

Front Matter

- Transmittal memorandum or letter—transmits the report from the writer to the person who requested the report or who will act on it. This letter may include information such as an identification of the report, the reason for the report, how and when the report was requested, problems associated with the report, or reasons for emphasizing certain items. Or it may communicate a simple message, such as: "Enclosed is the report on customer parking facilities, which you asked me on July 5 to investigate." (For further discussion of transmittal letters, see pages 450 and 592–593; for examples of letters of transmittal, see pages 348, 398, 452, 594–596.)
- Title page—presents an exact and complete title; usually includes the name of the person or organization for whom the report was prepared, the name of the person making the report, and the date. Arrangement of these items varies. See pages 319, 341, 349, 360, 397, and 565 for sample title pages.
- Table of contents—shows the reader the scope of the report, the specific topics (headings) that the report covers, the organization of the report, and the page references. Headings in the table of contents should be exactly as they appear in the report. See pages 399 and 566 for sample tables of contents.
- List of tables and figures—lists all tables and figures included in the report. Captions for tables and figures in this list are the same as those used within the report.
- Summary or abstract—presents the content of the report in a highly condensed form; a brief, objective description of the essential, or central, points of the report. This part may include an explanation of the nature of the problem under investigation; the procedure used in studying the problem; and the results, conclusions, and/or recommendations. The summary is usually no more than 10 percent of the length of the report, but it *represents* the entire report. (For further discussion and examples of abstracts, see pages 296–299, 350, 400, 401, 567.)

Report Proper

- Introduction—gives an overview of the subject; indicates the general plan and organization of the report, provides background information, and explains the reason for the report. More formal introductions give the name of the person or group authorizing the report, the function the report will serve, the purpose of the investigation, the nature and significance of the problem, the scope of the report, the historical background, the plan or organization of the report, a definition or classification of terms, and the methods and materials used in the investigation.
- Body—presents the information; includes an explanation of the theory on which the investigation is based, a step-by-step account of the procedure, a description of materials and equipment, the results of the investigation, and an analysis of the results. The body may also include graphics and other visuals.
- Conclusions and recommendations—deductions or conclusions resulting from investigation and suggested future activity. The basis on which conclusions were reached should be fully explained, and conclusions should clearly derive from evidence given in the report. The conclusions are stated positively and specifically and are usually listed numerically in order of importance; the recommendations parallel the conclusions. Conclusions and recommendations may appear together in one section, or they may be listed under separate headings.

End Matter

- Bibliography—lists the references used in writing the report, both published and unpublished material. (For a discussion of bibliography forms, see pages 547–556.)
- Appendix—includes supporting data or technical materials that contain information that supplements the text but if given in the text would interrupt continuity of thought. Appendix information is useful but should not be essential to reader understanding.

Other end materials might include a glossary or an index.

Examples of formal reports appear on pages 348–353, 397–406, and 565–578.

REMINDER

The report author chooses an appropriate format for the audience, purpose, and situation.

DESIGN OPTIONS IN REPORTS

Critical to the effectiveness of a report is the choice of design options to support reader needs. Document design includes choices of paper, binding, report elements and their order, graphics and other visuals, kinds and sizes of typefaces, uses of color, and so on.

There are also many options for page layout, placement of material (text and graphics), on the page. Ample white space is needed so that the text will be visually appealing and the reader can easily retrieve information. Options include lists, emphasis markers, boxes, sidebars, columns, fonts, and type sizes.

See pages 56–69, 72 for a detailed discussion of design options.

USE OF GRAPHICS AND OTHER VISUALS

Frequently, graphics and other visuals make information clearer, more easily understood, and more interesting. A report author must decide whether visuals will make the report more accessible for the reader, and if so, what kinds of visuals and what placements will be most effective. For a detailed discussion of graphics and other visuals, see Chapter 5.

TYPES OF REPORTS

Types of reports commonly used in the workplace include the observation report, the progress report, and the feasibility report. The following discussion examines each of these types in terms of purpose, uses, main parts, and organization. The examples of each type illustrate various report formats.

Observation Report

Purpose

The observation report records observable details. It may describe a particular location or site (sometimes called a field report); be a collection of information about an existing condition; or present results of experimentation, research, or testing. Other sources of information for the observation report may be personal observation, experience, or knowledgeable people. (See pages 520–522, for how to conduct an informational interview.)

Uses

The observation report can be used in many ways. For example, such a report may be important in estimating the value of real estate or the cost of repairing a house; establishing insurance claims for damage from a tornado or a blizzard; improving production methods in a department or firm; choosing a desirable site for a building, highway, lake, or computer lab; or providing an educational experience for a prospective employee or an interested client. The observation report may include the results of tests, such as a blood type. Some reports may become quite involved, not only describing the experiment, but also giving test results and applying the results to specific problems or situations. The observation report may or may not include recommendations.

Main Parts

Since the observation report has numerous uses and includes various kinds of information, it has no established divisions or format. The report may include sections on a review of background information, an account of the investigation or description of the observed activity or surroundings, an analysis and commentary, and conclusions and recommendations. An observation report on a visit to a company might include a description and explanation of physical layout, personnel, materials, and equipment; the activities that reflect the major function of the company; and evaluative comments.

An observation report that focuses on experimentation might include sections (and headings) such as Object (or Purpose), Theory (or Hypotheses), Method (or Procedure), Results, Discussion of Results, or Comments, and Conclusions/Recommendations. A longer report may include Appendixes and Original Data.

Organization

The organization of the report may vary depending on the subject of the report, the purpose, and the audience. The beginning of the report may state the purpose of the report; the specific site, facility, or division observed; and the aspects of the subject to be presented. The report may then continue with results of the investigation, conclusions, and recommendations. If the report focuses on experimentation, headings such as those listed above may be combined or rearranged to suit the author's and audience's needs.

REMINDER

An observation report presents description and analysis of observable details.

Student Examples: Observation Reports

Two examples of observation reports follow. The first (preceded by the completed Project Plan Sheet and the completed Organization Plan Sheet) is a student's report on a visit to the social work department of a hospital. The second report concerns the field inspection of a pond dam and uses a more formal report format, including a transmittal letter, title page, and summary.

PROJECT PLAN SHEET

FOR REPORTS

Audience

- Who will read the report?

  People considering going into Speech-Language Pathology and people thinking about going into Social Work

- How will readers use the report?

  Speech-Language Pathology majors and Social Work majors will use the report to better understand how social problems influence health conditions.

  The report will help readers to understand that the social, psychological, physical aspects of a person are interrelated. The report may help readers who are considering social work as a career to make a decision.

- How will your audience guide your report choices?

  I need to choose details that will be useful to individuals considering Speech-Language Pathology and Social Work as a career.

Purpose

- What is the purpose of the report?

  To report my observations of the Social Work Department at St. Dominic Hospital

- What need will the report meet? What problem can it help to solve?

  The report will give details about a hospital social work department. It can help readers trying to make a career choice.

Subject

- What is the report's subject matter?

  What goes on in a hospital Social Work Department

- How technical should the discussion of the subject matter be?

  Only slightly technical

- Do you have sufficient information to complete the report? If not, what sources or people can help you to locate additional information?

  No. An interview with one or more social workers at the hospital

- What title will most clearly reflect the report's subject and purpose?

  Hospital Social Work Department

Author

- Will the report be a collaborative or an individual effort?

  Individual

- How can the developer(s) evaluate the sources of the completed report?
  I will ask a social worker at St. Dominic to verify the content of the report.

Project Design and Specifications

- Are there models for organization of forms for reports?
  Yes
- In what medium will the completed report be presented?
  Written
- Will the report require graphics or other visuals? If so, what kinds and for what purpose?
  I will include examples of preprinted report forms that social workers use.
- What information design features can best help the report's audience?
  Headings and appendix

Due Date

- What is the final deadline for the completed report?
  December 4, 2001
- How long will the report take to plan, research, draft, revise, and complete?
  Two weeks (six class meetings)
- What is the timeline for different stages of the report?
  1st on Nov. 15, visit Social Work Department at St. Dominic Hospital; 2nd (Nov. 18), class meeting, Project Plan Sheet; 3rd (Nov. 20), informal oral report on site visit and interview; 4th (Nov. 22), Organization Plan Sheet; 5th (Nov. 27), preliminary draft of report, 6th (Dec. 4), final draft of report

ORGANIZATION PLAN SHEET

FOR REPORTS

Introduction

1. How can the introduction clearly indicate the report's subject, scope, and purpose?

   The introduction will state when and why I visited St. Dominic Hospital and the areas I observed.

Discussion

1. What are the main parts of the discussion?

   Physical Layout, Personnel, Materials and Equipment, Activities

2. In what sequence should the main parts appear?

   Sequence as listed in the overview statement in the introduction

3. What organizational patterns should be used to develop each part?

   Several patterns of organization have been selected, spatial, sequential cause-effect; the overall pattern is partition.

4. What information design elements indicate the report's sequence and organizing patterns?

   Headings

5. If graphics or other visuals are needed, where do they appear in the sequence?

   I will include a 4-page Psychological Assessment Record as an appendix.

Closing

1. How can the closing reinforce the report's purpose?

   I will conclude with a positive statement about how the social workers try to look at all aspects of a patient's well-being.

A Hospital Social Work Department

Jessica Barlow

English 1113 Section AL
December 4, 2001

A Hospital Social Work Department

On 1 December 2001, I visited the Social Work Department at St. Dominic Jackson Memorial Hospital. While there, I observed the social work referral process and talked to a social worker about the nature of her job. I also observed details of the department such as the physical layout, the personnel, the materials and equipment, and the activities of the social worker.

Physical Layout

The Social Work Department consists of a waiting area that contains chairs for the comfort of family members who are waiting to talk with a social worker. The secretary's desk is located in this area. Each social worker has a small individual office, which is necessary to provide privacy for interviews. A conference room is located on the first floor near the Social Work Offices. The conference room is used for staff meetings, as well as for support group meetings that are offered to the community.

Personnel

The Social Work Department consists of four full-time social workers, all of whom have a master's degree in social work. A social worker is available to work on an as-needed basis when hospital census is high or when regular staff members are unavailable. A secretary/receptionist performs clerical duties and greets family members. A director is responsible for clinical supervision of the social work and clerical staff.

Materials and Equipment

The Social Work area includes various pieces of office equipment, such as a copy machine, a fax machine, and a personal computer. There are numerous filing cabinets that contain patient information, as well as extensive resource information for the social workers. There are also office supplies such as pens, legal pads, and staplers.

Activities of the Social Worker

Consultations are initiated by physicians or other members of the health care team. These consultations are communicated to the Social Work Department through the hospital computer system. The social work secretary retrieves the consultation from the computer and relays it to the social worker who is assigned to the specific unit in which the patient is hospitalized. The social worker first reviews the chart, interviews the patient and/or family to make an assessment of the patient's psychosocial needs, and determines what intervention is needed to address the identified problems. The social worker communicates his/her assessment and recommends follow-up activities in the progress notes section of the patient's chart.

Health problems and social problems are interrelated; health problems affect social situations and social problems affect health. For example, a battered woman requires more than medication to address her life-threatening situation. People living in poverty who cannot pay for needed medications and lack adequate nutrition or heat for their homes will more likely benefit from their medical treatment if they also receive intervention for their social problems from their social worker.

Medical problems affect a person's ability to function. For instance, an elderly woman who lives alone and breaks her hip in a fall cannot be discharged from the hospital in three days without extensive discharge planning. The social worker suggests resource options to help her meet her needs. For example, short-term nursing home care may be required. If home care is desired, arrangements may be needed for home nursing and physical therapy visits, as well as medical equipment. For patients who are experiencing psychiatric problems, the social worker is often called on to complete "psychosocial assessments." This extensive information assists the psychiatrist in determining a diagnosis and helps the health care team to better the patient's situation. Appendix 1 is an example of a psychosocial assessment form.

Conclusion

This profession demands patience and good communication skills. Social workers are dedicated workers and each day they must be ready to listen to their patients and help them to plan their actions once they are discharged from the hospital. A visit to the Social Work Department at St. Dominic Jackson Memorial Hospital could broaden the perspective of Speech-Language Pathology majors who look forward to one day working alongside social workers as Speech-Language Pathologists.

Appendix 1. Psychosocial Assessment

Patient's Name__________________________ Admission Date______________ Date__________

I. Identifying Data: Age_____________ Sex__________ Marital Status____________________

Occupation: ____________________________ Patient's Living Situation: ___________________

Informant: ____
(NAME) (RELATIONSHIP)

II. History of Presenting Problems: (Precipitating Events, i.e. behavioral and interpersonal changes, losses, major symptoms, recent conflict with family members/others)

____
____
____
____
____
____
____
____
____
____
____
____
____
____

III. Psychiatric/Medical History:

Psychiatric Inpatient/Outpatient Treatment: ____________________________________

____
____
____
____
____
____
____

Patient	Room #	Physician	MR #	Acct #

Psychosocial Assessment Record

History of Substance Abuse: ____________________

Medical Problems: ____________________

Family History of Mental Illness/Substance Abuse: ____________________

IV. Family Background:

Parents (Place of birth, nature of relationships, home environment): ____________________

Child Development: ____________________

Siblings: (Nature of Relationships) ____________________

Patient	Room #	Physician	MR #	Acct #

Psychosocial Assessment Record

History of Physical/Sexual Abuse: ______________________________

V. Marital History: (Marriages, separations, problems) ______________________________

VI. Children: (Names and ages, nature of relationships) ______________________________

VII. Education/Vocational:

Schools and Academic Performance: ______________________________

Military: ______________________________

Disability: ______________________________

Work History: (Places worked, lengths of employment, nature of relationships with employer[s])

Patient	Room #	Physician	MR #	Acct #

Psychosocial Assessment Record

Financial Resources: ______________________________

____

____

VIII. Social:

Spiritual Orientation: ______________________________

____

Peer Relationships: ______________________________

____

Hobbies, Interests: ______________________________

____

IX. Patient's Strengths and Weaknesses: (Social Worker's Impressions)

Strengths: ______________________________

____

____

Weaknesses: ______________________________

____

____

X. Discharge Planning/Recommendations:

Family's Understanding of Patient's illness and Their Expectations of Treatment: ______________________________

____

____

____

Support Systems: (Home, community) ______________________________

____

____

Plan for Disposition: ______________________________

____

XI. Impressions: ______________________________

____

____

____

Social Worker

Patient	Room #	Physician	MR #	Acct #

Psychosocial Assessment Record

1135 Combs Street
Jackson, MS 39204
6 April 2001

Mr. Harry F. Downing
4261 Marshall Road
Jackson, MS 39212

Dear Mr. Downing:

Attached is the field inspection report of the pond dam on your property. The dam is considered to be in stable physical condition although some minor seepage and erosion were discovered. Recommendations for correcting these are included in the report.

It has been a pleasure to work with you on this project.

Sincerely yours,

Paul Kennedy

Paul Kennedy

FIELD INSPECTION REPORT

OF

HARRY F. DOWNING DAM

Paul Kennedy

6 April 2001

Observation Report, Summary

Summary

On 2 April 2001 an unofficial inspection of the dam of the Harry F. Downing Pond was conducted by Paul Kennedy.

This dam is considered to be in stable physical condition although some minor seepage and erosion were discovered. This conclusion was based on visual observations made on the date of the inspection.

FIELD INSPECTION REPORT
HARRY F. DOWNING POND
HINDS COUNTY, MISSISSIPPI
PEARL RIVER BASIN
CANY CREEK TRIBUTARY
6 APRIL 2001

PURPOSE

The purpose of this inspection was to evaluate the structural integrity of the dam of the Harry F. Downing Pond, which is identified as MS 1769 by the National Dam Inventory of 1973.

Description of Project

Location. Downing Pond is located two miles SE of Forest Hill School, Jackson, Mississippi, in Section 23, Township 6, Range 1 East (see Figure 1).

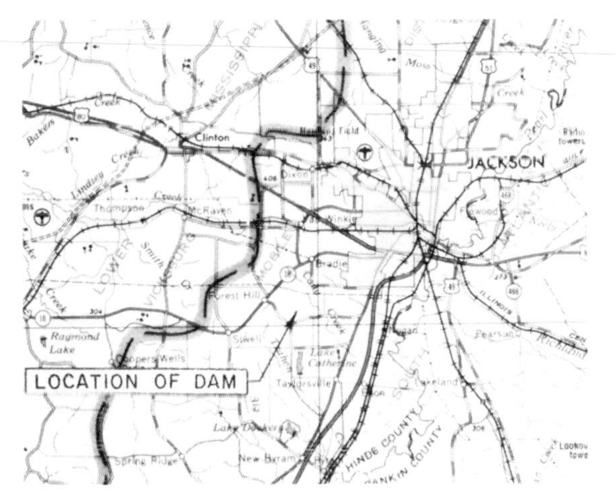

Figure 1. Location of Harry F. Downing Dam. (Map courtesy of U.S. Corps of Engineers Waterways Experiment Station at Vicksburg)

Hazard Classification. The National Dam Inventory lists the location of Downing Pond as a Category 3 (low risk) classification. Personal observation of areas downstream confirm this classification since no more than 25 acres of farmland would be inundated in the case of a sudden total failure of the dam.

Description of Dam and Appurtenances. The dam is an earth fill embankment approximately 200 feet in length with a crown width of 6 to 10 feet. The height of the dam is estimated to be 16 feet with the crest at Elevation 320.0 M.S.L. (elevations taken from quadrangle maps). Maximum capacity is 47 acre feet. The only discharge outlet for the pond is an uncontrolled overflow spillway ditch in the right (east) abutment. The spillway has an entrance crest elevation of 317.0 M.S.L. and extends approximately 150 feet downstream before reaching Cany Creek. The total intake drainage area for the pond is 30 acres of gently rolling hills.

Design and Construction History. No design information has been located. City records indicate that the dam was constructed in 1940 to make a pond for recreational purposes.

FINDINGS OF VISUAL INSPECTION

Dam. Apparently the dam was constructed with a 1V or 2H slope. This steep downstream slope is covered with dense vegetation, which includes weeds, brush, and several large trees. These trees range from 15 to 20 feet in height (see Photos 1 and 2). These trees have not likely affected the dam at the present time, but decaying root systems may eventually provide seepage paths.

A normal amount of underseepage was observed about halfway along the toe of the dam (see Photo 3). This seepage was not flowing at the time of the inspection but should be watched closely during high-water periods. The upstream face of the dam has several spots of erosion near the water's edge due to the lack of sod growth. Apparently topsoil was not placed after construction.

Overflow Spilway. The uncontrolled spillway shows no signs of erosion and is adequately covered with sod growth. (see Photo 4).

RECOMMENDATIONS

It is recommended that the owner:

1. Periodically inspect the dam (at least once a year).
2. Prevent the growth of future trees on the downstream slope.
3. Install a gauge and observe the flow of underseepage as compared to pool levels.
4. Fill areas of erosion and place topsoil and sod to prevent future erosion.

Photo 1. View looking east from left (west) abutment. Note large trees on dam at left.

Photo 2. View looking north at dam from inlet area. Note large trees on dam in background.

Photo 3. View looking west along toe of dam. Note seepage at left.

Photo 4. View looking upstream of uncontrolled overflow spillway ditch. Note adequate sod growth.

Progress Report

Purpose

The progress report gives information concerning the status of a project currently under way. Progress reports allow large businesses or organizations to keep up with what is happening within the business or organization. Supervisors, for example, might need to know if more workers will be needed to complete the project. Engineers might need to know if a design should be altered. Students might need to know if they should allot more time to a major project. A doctor might need to change the medicine or the therapy prescribed for a patient. The sales force for a company might need to change their promotion strategy. Your instructor may ask you to prepare a progress report as you work on a project.

Uses

Students and employees use progress reports to describe investigations to date, either at the completion of each stage or as requested by a supervisor. A student's progress report may signal the instructor that assistance or direction is needed. In industry and business, the progress report keeps supervisory personnel informed so that timely decisions can be made.

Progress reports answer a variety of questions. How much work has been accomplished toward completing the project? How much money has been spent? How much money is still available? Is it enough to complete the project? Is the project on schedule? Are changes needed in the project plan, performance specifications, method of collecting data, expenditures, or personnel? What unforeseen problems have arisen? The progress report should answer any questions about the progress (How far have we come?), the status (Where are we now?), and the completion of the project (What remains to be done?).

Main Parts

The progress report introduces the subject, describes work already completed on the project, discusses in detail the specific aspects that are currently being dealt with, and often states plans for the future. Unexpected developments or problems encountered in the investigation may be discussed in a separate section under a heading such as Problem or Unexpected Development or at the points in other parts of the report where they logically arise. A progress report might include recommendations for change in the plan or procedure. If such recommendations are made, they must be supported by reasons and an explanation of how the changes will affect the project. The recommendations would likely appear under the heading Recommendations. Recommendations would likely be presented at or near the beginning of a lengthy report, since they may be the most important information in the report and readers might otherwise miss them.

Organization

The three parts of the report (previous work, current work, future plans) form a natural, sequential order for presenting the information. Indeed, chronology or sequence is commonly used in organizing progress reports (see the report on pages 360–363 and 562–564). Another possible order for presenting information in progress reports is by activity (see the report on pages 355–359).

When progress reports are prepared for use within a business or organization, the content is often presented in memorandum form or on a preprinted form. Other commonly used forms are the business letter and the short report.

Student Examples: Progress Reports

Two examples of progress reports follow. The first describes progress on a research report. The second report was written by a student for her faculty advisor, who is recommending students for employment. Notice that both reports include specific data, not generalities.

REMINDER

A progress report presents the status of a project including work completed, current work, and work to be done.

Progress Report

Date: April 2, 2002
To: Dr. Katherine Staples
From: Kathy Judge
Subject: Progress on Research Report on Karate as a Physically and Psychologically Integrated Art

RESEARCH PERFORMED TO DATE

I have performed the following research to date:

- perused most books and magazine articles listed in bibliography of proposal
- selected other books to research
- made decision about which books and articles in bibliography to use for different sections of report
- copied possible quotations for citing, along with information for proper credit of quotation
- copied some graphics and drawings to aid in explanation of terms

OTHER WORK PERFORMED TO DATE

I have performed the following work on my project:

- have begun a rough draft of two sections of my report:
  the concept of ki
  the history of the Asian weaponless fighting arts
- have had a preliminary consultation with Master Kim Geary

WORK TO BE COMPLETED

- perusal of books recently found in the library, as well as those from the original bibliography not yet read
- rough draft of all sections
- preliminary layout including drawings, graphics, and explanatory figures
- consultation with Master Kim Geary to check accuracy of information
- typing and layout of final draft

CHANGES FROM ORIGINAL PLAN

I do not need to make any changes to my original plan; however, I am finding that the diversity of styles of martial arts makes it difficult to decide which one to focus on. I will focus on the principles that are common to all of them and use examples from the masters who are recognized universally as major figures in this art.

I will also include information from my own study of karate, including teaching methods that my teachers or I have come up with to promote the practice of karate as an integrated art.

REVISIONS IN COSTS

To date, I have spent $6.00 on copies. My original estimate was $15.00. I will only need about $5.00 more to complete copies of graphics and the final copies. I have not yet had to pay for parking since I go to the library at night. I have only spent about $9.00 on office supplies and do not expect to buy any more. My time breakdown has been 2 hours in the library and 10 reading the materials, as well as 1 1/2 hours doing the rough draft work I have already done. I had originally planned for 30 hours total, but I think I will need about 5 more before the project is completed (5 more for research and the rest for the writing of the report). I feel that my inexperience with graphics and layout will cause the time to go over the original estimate.

Revised breakdown in cost:

Copies	$ 11.00
Gas for auto	$ 10.00
Binding for report	$ 2.00
Other office supplies	$ 9.00
Coffee	$ 5.00
Time researching	$350.00
Total:	$387.00
Difference:	+$37.00

OUTLINE

I. Introduction
II. History
    A. Bodhidharma's Move to China
        1. Martial arts and Zen Buddhism
        2. Relationship between physical and spiritual
    B. Move from China to Okinawa and Korea
    C. Gichin Funakoshi's Move to Japan
        1. Standardization of Shoto-kan karate
        2. Karate as a lifestyle
III. Physical, Spiritual, and Psychological Aspects of Martial Arts
    A. The "Art" of Martial Arts
        1. The concept of "way"
            a. The whole person
            b. Modern training deficiencies
        2. The concept of "ki"
            a. Physical manifestations
            b. Process of attainment
        3. Yin Yang
            a. Integration of positive and negative forces
            b. Connection with Eastern religions
    B. Physical Training Enhanced
        1. Methods of training
            a. Kata (form)
            b. Self-defense
        2. Rank system
IV. Teaching the Whole Student
    A. Psychological Concepts
        1. Respect
        2. Humility
        3. Generosity
    B. Responsibility of Teacher in Training
    C. Responsibility of Student for Own Training
V. Learning from the Masters
    A. Proven Methods of Instruction
    B. Bringing an Ancient Art into Modern Times
VI. Practical Methods in Teaching Martial Arts in a Child's Class
    A. Practical Lesson Plans
    B. Methods of Maintaining Order
    C. Instilling a Sense of the "Art"
VII. Summary

ANNOTATED BIBLIOGRAPHY

Burns, D.J. (1977). An introduction to karate for student and teacher. Dubuque, IA: Kendall/Hunt Publishing Co.

I found this book to be a good source of the history of martial arts as well as insights into the connection between Zen and karate.

Cook, D. (1999, February). Ki energy: The universal force everyone talks about but few understand. Black Belt, 37, 81–83.

This article had a very good explanation of ki along with some good suggestions on how to harness the energy.

Draeger, D.F., and Smith, R.W. (1980). Comprehensive Asian fighting arts. Tokyo: Kodansha International.

This book discusses the history of martial arts in depth. It breaks down the martial arts by country and explains the differences as well as the similarities between them.

Egami, S. (1980). The heart of Karate-Do. Tokyo: Kodansha International.

This author was a student of Gichin Funakoshi, the founder of Shoto-kan karate. His insights into how to use inner strength (with the tale of having to start all over and learn from his young students) are particularly helpful.

Funakoshi, G. (1981). Karate-Do: My way of life. Tokyo: Kodansha International.

This is the "bible" of Shoto-kan karate. It reveals much in the way of the integrated art of weaponless fighting. It discusses all the qualities necessary to practice the art, as well as teaching methods and the link to Zen.

Kim, D., and Leland, T.W. (1978). Karate. Dubuque, IA: Wm. C. Brown Company Publishers.

This contains some history, discussion of modern martial arts philosophy, and behavior of students in training. It lists the unwritten laws for showing respect that all students should practice.

Mattson, G.E. (1974). The way of karate. Rutland, VT: Charles E. Tuttle Company.

This book discusses the relationship between Zen and karate. It also contains martial arts legends.

Mormon, H.E. (1998, May). Teaching in the 21st century. Black Belt, 36, 80–83.

This article discusses the importance of values in the teaching of martial arts.

Norris, C., with Hyams, J. (1988). *The secret of inner strength: My story.* Boston: Little, Brown and Company.

Chuck Norris, the actor, is a true martial artist. His book contains many suggestions in regard to building inner strength while training. It goes one step further and tells how to carry these traits into everyday living.

Pyung, K.S., as told to Fine, R. (1990). History of Cha Yon Ryu. Houston, TX: self-published.

The author of this book is my Grandmaster. This book contains history of this particular martial art, as well as qualities students should strive to achieve in their training. These qualities are represented on the school patch, which I will use as a graphic.

Webster-Doyle, T. (1990). Facing the double-edged sword: The art of karate for young people. Ojai, CA: Atrium Publications.

This is a comprehensive study of how to teach karate to children by emphasizing the psychological along with the spiritual.

Webster-Doyle, T. (1986). Karate: The art of empty self. Berkeley, CA: North Atlantic Books.

A series of reflections by the author into the practice of karate as an art.

MY COLLEGE PROGRESS REPORT
FROM
MISSISSIPPI STATE UNIVERSITY
AND
HINDS COMMUNITY COLLEGE:
PURPOSE, PROGRESS, AND PROJECTIONS

Debbie Davis
Dr. Pickett
March 23, 2001

My College Progress from Mississippi State University and Hinds Community College: Purpose, Progress, and Projections
Debbie Davis
March 23, 2001

Although I experienced starts and stalls in my higher education progress from 1977 until the present, I now have a definite goal. I want to be an interpreter for the deaf. I attended Mississippi State University from the Fall of 1977 until the Spring of 1979. In the fall of 1998, I enrolled at Hinds Community College. Unsure of what I wanted to do with my life, I decided to start in Business and Office Technology. After taking many business courses, I decided that being a secretary was not what I wanted for the rest of my working life.

In the Fall of 1999, I enrolled in Interpreter Training Technology. The program requires a minimum of 64 semester hours and 128 quality points. I will need one more year to complete the Interpreter Training classes. I will graduate in the Spring of 2001 with an Associate in Applied Science degree in Interpreter Training Technology.

1977 Fall Semester

During the fall semester of 1977, I completed the following courses with the stated hours, grades, and quality points from Mississippi State University.

Courses	Course Name	Hours	Grades	Quality Points
ENG 1103	English Comp	3	C	6
FLS 1113	Elementary Spanish	3	B	9
HI 1013	Early Western World	3	B	9
MA 1153	College Algebra	3	C	6
PS 1013	American Government	3	C	6
		15		36

My grade point average for the semester was 2.4 on a 4.0 scale. I took general courses because I was unsure of what I wanted to accomplish in college.

1978 Spring Semester

I completed these courses at Mississippi State University with the stated hours, grades, and quality points.

Courses	Course Name	Hours	Grades	Quality Points
FLS 1103	Elementary Spanish II	3	B	9
HI 1023	Modern Western World	3	B	9
MIC 1113	Elementary Micro	3	C	6
SO 1003	Intro to Sociology	3	D	3
		12		27

My grade point average for the semester was a 2.25 on a 4.0 scale.

1978 Fall Semester

Courses	Course Name	Hours	Grades	Quality Points
HE 1523	Art in Dress	3	C	6
HE 1701	Survey of Home Ec	3	A	12
PSY 1013	General Psychology	3	C	6
SO 1503	Marriage and Family	3	B	9
		12		33

My grade point average for the semester was a 2.75 on a 4.0 scale.

1979 Spring Semester

Courses	Course Name	Hours	Grades	Quality Points
EPY 1073	Science Public Health	3	B	9
MIC 1123	Psychology of Adoles	3	C	6
		6		15

My grade point average for the semester was 2.5 on a 4.0 scale. My grades show that I was uninterested in school while attending Mississippi State University. I believe I was too young and immature to appreciate the opportunity to go to school. I dropped out of Mississippi State University after this semester because I thought I wanted to be in the real world. I found out the real world was not as glamorous as I thought.

1998 Fall Semester

I returned to college this time at Hinds Community College on the Rankin Campus. I needed to learn a skill to help support my family. I decided I would enroll in Business and Office Technology.

Courses	Course Name	Hours	Grades	Quality Points
BOT 1013	Keyboarding I	3	B	9
BOT 1133	Information Processing	3	A	12
BOT 1213	Professional Devel	3	A	12
BOT 1313	Business Math	3	A	12
BOT 1413	Records Management	3	A	12
BOT 1713	Mechanics of Comm	3	A	12
		18		69

My grade point average for the semester was 3.8 on a 4.0 scale. I was proud of my success as a student. I proved I could go back to school, but I began to wonder if I had chosen the correct field of study.

1999 Spring Semester

Courses	Course Name	Hours	Grades	Quality Points
BOT 1113	Document Form/Proc	3	B	9
BOT 1143	Word Proc Appl	3	A	12
BOT 1433	Business Accounting	3	A	12
BOT 1813	Electronic Spreadsheet	3	A	12
BOT 2323	Database Mgmt	3	A	12
BOT 2813	Business Comm	3	A	12
		18		69

My grade point average for the semester was 3.8 on a 4.0 scale. I was reading the Hinds Community College handbook when I saw the information on Interpreter Training Technology.

I got very excited about learning a skill that I always dreamed of being able to use, sign language. I decided that was what I wanted to do with my life. In the fall of 1999, I transferred to the Raymond Campus of Hinds Community College and began my study in Interpreter Training Technology.

1999 Fall Semester

Courses	Course Name	Hours	Grades	Quality Points
BOT 2413	Computerized Acct	3	A	12
IDT 1113	Intro to Interpreting	3	A	12
IDT 1131	Exp/Recp Finger Spelling	3	A	12
IDT 1143	Foundations Deafness	3	A	12
IDT 1164	American Sign Lang I	3	A	12
		15		60

My grade point average for this semester was a 4.0 on a 4.0 scale. I decided that I would stay in this field.

2000 Spring Semester

I am currently enrolled in courses to earn an Associate in Applied Science degree specializing in interpreting. I am taking the following courses.

Courses	Course Name	Hours	Midterm Grade
ENG 1123	English Comp II	3	A
IDT 1173	Transliterating I	3	A
IDT 1174	American Sign Lang II	3	A
IDT 2323	Artistic Interpretation	3	A
SPT 1113	Oral Communications	3	A
		15	

During the Fall of 2000 and the Spring of 2001 I plan to take the following courses to complete my associate degree.

Courses	Course Name	Hours
IDT 2123	American Sign Lang III	3
IDT 2163	Sign to Voice I	3
IDT 2173	Interpreting	3
IDT 2183	Transliteration II	3
IDT 2153	Interpreting in Special Setting	3
IDT 2223	Educational Interpreting	3
IDT 2263	Sign to Voice II	3
IDT 2424	Practicum	4
		25

Upon completion of these courses, I will have more than the required 64 hours and the 128 quality points to obtain an associate degree.

My Future

I will graduate from Hinds Community College in May of 2001 as an interpreter with an Associate in Applied Science degree. When I complete my degree, I will be qualified to take the QA Interpreter Exam and enter the interpreting field in a choice of several positions.

Feasibility Report

Purpose

The feasibility report is a systematic analysis of what is possible and practical, of what can be accomplished or brought about. The feasibility report offers answers to questions such as: Should we do this? Which of these choices should I select?

Uses

Feasibility reports are frequently used in business, industry, government, and the corporate world. The data they present serve as the basis for making significant decisions. Should our company move its central headquarters from this geographical area to another part of the country? Should we rent, lease, or buy our equipment? What is the best location in this city for my fast-food franchise? Which investment company can meet the needs of my client? Which marketing strategy should I adopt? Whatever the situation, the feasibility report provides a complete, accurate analysis of the possibilities and presents recommendations.

Main Parts

Typically, the main parts of the feasibility report are:

Summary
Conclusions
Recommendations
Introduction: background, purpose, definition, scope (criteria)
Method for gathering information
Discussion

The main parts, and thus the headings, may vary, depending on the subject, audience, and other considerations.

Organization

The summary, conclusions, and recommendations may be combined. They usually come first in a feasibility study because these items—particularly the recommendations—are the focus of the report. Then follows a section that gives background information, explains the purpose of the study, defines and describes the subject, and explains the scope of the study. The scope of the study lists the criteria—that is, the standards by which an item is judged or a choice is made.

The lengthiest section of the report is the discussion; here, data are given with analysis and commentary on each possible solution or option, and all recommendations are substantiated. If you are asked to recommend a suitable site for dumping hazardous waste, for example, your recommendation will have far-reaching implications—financial, ecological, environmental, and human. Your analysis, therefore, of the feasibility of locating the dump site in a particular area must be thorough. A casual look at the areas is not a sufficient basis for a recommendation. You must collect as much relevant data as possible. Are there other similar sites used for dumping hazardous waste? How have persons living in the area been affected? How long have the sites been used for dumping? What are the similarities/differences between these sites and the proposed site? Is the proposed site accessible for transporting the waste at a minimum level of danger? At what cost? Is the land available? At what cost? You would have to answer these and many other questions before you could justifiably recommend for or against the proposed dumping site.

The discussion section of the feasibility report may be organized using one of the three main comparison-contrast patterns (point by point, subject by subject, similarities/differences). The most commonly used pattern is point by point, that is, criterion by criterion (a criterion is a standard by which an item is judged). The criteria, for instance, for selecting a new car might be cost, gasoline mileage, standard equipment, and warranty coverage. These criteria might then become headings in the report, each followed by a discussion of relevant data.

REMINDER

A feasibility report presents a recommended course of action based on a logical, organized analysis of possibilities.

Visuals can enhance the meaning and clarity of the feasibility report. If the report recommends purchasing new office furniture to complement the existing carpet and walls, color photographs would reinforce the recommendation. Tables and graphs are often used to compare facts and features. A table could clearly show the criteria—cost, gasoline, mileage, standard equipment, and warranty coverage—as each applies to individual cars considered for purchase. See Chapter 5 Visuals.

Student Example: Feasibility Report

An example of a feasibility report follows. In the study of compact sport pickup trucks, note how the data are concisely presented in a table.

TO: The Reverend Danny Wells
FROM: James David Wells
DATE: 6 April 2001
SUBJECT: FEASIBILITY REPORT OF AVAILABLE COMPACT SPORT PICKUP TRUCKS

Recommendation: While any of the pickup trucks looked at would be acceptable, I recommend the GMC Sonoma. It meets all of the criteria. The price of $17,445.00 is also within the established cost limit.

Purpose: The purpose of this feasibility study is to locate and purchase a compact sport pickup truck for towing trailers, transporting lawn and garden equipment, and moving music equipment to and from Pleasant Hill Baptist Church. The truck will be used by all members of the Wells family. The criteria established by the family for the truck are the following:

- Towing capacity within 5,000 lbs.
- Braking distance (55-0 mph-ft) within 180 ft.
- Acceleration (0-60 mph-sec) within 14 seconds
- Acceleration (standing 1/4-mile-sec) within 18 seconds
- Fuel economy with at least 16 miles per gallon
- Cost within $20,000.00

Method: I visited six local car retailers: East Ford Inc., Blackwell Chevrolet Inc., Wilson Dodge Inc., Fowler Buick GMC Inc., Mark Escude Nissan Inc., and Northpark Mazda Inc. I also looked at the February 2001 issue of *Sport Truck*. I found many possible pickup trucks for purchase, but I limited my selection to six for serious consideration.

Data: The six pickup trucks considered include a Chevrolet S-10, Dodge Dakota, Ford Ranger, GMC Sonoma, Mazda B4000, and a Nissan Frontier.

Data for the six pickup trucks are shown in the accompanying table.

Only two of the six pickup trucks had a towing capacity of 5,000 lbs. or more. The Dodge Dakota had a towing capacity of 6,800 lbs. while the GMC had a capacity of 5,500 lbs. The Mazda B4000 had a towing capacity of 4,000 lbs. The Chevy S-10, the Ford Ranger, and the Nissan Frontier all had a towing capacity of 2,000 lbs. Each of the pickup trucks stopped within 180 ft. from doing 55 mph except the Ford Ranger which stopped at a distance of almost 200 ft. Four of the pickup trucks accelerated to 60 mph within 14 seconds, the Nissan Frontier made it in 14.5 seconds, and the Chevy S-10 made it in 14.56 seconds. The Dodge Dakota, the GMC Sonoma, and the Mazda B4000 accelerated in the 1/4 mile within 18 seconds. It took well over 18 seconds for the Chevy S-10, the Ford Ranger, and the Nissan Frontier to accelerate in the 1/4 mile. Each of the pickup trucks got at least 16 miles per gallon except the Dodge Dakota, which got 12.3 mpg. With the exception of the Mazda B4000, each of the pickup trucks costs $20,000.00 or under.

Sport Pickup Trucks for Possible Purchase

Vehicle Make & Model	Towing Capacity	Braking Distance (55–0 mph)	Acceleration (0–60 mph)	Acceleration (standing 1/4–mile sec.)	Fuel Economy (mpg)	Cost
Chevrolet S-10	2,000	150	14.56	19.48	19.1	$15,005.00
Dodge Dakota	6,800	147	8.22	16.17	12.3	$20,000.00
Ford Ranger	2,000	198	13.18	19.05	18.5	$19,325.00
GMC Sonoma	5,500	166	10.30	17.38	19.4	$17,445.00
Mazda B4000	4,000	143	11.07	17.87	16.9	$21,360.00
Nissan Frontier	2,000	179	14.50	19.62	18.9	$18,580.00

GENERAL PRINCIPLES for Reports

- Reports serve practical, immediate needs by recording factual data or communicating data that are the basis for decision making.
- Effective reports are accurate, clear, concise, and objective.
- A responsible author makes sure that reports are complete and free from bias.
- Reports may be presented in various formats. These include special formats (such as a preprinted form, a memorandum, a letter, or an online template), conventional formats (informal and formal), and oral presentations.
- Formal reports follow a stylized format with front matter, the report proper, and back matter.
- The author selects appropriate design features and visuals according to the reader's needs and the purpose of the report.
- Regardless of the report type (observation, progress, feasibility, laboratory, trip, environmental impact), the questions that guide report preparation are:

  What is the purpose of the report?

  Who is the audience?

  How will the audience use the report?

CHAPTER SUMMARY

REPORTS

Reports are a firmly established part of the business world. A report conveys information that becomes a record of observations and activities or information on which decisions are based.

There are many ways to classify reports: by type, purpose, degree of formality, length, format, method of presentation. Regardless of classification, an effective report is accurate, clear, concise, and objective. Further, a responsible author ensures that the report is complete and unbiased.

Although much reporting is oral, most reports are also recorded in writing for ease in future reference and for efficient transmittal.

Commonly used types of reports are:

- Observation report—records and analyzes details about a site, condition, facility, or experiment; may include evaluative comments, conclusions, recommendations.
- Progress report—explains the current status of a project; includes a statement about the project, work completed, work to be done, problems encountered, possible changes in the project plan, and other details that show how things are coming along in completing a project.
- Feasibility report—analyzes what is possible and practical; includes detailed data about cost, time, alternative courses of action, possible repercussions, various factors to be dealt with in reaching the best decision for the proposed action.

In presenting reports, the author must consider not only the text of the report but also graphics and other visuals, and options concerning page layout and document design. Effective reports integrate textual and visual components to best meet the audience's needs.

ACTIVITIES

INDIVIDUAL AND COLLABORATIVE ACTIVITIES

13.1. Look up the word *report* in a standard desk dictionary and list all the meanings given. Indicate the meanings that you think are related to report writing in your chosen profession.

13.2. Interview at least two people who are employed in business, industry, government, or the corporate world. Find out the kinds of communications they use in the workplace. Determine which kinds of communications could be classified as reports. Present your findings in a one-page report.

13.3. Working in teams of three of four, collect at least three to five reports. Using the criteria of audience, purpose, accuracy, clarity, conciseness, objectivity, completeness, and nonbias, analyze the reports. Present the team's analysis in a one-page report. Choose one member of the team to show the sample reports to the other class members and discuss your team's analysis of the reports. Or have each team member discuss one report.

Preparing Reports

For Activities 13.4 –13.9

a. Review the procedure for preparing the particular type of report, pages 336–337, 354–355, 364–365.
b. Adapt and complete a Project Plan Sheet, page 375.
c. Using decisions from the completed Project Plan Sheet, adapt and complete an Organization Plan Sheet, page 376.
d. Using decisions from the completed Project Plan Sheet and the completed Organization Plan Sheet, write a preliminary draft.
e. Revise, incorporating design decisions.
f. Prepare a final draft.

Observation Reports

13.4. Assume that you have been experimenting with three brands of the same product or piece of equipment in an effort to decide which brand name you should select for your office, shop, or lab. Write a report of your observations.

13.5. Visit a business or industry, a government office or department, a service facility, or some other organization (or a major division of it) to survey the general operation. Write an observation report on your visit.

13.6. Write an observation report including conclusions and recommendations on one of the topics below, or on a similar topic of your own choosing.

a. Parking problems at a particular location or site
b. On-the-job training for persons in a particular field
c. Condition of a structure (building, bridge, water tower) or piece of equipment
d. Conditions at a local jail, prison, hospital, rehabilitation facility, mental institution, nursing home, or other public institution

e. Employment opportunities in your field
f. Services provided by an organization such as the Better Business Bureau, Chamber of Commerce, Red Cross, or Salvation Army

Progress Reports

13.7. Write a report showing your progress toward reaching a particular goal, such as completing a degree or receiving a certificate in your field, attaining a specific level of achievement in production or sales, or winning a particular prize.

13.8. Write a weekly progress report on a project to be completed within a specified length of time. Using the same project, write a monthly progress report on your work.

Feasibility Reports

13.9. Write a feasibility report including conclusions and recommendations on one of the topics below, or on a similar topic of your own choosing.
a. Purchase or rental of an item, such as an office, an automobile, or a computer system
b. Selection of a site for a business
c. Choice of a particular course of action, such as changing jobs or accepting a promotion that requires moving to another part of the country

Oral Reports

13.10. As directed by your instructor, adapt for oral presentation a report you prepared in one of the activities above. Ask your classmates to evaluate your presentation by filling in an Evaluation of Oral Presentations (see page 508).

READING

13.11. Study the following "1998 Drinking Water Quality Report," which appeared in the *Hinds County Gazette,* September 30, 1999, page 7. In small groups or as a class, answer the questions for discussion.

The following is the new *Consumer Confidence Report,* a requirement by the Mississippi Department of Health concerning the Town of Utica Water Quality. If you have any questions concerning this report, please contact Melissa Martin, Town Clerk, at 885-8718.

1998 Drinking Water Quality Report
Town of Utica
(PWS ID# 0250026)

Is my water safe?
Last year, as in years past, your tap water met all U.S. Environmental Protection Agency (EPA) and Mississippi State Department of Health drinking water standards. We vigilantly safeguard our water supply and once again we are proud to report that our system has not violated a maximum contaminant level or any other water quality standard. This report is a snapshot of last year's water quality. Included are details about where your water comes from, what it contains, and how it compares to standards set by regulatory agencies. We are committed to providing you with information because informed customers are our best allies.

Do I need to take special precautions?
Some people may be more vulnerable to contaminants in drinking water than the general population. Immuno-compromised persons such as persons with cancer undergoing chemotherapy, persons who have undergone organ transplants, people with HIV/AIDS or other immune system disorders, some elderly, and infants can be particularly at risk from infections. These people should seek advice about drinking water from their health care providers. EPA/Centers for Disease Control (CDC) guidelines on appropriate means to lessen the risk of infection by Cryptosporidium and other microbial contaminants are available from the Sage Water Drinking Hotline (800-426-4791).

Where does my water come from?
Our water comes from two wells both on Well House Road in the Town of Utica. Both of these wells draw water from the Catahoula Aquifer.

Source water assessment and its availability:
Currently, our source water assessment is being prepared by the Mississippi State Department of Health. When it is completed you will be notified and copies will be made available upon request.

Why are there contaminants in my drinking water?
Drinking water, including bottled water, may reasonably be expected to contain at least small amounts of some contaminants. The presence of contaminants does not necessarily indicate that water poses a health risk. More information about contaminants and potential health effects can be obtained by calling the Environmental Protection Agency's Safe Drinking Water Hotline (800-426-4791).

How can I get involved?
Our monthly town board meetings are held on the first Tuesday of each month at 7:30 p.m. at the Utica Town Hall. We encourage all citizens who have any questions or concerns regarding their water service or other public services that the town provides to meet with us. We ask that customers who have questions concerning their water bills or regarding disruptions in service or other technical concerns, please first contact the Town of Utica Water Department at the telephone number listed below. You may also e mail any comments or questions to us at uticams@aol.com.

Other information:
You may want additional information about your drinking water. You may contact our certified waterworks operator or you may prefer to log on to the Internet and obtain specific information about your system and its compliance history at the following address: http://www.msdh.state.us/watersupply/index.htm. Information including current and past boil water notices, compliance and reporting violations, and other information pertaining to your water supply including "Why, When, and How to Boil Your Drinking Water" and "Flooding and Safe Drinking Water" may be obtained.

Melissa Martin, Town Clerk
Gloria Wilson, Deputy Clerk
Leonard Graham, Public Works Supervisor
P.O. Box 335
110 White Oak Street
Utica, MS 39175-0335
(601)885-8718
UticaMS@aol.com

Water Quality Data Table

The table below lists all of the drinking water contaminants that we detected during the calendar year of this report. The presence of contaminants in the water does not necessarily indicate that the water poses a health risk. Unless otherwise noted, the data presented in this table is from testing done in the calendar year of the report. The EPA or the State requires us to monitor for certain contaminants less than once per year because the concentrations of these contaminants do not change frequently. Some of the data, though representative of the water quality, may be more than one year old.

Terms and Abbreviations used below

MCLG: Maximum Contaminant Level Goal: The level of a contaminant in drinking water below which there is no known or expected risk to health. MCLGs allow for a margin of safety.

MCL: Maximum Contaminant Level: The highest level of a contaminant that is allowed in drinking water. MCLs are set as close to the MCLGs as feasible using the best available treatment technology.

Contaminants (units)	MCLG	MCL	Your Water	Range Low	Range High	Sample Date	Violation	Typical Source
Inorganic Contaminants								
Nitrate [measured as Nitrogen] (ppm)	10	10	0.169	0.169	0.169		No	Runoff from fertilizer use; Leaching from septic tanks, sewage; Erosion of natural deposits
Nitrite [measured as Nitrogen] (ppm)	1	1	0.03	0.03	0.03		No	Runoff from fertilizer use; Leaching from septic tanks, sewage; Erosion of natural deposits
Radioactive Contaminants								
Alpha emitters (pCi/l)	0	15	1.9	1.9	1.9		No	Erosion of natural deposits
Beta/photon emitters #2 (pCi/l)	0	50	3.4	3.4	3.4		No	Decay of natural and man-made deposits. *EPA considers 50 pCi/l to be the level of concern for beta particles

Contaminant(s) (units)	MCLG	AL	Your Water	# of Samples> AL *	Sample Date	Exceeds AL	Typical Source
Inorganic Contaminants							
Copper (ppm)	1.3	1.3	0.333	0	12/31/96	No	Erosion of natural deposits; Leaching; Corrosion of household plumbing systems, from wood preservatives

*AL: Action Level; The concentration of a contaminant which, if exceeded, triggers treatment or other requirements which a water system must follow.

Units Description:

ppm: parts per million, or milligrams per liter (mg/l)

pCi/l: picocuries per liter (a measure of radioactivity)

a. Does the report fulfill its purpose, as stated in the third and fourth sentences: "This report is a snapshot of. . . . Included are details about. . . . "? Support your answer.
b. In what ways does the table explain and complement the report?
c. In what ways is the question-and-answer format appropriate or inappropriate for this report to the citizens of a small rural town?
d. What is the report author's attitude? In what ways does she invite citizen participation?
e. If you were a citizen of this town, how would you respond to this report?

CASE STUDY

13.12. The following case study concerns a travel report. Read the case study; in small groups or as a class, answer the questions for discussion.

A Resounding Report

There's no question about it. Barry Russell is dejected as he sits alone in the Water Regulatory Commission cafeteria. His friend Tran Nguyen decides to ask what it's all about.

"Can I join you, Barr?" he asks, as he puts his lunch tray down next to Barry's.

"Sure, Tran. But I'm afraid I'm not very good company today. I've really been raked over the coals by the director this morning, and I may even have to pay $473.27 in travel money!"

"What? You're not talking about that series of field inspections we had to make last month, are you?"

"That's it," Barry sighs. "I am in some deep and dirty water this time."

"But we were supposed to be reimbursed for every penny of that travel money," says Tran indignantly. "We even received a memo reminding us to submit travel reports listing our activities and itemizing expenses. Didn't you send in your report this time?"

"Oh sure. But Chief Batterton didn't like my report, Tran. She says that I didn't submit all of the receipts, and she was pretty peeved that I stayed a weekend over, even though I didn't claim any reimbursement. She also was upset about my record of our findings at the two test sites in Muleshoe. Both of them had been in violation of our water regulations for the last three inspections, and they were counting on accurate test results this time."

"Why, Barry, we were told that we could stay over the weekend if we sent in a memo in advance. And what was wrong with the results, anyway? You're the best quality inspector on the team!"

"I forgot to send in the memo advising Chief Batterton about the extra time. And my notes about the inspection were a little rough. In fact, I somehow misplaced my notes of the intermediate tap sampling data. Nobody at the facility had kept them, either. Chief Batterton says that we'll have to test again in Muleshoe. I also forgot the names of the people we discussed the zinc and copper problems with at Plant #2. And the chief says I'm not using the right template. She doesn't like my spelling, either."

Tran shakes his head sadly. "Those reports are public record, Barry," he says. "There could be some big troubles waiting for us in Muleshoe. I sure wish you'd taken better notes."

QUESTIONS FOR DISCUSSION

a. In what ways has Barry misunderstood the purpose of his travel report?
b. Who are Barry's audiences? What did they need to learn from his report?
c. In what ways has Barry's carelessness called his professionalism into question?
d. How has Barry's report also involved the Water Regulatory Commission?
e. What actions must Barry take to correct his problem?
f. What actions will the Water Regulatory Commission have to take?
g. What advice do you have for Barry? For Chief Batterton?

PROJECT PLAN SHEET

FOR REPORTS

Audience

- Who will read the report?
- How will readers use the report?
- How will your audience guide your report choices?

Purpose

- What is the purpose of the report?
- What need will the report meet? What problem can it help to solve?

Subject

- What is the report's subject matter?
- How technical should the discussion of the subject matter be?
- Do you have sufficient information to complete the report? If not, what sources or people can help you to locate additional information?
- What title will most clearly reflect the report's subject and purpose?

Author

- Will the report be a collaborative or an individual effort?
- How can the report developer(s) evaluate the success of the completed report?

Project Design and Specifications

- Are there models for organization or forms for reports?
- In what medium will the completed report be presented?
- Are there special features the completed report should have?
- Will the report require graphics or other visuals? If so, what kinds and for what purpose?
- What information design features can best help the report's audience?

Due Date

- What is the final deadline for the completed report?
- How long will the report take to plan, research, draft, revise, and complete?
- What is the timeline for different stages of the report?

ORGANIZATION PLAN SHEET

FOR REPORTS

Introduction

1. How can the introduction clearly indicate the report's subject, scope, and purpose?

Discussion

1. What are the main parts of the discussion?
2. In what sequence should the main parts appear?
3. What organizational patterns should be used to develop each part?
4. What information design elements indicate the report's sequence and organizing patterns?
5. If graphics are needed, where do they appear in the sequence?

Closing

1. How can the closing reinforce the report's purpose?

CHAPTER 14

Proposals: Using Facts to Make a Case

CHAPTER GOALS

This chapter:

- Defines the proposal
- Shows the ways in which proposals can offer responsible solutions to problems
- Describes different kinds of proposals and the occasions that call for them
- Lists and explains the typical parts of proposals
- Provides a procedure for developing effective proposals

INTRODUCTION

Every organization has needs—for goods and services, for professional assistance, or for change or improvement. How can organizations meet such needs? Who will help? What will happen? How can a company be sure that a course of action will work? What is the cost?

One way that organizations meet these needs and answer such questions is through the proposal, a persuasive report that offers a plan for meeting a need or solving a problem. The proposal typically is a well thought out plan that specifies a clearly described service, product, or process for a stated price. It may also be a proposed plan within an organization to recommend a change. This chapter describes the occasions for proposals, explaining typical proposal elements and showing ways in which proposals can be used effectively and ethically. The chapter also shows that proposals are persuasive documents, clear and well-developed discussions that respond thoughtfully to readers' needs.

WHAT IS A PROPOSAL?

A proposal is a document written to persuade the reader to accept a clearly described offer of a service, product, or process—to purchase equipment, hire additional employees, change suppliers of a service or materials, authorize work on a project, redirect the work flow on an assembly line. The proposal author typically offers to provide a particular service, product, or process for a stated price. The proposal reflects the author's understanding of both sides of a business agreement—the side that has a need and the side that meets it. A proposal can also make a strong, factual case for a needed change inside an organization. While such changes may not always require cash expenditure, they will ask the organization to adapt. The author of such a proposal must understand how change affects every person involved and show how it benefits the organization as a whole. The kind of proposal you write depends on your reader, your subject, and your purpose.

A proposal offers a complete plan for the reader's consideration. It can be as short as one page or as long as a bound manuscript with hundreds of pages. Both a one-page handout offering to provide housecleaning services and a 451-page bid for a new wastewater plant are examples of proposals. Each offers to provide a particular kind of service—to clean a home or design and build a fully operative wastewater plant to city specifications.

A proposal can be written by one person or a group. It can be written in a few hours or over several months. A letter from a student government association to the college president proposing changes in the class attendance policy or a memo from a sales associate to the department manager detailing the need for additional sales persons are examples of proposals that might be written rather quickly. In contrast, a multivolume document from an aircraft manufacturer proposal team bidding on a Department of Defense contract might require a team of writers and researchers to work for several months.

Proposals in the Workplace

In the workplace, proposals are an established way of doing business. Typically, a company or an individual announces a project, requesting other companies or individuals to respond with proposals, usually by a certain date. The proposals are evaluated, and the one that most clearly addresses the original project is chosen. Sometimes, a specific company is asked to submit a proposal for a specific project.

Following is a workplace proposal that The Austin Tree Specialists submitted to Edmond Snopes, owner of Contemporary Garden Design, which is working on the garden of the property at 121 Lucas Lane. Contemporary Garden Design asked The Austin Tree Specialists to assess the work that needs to be done to make the trees on the lot healthy and attractive.

The proposal is in letter form, addressed to Edmond Snopes, who requested it. The discussion of the trees is orderly and specific, keyed to a map of the lot indicating the position of each tree. Since the reader is not a tree specialist or a botanist, the discussion is nontechnical. This discussion proposes work for each tree, providing observations and concise background information to help the reader understand each recommendation and make an informed decision about it. The section on estimates summarizes work needed to be done and states the cost. The author explains when the work can take place and the kinds of insurance coverage that protect both workers and property owners from potential damages. Last, the terms of payment and the limits of service are clearly outlined.

This letter proposal is relatively short, nontechnical, and informal. It is clearly focused on the original request: work on trees at 121 Lucas Lane. The proposal uses headings to make the parts of the report easy to follow and review, keys discussion to the diagram of trees on the lot, and explains the terms of service and obligation for both parties. Last, the author, who is also the owner of The Austin Tree Specialists, is careful to write clearly, concisely, and logically. He knows that his company and his work will be judged on his ability to propose services and carry them out professionally. This proposal was accepted.

REMINDER

A proposal respects the needs of readers with accurate, complete, and helpful information. An effective proposal wins business and fosters strong business relationships.

P.O. BOX 50061
AUSTIN, TEXAS 78763
(512) 451-7363
FAX: (512) 451-7362

May 20, 1999

Edmond Snopes, Owner
Contemporary Garden Design
P.O. Box 1779
Round Rock, Texas 78621

RE: 121 Lucas Lane

Dear Edmond,

I've looked over the property at 121 Lucas Lane both with you and on a second return visit and have the following observations, recommendations and estimates for the homeowners to consider.

Observations and recommendations:
I've included a simple sketch of the yard with the trees labeled and numbered. Please refer to it as you read along.

The trees in general are in good health with but a few exceptions. In the front yard, starting at the street, the cedar elm (#1) needs minor pruning to remove the larger dead branches (3/4" in diameter and larger).

The three yaupons (#2,3,8) should be rounded over, reducing both their height and size by 18 to 24 inches. The goal with #2 and #3 will be to bring them closer to the same shape and size. "Rounding over" means just that. The plants will not have a real formal shape, just a rounded top. Number 8 should be brought down at least 36 inches to help it thicken up. Yaupons respond very well to being cut back. They are the favorite plant of Disney for making topiary art.

The mountain laurel recently planted by the edge of the drive is doing poorly. As you know, this plant can be difficult to transplant. When dug in the field, the tap root is usually cut. Mountain laurels usually have about a 50/50 chance of surviving transplanting. Watering should be done carefully. Too much (or too little) will kill it.

The cedars (ash junipers – #9,10,11) in the front should be left alone for the time being.

The small red oak (#6) and live oak (#4) need to be pruned to remove the small dead twigs (1/4" in diameter and larger) and lightly train. The Mexican plum (#5) needs no work.

The red oak (#7) needs to have the larger deadwood removed from its canopy (3/4" in diameter and larger).

The 2 cedar elms (#12,13) need very little work. Number 12 has a single dead branch tip (squirrel damage) and number 13 should be cleared away from the roof of the house.

The red oak, #14, needs no work.

Ash juniper #15 should have the low limb over the patio removed. No work is needed on #16 or #17.

The magnolia (#18) is suffering from the alkaline soils it finds itself growing in. A soil test should be run and then the tree should be fertilized as indicated by the soil test. This is a service we provide, and over the years we have successfully treated hundreds of magnolias in this condition. When the tree improves in health, the removal of the adjacent ash junipers may be warranted to give it both more room and light.

The shin oaks (#20) are struggling in the dense shade. Indeed, two of the 3 trunks are dead and should be removed. Perhaps all three should be eliminated.

No work is recommended on #19, #21, #22, and #23.

The red oak (#24) has been cabled. While the cabling was prudent, the method used to install the cables was poor. Replacement of the cables with correctly installed and placed cables will be needed in the future. In the meantime, these cables should be closely monitored by the homeowners.

Behind the gazebo, the small ash juniper labeled #25 should be removed. It serves no real purpose and will only damage the gazebo roof.

Cedar elm #28 should have its lowest limbs removed to increase light beneath the otherwise healthy tree.

Ash junipers #29 and #30 should have the lowest limbs removed to put more light on the yard. Ash juniper #32 should be cleared away from the roof of the garage.

Estimates:

Prune to remove large deadwood (3/4" in diameter and larger) from cedar elm #1, red oak #7
Prune to train red oak #6, live oak #4 removing small dead twigs 1/4 inch in diameter and larger, crossing limbs
Round over as stated above yaupons #2,#3,#8
Remove one dead branch tip in cedar elm #12

Clear cedar elm #13 and ash juniper #32 away from roof
Raise low limbs as mentioned above for ash junipers #15, #29, #30 and cedar elm #28
Remove 2 dead shin oaks (#20)
Perform soil test and fertilize magnolia appropriately (#18)

$1,595.00 + sales tax

Schedule:
This is a very busy time of the year. Once notified, we could begin the work in about 3 weeks.

Insurance:
As always, we are covered with general liability ($600,000.00) and a workmen's comp alternative plan ($5,000,000.00 per employee) to protect both your client and our employees.

Payment:
Payment is due in full on the day of completion. Any additional work the homeowners would like to have done would be in addition to this bid. We are always happy to accommodate any special requests while we are on the property. Additional work will be based on number of man hours worked in addition to the original estimate.

If you have any questions or would like to proceed with any or all of this proposal, please feel free to call.

Sincerely,

Patrick Wentworth
ISA Certified Arborist #TX-0019

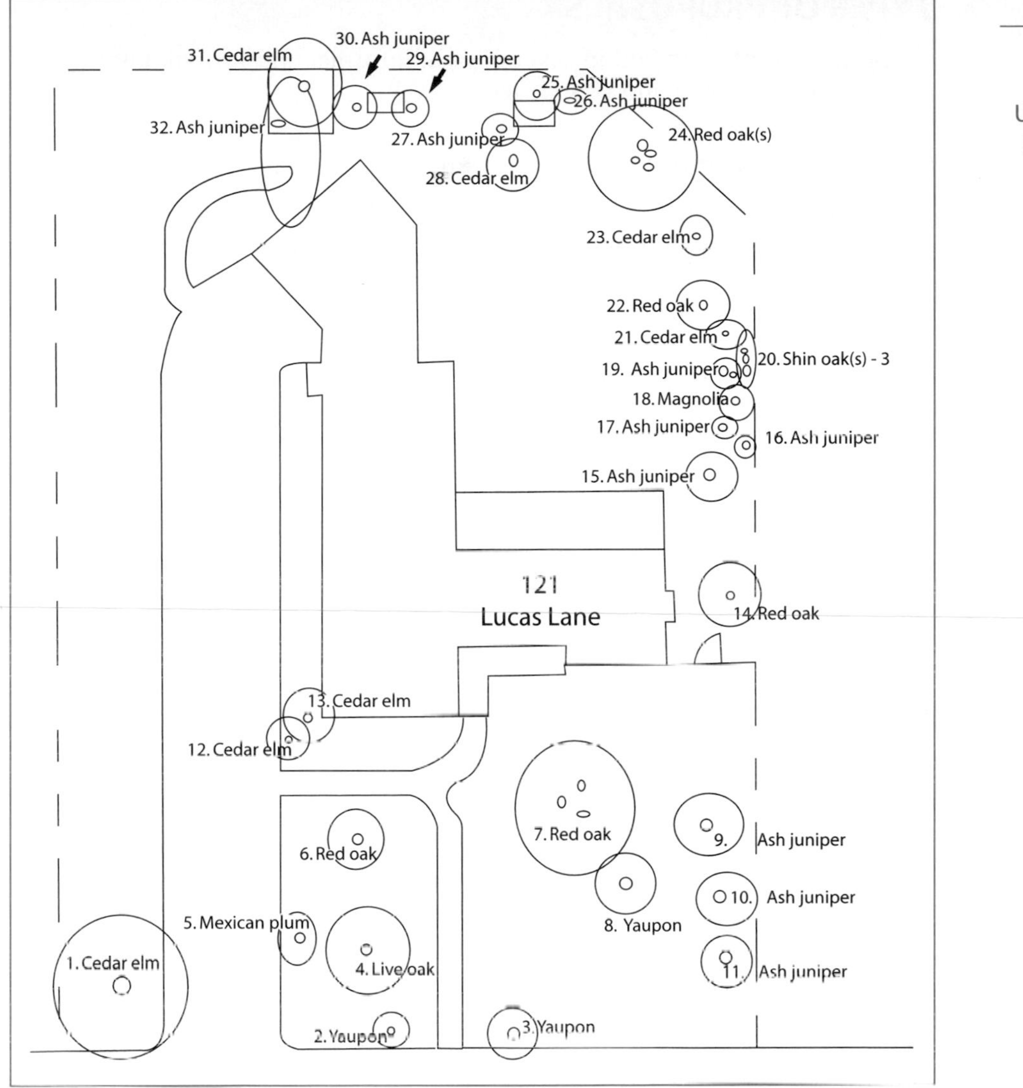
31. Cedar elm
30. Ash juniper
29. Ash juniper
25. Ash juniper
26. Ash juniper
32. Ash juniper
27. Ash juniper
24. Red oak(s)
28. Cedar elm
23. Cedar elm
22. Red oak
21. Cedar elm
20. Shin oak(s) - 3
19. Ash juniper
18. Magnolia
17. Ash juniper
16. Ash juniper
15. Ash juniper
121
Lucas Lane
14. Red oak
13. Cedar elm
12. Cedar elm
7. Red oak
9. Ash juniper
6. Red oak
10. Ash juniper
8. Yaupon
5. Mexican plum
1. Cedar elm
4. Live oak
11. Ash juniper
2. Yaupon
3. Yaupon

TYPES OF PROPOSALS

Proposals can be classified in several ways: informal and formal, internal and external, and solicited and unsolicited.

Informal and Formal

Depending on length and complexity, proposals can be classified as informal or formal. An informal proposal may be as simple as a brief statement of need and the criteria for meeting that need. An example of an informal proposal is the bid specification that describes cattle feed wanted for purchase for the farm at Hinds Community College (see pages 224–225). A formal proposal is typically longer and may include a letter of transmittal; an abstract; a title page; a table of contents; a list of graphics and other visuals; a lengthy discussion of budget, personnel, and schedule; and appendices.

Internal and External

Proposals can also be classified as internal or external according to origin. An internal proposal is initiated within a company and addresses an internal need. Typically, internal proposals recommend a change or improvement. For example, an employee at a large store realizes that the flow of customers could be greatly improved if the store layout were changed. The employee mentions the suggestion to a store manager and the two of them write a proposal, in memo format, to the district manager, who has the authority to accept or reject it. An external proposal is initiated outside the company or organization. The external proposal typically offers to provide goods or services to a client for a specific amount of money and within a certain amount of time. For example, a power company must find sources of fuel. The company seeks proposals from other companies who can supply the fuel, including the amount they can supply, within what time frame, with what kind of transportation, and at what cost.

Solicited and Unsolicited

A solicited proposal is submitted in response to a request for proposals (RFP). An organization identifies a need for goods, services, change, or improvement. Not knowing who can best meet their need, the company issues an RFP. An RFP can be as informal as a telephone call by a police department representative to ask for an estimate for repairing a patrol car or as formal as a government publication requesting bids for projects costing millions of dollars. Each request explains exactly what applicants should submit. The car repair estimate, for example, could be submitted on a form with space to describe the vehicle, the damage the mechanic has observed, the kinds of parts and services needed for repairs, the time the repairs will require, a breakdown of costs, a warranty, the business address, and signature of the repair shop owner. Such a proposal specifies what both the police department and the repair shop must agree to. It defines the limits of responsibility of the repair shop and sets forth the expectations of the police department.

Following is an example of an RFP from the town of Utica, Mississippi, seeking inspectors for two areas.

Request for Proposals for Rehab Inspectors and Asbestos Inspectors

1. The town of Utica, Mississippi, is requesting proposals for Certified Mississippi HOME Rehabilitation Inspectors to perform rehab inspections, work write-ups, and estimates.
2. The town of Utica, Mississippi, hereby notifies DEQ Certified Asbestos Inspectors that proposals for asbestos inspection services will be accepted as they relate to the implementation of the HOME Rehabilitation program.

Proposals must be received by 12:00 noon on Friday, November 26, 1999, in the Office of the Town Clerk, Town of Utica, Mississippi, 110 White Oak Street, Utica, Mississippi 39175. The prospective offer should provide the Town Clerk four (4) copies of the proposal in response to this Request for Proposal.

Proposals will be evaluated on the following criteria and relative importance: (1) Qualifications—25 points; (2) Experience—25 points; (3) Capacity—25 points; (4) Understanding—25 points. If you have any questions regarding this Request for Proposals, you may contact James Curtis Smith, Town of Utica HOME Project Administrator, 717 Thomas Lane, Madison, Mississippi 391101 or 601-856-2431.

The town of Utica reserves the right to reject any and all proposals

Reprinted by permission from Mary Ann Keith, editor, *Hinds County Gazette*.

A scholarship announcement is another type of RFP. Scholarship announcements usually list the kinds of information that review committees expect to see from applicants, including transcripts, personal and academic references, and applicant essays in response to particular questions. Often, forms are provided for applicants to complete. Successful scholarship applicants realize that they are being evaluated on more than their references and past academic performance. They are also judged by their ability to respond thoughtfully, clearly, correctly, and even neatly to all of the questions the scholarship RFP requires—and to submit the application before the deadline.

A grant funding announcement is also a type of RFP. Consider the following example of an RFP for grant funding.

The Kotmeister Foundation
1922 Madison Avenue
Suite 904
New York, NY 01162

NAME(S) OF PROGRAMS:
Travel Grant Award Program
International Language Study Program

TYPE
Direct stipends to undergraduate students to fund travel to professional or academic conferences in the student's major, to fund foreign language study programs overseas four to six weeks in duration, and awards to student research projects documented by reports.

ELIGIBILITY
Applicants must be enrolled full-time in an undergraduate degree program with a declared major.

FINANCIAL DATA
Travel grants reimburse conference fees, room, board, and travel for conferences in the applicant's area of study for up to $1,000.00.
International language study grants are awarded for up to $2,500 to pay tuition, travel, room, and board at accredited international language schools.

NUMBER OF AWARDS
23 Travel Grant Awards in 2000.
13 International Language Study Grants in 2000.

DEADLINE
None. Applicants should submit proposals outlining their request, documenting their field of study, and demonstrating how the grant will support their educational success at least 60 days before the date of the program they wish to attend or take part in.

Whether it asks for a well-documented, formal proposal or only a completed form, a published RFP gives clear guidelines about the form, length, and subject matter of proposals. Published RFPs typically appear in newspapers and journals.

The most frequently used source for U.S. government proposals is *Commerce Business Daily*. Following is an RFP from *Commerce Business Daily,* posted on CBDNET on November 16, 1999, and accessed on November 18, 1999. In this request for proposals, the Department of the Air Force is seeking a supplier for runway sand that meets the specifications outlined.

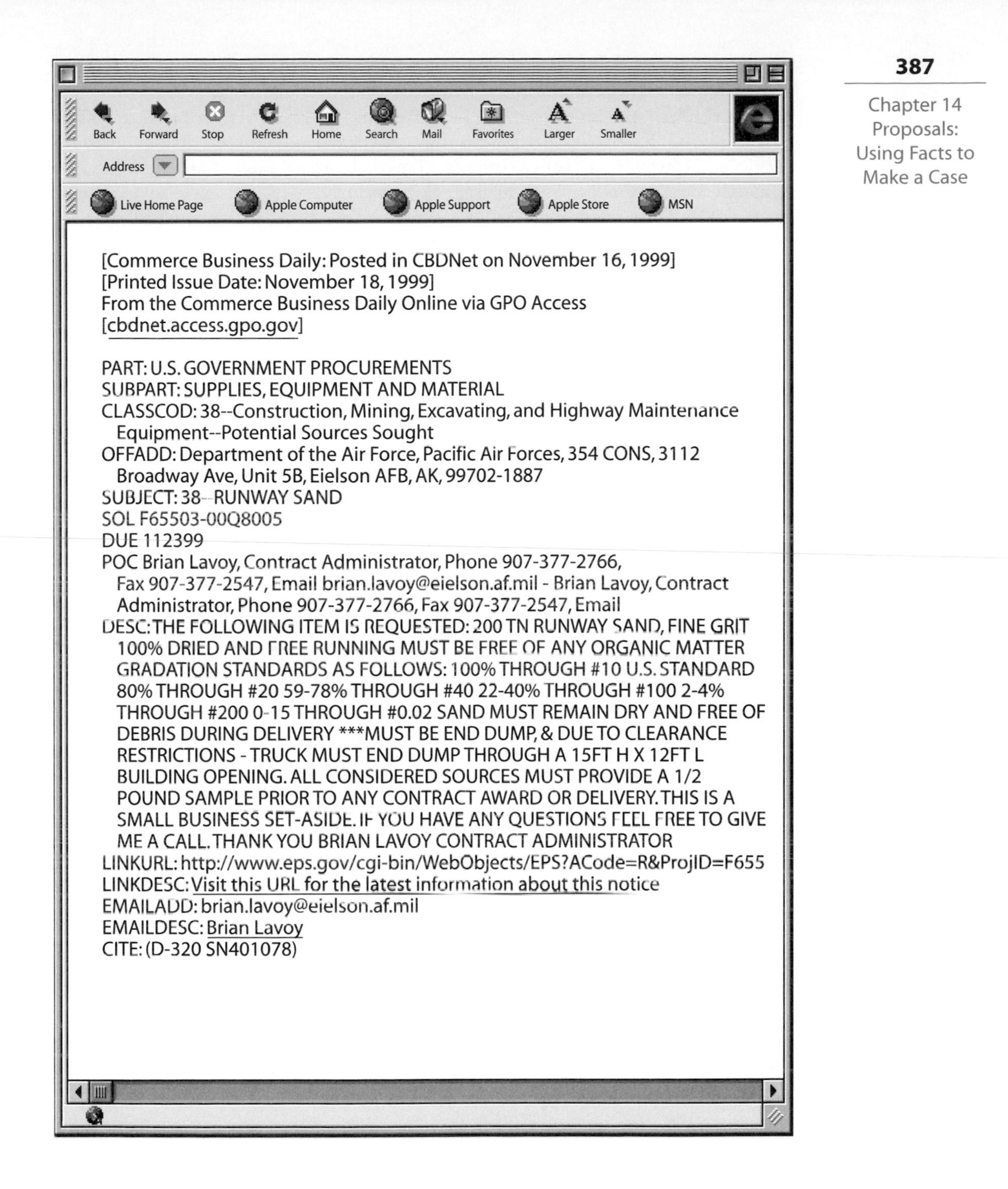

[Commerce Business Daily: Posted in CBDNet on November 16, 1999]
[Printed Issue Date: November 18, 1999]
From the Commerce Business Daily Online via GPO Access
[cbdnet.access.gpo.gov]

PART: U.S. GOVERNMENT PROCUREMENTS
SUBPART: SUPPLIES, EQUIPMENT AND MATERIAL
CLASSCOD: 38--Construction, Mining, Excavating, and Highway Maintenance Equipment--Potential Sources Sought
OFFADD: Department of the Air Force, Pacific Air Forces, 354 CONS, 3112 Broadway Ave, Unit 5B, Eielson AFB, AK, 99702-1887
SUBJECT: 38--RUNWAY SAND
SOL F65503-00Q8005
DUE 112399
POC Brian Lavoy, Contract Administrator, Phone 907-377-2766, Fax 907-377-2547, Email brian.lavoy@eielson.af.mil - Brian Lavoy, Contract Administrator, Phone 907-377-2766, Fax 907-377-2547, Email
DESC: THE FOLLOWING ITEM IS REQUESTED: 200 TN RUNWAY SAND, FINE GRIT 100% DRIED AND FREE RUNNING MUST BE FREE OF ANY ORGANIC MATTER GRADATION STANDARDS AS FOLLOWS: 100% THROUGH #10 U.S. STANDARD 80% THROUGH #20 59-78% THROUGH #40 22-40% THROUGH #100 2-4% THROUGH #200 0-15 THROUGH #0.02 SAND MUST REMAIN DRY AND FREE OF DEBRIS DURING DELIVERY ***MUST BE END DUMP, & DUE TO CLEARANCE RESTRICTIONS - TRUCK MUST END DUMP THROUGH A 15FT H X 12FT L BUILDING OPENING. ALL CONSIDERED SOURCES MUST PROVIDE A 1/2 POUND SAMPLE PRIOR TO ANY CONTRACT AWARD OR DELIVERY. THIS IS A SMALL BUSINESS SET-ASIDE. IF YOU HAVE ANY QUESTIONS FEEL FREE TO GIVE ME A CALL. THANK YOU BRIAN LAVOY CONTRACT ADMINISTRATOR
LINKURL: http://www.eps.gov/cgi-bin/WebObjects/EPS?ACode=R&ProjID=F655
LINKDESC: Visit this URL for the latest information about this notice
EMAILADD: brian.lavoy@eielson.af.mil
EMAILDESC: Brian Lavoy
CITE: (D-320 SN401078)

Unsolicited proposals are submitted to those who may need a service but who have not requested offers from anyone to perform it. The one-page handout offering cleaning services is an example of such a proposal. Individuals reading the handout may or may not have considered using cleaning services. However, a well-written and easy-to-read page that explains the kinds of cleaning services available, gives hourly costs for each service, and provides business references will be far more persuasive than just a business card with a name and telephone number.

Solicited or unsolicited, a proposal tries to answer the kinds of questions a customer might have before purchasing a service or product or implementing a process.

- Exactly what will be involved?
- How long will purchase or implementation take?
- Will the service, product, or process be reliable and efficient? How can I be sure?
- Will it make a difference?
- What is the cost?

REMINDER

Whether your proposal is unsolicited or solicited, be sensitive to your audience's needs and interests. Clear writing and complete information are good business.

Companies may seek new business with solicited or unsolicited proposals. In either case, a well-developed proposal speaks strongly in favor of the company that submits it. In order to win business, a proposal must define the reader's needs and answer any questions a prospective customer will have.

REMINDER

A proposal is a record of a promise you plan to keep and an agreement that you and your customer can understand and agree to.

QUALITIES OF EFFECTIVE PROPOSALS

Effective proposals have common qualities. They include factual information to show that you understand the audience's need or problem, that you have researched and planned a complete and reliable solution, and that you have considered the needs and expectations of your reader fully and responsibly. They also show that you are a person on whom others can depend to consider both sides of a bargain and to honor an agreement.

Effective proposals reflect an informed and comprehensive view of the need or problem they address. Writing such a proposal may require more research than simply reading the RFP. You may need to research the company your proposal will address.

- What is the size of the prospective customer's company?
- What are the company's resources and goals?
- Has the company made any recent changes?
- Have they funded other projects like the one you are proposing?
- Are copies of previously accepted proposals on file? (Accepted proposals are often public record, on file in libraries, or on the Web.)

As you develop a proposal, consider your own professional strengths. These strengths are details that demonstrate your experience and expertise.

- What can you legitimately offer to protect the interests of your customers or to demonstrate your professional expertise?
- Can you provide professional references?
- Do you have liability insurance, or can you offer warranties?
- Have you conducted similar projects that you can discuss with your customers?
- Are you a member of any professional or licensing organizations?

Last, consider the limits of your proposal.

- What can your reader reasonably expect of you, and what do you expect of your reader?
- When will you perform the service, and how long will it take?
- Will you use any consultants or special equipment?
- What fee do you expect, and what methods of payment are acceptable?
- Will you report to your customer during the service? If so, exactly when will you report?
- Will you provide related or follow-up services? If so, what fee will you charge?
- How will you notify your customer of any changes in procedure, materials, or time?
- Will changes affect the price of your service?

Put yourself in your reader's place. What would you want to see in a business arrangement that seems fair to both sides?

MAIN PARTS OF A PROPOSAL

Because proposals have various uses, may be solicited or unsolicited, may be formal or informal, and may be written by one person or a group, no set format exists. However, all proposals typically answer four basic questions:

1. What is the problem?
2. What is the solution?
3. How can the solution be implemented?
4. Why are you the best person, company, or organization for the job?

Depending on the purpose of the proposal and the audience's needs, you might include the following parts:

- Purpose or overview
- Problem (Service, Product, or Process)
- Proposed solution or plan (Who? What? How much? When?)
- Methods or procedure

- Comparison of current and proposed methods
- Equipment or material
- Cost
- Schedule
- Outline of the client's and the company's responsibilities
- Qualifications of the writer
- Recommendations or permission to implement the recommendations
- Conclusion

Proposals may also include supporting materials such as specifications; graphics and other visuals such as tables, charts, maps, photographs, line drawings; ancillary research data; or any other material that helps the reader to understand the proposal. The supporting material may appear within the proposal if it fits into the flow of the discussion. Usually, though, the supporting material will appear in an appendix.

Longer, more formal proposals may include supplementary material such as:

- Letter of transmittal
- Abstract
- Executive summary
- Title page
- Table of contents
- List of figures, illustrations, tables, etc.
- Appendixes

Remember that if you are writing a solicited proposal, you must follow the company's instructions for including supplements.

PLANNING AND WRITING A PROPOSAL

A proposal typically has three major divisions. The first (introduction) establishes that a problem exists and needs a solution; it may include any necessary background information. The second (discussion) offers a practical solution to the problem and builds a case for the solution; it includes methods, timetable, materials and equipment, personnel, and cost. The third (conclusion) summarizes the main points and recommends action.

As you plan your proposal, consider the answers to these questions:

- Who will read the proposal?
- Do the readers recognize that a problem exists, or must the proposal convince them?

- How much do the readers know about the background of the problem or circumstance?
- Who will implement the recommendations?

The Project Plan Sheet (see page 419) will guide you in identifying readers and their needs, interests, and preferences. Any proposal may be read by a diverse audience, including technicians, engineers, nontechnical readers, and clients. Determine who the main or primary readers will be and write the proposal to that audience. Supply needed information for other levels of readers through supplementary materials such as a glossary, an appendix, or a specialized explanation.

After identifying the audience, analyzing the audience's needs, and gathering information, select an appropriate format. For a brief proposal, you might use a memorandum or a business letter. In some instances, the form may be dictated by your instructor, employer, company, or RFP (request for proposal) specifications. In most proposals, you will organize the information in blocks or chunks of data with appropriate headings to make reading, understanding, and retrieving information as easy as possible. Use graphics or other visuals wherever they would enhance meaning. See Chapter 5 for a discussion of visuals.

Make specific suggestions for solving specific problems. Avoid vagueness. If you mean that you will complete the proposed plan in two weeks, do not write that you will complete the plan in a few weeks. If the cost to complete the proposal is $3,500, say so; do not write that completing the proposal will cost several thousand dollars. Make the readers feel that you understand their problem and that you are offering a well thought out solution. Do not ignore or attempt to hide possible limitations. If you promise something that you cannot deliver, you and your company could be liable in the event of failure. Promising more than you can deliver is both unethical and illegal.

REMINDER

Keep the following overriding principle in mind: write your proposal to show how the readers will benefit if they accept it.

The following steps can help you organize a proposal. Remember, however, that ultimately you must adapt your organizational plan to meet the needs of your audience and purpose; this may require rearranging, combining, omitting, or dividing steps.

- Write a summary. Note: You may want to revisit the summary after you have completed the proposal.
- Introduce the proposal.
  1. Come up with a title that clearly reflects the purpose and content of the proposal.
  2. Give an overview of the situation.
  3. Clearly state the problem.
  4. Describe, define, and explain the problem in detail.
     a. Show how and why the current situation or circumstance is unacceptable.

b. Convince the readers that you have thoroughly investigated the situation.
c. You may need to show that you have looked at significant records, observed activities, talked to appropriate persons, and researched any needed background information.
d. Include any useful supplementary material.

- Write the discussion section of the proposal.
  1. Explain how your plan will be implemented, including methods to be used.
  2. Set up a timetable for various phases of implementation.
  3. Identify any needed materials, equipment, personnel.
  4. Discuss needed facilities.
  5. Explain the cost in detail.
  6. Discuss the expected results and the probability of success.
- Write the closing section.
  1. Summarize the key points of the proposal.
  2. Recommend that the solution be implemented or, if appropriate, request permission to implement the proposed solution.
- Evaluate the proposal, considering audience and purpose. Add any appropriate supporting material.

REMINDER

Before your audience reads a word, they will *see* the proposal. The appearance must be pleasing and the information must seem accessible.

Presenting the Proposal

Be careful to present a proposal that is appealing and accessible. Use information design techniques such as fonts, point size, white space, headings, options for highlighting (listing, bold print, underlining, italicizing) to guide readers through the proposal. A neat, effectively presented proposal suggests that you have been attentive to the project and are concerned about the audience's needs.

Student Example: A Proposal

A student-written proposal for a Web page for Park Primary School appears on pages 394–406. A completed adapted Project Plan Sheet is included.

This proposal meets the needs and interests of its readers in many convincing ways. First, the proposal is clearly written and easy to follow, which is especially important for an audience of educators. From the stated title to the last page of works cited, the author has tried to show readers that every element of his plan has been carefully designed to meet their needs. The author introduces his subject and goals in the abstract, executive summary, and introduction. Readers can easily understand that the author has adapted his information to their particular school.

The author has also used information design to make his proposal easy to follow. The table of contents provides a clear guide with short and accurate headings. Each page provides ample white space, and the font is easy to read. Those who wish to read or study selected parts of the proposal can do so with ease.

What makes this proposal persuasive is the way in which the author demonstrates his expertise. He speaks directly and clearly to his readers without boasting. His discussion of the school and its students, of educational Web pages, of the technology needed to support a Web page, and of the design and subject matter he has adapted for Park Primary School all show what he has to offer clearly and thoughtfully. All of the author's research relates to his ideas and demonstrates an understanding of his diverse audience.

By working to understand his readers' needs and by using his expertise to propose a Web page design to meet those needs, the author shows how successful an unsolicited proposal can be.

PROJECT PLAN SHEET

FOR PROPOSALS

Audience

- Who will read the proposal?

  My main audience will be teachers and administrators at Park Primary School. The main audience will make the decision about what I propose. Parents, school board members, and teachers from other schools may be a secondary audience.

- How will readers use the proposal?

  My readers will use the proposal to determine whether they want a Web page for Park Primary and whether the features I have outlined are appropriate for the school. Secondary readers may use my recommendations as a basis for other educational Web pages.

- How will your audience guide your communication choices?

  I will need to consider all of the interests and needs of my main audience, being careful to discuss ways in which the proposed Web page presents Park Primary, the advantages it provides, and the educational advantages for students and teachers.

Purpose

- What is the purpose of the proposal?

  I need to develop the Web page project for Park Primary as part of my portfolio. If my proposal is accepted, I will be able to develop a professional credential while providing a public service.

- What need will the proposal meet? What problem can it help solve?

  My readers can use the proposal to learn what a Web page can do for the school and its students. If they understand the project's potential, they can determine whether the design I propose has the features their school needs.

Subject

- What is the proposal's subject matter?

  My proposal should cover my design and subject matter plans for the Park Primary School Web page. It should also show why I designed the page as I did, how the page relates to the school's needs, and whether Park Primary's existing technology can support what I propose.

- How technical should the discussion of the subject matter be?

  I want to make my discussion as easy to read and as nontechnical as possible so that it is understandable for all of my readers. I do not want to intimidate those who have little technical expertise.

- Do you have sufficient information to complete the proposal? If not, what sources or people can help you to locate additional information?

  I have already done most of the research by interviewing people at Park Primary and by reviewing information about educational Web pages. I will probably review other models and look for more information as I finish my plans for design and content.

- What title can clearly indicate the proposal's subject and purpose?

  Internet Home Page Structure and Content Recommendations for Park Primary School

Author

- Will the project be a collaborative or an individual effort?

  Individual

- How can the developer(s) evaluate the success of the completed project?

  I would like a peer reviewer who knows something about education and about Web pages to look over my revised draft to verify that I have considered all of the elements of a Web page that Park Primary would want to have.

Project Design and Specifications

- In what medium will the completed proposal be presented?

  Written

- Are there special features the completed proposal should have?

  I need to present the material so that my readers can easily locate and understand the information. Since the proposal is somewhat formal, I will include a letter of transmittal, a title page, a table of contents, and summaries as supplementary materials.

- Will the proposal require graphics or other visuals? If so, what kinds and for what purpose?

  I am considering including a title page with a picture of the entrance to Park Primary School and the image map listing the main categories of information.

- What information design features can best help the proposal's audience?

  Headings, clear fonts, and use of white space will make the text easy for readers to see, read, and skim.

Due Date

- What is the final deadline for the completed proposal?

  Since I am initiating the project, there is no deadline. However, I do not want the people I interviewed to wait too long for the proposal.

- How long will the proposal take to plan, research, draft, revise, and complete?

  Now that I have completed the interviews and the research, I need three weeks to plan, draft, and revise. I would also like to take a week for peer review and one last revision.

- What is the timeline for different stages of the proposal?

October 25–29	*Planning*
November 1–5	*Drafting*
November 8–12	*Revision*
November 15–22	*Peer Review and Revision*

INTERNET HOME PAGE STRUCTURE AND CONTENT
RECOMMENDATIONS FOR PARK PRIMARY SCHOOL

Brian Westbrook
18 November 1996

Prepared for Rebecca Green

Title I Coordinator

Park Primary School

PO Box 19772
Hot Springs, AR 98377
18 November 1996

Rebecca Green
Title I Coordinator
Park Primary School
220 Green Street
Hot Springs, AR 98377

SUBJECT: INTERNET HOME PAGE STRUCTURE AND CONTENT
RECOMMENDATIONS FOR PARK PRIMARY SCHOOL

Dear Ms. Green:

Here is my proposal for an Internet home page for Park Primary School.

I would like to thank you for granting time from your busy schedule to share with me your views about this project. This information was vital in determining your school's needs for its online presence. I would also like to thank all of those who responded to my e-mail questionnaire: A. Bisson, Barbara Campbell, Robert L. Clowers, Leni Donlan, Frank M. Flynn, Matt Freund, David Hoffman, Misty Joy Jones, Laura Lee, Gregg Legutki, Jo Lewis, Rich Rice, Charlotte Anne Robinson, and Jerry Wise.

Through our interview and my research of existing home pages, I have determined what I believe to be the best design for your school's home page. Structurally, the page should be technologically impressive while still retaining accessibility and speed. The content should be divided on a schoolwide basis into the categories of the school, the students, the faculty, and parent involvement. The school category would include the thematic projects that Park Primary uses and would be the main area of interest for the main audience of teachers at other schools.

Sincerely,

Brian Westbrook

Brian Westbrook

cc: Dr. Richard C. Raymond

Enc.: Proposal

i

TABLE OF CONTENTS

ii

ABSTRACT

Park Primary School wants to be the first school in Hot Springs to have a home page on the Internet. The school's goals for the page and its current hardware limitations are discussed in this report. Since Park uses thematic teaching among the classes, a schoolwide focus is recommended. The school should have a title page with several secondary pages for various aspects of the school such as their projects, the students, the faculty, and parent involvement. Until the school's own server is installed, the page can be posted at Geocities.

iii

EXECUTIVE SUMMARY

Working freelance as an HTML designer, I have created several pages for the World Wide Web. Along with my design experience, I also know how to publicize Internet sites. My pages have been linked from many sites including universities, movie studios, television networks, the Internet Movie Database, and Mensa. My pages have also received several awards.

I am designing a home page for Park Primary School free of charge as a showpiece for my portfolio and as a favor to Rebecca Green, Park Primary's Title I Coordinator. Park receives little respect from the other Hot Springs schools, and we would like to see it become the first local school with a home page.

Along with my own knowledge of Web publishing, I studied several school home pages and surveyed the designers of those pages. Though I saw no uniform structure for these pages, I was able to find elements that could be adapted to Park's home page. I also interviewed Ms. Green to determine the school's specific needs for their home page.

From that interview I concluded:

- Most school home pages include a mission statement, and many include school histories.
- The school has several extracurricular activities that can be highlighted.
- The technology needs to be impressive but still accessible via older equipment.
- The school's server is not yet operational, so they cannot host the page yet.

Based on my experience and research, I recommend:

- To reduce the size of the files, the home page should consist of a small title page with several secondary pages detailing various aspects of the school.
- The title page should contain a menu of options and the school's mission statement.
- The handbook, the calendar, the school newspaper, the ecology club, the post office, the faculty, and parent involvement should each have a separate page.
- Because of the thematic teaching, the focus should be schoolwide with an emphasis on the thematic projects.
- Student writing should be posted in plain text so the students can enter their own work without the teachers having to learn HTML programming.
- Since the school's own server is not yet operational, the page should be temporarily posted on Geocities. Because of space limitations at Geocities and download times, student artwork should be kept to a minimum.

1

INTRODUCTION

Park Primary School is in a predominantly low-income neighborhood, but the school's philosophy is that "all children can learn and all children can behave" (Coble 37). This statement shows that Park believes in not letting the economic level of its district interfere with the education of its students. Park hopes to be the first Hot Springs school with a home page on the Internet.

An article on *CNN Interactive* (Wilson) tells the success story of another underfunded school's home page. The high school, in Aberdeen, Mississippi, installed an Internet home page. By creating a local and familiar starting point for new users, they were able to get most of the population interested in the Internet. After a fundraising drive, Aberdeen was able to offer free public access for the entire town. Now the library has computers for patrons to use. They range in age from school children to 76-year-old Clara Thompson, who is learning to use e-mail. In Aberdeen, more than half of the families receive financial assistance from the federal government.

Park's economic level, however, does mean that few students' families have computers at home. There will also be very little use of the Internet in the classroom because of the school's lack of equipment. For these reasons, Ms. Green asked that this home page be aimed mainly at teachers at other schools, with Park's students and a few parents as secondary audiences.

Park Primary must decide what information should be offered to these three audiences and how that information should be presented. The report proposes a Web design to meet Park Primary School's needs.

DISCUSSION

Structure

- **Page Layout.** The biggest complaint of the Internet users is that many pages take too long to load. There are two main causes for this problem: slow servers and pages that are too large. Slow servers are caused by a narrow bandwidth, which is the opening through which data travels from computer to computer. Only the owners of the servers can increase the bandwidth, and therefore, controlling page size is the programmer's only option. To present a large amount of information, I recommend building a small title page (see Figure 1) with several secondary pages, each focusing on a different aspect of the school. For the menu on the title page I have designed an image map that will list the main categories of information (see Figure 2). For the sake of readers with nongraphical browsers, the same options can be offered in plain text below the image.

2

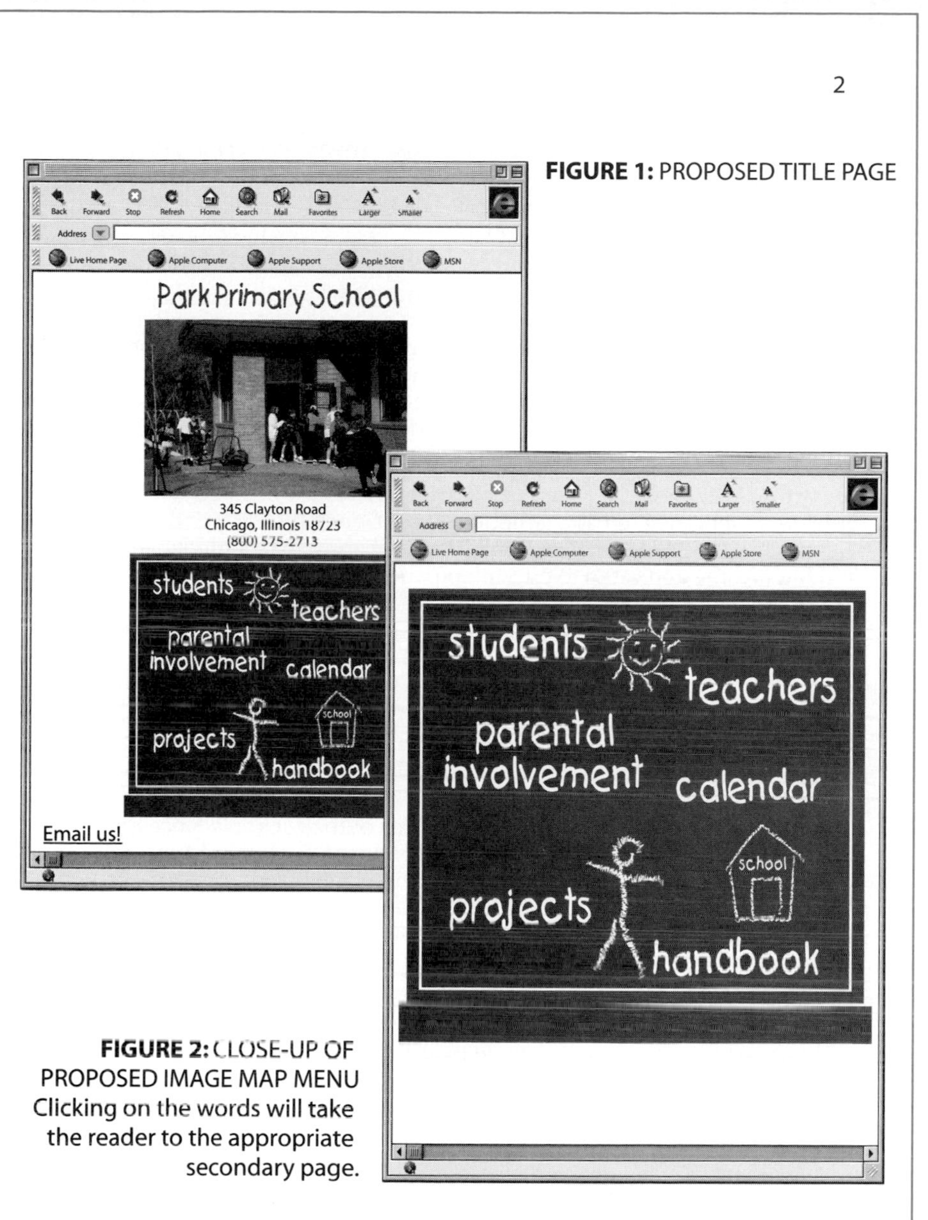

FIGURE 1: PROPOSED TITLE PAGE

FIGURE 2: CLOSE-UP OF PROPOSED IMAGE MAP MENU Clicking on the words will take the reader to the appropriate secondary page.

- **Division of the Information.** The three options are a schoolwide focus, grade by grade, and classroom by classroom. There are schools that use each, but the best grade school home page I have seen, Coyote Canyon Elementary School in Rancho Cucamonga, California, uses the schoolwide focus (Legutki). Like Park Primary, Coyote Canyon appears to be a small school, and a schoolwide focus lets the school present only its best features without having to use filler material

3

about its weaker areas. Since Park Primary uses thematic teaching units among its classes, this format would also allow those themes to be discussed as a whole rather than having to repeat information for each class or grade level.

- **Technology Level.** The page should incorporate new technology to attract the technologically savvy user who wants to see the net pushed to the limit, but without making the page inaccessible to those who are using older systems. Whenever information or a navigational option is given in a graphic, there should also be a text only version of the same. New programming options such as Java and Active X, two Web programming languages, have not yet become a standard and would be inaccessible for many users. They also offer nothing to this project that cannot be achieved by HTML, the standard programming language for home pages and the simplest.
- **Location.** Since the school's Internet server has not yet been installed, an outside server is needed as a temporary host. Geocitites (http://www.geocities.com/) offers free home pages up to 1 megabyte in size and has an area dedicated to education pages. The 1 MB limit (less than the size of a single high-density floppy disk) will affect the amount of material offered at first, but the page can be expanded when it moves to its permanent home on the school's own server.

Content

- **School.**

  Overview. During my research I conducted an e-mail survey of several school home page webmasters and studied their pages. Most of these schools had a mission statement, and many had histories of their schools. Since Park Primary is in the neighborhood where President Clinton lived as a child and replaced Ramble School, which Clinton attended (Green, interview), there is historical significance to the location. The school is also in a national park and is located near the hot springs that gave the town its name.

  Thematic Teaching. Because Park practices thematic teaching among the grade levels, it will be much easier to coordinate special-interest pages and links designed to reflect these themes on a schoolwide basis. For a unit on space exploration, the page could have links to NASA. For Arkansas history, there could be a link to the "Famous Arkansans" page. There should also be a search form for "Yahooligans," a search engine for children, so students can learn how to search the Internet.

  Documents. School documents such as those Ms. Green suggested (the handbook and the calendar) will be simple to convert to HTML. If the original text files are available, I can convert them directly. Otherwise it is only a matter of retyping them.

- **Students.**

  Writing. Student writing can either be converted to HTML or posted as unformatted text files. Either option would create an opportunity to teach basic keyboarding skills, but the age of the children will put much of the programming responsibilities on the teachers if HTML is chosen. However, in

4

the school newspaper, Ms. Green mentioned that some of the older children can type. Therefore, unformatted text would allow some students to enter their own work, or the work of classmates.

Artwork. Student artwork can either be created with one of several drawing programs (most computers have a basic drawing program preinstalled such as Paintbrush, which comes standard with Windows) or digitized from a hard copy. However, graphics files are usually large, and should be used sparingly where download speed or storage space is an issue.

Activities and Clubs. The student newspaper could be posted to the page, allowing readers a student's-eye view of what is happening at Park. The students can also post the minutes and activities of the ecology club, and an explanation of the school's internal post office system.

- **Faculty.** As Ms. Green requested in our interview, the page can feature biographies of each teacher, including education, the number of years they have taught, and any specializations. However, my research has shown that many webmasters have had trouble getting teachers to contribute material to their schools' home pages, either because of lack of time or apathy. There should be an e-mail link to each featured faculty member. Photos can be added when the page moves to its permanent site at the school.

- **Parent Involvement.** Park Primary has many programs for parental involvement. Though not many parents are likely to see this page, these programs should be listed for the occasional parent who would see them and for other schools to see. Along with the Parents and Teachers Together (PATT) program, there is a calendar of parent events such as parent-teacher conferences, "Donuts with Dad," and "Portfolio Picnics." This page should be closely linked to the calendar page.

CONCLUSIONS

- Most school home pages include a mission statement, and many include school histories.
- There are several school documents that can be easily posted.
- The school has several extracurricular activities that can be highlighted.
- Student artwork can be digitized or created on the computer, but graphics files are large.
- The faculty should be profiled.
- To make the home page easy to view, the pages need to be small.
- The technology needs to be impressive but still accessible to older equipment.
- The school's server is not yet operational, and so the school cannot host the page yet.

5

RECOMMENDATIONS

- The Park Primary School home page should consist of a small title page with several secondary pages detailing various aspects of the school.
- Because of the thematic teaching, the focus should be schoolwide with an emphasis placed on the thematic projects.
- The title page should contain a menu of options and the school's mission statement.
- Important school documents such as the handbook and calendar should be linked directly from the main page and, because of their size, will require their own pages.
- The school newspaper, the ecology club, and the post office should each receive their own page.
- Student writing should be posted in plain text so the students can enter their own work without the teachers having to learn HTML programming.
- The faculty should have a page listing their biographies and e-mail addresses.
- There should be a page devoted to parental involvement.
- Since the school's own server is not yet operational, the page should be temporarily posted on Geocities. Because of space limitations at Geocities and download times, student artwork should be kept to a minimum.

REFERENCES

Coble, Martha, et al. *Park Primary Parent Handbook.* Hot Springs School District. Hot Springs: Park Primary, 1996.

Green, Rebecca. Personal interview. 5 Nov. 1996.

Legutki, Gregg. *Coyote Canyon Elementary School.* 1996. Internet document. http://www.geocities.com/Athens/1051/.

Wilson, Dick. 21 Feb. 1996. "Small Towns Stretch Their Horizons Via the Web." *CNN Interactive,* Internet document. http://cnn.com/TECH/9602/internet_classroom/index.html.

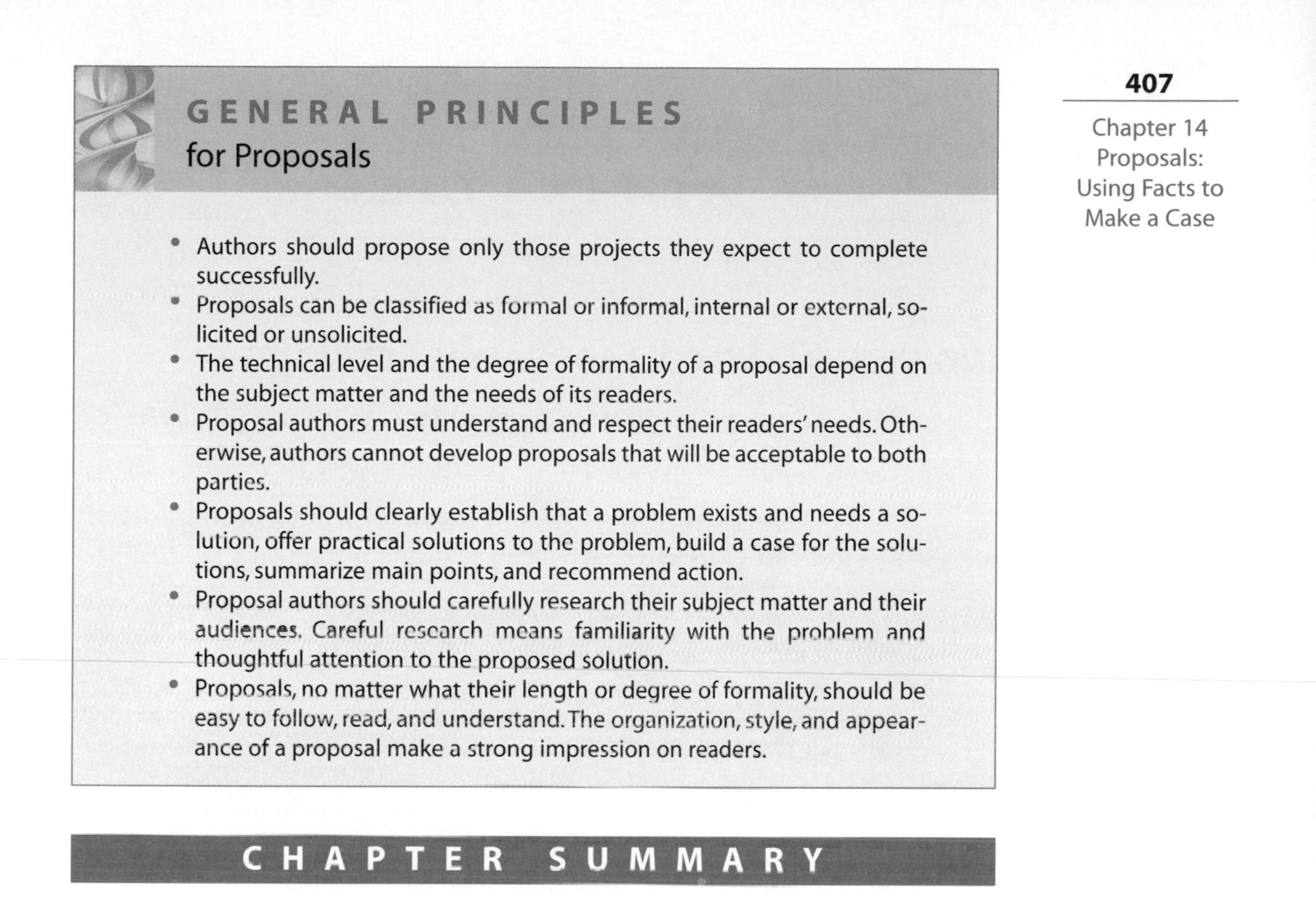

GENERAL PRINCIPLES for Proposals

- Authors should propose only those projects they expect to complete successfully.
- Proposals can be classified as formal or informal, internal or external, solicited or unsolicited.
- The technical level and the degree of formality of a proposal depend on the subject matter and the needs of its readers.
- Proposal authors must understand and respect their readers' needs. Otherwise, authors cannot develop proposals that will be acceptable to both parties.
- Proposals should clearly establish that a problem exists and needs a solution, offer practical solutions to the problem, build a case for the solutions, summarize main points, and recommend action.
- Proposal authors should carefully research their subject matter and their audiences. Careful research means familiarity with the problem and thoughtful attention to the proposed solution.
- Proposals, no matter what their length or degree of formality, should be easy to follow, read, and understand. The organization, style, and appearance of a proposal make a strong impression on readers.

CHAPTER SUMMARY

PROPOSALS

A proposal agrees to provide a particular service, product, or process. It reflects the author's understanding of both sides of a business agreement—the side that has a need and the side that meets it. The kind of proposal you write depends on your readers, your subject, and your purpose. A proposal offers a complete plan for the reader's consideration.

A proposal typically has three major divisions. The first (introduction) establishes that a problem exists and needs a solution; it may include any necessary background information. The second (discussion) offers a practical solution to the problem and builds a case for the solution; includes methods, timetable, materials and equipment, personnel, and cost. The third (conclusion) summarizes the main points and recommends action.

A proposal is a promise, and it can have legal implications. Making only those promises you can keep and defining those promises clearly for your readers can win contracts and build strong business relationships.

Proposals reflect the author's or company's professional expertise and business reputation. Well-written, fully developed proposals win acceptance. The persuasive power of a successful proposal lies in its ability to responsibly and fully meet the needs of its readers.

ACTIVITIES

INDIVIDUAL AND COLLABORATIVE ACTIVITIES

14.1. Work in groups of three or four students who have similar career goals. Discuss the following questions.

- What kinds of proposals do you anticipate writing in your career?
- How can proposals make changes inside a company, an organization, or a community?

Make a list of situations that may call for proposals.

14.2. Review the RFP from The Kotmeister Foundation on page 386. Then consider these questions:

- How would you advise applicants to prepare effective proposals in response to this RFP?
- What kind of research should candidates conduct?

Write a brief list of suggestions and strategies to help candidates write their proposals.

14.3. In groups of three or four, develop three problems you would like to solve—a service or product you would like to purchase, a rule or procedure you would like to have changed, or a community project you would like to conduct. For each project, write a short RFP. Review the sample RFPs on pages 385, 386, and 387.

For Activities 14.4–14.6

a. Review "Planning and Writing a Proposal" on pages 390–392.
b. Adapt and complete a Project Plan Sheet, page 419.
c. Following the steps for organizing a proposal, pages 391–392, plan the content and organization for the proposal.
d. Write a preliminary draft.
e. Revise the draft, incorporating design decisions.
f. Prepare a final draft.

14.4. Write a proposal in response to a research assignment in one of your classes. Your purpose will be to demonstrate to your instructor that you have defined the topic, planned the research, and located preliminary sources to meet the assignment. Consider the following questions:

- What headings will answer your instructor's questions about your proposed project?
- What kinds of information will persuade your instructor that the topic is sound and that the finished report will meet assignment requirements?

14.5. Identify a problem that needs solving or a situation that needs changing in your school, workplace, or community. Write a brief, unsolicited proposal that addresses the problem or the situation. Clearly identify the problem, give background information, state the methods proposed for solving the problem, and explain how accepting the proposal will result in improvement. Present the proposal using effective information design. You may work alone or in a team with two or three other students.

14.6. Write a proposal recommending a way to raise money for an organization. Identify primary and secondary audiences. Research the plan carefully. Then present a detailed, well thought out proposal.

READING

14.7. In "Why Your Last Technical Proposal Failed," technical communication educator and consultant Dr. Edmond Weiss describes the problems that keep proposals from being successful. Read the selection. In small groups or as a class, respond to the questions for discussion that follow.

Why Your Last Technical Proposal Failed

EDMOND H. WEISS

If you're like most technical communicators, your first technical proposal is a baptism by fire. Without training or orientation, you're tossed a nearly impenetrable request for proposals (everyone will call it an "RFP" and not tell you what that means). You're attached to a team that needs to do a month's worth of work in about two weeks. And, after a marathon of late nights and weekends (known to experienced proposal writers as "crash time"), you submit a proposal that fills you with pride—and loses.

There is no such thing as a good proposal that fails. Even when a competition is "wired" for someone else, the good proposal writer knows about it. That's why companies that live by proposals usually conduct postmortems on proposals that fail. The process isn't over until you believe you deserved to lose.

The ten items listed below are, in my experience, the ten most common mistakes made by *offerors*. (In the world of contracting, a proposal is also called an offer. Therefore, an "offeror" is one who responds to an RFP with a proposal.)

1. YOU CHOSE THE WRONG RFP.

Even if the competition was fair, you may not have had a chance of winning. Your rivals may simply have been far better qualified—something your management should have known. There are also cases where the playing field is not level, where one or two firms have to win. Again, your manager should have recognized these cases.

The decision to write a proposal is a high-stakes bet. In research and development companies especially, the money sunk into proposal writing is often the company's capital. Yet most companies, especially small ones, have no formal scheme for evaluating proposal opportunities. Unwise companies pick projects on impulse, the way one might pick a horse. Often, managers will infuriate their employees by waiting until the last moment to make a decision.

There are hundreds of ways to build a bid/no-bid model, and wise companies nearly always apply them in some form. (You can even make the gambling metaphor work by talking in terms of your company's *track record*. Monitoring your track record will tell you whether your model needs to be revised.)

A formal bid/no-bid decision model includes objective criteria and a threshold of acceptable risk. In addition, the timeliness of the project must be considered. Once you've got a good decision model, a prospective proposal can be *handicapped*—systematically appraised and assigned a probability of paying off.

Criterion

Most successful companies isolate several important factors in considering a proposal:

- The company's current relationship with the customer
- The number and strength of the competitors
- The probability that the project will not be awarded at all or that it has been virtually promised to another firm
- The closeness of fit between what is asked for in the RFP and what the firm can offer

Smart companies also evaluate the *attractiveness* of the project:

- Potential profits
- Compatibility with company plans
- Possible interference with other projects or proposals

Risk Thresholds

It is possible to predict what will happen to a proposal. Sometimes one will be written in desperation, or merely to make a company known in a new market. These are proposals that, logically, should not have been written at all. But sometimes these casual, ill-considered, improbable proposals find themselves alone on the customer's desk and result in a contract.

Remember, though, that a proposal is an investment. In general, it is best to plan conservatively and not squander company resources.

. . .

2. YOU WERE NOT *TRACEABLE*.

Most proposals are read by two classes of reviewers: those who wish to know what you can do and those whose only objective is to disqualify you and reduce the number of competitors. This latter group is typically the first bank of readers. They are administrative types whose task is to reduce the number of competitors to a manageable list. To this end, they are hoping that you will be in some way unresponsive, that you will leave out one of the many forms, fail to provide the right number of copies, or otherwise allow them to mark you down for an infraction on their checklists.

Those checklists are important. If the reviewers believe your response lacks an item on that list—even if you are sure it is there—your proposal may be disqualified.

If the reviewers cannot disqualify you for obvious errors of compliance or packaging, they will look inside the document to see whether you responded to every item in the statement of work (SOW). Here is where successful proposal writers have developed the craft of *traceability,* making sure that the proposal contains everything asked for and that the reviewers can find it without a struggle.

. . .

3. YOU WERE INDISTINGUISHABLE.

One well-known proposal consultant says that there is but one main question to be answered in a proposal: Why would the customer, given several perfectly acceptable competitors, choose your proposal above that of the others?

The only widely accepted theory of proposal writing—known as *win theory*—is that proposal success depends on finding some exploitable difference between you and your competitors and emphasizing it at every opportunity in your writing and negotiations. Instead of believing that your reader will be eager to read your proposal and grateful for your solution to their problems, it is better to presume that the potential customer would find it much more justifiable to choose one of the competitors.

. . .

What if you cannot define an exploitable difference? What if, because of the nature of the project or the particular mix of competitors, there are no differences that matter? What if, for example, a city needs an environmental study and there are eight firms, equally competent and comparably priced, that could do the job?

My advice in this situation—although many would disagree—is to pass up this proposal and invest your time and money in something better than a seven-to-one shot.

4. YOU SHOWED NO SUPERIOR INSIGHT.

To return to the example above, it is quite difficult to imagine a community in which eight consulting firms have an equal chance of winning a city contract—even if that is the official story. Proposals do not begin business relationships; they consummate them. Authors of winning proposals know well in advance that RFPs are coming. In major military and aerospace procurements, most of the proposal is written *before* the RFP arrives. There is nothing illegal or even improper about this procedure. On the contrary, developing and maintaining trusting relationships between potential customers and vendors is the main dynamic of technical marketing.

Among the most important criteria used to evaluate technical proposals is the offeror's understanding of the problem. To inexperienced engineers and scientists, this seems to be a technical criterion, an assessment of the power of the analysis in the technical component of the proposal. In fact, though, understanding of the problem usually means understanding of the client.

. . .

Put another way, a winning proposal nearly always reveals a deeper knowledge of the customer than can be learned from the RFP or other easily available public documents. This is known in the trade as customer intelligence. A thorough "understanding of the problem" usually entails familiarity with events, documents, and issues not in the public domain, even when bidders are not supposed to have access to that information. Short of outright espionage, the more inside knowledge evident in a proposal, the better.

5. YOU FORGOT TO SELL.

The objective of selling is to define how the features of one's own proposal differ from the features of the competitors' proposals, and then to prove that these differences yield important benefits to the customer.

. . .

Every section of a well-made proposal teaches, proves, or sells what is unique to the proposal. At Hughes Aircraft, the publications engineers invented the idea of breaking immense technical proposals into two-page chunks, "small, illustrated essays with a single theme." These *two-page spreads* begin with a thematic heading, followed by a summary overview, and then all the text and pictures or other exhibits needed to develop the theme to make the point.

6. YOU TURNED YOUR AUTHORS LOOSE.

The worst mistake an organization can make at proposal time is to turn loose a handful of technical staff, each with a piece of the RFP and each responsible for a substantial part of the document.

Before proposals are written they must be closely planned and modeled, with detailed thematic outlines. A small group of sales professionals must define the main selling theme and the subsidiary themes and ensure that every one of these themes, as well as every item in the statement of work, is traceable within the table of contents.

. . .

Unfortunately, many of the people working on a proposal have been conscripted; they dislike the assignment, especially if it is done after hours, and particularly if writing has been a recurring career problem for them. Too often, their goal is simply to cover some pages with appropriate-looking material. It is crucial that all proposal authors, especially the reluctant ones, write to specifications prepared by senior employees. Every page will then contribute to the sale.

7. YOU HEDGED TOO MANY PROMISES.

Large stretches of a . . . proposal consist of promises: schedules, resources, personnel. . . . Not surprisingly, then, successful proposals come from organizations that are reputed to keep their promises. Nothing impresses a prospective client more than the unqualified pledge that certain people will be committed, for a specific share of their time, to the proposed project. Similarly, nothing makes a prospect more wary than hedged promises and evasive commitments.

Of course, everyone who works in a proposal-driven business understands the dilemma of uncertain resources. A proposal is at best a strong possibility of work; at any time several proposals, pledging the same resources, may be out for consideration. What will happen if the offeror is especially successful and key people are overcommitted?

To protect against this eventuality, most companies will draw back, promising, for example, that the project will be led by Dr. Jones "or a person of equal qualifications." Other firms will not even go that far; they provide a list of employees and promise only to staff the project from among the people on the list—or perhaps others equally qualified.

. . .

Customers are not fools. They have been burned by empty promises in earlier proposals and can spot the hedged bet at great distance. But they also know that, downstream, predictions and promises, even those made in earnest, will almost always need adjustment. Everyone knows that any project is subject to a host of such unpredictable factors as strikes, shortages, weather, illness, and death. But customers will expect a serious promise from an offeror with a reputation for keeping promises. And most also expect enforceable guarantees and protections.

8. YOU PUBLISHED YOUR FIRST DRAFTS.

First drafts are not nearly good enough. A main difference between amateur and professional writers is that amateur writers think the first draft is a finished document, while professionals view it as a raw approximation of the finished work.

Engineers, scientists, analysts, and other technical staff often protest that there is not enough time to edit and revise their drafts. Don't fall into this trap. The writing that goes into a proposal should be clear and robust. It should also be precisely edited.

Clear Writing

Too often, technical professionals do not realize that simple language is better. Indeed, most university-educated workers think that the more difficult and long-winded the text, the more "impressive" it is to the reader.

. . .

Nonetheless, a proposal will be more effective if it can be easily understood.

. . .

Robust Writing

Robust writing is crisp text that engages and holds the attention of the reader. Sentences are short; grammar is uncomplicated. Words are familiar to the reader and new ideas are clearly explained.

Robust writing is invisible to the reader. The reader is unaware of the style, does not notice the word choice or organization, and certainly never stops to ponder grammar or usage. Robust writing creates the illusion that ideas are leaping directly from the writer to the reader.

Editing

One of the hardest things for a technical communications consultant to accept is that clients may consider editing largely a cosmetic activity.

Good writing is the result of good editing. (No first draft ever meets the standard.) Any company that wants to win its fair share of proposals will insist that a nearly finished document be reviewed by a content editor, someone who writes considerably better than the average college graduate, and who has enough time for a thorough inspection of the document.

9. YOU OVERBURDENED YOUR READERS.

Too many engineers and technical staff expect their proposals to be greeted eagerly and studied closely. They expect important points to be seen and understood—even if they're buried in the proposal. To communicate successfully, however, it is best for the proposal writer to presume that the reader is inattentive, forgetful, and error-prone, more likely to skim a passage than to read it closely.

Effective proposal writers try to reduce the burden on their readers. The best style of writing is the one that communicates its meaning with the least possible effort for the reader.

Along with thematic titling, there are other devices that lighten the reader's load.

- Incorporating frequent summaries and overviews
- Using lists and tables instead of paragraphs, when appropriate
- Placing charts and other exhibits so that they can be seen when they are discussed, either on the same page or the facing page. (When readers are told to "See Figure 1" they should be able to see it.)
- Leaving lots of white space on the page, with frequent side headings and limiting the width of text columns to five inches, unless these practices are prohibited in the RFPs
- Using easy-to-read typefaces and font sizes
- Highlighting key words, phrases, and brief passages with italics, boldface, or other devices

10. YOU OVERPROTECTED YOUR DOCUMENT.

The purpose of a test is to find bugs, to make the thing being tested fail. But the natural inclination of a team of writers that has been working for weeks on a frantically constructed proposal is to try to conceal any flaws from the people who cannot appreciate how hard the team has worked.

. . .

Proposals must be tested in earnest, that is, read hard by an individual or group not associated with the writing team, whose mission is to misunderstand and find fault with the document. Some firms call the final readers the "killer team," the "red team," or some similar name. Whatever the reviewing team is called, this final reading must be independent of the authors, who will learn over time not to take the team's criticisms personally.

Written by Edmond Weiss and reprinted with permission from *Intercom,* the magazine of the Society for Technical Communication, Arlington, VA, U.S.A.

QUESTIONS FOR DISCUSSION

a. What useful strategies does Weiss suggest for teams who develop proposals? How does teamwork differ from the work of an individual proposal author?
b. What problems make a proposal "indistinguishable"? How can proposal authors overcome these problems?
c. What "sells" a proposal, according to Weiss?
d. What are the risks of careless writing, incomplete research, and hedged promises?
e. What kinds of review should proposals undergo before they are submitted? Why are peer reviews valuable to proposal authors?

CASE STUDY

14.8. In a small group or in a short written assignment, discuss the ways in which the following case study illustrates what a proposal can do for its author, its readers, and the author's organization.

You're Proposing What?

EcoTech Publications is a large company that designs and prepares documents for local businesses. Since the company believes in promoting an efficient, comfortable, and safe workplace, its directors are always looking for ways to streamline production, to create a safe and comfortable work environment, and conserve and share resources. EcoTech's administration has set up a generous incentive for any employee proposals for projects and changes that meet these important company goals.

The Board of Directors of EcoTech is considering in-house proposals at its monthly meeting.

"We have only one proposal this month," announces Brenda Baron, the Board Chair. "We have all read it. It's Sally Twintree's request for a new work station."

Gary Edwards chuckles. "It's certainly the shortest one I've ever read," he says. "All Sally suggests is that she can work faster and more efficiently if she has a new work station with more power. However, she doesn't explain exactly what her job is, what kind of computer she needs, or how it will help EcoTech in general. That's pretty short-sighted."

"I certainly have more questions to ask," says Jean Smyles. "Just how much does a new system cost? This seems like a suggestion that will please Sally, but that really isn't enough for us to consider it. Didn't we publish some guidelines about proposals? This one is no more than a self-seeking suggestion."

"Sure, we published a very clear RFP," sighs Brenda. "It asks employees to explain their ideas fully, to give cost estimates, and to show how proposed ideas affect other EcoTech employees. We even listed our company goals and asked for each proposal to show how new ideas were related to them. We also listed every element we wanted an EcoTech proposal to have. And we even asked the questions each proposal should address, including expenses. I guess Sally didn't send for our guidelines."

"Those are great proposal guidelines! And some of the proposals have been great, too," says Gary. "Do you remember the one about flextime work hours? The author showed that EcoTech could meet the needs of over 40 percent of its employees by setting up a flextime schedule that cost us nothing to establish. That project's evaluation certainly showed that the idea worked. That proposal author helped *everybody* at EcoTech."

Jean remembers another good proposal. "We really liked that one about color-coded recycling bins in production areas. The bins made it easy to separate paper and other materials. We also started donating some collected drawing materials to Smithtown Elementary School down the street. The author really thought out an easy and inexpensive way to save resources and to help the school. Everybody won with that proposal."

"I guess Sally Twintree just didn't see more than her own side of things," says Brenda. "She is only one of 332 employees, and she forgot to think further than her own needs. Even so, most of our production work is done with computers. Maybe she has some ideas about new equipment and software that could help everybody. Jean, would you please call Sally and explain the information we need to see so that we can understand her proposal?"

"All right," Jean agrees. "I'll call Sally. She needs to understand that a proposal must explain ideas fully and to think about more than her side of things. Maybe I can show her some of our business proposals as examples and some of the RFPs we responded to, as well. We'd never get a proposal accepted if we only explained EcoTech's side of a business deal."

QUESTIONS FOR DISCUSSION

a. Who are Sally's readers? What assumptions has Sally made about them? What would you explain to Sally Twintree about her proposal and its readers?
b. What questions does Sally need to answer before she submits her proposal?
c. Who are some of the other employees at EcoTech who would potentially be affected by Sally's proposal? Has Sally considered these other employees?
d. What kinds of research does Sally need to conduct to show that her proposal should be funded?
e. How would you tactfully explain to Sally that her needs and ideas are important to EcoTech, but that she needs to explain them in a different way in order to get what she wants?
f. Are EcoTech's cash incentives for accepted in-house proposals good investments for the company? In what ways?

PROJECT PLAN SHEET

FOR PROPOSALS

Audience

- Who will read the proposal?
- How will readers use the proposal?
- How will your audience guide your communication choices?

Purpose

- What is the purpose of the proposal?
- What need will the proposal meet? What problem can it help solve?

Subject

- What is the proposal's subject matter?
- How technical should the discussion of the subject be?
- Do you have sufficient information to complete the proposal? If not, what sources or people can help you to locate additional information?
- What title can clearly identify the proposal's subject and purpose?

Author

- Will the project be a collaborative or an individual effort?
- If the project is collaborative, what are the responsibilities of each team member?
- How can the developer(s) evaluate the success of the completed project?

Project Design and Specifications

- Are there models for organization or format for proposals?
- In what medium will the completed proposal be presented?
- Are there special features the completed proposal should have?
- Will the proposal require graphics or other visuals? If so, what kinds and for what purpose?
- What information design features can best help the proposal's audience?

Due Date

- What is the final deadline for the completed proposal?
- How long will the proposal take to plan, research, draft, revise, and complete?
- What is the timeline for different stages of the proposal?

CHAPTER 15

Correspondence: Sending and Responding to Messages

CHAPTER GOALS

This chapter:

- Defines the nature of business correspondence
- Explains reader-centered emphasis
- Describes audience considerations
- Describes the author's responsibilities
- Discusses various media for transmitting correspondence
- Explains the differences between memorandums and letters
- Describes the conventions, formats, styles, and occasions for memorandums
- Describes the conventions, formats, styles, and occasions for letters
- Explains the particular requirements of inquiry letters, good news and bad news letters, claim and adjustment letters, and transmittal letters

INTRODUCTION

Writing effective memorandums and letters, according to employers, is one of the major skills that an employee needs. Employees must be able to handle workplace correspondence, such as communicating with other employees, making inquiries about processes and equipment, requesting specifications, making purchases, responding to customer questions, answering complaints, and promoting products. Since many business transactions are conducted in part or wholly by correspondence, knowledge of how to handle correspondence in the workplace is essential in the daily activities of the successful professional.

This chapter is designed to assist you in making appropriate choices as you handle correspondence through memorandums and letters, whether in print or nonprint media.

WHAT IS CORRESPONDENCE?

Correspondence in the workplace involves sending and receiving messages to get work done. The messages are direct responses to particular problems and needs. The technician for computer networking, for example, discovers that the cables received are not the cables ordered. To address this problem, the technician quickly makes several decisions: Telephone the vendor? No, that takes too much time, plus there would be no written record. There is also the real possibility of misunderstanding oral messages, particularly if the person receiving the call must relay the message to someone else. E-mail the vendor? Yes, the packing slip shows an e-mail address as well as a fax number. Within a few minutes, the technician sends the complaint message and receives a response. The correct cables will be shipped immediately via overnight carrier and the wrong cables are to be returned by the same delivery service.

Memorandums and letters are records of business transactions that may detail specifics such as who is responsible for what, due dates, or responses to requests or claims. Because more than one message is frequently required in settling a matter, easily accessible copies are a necessity for both the sender and the receiver. Typically, accessible copies are stored in print (on paper) or electronically (on disk or hard drive).

READER-CENTERED EMPHASIS

Since workplace correspondence is directed to a named reader, using a courteous, reader-centered tone is essential. To create such a tone, address the reader directly and helpfully. Stress the "you" aspects to focus on qualities of the addressee. Minimize references to "I," "me," "we," "our" to avoid sounding writer-centered. Compare these responses to a client's order:

Poor: We were pleased to receive your order for ten microscopes. We have forwarded it to our warehouse for shipment.

Improved: Thank you for your order for ten microscopes. You should receive the shipment from the warehouse within two weeks.

The first response is writer-centered; the second is reader-centered.

The "you" emphasis helps the writer to take a positive approach, even when the response is not exactly what the reader wants to hear. Study these two versions of a negative response:

Poor: We regret that we cannot fill the order for ten microscopes by December 1. It is impossible to get the shipment out of our warehouse because of a rush of Christmas orders.

Improved: Your order for ten microscopes should reach you by December 10. Your bill for the microscopes will reflect a 10 percent discount to thank you for accepting a delayed shipment caused by a backlog of Christmas orders.

The negative situation, rephrased as a positive situation, is much more likely to satisfy the customer.

Part of the reader-centered approach is considering how the reader may perceive both the message and the writer. Natural, direct wording conveys professionalism and clear thinking. Consider the following example:

Poor: Pursuant to your request of October 10 that I present my in-depth knowledge about Thanksgiving observance by our neighbors to the north, at the November convening of the DECA membership, I regret to inform you that my impending holiday tour will prevent a positive response to same.

Although the writer is trying to sound impressive, the result is stilted, awkward, and confusing. The same information can be clearly and naturally worded.

Improved: Thank you for your invitation of October 10 to present the program at the November DECA meeting on the Canadian observance of Thanksgiving. I must decline, however, because I will be on a European vacation.

REMINDER

Effective correspondence is reader-centered, focused on the needs of the person addressed.

As you compose correspondence, keep your reader in mind. How will the memorandum or letter sound to the reader? Will it favorably impress the reader? Is the information presented in a natural, direct way so that the content is clear?

AUDIENCE CONSIDERATIONS

The thoughtful correspondence writer studies the audience's needs, considering the audience's culture, technical knowledge, and organizational distance.

The "you" approach and directness, favored in correspondence in the United States, for instance, may cause discomfort and misunderstanding for recipients from other cultures. A tone that intimates a "pat on the back" may cause one individual to beam with pride and another to cringe. The formal tone of European business correspondence seems flowery to Americans. Conversely, American correspondence seems abrupt and rude to Europeans.

The recipient's degree of technical knowledge influences the choice of technical terminology. In a memo to another health professional, a doctor uses the term "myocardial infarction" but for the family uses "heart attack."

Organizational distance refers to employees' official relationships within a company. Each employee has a job title, a job description, and an official status within the company (typically reflected in an organization chart—see pages 100–101). Various job titles and responsibilities influence wording in correspondence within and outside the company.

THE AUTHOR'S RESPONSIBILITIES

The author of a memorandum or letter is responsible for treating recipients with respect and courtesy, building and maintaining positive working relationships, and following ethical practices in all matters of correspondence.

Respect

The recipient should be addressed appropriately and professionally, and the tone of the communication should be cordial and sincere. Tone is conveyed not only through word choice, but also through attention to the appearance of the communication. A sloppy letter with typos and unplanned sentences conveys a lack of respect for the reader and the message, "Neither you nor this letter is worth my time."

Courtesy

Since much of the world's work is carried out through correspondence, it is just good business to be courteous. Showing appreciation, giving praise, and personalizing the communication reflect courtesy.

Positive Working Relationships

Positive results are usually accomplished through positive collaborative efforts, and these efforts are typically recorded through letters and memorandums. Practice sensitivity to the recipients' feelings, intentions, and cultural mores. Avoid combative and threatening wording. In all correspondence, foster open, aboveboard relationships with others.

Ethical Practices

Unethical practices in correspondence matters are a hot topic in the workplace. Efficient, speedy, high-tech machines (fax machines, photocopiers, computers) encourage misuse of correspondence. Basic don'ts include: (1) Do not copy confidential correspondence without permission, (2) Do not read, forward, or in any other way share confidential correspondence without permission, and (3) Regardless of the forms (memorandum, letter, or e-mail) and its method of delivery, respect the privacy of the correspondence.

MEDIA FOR ELECTRONIC CORRESPONDENCE

Media for correspondence delivery have changed dramatically in the last twenty years. Until a few years ago, courier and postal services handled the delivery of correspondence. Then other mail-handling businesses evolved, such as United Parcel Service (UPS), Federal Express (FedEx), and Airborne Express (ABX). While regular mail delivery takes anywhere from one to several days, these companies guarantee same-day, next-day, or two-day delivery, depending on the mail service and the geographical area.

Electronic correspondence has made delivery even faster. Common electronic delivery systems, which typically require only a few seconds for transmission, include facsimiles (faxes), voice mail, and electronic mail (e-mail).

When both the sender and the receiver have fax capabilities (a fax machine/computer), a letter, memorandum, or other printed material can be sent through the machine (via telephone lines) in a matter of seconds; shortly thereafter, a reply can be sent. Faxed messages typically include a fax cover sheet and the message page(s).

Using a computer, electronic typewriter, or even handwriting, you can design and print cover sheets as you need them. Below is an example of a department fax cover sheet.

505 East Main Street
PO Box 1264
Raymond, MS 39154
Phone: (601) 857-3349

Hinds Community College
Department of English
Raymond Campus
FAX (601) 857-3648

Fax

To:	**From:**
Fax:	**Date:**
Phone:	**Pages:**
Re:	**CC:**

☐ **Urgent** ☐ **For Review** ☐ **Please Comment** ☐ **Please Reply** ☐ **Please Recycle**

***Comments:**

Companies may have a cover sheet designed and printed or photocopied in large quantities if employees use a fax machine often. Basic information on the fax cover sheet includes:

- name of the person to whom the message is sent
- name of the person sending the message
- date
- subject

Other items often included are:

- number of pages
- sender's address
- sender's telephone number
- sender's and receiver's fax numbers
- a telephone number and a person to call if the fax is received at a wrong location or if not all pages go through
- space for comments or a message

Fax cover sheets have no standard design or content.

An example of a filled-in fax cover sheet is shown on page 427.

Notice that the date and time of transmission, the sender's name and fax number, and the page number are printed at the top of the received cover sheet.

With voice mail, callers can leave messages in an electronic mailbox. Individuals who have an electronic mailbox can retrieve messages from any touchtone telephone at any time. Voice mailbox owners can also leave messages to themselves (such as reminders of meetings and appointments). Also, unlike some answering machines, from which anyone can retrieve messages, a voice mailbox is confidential; messages are typically accessed using a password or code.

E-mail requires both the sender and the receiver to have computers and a connection through a local area network (LAN) or an online service. With a LAN, individuals can communicate, for example, between computers within a building. With an online service, individuals can communicate between computers anywhere in the world. Communications may be sent to one person or to any number of people using options such as listservs, chat rooms, forums, and bulletin boards. E-mail communication can take place in minutes.

MAY.-07 01 (MON) 11:11 G. PARTNERSHIPS TEL:601 - 352 - 4752 P.001

FAX TRANSMITTAL FORM

DATE: *5/7/01* TIME: *11:05 a.m.*

* * PLEASE DELIVER THE FOLLOWING INFORMATION TO:

NAME: *Mitzi Reed*

DEPT: *Honors Program*

FAX: *(601) 857-3213*

* * FROM: *Joshua C. Johnston III, M.D.*

FAX: *(601) 355-1358*

TOTAL # OF PAGES: *3* (INCLUDING COVER SHEET)

REMARKS: *These are the dates on which James C. Salmon visited our office and was seen by Kirk C. Smith, M.D.*

* * IF YOU DO NOT RECEIVE ALL PAGES, PLEASE CALL: (601) 355-4752

This facsimile transmission is intended only for use of the individual or entity to which it is addressed and may contain confidential information belonging to the sender which is protected legally by the physician-patient privilege and/or other privilege provided by law, including but not limited to, the privilege provided pursuant to M.C.A. #41-9-67 and #13-1-21. If you are not the recipient or agent responsible for delivering the transmission, you are hereby notified that any dissemination, distribution, copying or use of this transmission is strictly prohibited. If you have received this transmission in error, please immediately notify the sender by telephone to arrange for return of the transmission.

Charles H. Long, M.D.
Reed B. Hughes, M.D.
Isabelle M. Walker, M.D.

GASTROINTESTINAL
PARTNERSHIPS

Joshua C. Johnston, III, M.D.
Kirk C. Smith, M.D.
Mckenzie A. Randolph, M.D.

1425 North State * Suite 100 * Jackson, MS 39202

601/355-1356

Below is an example of an e-mail transmission from one person to a group. Note that the headers at the end of the message give technical data about who, to whom, what, when, why, and how—should this message require tracing or retrieval.

Subj: Re: [tyca-ec] TYCA wins
Date: Thursday, June 17, 1999 8:41:01 AM
From: BODMER@gwmail.nodak.edu
To: tyca-ec@serv1.ncte.org
bcc: PickettHCC

Congratulations to the new officers of TYCA. Georgia brings a strong, steady voice with the experience and history of the organization, and T. Ella represents those new voices we want to hear on the Executive Committee. As we bring our new voices into the conversation, we will see new strength and growth as the organization continues to evolve to meet the needs of the membership.

Paul

---------------------------- Headers --

Return-Path: <tyca-ec-owner@serv1.ncte.org>
Received: from rly-yb02.mx.aol.com (rly-yb02.mail.aol.com
[172.18.146.2]) by air-yb05.mail.aol.com (v59.51) with SMTP; Thu, 17 Jun 1999 10:41:01 -0400
Received: from serv1.ncte.org (serv1.ncte.org [208.223.98.5]) by rly-yb02.mx.aol.com (vx) with SMTP; Thu, 17 Jun 1999 10:40:55 -0400
Received: (from majordomo@localhost) by serv1.ncte.org (8.8.7/8.8.7) id JAA14516 for tyca-ec-outgoing; Thu, 17 Jun 1999 09:43:07 -0500
X-Authentication-Warning: serv1.ncte.org: majordomo set sender to owner-tyca-ec@serv1.ncte.org using -f
Message-Id: <s768c27d.065@gwmail.nodak.edu>
X-Mailer: Novell GroupWise 4.1
Date: Thu, 17 Jun 1999 09:37:13 -0600
From: Paul Bodmer <BODMER@gwmail.nodak.edu>
To: tyca-ec@serv1.ncte.org
Subject: Re: [tyca-ec] TYCA wins
Mime-Version: 1.0
Content-Type: text/plain
Content-Disposition: inline
Sender: owner-tyca-ec@serv1.ncte.org
Precedence: bulk
Reply-To: tyca-ec@serv1.ncte.org

6/17/99 America Online : PickettHCC Page 1

Much e-mail is transmitted by special systems that create multipart addresses combining geographical and conceptual information. One such system is the Domain Name System (DNS) used by the Internet, the largest electronic information exchange. The DNS includes a six-part address. Parts 1–4 and Part 6 tell who and where; Part 5 tells what:

1. Identification of the user or organization (who)
2. @ symbol (connects the who with the where identification)
3. Subdomain (department, building)
4. Domain (name of institution, business)
5. Type of organization (edu—education, com—commercial, gov—government, mil—military)
6. Country (U.S. addresses usually omit the country segment)

A sample address in the Domain Name System is <aalaster@hinds.cc.ms.us>.

Individuals may chose to use other international online services. In Somerset, England, for instance, a forensics specialist subscribes to a service that requires only the user name and the server name. His e-mail address is <mblenkinsop@virgin.net>.

Changes in transmission methods of workplace correspondence are inevitably accompanied by other changes. Instantaneous communication encourages paying less attention to formalities and more to answering immediately.

CHOICE OF CORRESPONDENCE MEDIUM: MEMORANDUM (MEMO) AND LETTER

Memorandums are used for in-house correspondence, that is, communications between persons within the same company. The memo as a communication within a company often addresses persons on a first-name basis and assumes much shared information. The correspondents may be in the same building or in different branch offices of the company. Memos are used to convey or confirm information. They serve as records for transmittal of documents, policy statements, instructions, minutes of meetings, and the like. Although the term *memorandum* was once associated with a temporary communication, usage of the term has changed. Today, a memorandum is regarded as a communication that makes needed information immediately available or that clarifies information.

Letters, on the other hand, are typically used for communication with persons outside the company or for formal correspondence within a company. Letters follow a conventional format with complete names, addresses, and dates in the heading and inside address. Letters also typically follow the conventions of a salutation and complimentary close plus a handwritten signature above the keyboarded name of the writer. Although the body of a letter may be similar in content to the body of a memorandum, the letter tends to be more formal in appearance and content.

Memorandums: Conventions, Formats, Styles, and Occasions

The memorandum, unlike the business letter, has only two regular parts: the heading and the body. The formalities of an inside address, salutation, complimentary close, and signature are usually omitted. (Some companies, however, prefer the practice of including a signature, as in the memorandum on page 431.) The memorandum may be initialed in handwriting, as illustrated in the memorandum on page 432. This initialing (or signature) indicates official verification of the sender. As in letters, an identification line (see pages 431, 441, 442, and 444), a copy line (see pages 398 and 431), and an enclosure line (see pages 398, 432, 451, 594, 595, 596) may be used, if appropriate.

See also the memorandums in Chapter 13 Reports, pages 331, 355–359, and 366–367.

Heading

The heading in a memorandum is typically a concise listing of:

- *To* whom the message is addressed
- *From* whom the message comes
- *Subject* of the message
- *Date* of the message

For ease in reading, the guide words To, From, Subject, and Date usually appear on the memorandum. These guide words are not standardized as to capitalization, order, or placement at the top of the page. Most companies use a memorandum form, printed with guide words and the name of the company or the department or both. It is perfectly permissible, however, to make your own memorandum form by simply keyboarding the guide words.

Body

The body of the memorandum is the message. It is composed in the same manner as any other business communication. The message should be clear, concise, complete, and courteous. If internal headings will make the memo easier to read, insert them (see Headings, pages 58–60).

In the following two examples of memorandums, note how the messages serve different purposes. The first memorandum urges cooperation in arranging for contractors and regulatory agencies to view prototype testing. The second memorandum announces information for fall classes given by a professional organization within the insurance industry.

Letters: Conventions, Formats, Styles, and Occasions

Letter writing in the workplace follows conventional standards. Failure to follow these standards shows poor taste, reflects the writer's lack of knowledge, and invites an unfavorable response. There is no place for unusual or "cute" stylistic expressions in a business letter. (A specialized type of business letter, the sales letter, sometimes uses attention-getting gimmicks. Follow-up sales letters, however, tend to conform to standard practices.)

TILLMAN MARSHALL
Marine Company

OFFICE MEMORANDUM

Date: July 5, 2001
To: Earlene Beckwith
From: Don C. Eckermann *Don C. Eckermann*
Subject: Elevating Unit Tests, Customer Visits, and Regulatory Agency Involvement

The new Tillman Marshall elevating system and the new mobile offshore drilling rig designs continue to be a major undertaking for our company.

I believe that marketplace acceptance can be accelerated by arranging for selected drilling contractors to view the prototype testing at Longview in August 2002. Similarly, the necessary regulatory agency approvals can be expedited by inviting selected agency representatives to the prototype testing.

These visitors will be important people on an important mission, particularly to us. These visitors will convey their experiences and observations to the marketplace.

I seek your help in planning the proposed visits for targeted drilling contractors and for regulatory agency representatives.

Within the next two weeks please send me
1. a list of selected contractors you think we should invite
2. a list of all the regulatory agencies that will be involved in the production and marketing of the elevating systems

ljb

c:	Longview:	New Orleans:	Houston
	Rudy Harrison	Will Trimble	Pharr Ingahorn

Paper
Choose good quality paper. Most companies use bond (at least 20 pound) letterhead stationery.

Appearance
The general appearance of a letter is very important. A letter that is neat and pleasing to the eye invites reading and consideration more readily than one that is unbalanced or crammed on the page. The letter should be like a picture,

Memorandum

Women in Insurance
P.O. Box 31580
PORTLAND, OREGON 97233-0294
(503) 932-5990 FAX (503) 932-5981

DATE: July 9, 2001
TO: All Education Coordinators and IIA Students
FROM: Raja Rojillio, CPIW Chair RR
Katina Brewer, Cochair
SUBJECT: Fall 2001 Registration

We are pleased to enclose registration materials for our fall classes.

Additional Information:

STUDY MATERIALS Students will be responsible for obtaining study materials from the Insurance Institute, other publishers, or their employer. For your convenience, enclosed are excerpts from the Institute's Key Information Booklet. Please order the materials as soon as possible so they will arrive in time for the first class.

TUITION The class schedule lists the tuition fees applicable to each course. Please note the reduced fees for WIN members.

CLASSES Classes will begin the week of August 19, 2001, at the time and location shown on the schedule. As always, our classes are subject to enrollment minimums. If the number of registrants for a class falls short of the minimum, we will provide a list of the other applicants to assist in forming a study group.

We encourage your participation in our education program and welcome any suggestions that you may have.

Enclosures

framed on the page with margins in proportion to the length of the letter. Allow at least a $1\frac{1}{2}$-inch margin at the top and bottom, and at least a 1-inch margin on the sides. Short letters should have wide margins and be appropriately centered on the page. As a general rule, single space within the parts of the letter; double space between the parts of the letter and between paragraphs. (See the letter on page 434 for proper spacing.)

Format

Although there are several standardized layout formats for letters, most companies prefer the block format (see sample letters on pages 441, 442, 444, 446, 452, 594, 595, 596). The modified block is an older format that some users consider friendlier and easier to read (see sample letters on pages 348, 451). Another format gradually gaining favor is the simplified block format.

Block Format. This format is distinguished by all parts of the letter being even with the lefthand margin. This format is easy to keyboard because no tabbing is needed. Open punctuation is sometimes used; that is, no punctuation follows the salutation and complimentary close.

Modified Block Format. This format is distinguished by several parts of the letter—the inside address, salutation, and paragraphs (optional indenting)—being aligned with the lefthand margin. The heading, complimentary close, and signature align five spaces to the right of center page. Paragraphs may or may not be indented. Open punctuation is sometimes used; that is, no punctuation follows the salutation and the complimentary close.

Simplified Block Format. In the simplified block format recommended by the American Management Society (AMS), the salutation and the complimentary close are omitted. A subject line is almost always used. Like the block format, all parts begin at the lefthand margin.

REMINDER

Workplace correspondence follows conventional standards.

Parts of a Business Letter

Regular Parts of a Letter. The parts of a business letter follow a standard sequence and arrangement. The six regular parts in the letter include: heading, inside address, salutation, body, complimentary close, and signature (these are illustrated in the letter on page 434). In addition, there may be several special parts in the business letter.

1. *Heading.* Located at the top of the page, the heading includes the writer's complete mailing address and the date, in that order, as shown below. As elsewhere in standard writing, abbreviations are generally avoided. For generally accepted uses of abbreviations, see Appendix 2, pages 621–624. In writing the state name, use the two-letter abbreviation recommended by the postal service. (See the Two-Letter Abbreviations for States box on page 435.) Note that the heading does *not* include the writer's name.

Route 12, Box 758	704 South Pecan Circle
Elmhurst, IL 60126	Hanover, PA 17331
July 25, 2001	9 April 2001

Regular Parts of a Typical Business Letter and Their Spacing

Letter	Spacing	Part
	At least $1\frac{1}{2}$ inch margin at top	
80293 Hwy 32 South Elmhurst, IL 60126 25 July 2001	Two or more keyboard returns	**HEADING**
Ronald M. Benrey Electronics Editor, *Popular Science* 355 Lexington Avenue New York, NY 10017-0127	Content same as address on envelope	**INSIDE ADDRESS**
Dear Mr. Benrey:	Double space Followed by colon	**SALUTATION**
	Double space	
In an electronics laboratory I am taking as a part of my second-year training at Midwestern Technical College, Elmhurst, Illinois, I have developed a six-sided stereo speaker system. The speaker system is inexpensive (the materials cost less than $50), lightweight, and quite simple to construct. The sound reproduction is excellent.	at least 1-inch margin on sides	**BODY OF LETTER**
	Double space	
My electronics instructor believes that other stereo enthusiasts may be interested in building such a speaker system. If you think the readers of *Popular Science* would like to look at the plans, I will be happy to send them to you for reprint in your magazine.		
Sincerely yours,	Double space Capitalize first word, comma after close	**COMPLIMENTARY CLOSE**
Thomas G. Stein Thomas G. Stein	Four keyboard returns	**SIGNATURE AND TYPED NAME**
	At least $1\frac{1}{2}$ inch margin	

Many firms use *letterhead stationery* that has been printed especially for them, with their name and address at the top of the page. Some firms have other information added to this letterhead, such as the names of officers, a telephone number, a fax number, or a logo.

Two-Letter Abbreviations for States

Alabama	AL	Montana	MT
Alaska	AK	Nebraska	NE
Arizona	AZ	Nevada	NV
Arkansas	AR	New Hampshire	NH
California	CA	New Jersey	NJ
Colorado	CO	New Mexico	NM
Connecticut	CT	New York	NY
Delaware	DE	North Carolina	NC
District of Columbia	DC	North Dakota	ND
Florida	FL	Ohio	OH
Georgia	GA	Oklahoma	OK
Hawaii	HI	Oregon	OR
Idaho	ID	Pennsylvania	PA
Illinois	IL	Rhode Island	RI
Indiana	IN	South Carolina	SC
Iowa	IA	South Dakota	SD
Kansas	KS	Tennessee	TN
Kentucky	KY	Texas	TX
Louisiana	LA	Utah	UT
Maine	ME	Vermont	VT
Maryland	MD	Virginia	VA
Massachusetts	MA	Washington	WA
Michigan	MI	West Virginia	WV
Minnesota	MN	Wisconsin	WI
Mississippi	MS	Wyoming	WY
Missouri	MO		

Note that each abbreviation is written in uppercase (capital) letters and that no periods are used.

On letterhead stationery, the inside address is already printed on the paper; you add only the date to the heading. (See letters on pages 441, 442, 444, 451, 452.) Letterhead paper is used for the first page only.

2. *Inside Address.* The inside address is placed flush (even) with the left margin and is usually three spaces below the heading. It contains the full name of the person or firm being written to and the complete mailing address, as in the following examples:

Ronald M. Benrey	Kipling Corporation
Electronics Editor, *Popular Science*	Department 40A
355 Lexington Avenue	P. O. Box 127
New York, NY 10017–0127	Beverly Hills, CA 90210

Preface a person's name with a title of respect (for example, Ms., Mrs., Mr.) if you prefer; and when addressing an official of a firm, follow the name with a title or position. Write a firm's name in exactly the same form that the firm itself uses. Although finding out the name of the person to whom a letter should be addressed may take some time, it is always better to address a letter to a specific person rather than to a title, office, or firm. In giving the street address, be sure to include the word Street, Avenue, Circle, and so on.

3. *Salutation.* The salutation, or greeting, is two spaces below the inside address and is flush with the left margin. The salutation typically includes the word *Dear* followed by a title of respect plus the person's last name or by the persons' full name: "Dear Ms. Badya:" or "Dear Maron Badya." In addressing a company, "Dear Davidson, Inc.:" or simply "Davidson, Inc.:" are acceptable forms.

   Americans usually follow the salutation with a colon. Other practices include using a comma if the letter is a combination business-social letter, and in the modified and full block forms, omitting the punctuation after both the salutation and the complimentary close.

4. *Body.* The body, or the message, of the letter begins two spaces below the salutation. Like any other composition, the body of a letter is structured in paragraphs. Generally, it is singled space within paragraphs and double spaced between paragraphs.

Second Page. For letters longer than one page, observe the same margins as used for the first page. Be sure to carry over at least two lines of the body of the letter to the second page.

Although there is no one conventional form for the second-page heading, it should contain (a) the name of the addressee (the person to whom the letter is written), (b) the page number, and (c) the date. The following illustrate two widely used forms:

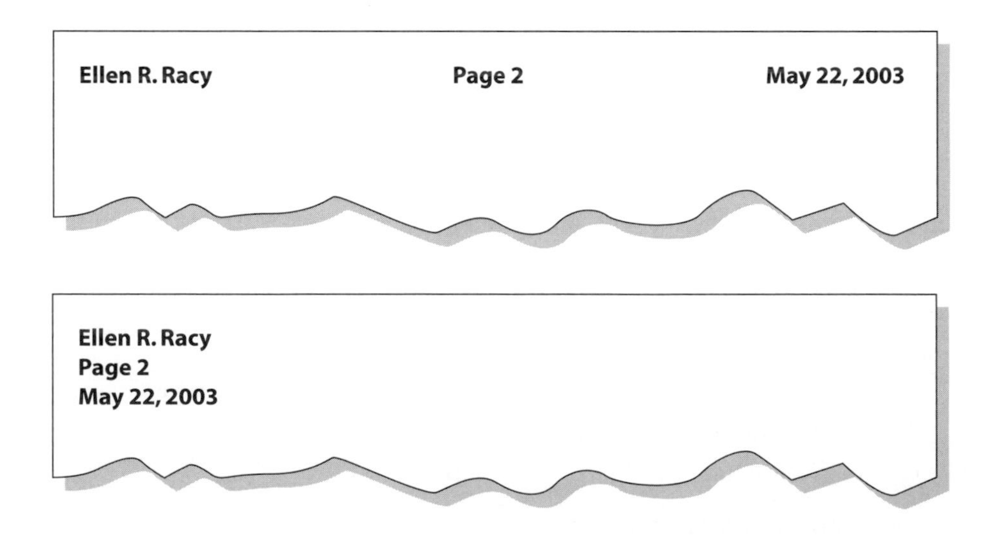

5. *Complimentary Close.* The complimentary close, or closing, is two spaces below the body. It is a conventional expression, indicating the formal close of the letter. "Sincerely" is the most commonly used closing. "Cordially" may be used when the writer knows the addressee well. "Respectfully" or "Respectfully yours" indicates that the writer views the addressee as an honored individual or that the addressee is of high rank.

   Capitalize only the first word, and follow the complimentary close with a comma. (In using open punctuation, the comma is omitted after both the salutation and the complimentary close.)

6. *Signature.* Every letter should have a legible, handwritten signature in ink. Below this is the keyboarded signature. If the letter is handwritten, print the signature below the handwritten signature.

   If a writer's given name (such as Dale, Carol, Jerry) does not indicate whether the writer is male or female, he or she may want to include a title of respect (Ms., Miss, Mrs., Mr.) in parentheses to the left of the keyboarded signature. In a business letter, a married woman uses her own first name, not her husband's first name. Thus, the wife of Jacob C. Andrews signs her name as Thelma S. Andrews. If she prefers, she may type her married name in parentheses (Mrs. Jacob C. Andrews) below her own name.

   The name of a firm may appear as the signature as well as the name of the individual writing the letter. In this case, responsibility for the letters rests with the name that appears first.

   Following the keyboarded signature, there may be an identifying title indicating the position of the person signing the letter; for example: Estimator; Buyer, Ladies Apparel; Assistant to the Manager, Food Catering Division. (See pages 441, 442, 444, 451, 452.)

Special Parts of a Letter In addition to the six regular parts of the business letter, sometimes special, or optional, parts are necessary. The main ones, in the order in which they would appear in the letter, are:

1. *Attention Line.* When a letter is addressed to a company or organization rather than to an individual, an attention line may be given to help in mail delivery. An attention line is not used when the inside address contains a person's name. Attention lines directed to Sales Division, Personnel Manager, Billing Department, Circulation Manager are typical. The attention line contains the word *Attention* (capitalized or all capitals, and sometimes abbreviated) followed by a colon and name of the office, department, or individual.

   Attention: Personnel Manager *or* ATTN: Personnel Manager

2. *Subject or Reference Line.* The subject or reference line saves time and space. Typically it consists of the *Subject* or *Re* (a Latin word meaning "concerning") followed by a colon and a word or phrase of specific information, such as policy number, account number, or model number.

   Subject: Policy No. 10473A *or* Re: Latham VCR Model 926

The position of the subject line is not standardized. It may appear to the right of the inside address or salutation; it may be centered on the page several spaces below the inside address; it may be flush with the left margin several spaces below the inside address; it may even be several spaces below the salutation. (See pages 398, 441, 446, 451, 595.)

3. *Identification Line.* When the person whose signature appears on the letter is not the person who keyboarded the letter, an identification line is needed. Current practice is to include only the initials (in lowercase) of the typist. The identification line is two spaces below the signature and is flush with the left margin of the letter. (See pages 431, 441, 442, 444.)
4. *Enclosure.* When an item (pamphlet, report, check) is enclosed with the letter, place an enclosure line two spaces below the identification line and flush with the left margin. If there is no identification line, the enclosure line is two spaces below the signature. The enclosure line can be written in many ways and gives varying amounts of information, as illustrated on pages 398, 432, 451, 594, 595, 596 and in these four examples.

   Enclosure

   Enclosures: Inventory of supplies, furniture, and equipment
   Monthly report of absenteeism, sick leave, and vacation leave

   Encl: Application of employment form

   Encl. (2)

5. *Copy.* When you send a copy of a letter to another person, place the letter *c* (usually lowercase), or the word *copy* followed by a colon and the name of the person or persons to whom you are sending the copy one space below the identification line and flush with the left margin of the letter. (See pages 398, 431.) If there is no identification line, the copy notation is two spaces below the signature and even with the left margin.

   c: Mr. Jay Longman

   copy: Joy Minor

The Envelope

The U.S. Postal Service sets standards for envelope sizes and addresses for mail in this country. These standards allow the use of an OCR (Optical Character Reader) to sort mail electronically.

Envelope sizes regularly used by businesses are classified as All-Purpose (3 5/8 x 6 1/4 inches), Executive (3 7/8 x 7 1/2 inches), and Standard (4 1/8 x 9 1/2 inches). Larger envelopes are available in three standard sizes: 6 1/2 x 9 1/2 inches, 9 x 12 inches, and 11 1/2 x 14 5/8 inches.

Regular Parts on the Envelope. The two regular parts on the envelope, the outside address and the return address, are illustrated on page 439.

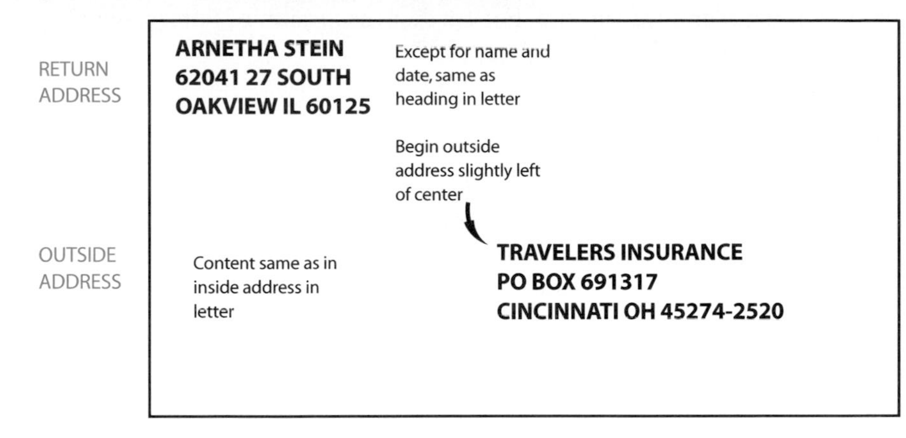

1. *Outside Address.* The content of the outside address on the envelope is identical to the content of the inside address. The postal service prefers single spacing, all uppercase (capital) letters, and no punctuation marks for ease in sorting mail using an OCR. For obvious reasons, the address should be accurate and complete. The postal service encourages using the nine-digit zip code to facilitate mail delivery.
2. *Return Address.* Located in the upper lefthand corner of the envelope, not on the back flap, the return address includes the writer's name (without "Ms.," etc.) plus the address as it appears in the heading. The zip code should be included.

Special Parts on the Envelope. In addition to the two regular parts of the envelope, sometimes a special part is needed. The main ones are the attention line, the personal line, and mailing directions.

1. *Attention Line.* An attention line may be used when a letter is addressed to a company rather than to an individual. The wording of the attention line on the envelope is the same as that of the attention line in the letter. On the envelope, the attention line is written directly above the first line of the address.
2. *Personal Line.* The word *Personal* or *Confidential* (capitalized and usually italicized or underlined) indicates that only the addressee is to read the letter. The personal line, aligned with the left margin of the return address, appears three spaces below the return address.
3. *Special Mailing Directions.* Mailing directions such as SPECIAL DELIVERY, REGISTERED MAIL, CERTIFIED MAIL, or PRIORITY MAIL are typed in all capital letters below the stamp.

TYPES OF LETTERS

There are many types of letters and numerous books devoted wholly to discussions of letters. This section discusses some of the common types of letters written by businesses and individuals: inquiry, good news and bad news, claim, adjustment, and transmittal. For further discussion of letters, see Chapter 13 Reports page 333; see also pages 592–593.

Letter of Inquiry

A letter to the college registrar asking for entrance information, a letter to a firm asking for a copy of its catalog, a letter to a manufacturing plant requesting information on a particular product—each is a letter of inquiry, or request. To write a letter of inquiry, follow these guidelines:

1. State what you want clearly and specifically. If asking for more than two items or pieces of information, use an itemized list.
2. Give the reason for the inquiry, if practical. If you can clearly show a direct benefit to the company or person addressed, you increase your chances of a reply.
3. Include an expression of appreciation for the addressee's consideration of the inquiry. Usually a simple "thank you" is adequate.
4. Include a self-addressed, stamped envelope with inquiries sent to individuals who would have to pay the postage themselves to send a reply.

REMINDER

An inquiry letter states clearly, specifically, and courteously what the writer wants.

On page 441 is a letter of inquiry written from a company handling student loans to a parent regarding the status of her son's application materials.

Good News and Bad News Letters

Writing letters that convey good news or bad news is a common responsibility in the workplace.

Good News Letters

A good news letter is pleasant for both the writer and the recipient. People like good news. The letter might be congratulations for a job promotion, an announcement of a person's winning entry in a graphic design contest, or a positive response to an employee's request for new office furniture.

Whatever the occasion, the goods news letter typically has three parts:

1. Statement of the good news
2. Clarifying details
3. Cheerful closing

REMINDER

A good news letter directly states the good news with clarifying details.

The reader-centered approach is most important in the good news letter. The "you" approach comes naturally in the writer's focus on the recipient's accomplishment. A warm, sincere, personable tone helps to make the letter special.

A good news letter appears on page 442.

SOUTHERN EDUCATORS
LIFE INSURANCE COMPANY
4672 Knightsbridge NW Richmond, Georgia 30093-3239
Telephone: (404) 449-5267 Toll Free: (800) 221-1012

July 26, 2001

Mrs. Patricia A. Woodson
350 Byrd Road
Florence, NC 28560

RE: 78689, Bruce Eugene Woodson, Jr.

Dear Mrs. Woodson:

Recently we sent you forms and instructions for your son's convenience in applying fc
Guaranteed Student Loan (Stafford Loan).

You were to fill out certain sections and then send the forms to the college or univers
that your son will attend. The college or university was to complete the forms and the
send them to us for processing.

We have not received your son's completed forms.

If you have filled in the forms and sent them to the college or university, urge the schc
to send the forms to us immediately.

If we can assist you, please contact us. Our office hours are 8:30 A.M. to 4:45 P.M., Monc
through Friday. Our toll-free number is (800) 221-1012.

Sincerely,

Lakisa Khanta

Lakisa Khanta
Student Loan Department

GSL/S38

Good News Letter in Block Format

STATE OF MISSISSIPPI
DEPARTMENT OF ECONOMIC AND COMMUNITY DEVELOPMENT

JAMES B. HEIDEL
EXECUTIVE DIRECTOR

April 23, 2001

Honorable Jimmy Moore
President
Prentiss County Board of Supervisors
Post Office Box 477
Booneville, Mississippi 38829

Dear Mr. Moore:

I am pleased to inform you that your application for Emergency Shelter Grant funds in the amount of $100,000 has been approved by the Mississippi Department for Economic and Community Development.

Through this application for funding, Prentiss County has demonstrated local commitment in addressing the community development needs of Mississippi. I commend you for this initiative and for helping to improve the quality of life in your community.

Please contact Mrs. Alice A. Lusk, director for the Community Services Division, Department of Economic and Community Development, if you have any questions. Her mailing address is Post Office Box 849, Jackson Mississippi, 39205, or you may call her at (601) 359-3179. Mrs. Lusk's office will be in contact with you soon regarding your contract.

Sincerely,

James B. Heidel

James B. Heidel
Executive Director

JBH:jnt

1200 WALTER SILLERS BUILDING • POST OFFICE BOX 849 • JACKSON, MISSISSIPPI 39205-0849 • TEL: (601) 359-3449 • FAX: (601) 359-2832

Bad News Letters

Writing letters that convey a bad news message can be challenging. The task becomes more manageable if the writer assumes the role of the recipient and asks "What could make me feel more comfortable even in my disappointment?" First off, no one likes to be slapped in the face with "Too bad you didn't get the job. Better luck next time." Soften the blow of disappointment; start with statements that give the recipient positive reinforcement. Then move on to the particulars of the bad news.

Thus, the approach in the bad news letter is indirect, in contrast with the direct approach in the good news letter. The bad news letter typically follows this four-part sequence:

1. Begin with a buffer that softens the disappointment. Establish a common ground of agreement and accomplishment.
2. State the bad news tactfully and sensitively.
3. Give specific details that clarify the reasons for the bad news.
4. Close with an upbeat look to the future.

An example of a bad news letter appears on page 444.

REMINDER

A bad news letter attempts to soften the disappointment using tact, sensitivity, and an upbeat closing.

Claim and Adjustment Letters

Claim and adjustment letters are in some ways the most difficult letters of all to write. The claim letter describes the problem; the adjustment letter is the response.

Claim Letters

About one out of four purchases of services or goods by individuals results in a problem, according to the U.S. Office of Consumer Affairs. Therefore, it is wise to be knowledgeable about writing a claim letter.

Frequently, writers of claim letters are angry, annoyed, or extremely dissatisfied, and their first impulse is to express those feelings in a harsh, angry, sarcastic letter. But the claim letter that antagonizes the reader is not likely to result in positive action. Thus, in writing a claim letter, it is important to be calm, courteous, and businesslike. Assume that the reader is fair and reasonable and will consider all information presented. Include only factual information, not opinions; and keep the focus on the real issue, not on personalities.

In writing a claim letter:

1. Identify the transaction (what, when, where, etc.). Include copies (not the originals) of the substantiating documents—sales receipts, canceled checks, invoices, and the like.
2. Explain specifically what the problem is. Regardless of how angry or inconvenienced you may feel, don't use obscenities, threats, or libelous statements.
3. State the adjustment or action that you want taken, such as repair, replacement, exchange, or refund.

Bad News Letter in Block Format

STATE OF MISSISSIPPI
DEPARTMENT OF ECONOMIC AND COMMUNITY DEVELOPMENT
JAMES B. HEIDEL
EXECUTIVE DIRECTOR

April 23, 2001

Dr. Charles Main
Executive Director
Pine Belt Mental Healthcare Resources
Post Office Box 1030
Hattiesburg, Mississippi 39403

Dear Dr. Main:

The Community Services Division has completed its review of the FY 2001 Emergency Shelter Grants program (ESG) applications submitted for funding.

The limitations on ESG funds make the selection process very competitive. Based on the review and the rating process conducted by our staff, your Emergency Shelter application did not rate within the funding range.

Thank you for your interest in the Emergency Shelter Grants program. We look forward to working with you in the future.

Sincerely,

Alice A. Lusk

Alice A. Lusk, P.C.D.
Director
Community Services Division

AAL: jnt

1200 WALTER SILLERS BUILDING • POST OFFICE BOX 849 • JACKSON, MISSISSIPPI 39205-0849 • TEL: (601) 359-3449 • FAX: (601) 359-2832

4. Keep a record of all actions. Keep a copy of every letter you send. When speaking with someone over the telephone, record the person's name along with the date, time, and outcome of the conversation.
5. Send the letter by certified mail. Certified mail costs a few dollars extra, but you will have proof that the letter was delivered.
6. Remember that reputable companies want customers to be satisfied and that most respond favorably to justifiable complaints.

A claim letter written by a customer who believes she has not been treated fairly in trying to replace a dress with a defective zipper appears on page 446. Note the detailed explanation that includes names of personnel, dates, the problems, and the actions she wants the store manager to take. Preceding the claim letter are the Project Plan Sheet and the Organization Plan Sheet that the writer adapted and completed in preparing the letter.

REMINDER

A claim letter focuses on complete, substantiated information and reasonable expectation of adjustment.

Claim Letter in Block Format

2708 Bryonhall Drive
Austin, TX 78745
December 7, 2001

Mr. Larry Watkins, Manager
Mirabelle's Department Store
165 E. Main Street
Austin, TX 78701

DRESS WITH DEFECTIVE ZIPPER

Dear Mr. Watkins:

On December 3, 2001, I purchased a party dress for my daughter from your selection of Elegant Christmas Frocks displayed in your preteen department. I was assisted by Aileen Dawes in making my selection. Since my daughter was not with me, I asked Ms. Dawes if it would be possible to return the dress if it did not fit her or if she did not like it. Ms. Dawes assured me there would be no problem.

I took the dress home and had my daughter try it on that evening. As I zipped it up, the zipper came loose from the dress. I inspected the zipper and noted that the seam to which the zipper should have been sewn was too narrow and was frayed in parts.

On December 5, I returned the dress to your store. Ms. Dawes was not working that day, but another salesperson, Christine Mays, told me that I could not return the dress because it was damaged. She referred me to her supervisor, Mary Kaiser, when I protested that the dress was defective. Ms. Kaiser said that store policy prohibited the return of damaged goods.

I would like to protest this decision. I have been a Mirabelle's customer for many years and have come to associate your store with exceptional quality and outstanding service. This is the first item I have purchased that did not meet the standards I expect from Mirabelle's. Obviously the garment was not properly sewn. It is an expensive garment, and all I am asking is to have it replaced with one that is not defective.

I'm sure that the Mirabelle's return policy for damaged goods was not meant to include items that are defective. I would like to discuss this matter with you at your convenience in order to work out a fair adjustment. My telephone number is 555-0467; please call me with a time that I can meet with you.

Sincerely,

Kathy Judge

Kathy Judge

PROJECT PLAN SHEET

FOR CORRESPONDENCE

Audience

- Who will read the correspondence?

  Larry Watkins, Manager of Mirabelle's; perhaps sales staff

- How will the reader use the correspondence?

  Larry Watkins in deciding whether to permit the return of the dress

- How will your audience guide your correspondence choices?

  The facts should be presented without bias and without condemnation of staff, who said they were following store policy. All aspects of the claim must be accurate because sales staff are likely to be consulted for verification.

Purpose

- What is the purpose of the correspondence?

  To convince the store manager that the dress should be exchanged

- What need will the correspondence meet? What problem can it help solve?

  The store manager will want all pertinent details before making a decision about the return of the dress. He may want all of this information before responding to the request for an appointment. The factual information concerning the purchase and the attempted return of a dress will hopefully persuade the store manager to allow the return.

- What reference line (subject line) will most clearly reflect the correspondence's purpose?

  DRESS WITH DEFECTIVE ZIPPER

Subject

- What is the correspondence's subject matter?

  Appointment with the store manager to work out a fair adjustment

- How technical should the discussion of the subject matter be?

  Specific details should be included—in lay person's terminology—concerning the purchase and attempted exchange of the dress.

- Do you have sufficient information to complete the correspondence?

  Yes

Author

- Will the correspondence be a collaborative or an individual effort?

  Individual

- How can the correspondent(s) evaluate the success of the correspondence?

  The letter will be successful if the store manager and I can work out a fair adjustment.

Project Design and Specifications

- Are there models for organization or forms for correspondence?

  Yes
- In what form (memo, letter) and format (block, modified block) will the correspondence be presented?

  Letter in block format
- Will the correspondence require special parts, such as enclosures? If so, what kinds? For what purpose?

  Subject line, to state the reason for the letter

Due Date

- What is the deadline for the correspondence?

  Today, if my daughter is to get full benefit from the party dress for the Christmas season
- How long will the correspondence take to plan, research, draft, revise, and complete?

  About two hours

ORGANIZATION PLAN SHEET

FOR CORRESPONDENCE

Introduction

1. How can the opening paragraph clearly indicate the correspondence's subject, scope, and purpose?

   State the date, department, name of sales assistant, and return agreement.

Discussion

1. What are the main parts of the body?

   A narrative description of each event from my daughter trying on the dress at home on Dec. 3 through attempting to return the dress on Dec. 5

2. In what sequence should the main information appear?

   The order in which the events occurred

3. What organization patterns should be used to develop each paragraph?

   Chronological

Closing

1. How can the closing reinforce the correspondence's purpose?

   Ask for a face-to-face meeting to work out a fair adjustment.

Adjustment Letters

In writing an adjustment letter:

1. Respond to the claim letter promptly and courteously.
2. Refer to the claim letter, identifying the transaction.
3. Clearly state what action will be taken. If the action differs from that requested in the claim letter, explain why.
4. Be fair, friendly, and firm.

Following on page 451 is an adjustment letter in response to the claim letter on page 446.

Transmittal Letters

The transmittal letter (or memorandum for internal transmission) is a silent courier accompanying a document, typically a report or proposal, from sender to recipient. The sender is usually the writer of the document and the recipient is usually the person who requested the document or the person who will act on the document. The transmittal letter identifies, introduces, fills in background information, describes problems, and provides contact information on the material being transmitted.

A transmittal letter typically includes these parts:

1. Statement that a document (give the title) is enclosed
2. Identification of the document, such as who prepared it, why, when, and for whom
3. Brief summary of the document, with reasons for emphasizing certain points
4. Description of problems arising in preparing the document
5. Any additional information that would be helpful in understanding the context
6. Expression of appreciation to others who helped with the document to the person who requested it
7. Contact information, should questions arise

REMINDER

A transmittal letter identifies, introduces, fills in background information, describes problems, and provides contact information on the material being transmitted.

An example of a transmittal letter accompanying a reference book sent to pertinent Youth Court agencies appears on page 452. The book compiler gives background information concerning child advocacy, explains the book's purpose, and recognizes contributors to the book's publication.

Other examples of transmittal letters appear on pages 319, 348, 398, 452, 594–596.

Mirabelle's Department Store

165 East Main Street • Austin, Texas 78701
Telephone 512-551-9103
Fax 512-551-9100

December 10, 2001

Kathy Judge
2708 Bryonhall Drive
Austin, TX 78745
SUBJECT: Confirmation of telephone conversation

Dear Mrs. Judge:

As agreed in our telephone conversation earlier today, you will bring the dress to my office on December 13 at 4:00 p.m.

I believe that this matter can be resolved quickly. Please accept my personal apologies for our staff's misunderstanding in applying the terms "defective" and "damaged" to the dress you are returning.

You are a valued customer, Mrs. Judge, and we appreciate your shopping in Mirabelle's Department Store for the past ten years. Enclosed is a voucher for a 15 percent discount on your next purchase of a garment at Mirabelle's.

Sincerely,

Larry Watkins

Larry Watkins, Manager

Enclosure: discount voucher

Transmittal Letter in Modified Block Format

MISSISSIPPI COMMITTEE
FOR PREVENTION
OF CHILD ABUSE, INC.

CHILDREN FIRST
INCORPORATED

July 1, 2001

Dear Associate:

The Mississippi Youth Court Desk Book is a joint publication with Children First Incorporated and the Mississippi Committee for Prevention of Child Abuse.

Children First Incorporated was founded in 1991 with the mission to improve services to and information on victimized children. This mission is primarily carried out through our CASA programs, which provide the voice of a volunteer advocate on behalf of an abused child. Additionally, Children First Incorporated is engaged in several partnerships including the Mississippi College School of Law and the Children's Legal Clinic–Child Advocacy Project and the Mississippi Supreme Court's Youth Court Improvement Project. All three of these partners were of strategic importance in the development and publication of the 2001 *Law Desk Book*.

The Mississippi Committee for Prevention of Child Abuse was founded in 1984 with the mission of educating and promoting awareness in the field of Child Abuse. MCPCA has been active on the front lines of the battle to protect Mississippi children from harm. MCPCA additionally provides much needed financial and other support to two of the state's child abuse shelters, giving children access to protection any time of the day or night. One of these shelters is the Natchez Children's Home, which has served children's needs for over a century.

One program that MCPCA is extremely proud to be a part of is Survival, Inc. Survival, Inc., a MS Corporation, is a nonprofit 501(c) (3) organization dedicated to the empowerment of survivors, family members, and loved ones whose lives have been forever changed by violent crimes. This program offers individual, system, and educational support. The program has also implemented a 24-hour, statewide crisis line.

MCPCA publishes *Voices,* a social work newspaper. This publication reaches 17,000 professionals and advocates. *Voices* provides the reader with updates on legislation, child abuse cases, conferences and victim's rights.

We are very excited to present you with the third edition of the *Mississippi Youth Court Desk Book* prepared in cooperation with our partners (*Administrative Office of Courts, Mississippi College School of Law and the Children's Legal Clinic, Inc.*) and funded in part through a grant from the Mississippi Bar Foundation IOLTA Project. This book is designed to put key Youth Court information under one cover and thus provide professionals with easy access to up-to-date information, including: Statutes, Supreme Court Special Orders, Dispositional Resources, and General Reference Information.

If you have any input for future editions, please let us know. Our special thanks go to Nick Clark and his staff for the printing and binding of the book. It is our hope that this book provides you with useful information as you protect Mississippi's children.

Sincerely,

Dana Stanley
Dana Stanley
Executive Director
MS Committee for Prevention of Child Abuse

Jeffrey D. Johns
Jeffrey D. Johns
Chief Executive Officer
Children First Incorporated

GENERAL PRINCIPLES for Correspondence

- In sending correspondence, the writer chooses a delivery system (U.S. postal services, express land and air services, or electronic services) that best meets the needs of sender and recipient.
- The use of electronic systems such as faxes, voice mail, and e-mail speed delivery, but may pose problems of confidentiality and professional etiquette.
- Correspondence sent through the postal service should follow postal service guidelines for envelopes to allow electronic sorting (using all uppercase letters, omitting punctuation, and using the two-letter abbreviations for states).
- Memorandums are typically used for written communication between persons in the same company. Memorandums are used to convey or confirm information.
- An effective memorandum or letter is neat and pleasing to the eye, and follows a standardized layout format. The communication uses conventional parts, and, when needed, special parts, all following a standard sequence and arrangement.
- An effective memorandum or letter is well organized, stresses a positive approach, uses natural wording, and is concise.
- Effective correspondence is reader-centered, is sensitive to cultural differences, and follows ethical practices.
- Whether a letter is an inquiry, good news or bad news, claim or adjustment, transmittal, or some other type, each requires special attention to purpose, audience, and inclusion of pertinent information.

CHAPTER SUMMARY

CORRESPONDENCE

Knowledge of how to handle correspondence in the workplace through memorandums and letters is essential in the daily activities of the successful professional. Effective correspondence writers are considerate of the audience's culture, technical knowledge, and organizational distance. Their correspondence shows respect and courtesy, encourages positive working relationships, and follows ethical practices.

Delivery systems of correspondence have changed dramatically over the last twenty years. Electronic transmission of correspondence, for instance, has made communication instantaneous via fax machines, voice mail, and e-mail.

As in any communication project, the correspondence writer is a problem solver, answering questions such as these:

- Is a memorandum or a letter needed?
- What delivery system is most appropriate?
- What is the purpose of the communication?
- What are the reader's needs?
- What information is needed?
- How can I be sure that the communication is successful?

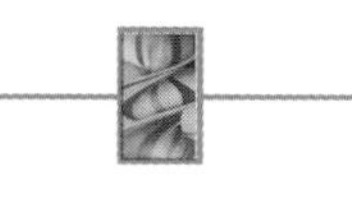

ACTIVITIES

INDIVIDUAL AND COLLABORATIVE ACTIVITIES

15.1. Collect at least five memorandums, letters, and e-mails of various types used in the workplace and bring them to class. For each communication, answer the following questions.

a. What layout format is used?
b. Is letterhead stationery used? If so, what information does the letterhead contain? (Remember that letterhead stationery may contain information at the bottom or sides as well as at the top of the page.)
c. Is the communication neat and pleasing in appearance? Explain.
d. What special parts of a memorandum or letter are used?
e. What is the purpose of the communication?
f. Is the communication reader-centered? Explain.

In small groups, analyze each memorandum and letter. Select the most effective one. Share it with the entire class, explaining why each example is effective.

15.2. Select one of the communications you collected for 15.1. Evaluate the communication according to the General Principles for Correspondence and any special instructions for this type of communication. Hand in both the communication and the evaluation.

15.3. Examine the content and form of the following body of a request letter addressed to *Popular Science Digest*. Point out items that make the request unclear.

> In regard to your article a while back on how to make a home fire alarm system in *Popular Science Digest*, which was very interesting. I would like to obtain more information. I would also like to know the names of people who have had good results with same. Give me where they live, too.

15.4. Rewrite the body of the letter shown in 15.3.

15.5. Read the following memorandum from a supervisor to his staff. As you read, assume that you are a member of the staff. Jot down your reactions to the memo; then answer the following questions:

TO: All Support Staff
FROM: Hevard Johnson
DATE: November 6, 2001
SUBJECT: Miscellaneous gripes and complaints

The following is a list of friendly reminders of things that either your mamas didn't teach you or that you forgot at some point in your checkered past. Most of these gentle prods toward respectable behavior apply to everyone. Others, however, may not apply to our subcontractors. I do, however, request that each of you take a moment to familiarize yourselves with the contents of this list and heed the messages found therein.

• Time sheets are due on either my or Dean's desk by **9:00 a.m.** on the 15th of each month (or the last working day before the 15th) and on the last working day of each month. I am extremely tired of having to go around searching, begging, and pleading for time sheets. You also signed an oath during orientation that you would keep these time sheets filled out on a daily basis. I think many of you lied.

• Our office opens at **8:00 a.m.** I have checked and I am quite certain that nobody changed that to 8:15.

• **Attention all smokers!!** We now have ash trays mounted beside each of the four entrances to the building. Use them! I'm really, really sick and tired of picking your butts up in the parking lot! You have to walk right past the new ash trays to get back into the building after you have thrown your butt on the ground. Have some pride and common courtesy.

• The "aluminum" recycling bin in the breakroom is for **aluminum.** What a shock! How could I expect anybody to figure that one out without written directions. There are trash cans for the other trash. Tours can be arranged if you are still having difficulty with this concept.

• As you are casually staggering down the hallways, up the stairs, or otherwise meandering through the buildings, put your pens or pencils in your pockets, behind your ears, or in some other convenient receptacle. Apparently it was only a pipe dream that you would all learn to walk straight enough to not make huge black marks on the walls. I guess that I will have to use our bonus money to repaint, or make you promise to use it on grace and poise lessons.

• Eating at your desks is now forbidden. Hansel and Gretel don't work here, so there is no need for the continuous food trails leading through the building. All eating done within the confines of 939 North Washington Street shall henceforth be conducted solely in the break room. Yes, the training room is off limits too!

• For your own safety—DO NOT stand anywhere near the back door at 4:55. You could be killed in the stampede of people leaving early. I find it strange that people's watches are so cheap that they can gain ten minutes in one day. The same watch can be five minutes late in the morning, yet be five minutes early in the afternoon.

This list is just the tip of the iceberg. We all need to exercise a little common sense and a little common courtesy rather than just being common. Think about the other people who work here and respect them and their property. We are all responsible adults here; let's behave like it.

a. What is your overall reaction to the memo? Would you be inclined to change your actions? Explain your answer.
b. State your reactions to phrases in the introduction such as
   - "your mamas didn't teach you"
   - "checkered past"
c. List other objectionable phrases. Why are they objectionable? How can concerns about the use of such language be addressed?
d. Revise the memorandum, as assigned:
   - Rewrite the introduction.
   - Rewrite any one section.
   - Rewrite the closing.

For Activities 15.6–15.17

a. Review the procedure for preparing a memorandum or letter, pages 429–439. (Review procedures for particular types of correspondence, pages 440–452.)
b. Adapt and complete a Project Plan Sheet, pages 465–466.
c. Using decisions from the completed Project Plan Sheet, adapt and complete an Organization Plan Sheet, page 467.
d. Using decisions from the two completed plan sheets, write a preliminary draft.
e. Revise, incorporating design decisions.
f. Prepare a final draft.

15.6. Assume that you are a department supervisor. The date and the location of the next monthly meeting have been changed. You need to send this information to the sales staff. Write the communication.

15.7. Assume that you are an employee in a company. You have an idea for improving efficiency that should lead to a larger margin of profit for the company. Present this idea in writing to your immediate supervisor. State the idea clearly and precisely, and give substantiating data.

Inquiry Letters

15.8. Prepare a letter to a person such as a former employer or a coworker asking permission to use the person's name as a reference in a job application.

15.9. Write a letter to the appropriate official at your college requesting permission to take your final examinations a week earlier than scheduled. Be sure to state your reason or reasons clearly and effectively.

15.10. Write a good news letter to a coworker congratulating the person on winning second place for an original design in a graphic arts competition.

15.11. As office manager of a 30-employee small business, you are hiring an assistant office manager. Two present employees have applied for the position. Write a good news letter to Manual Cortes, who has been selected for the position.

15.12. For the situation in 15.11, write a bad news letter to Hua Ling, who was not selected for the position.

Claim and Adjustment Letters

15.13. Assume that you ordered an item, such as a pair of running shoes, a set of wheel covers, or a computer memory chip, and the wrong size was sent to you. Write a claim letter requesting proper adjustment.

15.14. Write a claim letter requesting an adjustment on a piece of equipment, a tool, an appliance, or a similar item that is not giving you satisfactory service. You have owned the item for two months and it has a one-year warranty.

15.15. Write an adjustment letter in response to 15.13 or 15.14.

Transmittal Letters

15.16. Write a transmittal letter or memorandum to accompany a progress report for Project Double Speed. The Project is behind schedule and cost overruns are mounting. (You are not asked to write the progress report.)

15.17. Write a transmittal letter to accompany your résumé in applying for a job. (You are not asked to write the résumé.)

READING

15.18. Read the newspaper article, "Ethics in the Computer Age," by Nell Luter Floyd, which includes discussion of ethical considerations in cyberspace correspondence.

Ethics in the Computer Age

NELL LUTER FLOYD

When Heather Osborne wants to know how the stock market is performing, she simply checks the Internet. Never mind that she's at work, and it may not be pertinent to the job at hand. It's OK.

"Our feeling is that Internet and e-mail provide advantages that would outweigh any disadvantage from employees using them excessively," said Osborne, executive vice president at Godwin Group.

A Matter of Degree

BAD

Survey respondents thought it was unethical to:

- Sabotage systems/data of current coworker or employee—96 percent
- Access private computer files without permission—93 percent
- Use office equipment to shop on the Internet for personal reasons—54 percent

NOT SO BAD

Fewer respondents labeled the following actions unethical:

- Play computer games on office equipment during work hours—49 percent
- Use office equipment to help you children/spouse do school work—37 percent
- Use company e-mail for personal reasons—34 percent
- Use office equipment for personal reasons—29 percent

Source: American Society of Chartered Life Underwriters and Chartered Financial Consultants and Ethics Officer Association.

The Jackson marketing and advertising agency has no written policy about Internet or e-mail usage but relies on employees to use the tools judiciously.

Contrast that with Raytheon Aerospace Co. in Madison, a firm that specializes in aviation logistics support services. It monitors e-mail messages and Internet usage.

Raytheon Human Resources Director Bill Grosvenor recalls an employee at Raytheon who accidentally typed the wrong ending on an Internet address and wound up at a sexually oriented site in cyberspace.

"The employee backed out after seven or eight seconds and immediately notified the supervisor of the mistake," Grosvenor said of the employee who was innocent of misconduct.

While even the most virtuous among us has probably been guilty of goofing off at work or taking home an occasional pen or pencil, what's different today is that technology may be at the heart of wrong-doing.

Serious infractions can be grounds for firing. For example:

- Telephone Electronics Corp. cited unauthorized long-distance telephone and facsimile transmissions from company headquarters to Russia as one reason—along with allegations of poor job performance—for firing an employee in 1994, according to U.S. District Court records in Jackson. The employee's lawsuit that claimed violations of the Americans with Disabilities Act was dismissed July 24.
- Employees of Unisys downloaded child pornography using computers at the University of Mississippi Medical Center where

they were under contract to make repairs. The company did not renew their contracts, and both were sentenced in federal court earlier this year.

The issue isn't new but the age-old one of theft of time or goods, said Yolanda Estes, assistant professor of philosophy at Mississippi State University in Starkville, where she teaches courses in business ethics and professional ethics. "It just happens that the issue isn't as tangible as it once was."

How technology contributes to on-the-job pressure and increases the risk of unethical and illegal business practices is the heart of a survey this year by the American Society of Chartered Life Underwriters and Chartered Financial Consultants and the Ethics Officer Association.

"We wanted to get a pulse from American workers as to what things they believed to be unethical using new technology," said Amanda Mujica, director of communications for the Ethics Officer Association in Belmont, Mass.

"It's the first study I've seen that has looked at: 'What do you think is wrong?' and 'Have you ever done that?'"

Forty-five percent of workers said they have committed at least one of a dozen actions over the past year that are either unethical or fall into a hazy area, according to the survey of 726 workers.

"It's interesting that we have a consensus of what people think is wrong, yet they're many of the things American workers are doing," Mujica said.

The ethical abuses range from serious—4 percent of workers said they have done something to sabotage the computer system or data of their company or coworkers—to seemingly insignificant—13 percent of workers say they have used company computers to shop the Internet.

ROUGH NUMBERS

Other actions include: 5 percent who listened to a private cellular phone conversation; 6 percent who said they accessed private computer files without permission; 13 percent who copied company software for personal use; and 11 percent who logged on at work to search the Internet for another job.

Art Pullum, president of AAP Staff Services, Inc., a Jackson employment service that employs about 200 persons, said he's not surprised by survey findings that some workers use resources at work to conduct personal business.

"I've been in business for over 20 years, and as situations change, people change," said Pullum, who believes at least 15 percent of full-time workers get paid for 40 hours of work a week but don't work that many hours.

Employees without computers at home often feel justified in using them at work for personal business, said Craig Lowery, associate professor of computer science at Mississippi College. "The employees say they can't afford it and it won't hurt anything, so it's OK to browse the Web. Some employers don't mind. Others do, depending on how much the company relies on it."

Some employees don't think it's fair to use the computer at work for personal reasons.

"If we're on the clock and faithfully giving 100 percent to the job, there should not be idle time," said Carolyn Mayberry, customer service specialist at Deposit Guaranty National Bank in Ridgeland.

At AAP Staff Services, employees with e-mail and Internet privileges must use those tools to conduct business during the work day, Pullum said. After hours, it's fine for an employee to use the Internet to help a child with homework, he said.

TO A LIMIT

Employees at New Horizons Computer Learning Center in Ridgeland are welcome to shop for most anything on the Internet because they teach a course in how to use the Internet, but it's not OK to devote a large amount of time to research for one's child or spouse, said Robert Peet, operations manager at the center.

"If they're not doing things work-related and conducive to business issues at my place of business then they run the risk of being unemployed." At Horizons—like most businesses—pornographic sites are off-limits, no matter what. "I'm not going to be responsible for some sexual harassment lawsuit if a female employee walks by and a male employee is at some perverse site."

The trend, Lowery said, is that more companies are cracking down on personal use of the Internet, e-mail, and computer. Many employees fail to realize a quick game of Solitaire or a few minutes sending personal e-mail messages could tie up a company's computer network. Downloading a game off the Internet could introduce a detrimental code or virus that could bring a system down.

PERMISSION TO PLAY

Other employees have permission to play games at certain times.

When Beverly Mercier, secretary of First Baptist Church of Jackson Counseling Center, relieves the switchboard operator for an hour at lunch, she plays a computer game called Free Cell between calls. "I take that time for something I like to do."

Employees often fail to realize a business has the right to read e-mail messages its employees send, Lowery said. "That's why you should use your e-mail at work for business purposes. Get e-mail from another place for personal reasons."

At Raytheon, e-mail is strictly for business.

"There are occasions when you have no control over someone sending you a personal e-mail into the system," Grosvenor said. "It's a reasonable response to say, 'I got your message, and you can e-mail me at home.'"

WHAT THEY FOUND

Key findings from *Technology and Ethics in the Workplace: The Ethical Impact of New Technologies on Workers:*

- Employees see a number of new technology-related problems facing society. Primary concerns: Availability of dangerous and offensive materials on the Internet (76 percent), invasion of privacy by the government (76 percent), invasion of privacy by businesses (75 percent) and the loss of person-to-person contact.
- One in six employees agree with the statement: Traditional standards of right and wrong are no longer relevant. One-third agree or are ambivalent. This sentiment increases with age. Only 7 percent of those ages 30–34 agree, but more than 21 percent of survey respondents over the age of 55 say that traditional standards of right and wrong are no longer relevant.
- When asked if they feel they are addicted to the Internet, 2 percent said yes. The response rate did not vary by gender or age. When asked if others they work with are addicted to the Internet, 50 percent said yes. Sixty-five percent of those who work in large organizations say others they work with are addicted.

FLAG ON THE PLAY

Raytheon flags inappropriate Web sites, and a warning pops up if an employee tries to access one, Grosvenor said. The 350 employees at the Madison site know that, as do its 5,000 employees worldwide, he said.

Technology has expanded the horizons of what employees can do, and most large companies realize it's necessary to set limits, communicate what they are and enforce them.

Estes said her students are divided on whether businesses should monitor computer usage by employees. Half think so, and half think it's an invasion of privacy.

One solution is for business to focus on productivity.

"That would solve some problems because you wouldn't have to set the company up as big brother monitoring the computer use."

Human nature can't be changed, Grosvenor said, but businesses can outline what they expect to work.

"What employees do on their own time, we may not be able to control, but we do have a say when they're at work."

SURVEY QUESTIONS

People differ on what constitutes unethical behavior. Here are the actions 726 workers evaluated as to right or wrong as part of a study sponsored by the American Society for Chartered Life Underwriters and Chartered Financial Consultants and Ethics Officer Association.

The study asked workers to indicate whether they engaged in any of the following actions involving technology in the workplace in the past year (percentage in agreement follows the statement):

- Used company e-mail for personal reasons (29 percent).
- Used office equipment to network/search for another job (11 percent).
- Listened to a private cellular phone conversation (5 percent).
- Wrongly blamed an error you made on a technological glitch (14 percent).
- Copied the company's software for home use (13 percent).
- Accessed private computer files without permission (6 percent).
- Sabotaged systems/data of former employer (4 percent).
- Sabotaged systems/data of current co-worker or employer (4 percent).
- Used office equipment to shop on the Internet for personal reasons (13 percent).
- Played computer games on office equipment during work hours (30 percent).
- Visited pornographic Web sites using office equipment (5 percent).
- Used office equipment to help your children/spouse do school work (19 percent).
- Used new technologies to unnecessarily intrude on co-workers' privacy such as paging during dinner (6 percent).
- Created a potentially dangerous situation by using new technologies while driving (19 percent).
- Made multiple copies of software for office use (11 percent).
- Used office equipment for personal reasons (62 percent).

Source: *Technology and Ethics in the Workplace: The Ethical Impact of New Technologies on Workers*

The Clarion-Ledger [Jackson, Miss.] 2 Aug. 1998: F1+.

QUESTIONS FOR DISCUSSION

a. List three ethical problems in correspondence that employees and employers may face. Suggest ways to deal with these problems.
b. State two correspondence ethical concerns (not mentioned in the article) that you foresee as major workplace issues with the increasing capability of computers and other office machines. Suggest ways to deal with these issues.

15.19. In class discussion, or as a short written assignment, respond to the following case study by identifying the ways in which the case illustrates the nature of business correspondence.

Just the Fax

Joan Whitmore is startled by a loud hoot from the office next door. She jumps to her feet to find out what's wrong. Luella Shiflet, her boss, is standing in the hall, livid with rage. She is holding a letter.

"I have never been treated with such sarcasm, in print and on letterhead, no less!" snaps Luella. "Who does this twerp think he is? Doesn't he realize that we are a woman-owned business? The name on our letterhead, Woman Business Associates, should really give a clue."

"Whoa there, chief," says Joan. "Slow down a minute. You'll live longer."

She guides Luella back into her office. "Here, sit down and have a glass of water. Then you can explain what all the fuss is about."

Luella slowly counts to ten and sips her water. "Thanks, Joan. That's better. This letter is from some idiot who is interested in working with us on that big proposal we got, the one we wanted to see about handling with another firm. He read our request for proposals on our company Web page, and he has made a bid. It's a pretty good bid, actually."

"So what's the problem? I'd expect you to be pleased, not peeved."

"Maybe, but listen to this, Joan. He starts out by calling us 'mesdames'."

"Hmmm. That's pretty old-fashioned and stuffy, not to say sexist. But it's the offer that counts, isn't it?"

"But wait, Joan. There's more. He sounds like a pompous windbag. 'Pursuant to your request' and 'kindly be advised' and 'should the ladies of your outstanding commercial concern desire competent support from one long experienced.' He goes on and on for four pages. I can spot the bottom line, all right. But it took me two readings to figure out just what he had proposed when a simple list would have explained it all. I don't want to deal with anybody who can't respond clearly to a request for bids."

"You're probably right, Luella. Besides, just the little you've read to me convinces me that he'll be a real pain to work with. The way he writes tells it all. He's one of those know-it-alls who won't be open, won't meet deadlines, and will patronize us like crazy. Want me to write him a polite kiss-off letter? Or does he even deserve one?"

QUESTIONS FOR DISCUSSION

a. Are Luella and Joan justified in their conclusions about the writer of the letter that Woman Business Associates received? Why or why not?

b. Is the author of the letter necessarily guilty of sarcasm or sexism?
c. What tone does the author of the letter think he has achieved? How does this tone sound to his audience?
d. Are Luella and Joan acting wisely in rejecting the bid because they objected to the style of the letter?
e. Is this a good decision for Woman Business Associates? Why or why not?
f. Woman Business Associates advertised an interest in receiving bids by posting the request on the Web. Is it ethical for the company not to reply to all bidders?
g. What advice do you have for Joan and Luella?
h. What advice do you have for the author of the letter?

PROJECT PLAN SHEET

FOR CORRESPONDENCE

Audience

- Who will read the correspondence?
- How will the reader use the correspondence?
- How will your audience guide your correspondence choices?

Purpose

- What is the purpose of the correspondence?
- What need will the correspondence meet? What problem can it help to solve?
- What reference line (subject line) will most clearly reflect the correspondence's purpose?

Subject

- What is the correspondence's subject matter?
- How technical should the discussion of the subject matter be?
- Do you have sufficient information to complete the correspondence? If not, what sources or people can help you to locate the additional information?

Author

- Will the correspondence be a collaborative or an individual effort? If the correspondence is collaborative, what are the responsibilities of each team member?
- How can the correspondent(s) evaluate the success of the correspondence?

Project Design and Specifications

- Are there models for organization or forms for correspondence?
- In what form (memo, letter) and format (block, modified block) will the correspondence be presented?

- Will the correspondence require special parts, such as enclosures? If so, what kinds? For what purposes?
- What information design features can best help the correspondence's audience?

Due Date

- What is the deadline for the correspondence?
- How long will the correspondence take to plan, research, draft, revise, and complete?
- What is the timeline for different stages of the correspondence?

ORGANIZATION PLAN SHEET

FOR CORRESPONDENCE

Introduction

1. How can the opening paragraph clearly indicate the correspondence's subject, scope, and purpose?

Discussion

1. What are the main parts of the body?
2. In what sequence should the main information appear?
3. What organization patterns should be used to develop each paragraph?
4. What information design elements indicate the correspondence's sequence and organizing patterns?

Closing

1. How can the closing reinforce the correspondence's purpose?

PART V

Oral Communication

Part V of *Technical English* discusses an important means of presenting all of the types of communication discussed in Part IV. **Chapter 16 Oral Communication: Saying It Simply** provides practical advice for success in one of the most feared workplace situations: public speaking. This chapter demonstrates that oral communication, like other types of communication, can be successful if communicators consider their audiences, goals, and communication tools. It also provides a procedure to make oral communication easy to plan and deliver.

CHAPTER 16

Oral Communication: Saying It Simply

CHAPTER GOALS

This chapter:

- Defines oral communication
- Examines differences between speaking and writing
- Differentiates between informal, quasiformal, and formal presentations
- Specifies the speaker's responsibilities
- Offers suggestions for coping with nervousness
- Examines the purposes of oral communication
- Explains how to prepare an oral presentation
- Explains procedures for delivering an oral presentation
- Discusses the importance of developing and scripting visuals
- Shows how to evaluate an oral presentation

INTRODUCTION

Most people communicate more through speaking than they do in any other way. Speaking to others at work, greeting customers, sharing project ideas with team members, talking on the telephone—these kinds of informal communication occur frequently. Such informal occasions involve an audience of only one person or a small group and seldom require prepared comments.

Occasions that require prepared comments, a scripted presentation, or a formal speech make many people nervous. In fact, various polls show that most people dread standing before an audience and speaking more than death.

The information in this chapter presents a systematic approach to planning and delivering oral presentations. This approach can help you develop the self-confidence and the knowledge needed for effective oral expression.

WHAT IS ORAL COMMUNICATION?

Oral communication—**spoken** communication—is an exchange between speaker(s) and listener(s) and can take place in many settings and through various media.

Informal oral communication, for example, occurs daily in the workplace when people greet each other face to face with "How are you today?" or "Have a nice weekend."

Oral communication in the workplace may be a telephone conversation with a competitor down the street, a voice mail message from the office in Singapore, or an interactive video conference with team members in the United States, Canada, and Mexico.

Oral communications often become records. For instance, calls on the telephone—a much used medium—may be logged by time, date, caller, and message. Oral communications, particularly those transmitted electronically (including telephone conversations), may be stored on tape, disk, or digital recording (using a computer chip).

SPEAKING AND WRITING

Speaking and writing have much in common because they are both forms of communication based on language. Speaking differs, however, in several important ways:

1. *Level of diction.* Speaking typically requires a simpler vocabulary and shorter sentences.
2. *Amount of repetition.* More repetition is needed in speaking to emphasize and summarize important points.
3. *Kinds of transitions.* Transitions from one point to another must be more obvious in speaking. Transitions such as *first, second,* and *next* signal movement that is often conveyed on the printed page through paragraphing and headings.
4. *Kind and size of visuals.* Speaking lends itself to the use of exhibits and projected materials; some kinds of flat materials such as charts, drawings, and maps must be constructed on a large scale.

Spoken and written presentations differ in level of diction, amount of repetition, kinds of transitions, and use of visuals.

INFORMAL, QUASIFORMAL, AND FORMAL PRESENTATIONS

Oral communication can be informal, quasiformal, or formal. The term *informal* describes nonprepared speech while *formal* describes well-planned, rehearsed speech. Quasiformal includes characteristics of both informal and formal presentations.

Informal Presentations

Most of us spend a large percentage of each day in informal communication, talking with friends, family, and coworkers. In the workplace, you may be asked to share views or knowledge about projects, places, or plans within the company or with an external group, such as a service organization, a civic club, or a group of clients. You may be asked at a meeting to respond impromptu, or you may be asked ahead of time but make little or no preparation. In these situations, you share information through informal oral presentations.

Quasiformal Presentations

Committee meetings, discussion groups, and team planning sessions are examples of settings for quasiformal presentations. In these settings, the sharing of ideas and information focuses on specific purposes. The leader (usually selected by job description, appointment, or group consensus) sets the agenda and conducts the meeting. Discussions may include impromptu group sharing when ideas pop into participants' heads, oral reports from subcommittees, or a formal report, for example, from a task force asked to study a problem and make recommendations. Degrees of formality in quasiformal settings are blurred. The overriding purpose is to conduct the business at hand in a professional manner.

Formal Presentations

Formal oral communication involves a great deal of preparation and attention to delivery. Professionals are often asked to share their views and knowledge of their fields. The deliberate, planned, carefully organized and rehearsed presentation of ideas and information for a specific audience and purpose constitutes formal communication.

Formal presentations may be categorized according to the speaker's mode of delivery as:

- extemporaneous
- memorized
- read from a manuscript

In the extemporaneous mode—the most often used of the three modes—the speaker refers to brief notes or an outline, or simply recalls from memory the points to be made. In this way, the speaker is able to interact with the audience and convey sincerity and self-assurance. In the memorized mode, the speaker has written out the speech and committed it to memory, word for word. The memorized speech tends to lack spontaneity and an at-ease tone; also, the speaker must face the very real possibility of forgetting what comes next. In the third mode of delivery, the speaker reads from a manuscript. While this type of delivery may be needed when exact wording is required in a structured situation, the manuscript speech has serious limitations. It is difficult for the speaker to show enthusiasm and to interact with the audience, delivery is usually stilted, and the audience may soon become inattentive.

REMINDER

Most formal oral presentations are extemporaneous: The speaker refers to notes or an outline, but does not memorize or read the speech from a script.

THE SPEAKER'S RESPONSIBILITIES

A speaker is expected to fulfill certain responsibilities. If a speaker begins, for instance, by apologizing for forgetting to request a screen for projecting visuals, the audience may become wary and begin to wonder what else the speaker may have forgotten. Or if a speaker is allotted fifteen minutes but talks on for another twenty minutes, the audience may become agitated. Such situations can be avoided if the speaker attends to these responsibilities:

- checking the room and the equipment conditions
- stating how the allotted time will be used
- adhering to the time limit
- following ethical practices

Fulfilling these responsibilities helps to ensure a successful oral presentation.

Room and Equipment Conditions

Before an oral presentation, check the room for conditions such as appropriate temperature, seating arrangement, and ease of moving furniture (for instance, for small group discussions). Check the equipment to be sure that all is working and in place. A burned out bulb in an overhead projector or lack of an extension cord, for example, could undermine a well-planned presentation.

Use of Allotted Time

Sharing your plan for the oral presentation with the audience helps to establish rapport between you and your audience. If you welcome questions during the presentation, for instance, or if you want the audience to hold questions until the presentation is over, let the audience know. And if you can talk with people individually at the close of the presentation, make the audience aware of your availability.

Time Limit

Sticking to time limitations is very important. Unfortunately, many speakers cancel out careful preparation and good delivery by not adhering to specified time limits. A speaker given four to six minutes who speaks for only one or two minutes may be perceived as not having adequately developed the subject. On the other hand, a speaker given thirty-five minutes is often praised for not using the full time allotted. (Studies show that attention span for adults wanes after about twenty minutes.)

REMINDER

Know how long your speech will be and adjust it to fit within the time limit.

Ethical Practices

There is power in the spoken word. Whether in coffee-break discussions, committee meetings, or formal oral presentations, individuals in the workplace must remember ethical considerations in speaking situations.

Informed speakers:

- Recognize the power they have over listeners. Listeners expect the speaker to respect them as thinking, intelligent human beings capable of making sound choices when given accurate, full, unbiased information. An audience may be present because attendance is expected, the speaker is a well-known and respected civic leader, or the speaker is a company executive—whatever the situation, the speaker should not take advantage of a captive audience to push a hidden agenda.

- Recognize the power of ideas. In presenting information and ideas, the informed speaker gives needed background and context, taking into consideration the needs and interests of listeners. Often, a speaker supports a point and gives it credibility by referring to respected authorities and quoting them. Such quotations—in fairness—must be given in context. For example, a report about a United States government official who said that he wished he had been a better student of Latin so that he could better communicate with citizens in Latin America was unfair and unethical. In the full text, the speaker had referred to the Latin origin of Romance languages, such as Spanish, and how knowledge of Latin helps in learning modern languages derived from it.
- Avoid plagiarism. Informed speakers give credit where credit is due in presenting ideas and information. To do otherwise is unethical.
- Plagiarism is implying that another person's ideas or words are your own. Plagiarism may be unintentional or intentional. To avoid the problem of plagiarism:
  1. Indicate when you are quoting and give the source. Example: Peter J. Prestillo, Ford Motor Company, poses an insightful question: "We must ask ourselves, how willing are we to make education and training a top priority in our organizations?"
  2. Give paraphrased material in your own style and language, and credit the source. Example: According to Peter J. Prestillo, Ford Motor Company, international organizations must address the issue of their commitment to education and training as a top priority.

REMINDER

In any presentation, oral or written, give credit for ideas and information that are not your own.

COPING WITH NERVOUSNESS

Nervousness is a normal reaction to a strange, uncertain, or unfamiliar situation. Fortunately, nervousness is temporary. The sweaty hands, shaking knees, trembling voice, short breaths, queasy stomach, and light-headedness are emotional and physical reactions to being in the speaker's spotlight and fearing that the audience might not be receptive or that no words will come out when it is time to speak.

These tips can help to ease the anxiety:

- Prepare note cards and visuals, and rehearse again and again.
- Think positively: You have important things to say.
- Take several deep breaths and exhale slowly.
- Establish eye contact with the audience immediately. Smile. Make a comment such as "Thank you for being here today." This acknowledgment of the audience can help you feel at ease.
- If you have a mental block, be prepared to show a visual immediately. The visual, perhaps a slide or an overhead transparency or a poster, might show the main points of the presentation. The visual shifts the eyes of the

audience from the speaker to the visual. These moments can provide time for you to regain your composure, as the audience studies the visual.

- Know the introduction so well that you could recite it in your sleep. Once you get started speaking, usually you can continue.
- Remember that your audience is empathetic; your audience understands nervousness and wants you to do well.

REMINDER

Thorough preparation can help a speaker cope with nervousness.

PURPOSES OF ORAL PRESENTATIONS

Oral communication's major purpose can be to entertain, to persuade, or to inform. Each type of communication could be presented formally, quasiformally, or informally.

To Entertain

Oral communication that is meant to entertain is intended to provide enjoyment for listeners. Few individuals have occasion to communicate solely for the purpose of entertaining except on an informal basis with friends and relatives.

To Persuade

Communication that is meant to persuade seeks to affect the listeners' beliefs or actions. On the job you may find yourself responsible for persuading supervisory personnel or customers or employees to change a method or a procedure, to hire additional personnel, to buy a certain piece of machinery or equipment, and so forth. Whether presenting an idea, promoting a plan, or selling a product, the same basic principles of persuasion are involved.

The art of persuasion can be summed up in two sentences: Present a need, want, or desire of the audience. Show the audience how your idea, service, or product can satisfy that need, want, or desire.

Basic Considerations in Persuasion

If you are to prepare and present a persuasive speech effectively, you must be aware of several factors: the audience's needs, wants, desires and the kinds of appeals you can make.

Audience's Needs, Wants, and Desires. The basic needs of existence are few: food, clothing, and shelter. In addition to these physical needs, people have numerous wants and desires, including:

- *Economic security.* This includes a means of livelihood and ownership of property and material things.

- *Recognition.* People want social and professional approval. They want to be successful.
- *Protection of self and loved ones.* The safety and physical well-being of self, family, and friends are important to most people.
- *Aesthetic satisfaction.* People desire pleasant surroundings and experiences that are pleasing to the senses.

Consideration of the audience's needs, wants, and desires is essential in order to present an idea, plan, or product effectively. For example, if you were to try to persuade your employer to purchase new computers, you would probably appeal to the employer's economic and aesthetic desires; that is, you might emphasize the time that could be saved and the improved appearance of keyboarded material. Or if you were a supervisor impressing upon a worker the importance of following dress regulations, you would stress protection (safety) of self and the possible loss of economic security for the worker and his or her family should injury or accident occur. Or if you were to persuade a coworker to take college courses in the evening, you would point out the recognition and economic security aspects—gaining recognition and approval for furthering education and skills and the possible subsequent financial rewards.

Appeals—Emotional, and Rational, Direct and Indirect. Persuasion is the process of using combined emotional and rational appeals and principles. Emotional appeals are directed toward feelings, inclinations, and senses; rational appeals are directed toward reasoning, logic, or intellect. Many times, emotional appeals carry more weight than rational appeals.

The most satisfying persuasion occurs when people make up their own minds or direct their own feelings toward a positive reception of the idea, plan, or product—without being told to do so. Indirect appeals, suggestions, and questions are usually much more effective than a direct statement followed by proof.

The Persuasive Presentation

A persuasive presentation involves four steps or stages: opening, intensifying need and desire, supporting proof, and closing. For timing in moving from one step to another, you must use your judgment by constantly analyzing conditions and audience mood.

In approaching a single listener or a small group of listeners, be sincere and cordial. A firm handshake should set a tone of friendliness. It is also natural to exchange a few brief pleasantries (How are you? How is business? Beautiful weather we're having.) before getting down to business. With a large group of listeners, such a personal approach is difficult. You can, however, be sincere and cordial, and you can often speak with audience members individually following your presentation.

Opening. In the opening, the listener's attention is captured. Thus, the opening should immediately spark the listener's interest and should present the best selling points. This may be done directly or indirectly.

Direct Our new Mega Value CD player provides a clearer, more authentic sound than any other CD player on the market.

Indirect Do your customers ask for CD players that give a closer, clearer, more authentic sound?

Intensifying Need and Desire. Once the audience's attention is captured, each main selling point is developed with explanatory details. Both emotional and logical appeals show how the product or service will help satisfy one or more of these basic desires: economic security, recognition, protection of self and loved ones, and aesthetic satisfaction.

At this stage, and throughout the presentation, the listener may raise objections. The best way to handle these is to be a step ahead of the listener; that is, be aware of all possible objections, prepare effective responses, and incorporate them into your presentation.

Supporting Proof. If a persuasive oral presentation is to be successful, the speaker must provide evidence and proof. Evidence and proof require logical reasoning. Information must be relevant and must be examined on the basis of reason, not on emotion or preconceived ideas.

Valid critical thinking is based on logical reasoning, as in Sam's case on page 480. Sam asks, "Why didn't I get the 150 bucks I asked my parents for to get new front tires on my car?" To be logical, a conclusion must be based on reliable, relevant, and sufficient evidence, and on an intelligent analysis of the evidence. Reliable evidence is the proof that can be gathered from trustworthy sources: personal experience and knowledge, testimony of reliable individuals, reference works, reliable Internet sources, examples, statistics, and the like. Relevant evidence is information that directly influences the situation. Sufficient evidence means enough or adequate information with no omission of significant facts that would alter the situation. Evidence that is reliable, relevant, and sufficient must be analyzed intelligently. The meaning and significance of each individual piece of information and of all the pieces of information together must be considered if plausible conclusions are to be reached.

The speaker conveys proof and evidence not only through words but also, and often more importantly, through demonstrations and exhibits.

REMINDER

Logical reasoning is based on evidence that is reliable, relevant, sufficient, and intelligently analyzed.

Closing. In closing a presentation, it is wise to assume that the audience will accept the idea or plan or will buy the product. The following suggestions reflect such an attitude: *When may we begin using this procedure? Which model do you prefer?* Reaffirmation of how the product or service will enhance the listener's business often helps to conclude the deal and to reinforce his or her satisfaction. If you detect a negative attitude, avoid a definite "no" by suggesting further consideration or a trial use of the product and another meeting at a later date.

Scenario: Why Sam Didn't Get 150 Bucks

1. My parents don't love me.
2. They don't have the money.
3. They didn't receive my voice mail message.
4. They mailed me a check but something happened to the letter.
5. They just don't realize what bad shape those front tires are in.
6. They think I should pay my car expenses out of the money I earn from my part-time job.

Sam turns over each possibility in his mind. He immediately dismisses possibility 1 (My parents don't love me). He knows this was only a childish reaction and that love cannot be equated with money. Sometimes parents who don't love their children send them money; and sometimes, for very good reasons, parents who love their children don't send them money.

There is a real possibility Sam's parents didn't have the $150 to spare (possibility 2). But Sam reasons that if this is the cause, they would have asked him to wait a while longer or at least made some response.

Perhaps his parents did not receive his voice mail request (possibility 3). His parents, however, check their voice mail frequently.

Maybe they mailed him a check but something happened to the letter (possibility 4). The letter could have gotten lost in the mail. To his knowledge, however, no letter sent to him has ever gotten lost.

Possibility 5 (They just don't realize what bad shape those front tires are in) could certainly be a true statement. But Sam realizes that his parents could be aware of the condition of the tires and still have reason not to send the money.

The more Sam considers possibility 6 (They think I should pay my car expenses out of the money I earn from my part-time job), the more this seems like the reason. After all, Sam had promised his parents that if they would help him buy a car, he would keep it up. Furthermore, in his excitement of owning a car at long last, he had cautioned his parents to ignore any requests for car money, no matter how desperate they sounded.

So as Sam carefully considers each possibility, he comes to the conclusion that the reason he did not receive the money is that his parents expect him to pay for his car expenses.

To Inform

Of the general purposes of oral communication—to entertain, to persuade, and to inform—the informative purpose is most common in the workplace. In communicating procedures, reports, proposals, and other materials (see Chapters 12–15 in Part IV), the speaker's major goal is to *inform* the audience.

Giving informative oral presentations is a significant aspect of workplace communication responsibilities. The focus of the remaining sections of this chapter—preparing an oral presentation, delivery, and visual materials—are geared to the informative speech. The basic principles discussed in these sections are, however, applicable to any speech situation. For instance, the steps in preparing an oral presentation are essentially the same, whether the purpose is to entertain, to persuade, or to inform.

PREPARING AN ORAL PRESENTATION

Preparing an oral presentation involves these steps:

- determining the specific purpose
- analyzing the type of audience
- gathering the material
- organizing the material
- determining the mode of delivery
- outlining the speech
- preparing visual materials
- rehearsing

Determining the Specific Purpose

Typically, the general purpose of a workplace speech is to inform; sometimes the general purpose is to persuade, or, occasionally, to entertain. You must determine the *specific* purpose if the speech is to be effective. Establish the reason for the speech and determine who will use the information. Give data that completely, accurately, and clearly present the subject, and then analyze and interpret the data thoroughly and honestly. You can then make recommendations accordingly.

Analyzing the Type of Audience

Your speech, if it is to be effective, must be designed expressly for the knowledge and interest level of the audience. Adapt your vocabulary and style to the particular audience. For instance, if you were to report on recent applications of laser technology, your reports to a group of nurses, to a group of engineers, to a college freshman class of physics students, or to a junior high science club would differ considerably. Each group has a different level of knowledge and a different interest in the subject.

Gathering the Material

You gather material primarily from three sources: interviews and reading, field investigation, and laboratory research.

The extent to which you use one or more of these sources depends on the nature of the speech. A report in history, for instance, may simply call for reading certain material in a book. An investigation of parking facilities in a particular location may call for personal interviews plus on-site visits. An analysis of the hardiness of certain shrubs when exposed to sudden temperature changes may involve both field investigation and experimental observation.

Organizing the Material

To organize the material, select the main ideas, but do not exceed three or four. (Remember that your audience is listening, not reading.) Arrange supporting data under each main idea, using only the data necessary to develop each main idea clearly and completely.

After the main body of material is organized, plan the introduction. Let the audience know the reason for the speech, the purpose, the sources of data, and the method or procedure for gathering the data. Then state the main ideas to be presented. The function of the introduction is to set an objective framework so that the audience will accept the information as accurate and significant.

Next, plan the conclusion. It should contain a summary of the data, a summary of the significance or interpretation of the data, and conclusions and recommendations for action or further study.

A suggested outline for the introduction, body, and conclusion is given in the Outlining the Speech section below. (For the organization of a persuasive speech, see pages 478–479.)

Determining the Mode of Delivery

Once you have analyzed the speaking situation and gathered and organized the material, you can determine the appropriate mode of delivery. Is it more appropriate to speak extemporaneously, to recite a memorized speech, or to read from a manuscript? (Of course, you may have been told which mode to use; thus the decision has already been made for you.) The memorized speech is most appropriate in situations such as competing in an oratorical contest or welcoming an important visiting dignitary. Reading from a script is most appropriate if presenting a highly technical scientific report, giving a policy speech, or the like. For most other situations, extemporaneous speaking is the most appropriate.

Outlining the Speech

In most cases, a good outline will serve to structure your speech. A suggested outline form follows.

Introduction

- I. Reason for the speech
  - A. Who asked for it?
  - B. Why?
- II. Purpose of the speech
- III. Sources of data
- IV. Method or procedure for gathering the data
- V. Statement of main ideas to be presented

Body

- I. First main idea
  - A. Subidea
    - 1., 2., etc. Data

B., etc. Subideas
1., 2., etc. Data
II., III., etc. Second, third, etc., main ideas
A., B., etc. Subideas

Conclusion
I. Summary of the data
II. Summary of the significance or interpretation of the data
III. Conclusions and recommendations for action or further study

If you plan to present a memorized speech or read from a script, write out the speech. Give special care to manuscript form and to the construction of visuals if you are to distribute copies of the speech (copies should be distributed *after* the oral presentation, not before or during it).

Preparing Visual Materials

Carefully select and prepare visuals to help clarify information and to crystallize ideas. For a full discussion of visuals, see the section in this chapter "Developing and Scripting Visuals" and Chapter 5 Visuals.

Rehearsing

For an extemporaneous speech: From your outline, make a note card (3- by 5-inch) of the main points that you want to make. Indicate on the card where you plan to use visuals. Rehearse the entire speech several times, using only the note card (not the full outline). Get the ideas and supporting data and the order in which you want to present them fixed in your mind.

For a memorized speech: Commit to memory the exact wording of the script. As you practice the speech, put some feeling into the words; avoid a canned, artificial sound.

For a speech read from a manuscript: Just because you are to read a speech doesn't mean you shouldn't practice it. Go over the speech until you know it so thoroughly that you can look at your audience almost as much as you look at the script. Number the pages so that they can be kept in order easily. Leave the pages loose (do not clip or staple them together); you can then unobtrusively slide a finished page to the side or to the back of the stack.

REMINDER

Thorough preparation of an oral presentation includes determining the purpose, analyzing the type of audience, gathering and organizing the materials, determining the mode of delivery, outlining the speech, preparing visual materials, and rehearsing.

Some speakers find it helpful to videotape their speech once or twice while rehearsing and then play back the tape for an objective analysis of their strengths and weaknesses. Rehearsing your presentation several times is every important; rehearsing gives you self-confidence and prepares you to stay within the time allotted.

STUDENT EXAMPLE: ORAL PRESENTATION

Examine the following completed Project Plan Sheet and transcript of a student-prepared oral presentation, "How Alexander Fleming Discovered Penicillin."

PROJECT PLAN SHEET

FOR AN ORAL PROCESS EXPLANATION

Audience

- Who will listen to the process explanation?

  A lay audience of adult volunteers at a hospital
- How will hearers use the process explanation?

  For general understanding of how a major antibiotic was accidentally discovered
- How will your audience guide your communication choices?

  The audience, a group of volunteers, will need simple language presented in an informal style.

Purpose

- What is the purpose of the explanation?

  To explain the discovery of penicillin
- What need will the explanation meet? What problem can it help to solve?

  The explanation will supply background information for hospital volunteers on a major discovery in the medical field.

Subject

- What is the explanation's subject matter?

  How penicillin was discovered
- How technical should the discussion of the subject matter be?

  Since the explanation is intended for a lay audience, it should be nontechnical.
- Do you have sufficient information to complete the subject? If not, what sources or people can help you to locate the additional information?

  No. I will look at print and online reference sources.
- What title can clearly identify the project's subject and purpose?

  How Alexander Fleming Discovered Penicillin

Author

- Will the explanation be collaborative or an individual effort?

  The project will be an individual effort.
- How can the developer(s) evaluate the success of the completed explanation?

  Maybe I can ask a physician or nurse to look over my finished draft.

Project Design and Specifications

- Are there models for organization or format for the explanation?

  Yes. Printed speeches are available from a variety of sources.

- In what medium will the completed explanation be presented?

  Oral presentation

- Are there special features the completed explanation should have?

  The finished explanation should be especially clear and easy to understand. Complex sentence structure and technical terms will confuse my audience.

- Will the explanation require visuals or other graphics? If so, what kinds and for what purpose?

  Prescription vial of penicillin tablets, a prescription tube of penicillin ointment. A list of stages written on a chalkboard, dry erase board, or a page from a newsprint pad as each stage is introduced. A petri dish. A piece of moldy cheese and a slice of moldy bread.

- What information design features can best help the explanation's audience?

  Use of real objects; outline of stages in the discovery process

Due Date

- What is the final deadline for the completed explanation?

  I will give the presentation at the hospital volunteers' monthly meeting on 15 June.

- How long will the explanation take to plan, research, draft, revise, and rehearse?

  Research will take the longest time. I also need to allow at least a week for showing the speech to my final reviewer and another week to make any changes and final revisions. I need to allow time to gather visuals and to practice delivering the speech.

- What is the timeline for different stages of the explanation?

  I'll plan an outline during one day. I'll spend one week locating sources and taking notes. I'll spend two days on drafting and two days on revision, allowing my reviewer two additional days to read the explanation and contact me with suggestions.

ORAL PRESENTATION
How Alexander Fleming Discovered Penicillin

Title clearly reflects presentation's purpose.

[Marginal notes that show the organizational plan and suggest ways to use visuals in the presentation have been added.]

Introduction of subject, purpose, and significance of the process **(Directions for the speaker's use of visuals are shown in bold.)** **Hold up a vial of penicillin tablets and a tube of penicillin ointment.**

Many fortunate discoveries have resulted from accidents. We call such discoveries serendipitous. One such serendipitous discovery happened in September 1928, resulting in a miracle antibiotic drug. The drug was first used widely in World War II among the Allied military personnel in the treatment of infections, wounds, and disease. This miracle drug proved also to be very effective against the organism that causes the venereal disease syphilis. Today the drug is commonly available in tablets, capsules, ointments, and drips, as well as by injection. The miracle drug is penicillin. For over fifty years this drug has helped to ease pain, facilitate healing, even save lives—for millions of people around the world.

List of main stages As each major stage is introduced, write the stage for the audience.
1. Experimentation with antibacterial agents
2. Analysis of the mold
3. Identification of the mold
4. Extraction of the essential compound

How did an accidental discovery by Sir Alexander Fleming lead to the development of penicillin? These events occurred in four stages: experimentation with antibacterial agents, analysis of the mold, identification of the mold, and extraction of the essential compound.

Introduction and development of Stage 1:

Stage 1 Experimentation with Antibacterial Agents

The young Alexander Fleming, M.D., from Scotland, specializing in bacteriology at the University of London, was interested in developing antibacterial agents from nature.

Having made a special study of septic wounds while serving as a doctor in World War I, he continued his research and experimentation. While Fleming was working on a series of experiments with *Staphylococcus aureus,* a fortunate accident occurred. Fleming had left exposed to the open air some culture plates (also called petri dishes) of staphylococci.

A few days later, when he returned to the laboratory, he discovered that one of the bacteriological plates had become contaminated with a mold from the air, possibly from an open window.

Fleming probably admonished himself to be more careful. He started to discard the plate. But the trained eye of the scientist noticed something different about the mold.

Introduction and development of Stage 2: Hold up a petri dish.

Stage 2 Analysis of the Mold

Fleming carefully analyzed the conditions surrounding the mold. He noticed that the bacteria had failed to grow in the area of the mold. Thus, he reasoned, some unknown substance in the mold had killed the bacteria. Also, he noticed tiny drops of liquid on the surface of the mold.

Though Fleming had tainted his staphylococci experiment, he had stumbled on a discovery that would revolutionize the treatment of wounds and illness in human beings.

Introduction and development of Stage 3:
Hold up a piece of moldy cheese and a slice of moldy bread.

Stage 3 Identification of the Mold

Fleming continued to study and analyze the mold. This mold, by the way, was green and was very similar to the ordinary fungus growth we see on cheese and bread when they get a little old.

Eventually Fleming identified the mold as belonging to the genus *Penicillium,* and he called the unknown substance penicillin. The tiny drops of liquid on the surface of the mold—the substance that Fleming called penicillin—was the chemical that had destroyed the neighboring bacteria.

Introduction and development of Stage 4:

Stage 4 Extraction of the Essential Compound

Some ten years after Fleming had made his accidental discovery of penicillin, two fellow scientists—Howard Walter Flory and Ernst Boris Chain—took up Fleming's work. They extracted the essential compound from the liquid in which penicillin grows.

Knowledge of the drug and experimentation beyond the laboratory grew quickly. By early 1941, amazing results were being obtained in the treatment of infections in human beings. And later in 1941, penicillin began to be mass produced in the United States.

Closing

Closing highlights Fleming's recognition for his discovery of penicillin.

Alexander Fleming received widespread recognition and honor for his discovery of penicillin. He was knighted—hence, *Sir* Alexander Fleming. And in 1945, Fleming, with his two coworkers, was awarded the highest honor in his field—the Nobel Prize in medicine.

DELIVERING AN ORAL PRESENTATION

A major factor in oral communication is effective delivery, or *how* you say what you say. When giving a speech, observe the following.

- *Display poise and self-confidence.* Walk to the podium with poise and self-confidence. From the moment the audience first sees you, give a positive impression. Even if you are nervous, the appearance of self-confidence impresses the audience and helps you to relax.
- *Capture the audience's attention.* Get the audience's attention and interest. Begin your speech forcefully. Opening techniques include asking a question, stating a little-known fact, or making a startling assertion (all, of course, should pertain directly to the subject at hand).
- *Make eye contact.* Look at the audience. Interact with the audience through eye contact, but without special attention to particular individuals. Spend as little time as possible looking at your notes, and try not to stare at the floor or the ceiling, over the heads of your audience, or out the window.
- *Avoid shifting modes of delivery.* Stick to an appropriate mode of delivery. If, for instance, your speech should be extemporaneous, don't read a script to the audience.
- *Show zest in your speaking.* Put some zest in your expression. Relax, be alive, and show enthusiasm for your subject. Avoid a monotonous or "memorized" tone and robot image. Have a pleasant look on your face.
- *Speak clearly.* Get your words out clearly and distinctly. Make sure that each person in the audience can hear you. Follow the natural pitches and stresses of spoken language. Speak firmly, dynamically, and sincerely. Enunciate distinctly, pronounce words correctly, use acceptable grammar, and speak on a language level appropriate for the audience and the subject matter.
- *Adjust your voice.* Adjust the volume and pitch of your voice as needed for emphasis of main points or because of distance between you and the audience, the size of the audience, the size of the room, and outside noises. Be certain everyone can hear you.
- *Vary your speaking rate.* Vary your rate of speaking to enhance meaning. Don't be afraid to pause; pauses allow time for an idea to become clear to the audience and emphasize important points.
- *Stand naturally.* Stand in an easy, natural position, with your weight distributed evenly on both feet. Bodily movements and gestures should be natural; well-timed, they contribute immeasurably to a successful presentation.
- *Avoid mannerisms.* Avoid mannerisms such as toying with a necklace or pin, jangling change, or repeatedly using an expression such as "you know," "like," or "uh."
- *Show visuals with natural ease.* For specific suggestions, see Showing Visuals on pages 495–496.

- *Close appropriately.* Close the presentation—do not just stop speaking. Your speech should be a rounded whole, and the close may be indicated through voice modulation and a simple "Thank you" or "Are there any questions?" (See the closing in the speech on page 493.)

EXAMPLE: A SPEECH BY A COMPANY EXECUTIVE

The speech on pages 490–493 was delivered in a formal setting to other executives. As you read the speech (shortened here due to space limitations), note how the speaker does several things. While laying out the problems of insufficiently educated employees in the automotive industry, the speaker gives specific information about how his company historically has responded to consumer and employee needs and then challenges the audience to invest in human capital.

DEVELOPING AND SCRIPTING VISUALS

Visual materials can significantly enhance your oral presentation. Impressions are likely to be more vivid when you use visuals. Showing rather than telling an audience something is often clearer and more efficient. Showing *and* telling may be more successful than either method by itself. For instance, a graph, diagram, or demonstration may present ideas and information more quickly and simply than can words alone.

In brief, visual materials are helpful in several ways. They can convey information, supplement verbal information, minimize verbal explanation, and add interest.

See Chapter 5 Visuals for a discussion of various types of visual materials for both oral and written communications.

Types of Visuals in Oral Presentations

Technology has revolutionized the creation and delivery of visuals. Digital imaging technology, scanners, color printers, computer software programs, and telecommunications, for instance, have vastly changed the production and presentation of visuals. Some types of visuals that only a few years ago were considered the realm of professional drafters or commercial artists can now be produced at a desk in the workplace or home.

Regardless of how they are produced, visuals for use with oral presentations can be grouped into three types: flat materials, exhibits, and projected materials. A brief survey of these types can help you determine which visuals are most appropriate for your needs. The oral presentation, "How Alexander Fleming Discovered Penicillin," on pages 486–487 incorporates several types of visuals.

Flat Materials

Included in flat materials are two-dimensional materials such as dry erase boards, bulletin boards, magnetic boards, handouts, posters, charts, maps, and scale drawings.

The War for Talent
Preparing Our Workforce for the Next Millennium

Address by Peter J. Prestillo, Vice Chairman and Chief of Staff, Ford Motor Company
Delivered to the University of Michigan's Management Briefing Seminar,
Traverse City, Michigan, August 4, 1999

Introduction

I'm going to cover a topic that's critical to our industry—preparing our workforce for the next millennium. It's not news that the automotive industry is experiencing a severe talent shortage. Who knows better than all of us in this room? We're living the consequences—a lack of flexibility and adaptability in our workers and a higher than usual turnover. Both add tremendous costs to the bottom line.

We can, however, affect the degree to which these will continue to plague us. Only we can stop this unproductive and costly downward spiral.

We must ask ourselves, how willing are we to make education and training a top priority in our organizations? How far back into the educational pipeline are we willing to reach to help our children learn the fundamental employability and life skills that so many students today simply do not have?

I'll begin my remarks by briefly laying out the scenario: How bad is it and why? Then I'll look at the education and training needs of our workforce. And I'd like to conclude with a word about Ford's efforts in this area—both our historical commitment and what we're doing today.

Current Scenario

5 As I just said, our industry is on the brink of a potentially devastating shortage of talent—on the plant floors, in our salaried ranks and at our dealerships. At every level, in all job categories—we need more good people, and we need them today.

What's significant about this problem is its reach. The labor shortage is affecting OEMs [Original Equipment Manufacturer], suppliers, and dealers—not just here in Michigan— but around the world. . . .

The auto industry is one of the greatest consumers and producers of new technology. Microchips control nearly all systems on today's vehicles. They provide the logic behind the proper operation of such functions as emission controls and fuel management, antilock brakes, airbags, dynamics and ride control, sound systems, cruise control, and on and on.

Great technological leaps are being made with each model year. Take our new Lincoln LS, for example. Its 16 chips can process 40 million instructions per second. The Powertrain Control Module alone has five times the processing capability of previous generations.

We need to do a better job of informing the outside world that our industry offers great opportunities for forward thinkers.

10 In fact, today's hourly workers at Ford are required to demonstrate a high level of literacy and must understand statistics, metrics, and elementary Boolian Algebra, among other subjects. Yet, the negative image lingers.

As I said a moment ago, this war for talent is being felt in the managerial ranks as well as on the plant floor. . . .

This seeming disdain for manufacturing is not unique to the United States. It's a worldwide problem. According to one European survey, only 8 percent of university graduates there aspire to careers in engineering or manufacturing.

Education and Training Needs

I believe one of the reasons our industry is so affected by this talent shortage is, in part, a result of the shift in societal values. Our economy today is rewarding "ideas" the way it once rewarded the production of things. . . .

The world is moving from valuing a manufacturing-driven economy to valuing a knowledge-based economy. Even in manufacturing, that means people—our human resources—are our capital—our "intellectual capital." And, as everything else in our business these days, competition for that capital is ruthless.

15 No one can argue that the world is a much different place than it was a century ago when manufacturing was redefining the economy. Our products are different. Our workplaces are different. Our employees must be different—they must be prepared. And it's our responsibility to see to it that they are. . . .

Today's workplaces need people who are flexible and adaptable, who have an intuitive ability to solve problems and work in teams, who are independent, creative thinkers—and can communicate effectively. We need people who are trained to learn—because we don't know what they will need to know over the course of their careers.

Unfortunately, our educational system is not uniformly producing this type of student. And when they do, those top students often don't want to work in manufacturing.

Under this scenario, we cannot ensure an adequate consumer base either. As we know, cars and trucks can be expensive. If our consumers do not have sufficient skills to get good employment, how will they be able to afford our products? Education is the key answer.

A consistent finding in studies related to labor and employment is that the higher the level of educational achievement, the higher the earnings. And with time, the pay gap is getting wider.

20 In 1980, the pay gap between high school and college graduates was 50 percent. In 1998, it jumped to 111 percent.

Meanwhile, what are we—as an industry—going to do about the modern-day disconnect between education and business? It's time to stop the hand-wringing and roll up our sleeves. It's time to get involved—really involved. It's not enough to give money—though that's important too.

We need to redouble our efforts and touch these students—grab them, really, and interest them—long before they graduate. Working with colleges and high schools is no longer sufficient. We must reach further back into the pipeline—into the middle, elementary, even preschools.

Our expertise is needed in creative learning environments that foster the development of the type of schools we need. This means getting involved in curriculum planning, on local school boards—wherever education policy is made.

For us at Ford, it means creating schools, which we did with the Henry Ford Academy. I'll tell you more about that in a moment.

25 If we want the best and the brightest students, we have to do two things: make sure they're prepared for the business world—and then attract them to our industry.

Ford's Commitment to Education

At Ford, our commitment to education is more than good social responsibility. It is very much tied to attracting and retaining the best employees. Let me explain.

As parents and grandparents, our employees are deeply concerned about the education of their children. Traditionally, a "good job" at Ford—and all manufacturers, really—has been defined by wages and benefits. Today, the definition of a good job is being expanded by Ford employees to include the company's performance on social issues–especially education....

As early as 1914, Henry Ford established a school for immigrant workers who spoke little or no English....

Henry Ford knew instinctively then what exhaustive academic research and testing are showing us today: that technical and scientific skills of employees are as crucial as the attitudes and behaviors taught in his schools, which we know today as employability skills. They include being responsible, disciplined, and able to work with others to get things done. Having these skills was—and still is—a requirement to be a productive member of the workforce.

30 The attitudes and behaviors that he taught in his schools were basic employability skills. Today we call them "life skills."...

The world knows Henry Ford as the man who lowered the price of a car so that the average person could buy one. The Model T went from $950 in 1909, the year after it was introduced, to $360 in 1916—and sales skyrocketed from 12,000 to 577,000.

But he also was interested in the other half of the equation—creating a consumer class that could afford to improve their "lot in life." By introducing the five-dollar day, he did just that. He made the link between mass consumption and mass production.

In declaring the five-dollar day, Henry Ford was doing something else as well. Perhaps something even more important. He was recognizing—and rewarding—his workers as the backbone of his company....

And always, from the time Ford was assembling Model Ts in Highland Park to today, providing educational opportunities to our communities has been a priority....

35 We believe we're breaking new ground in our partnerships with public education around the world.

I mentioned the Henry Ford Academy a moment ago. Let me tell you about it. It's a charter high school being operated in conjunction with Henry Ford Museum.... The curriculum emphasizes technology, communications, and critical thinking. Ford employees participate in the classroom and in the design of the curriculum.

The Ford Academy of Manufacturing Sciences'... students are studying manufacturing disciplines in real-world situations. This fall, we will adapt the curriculum for the Internet, allowing teachers to retrieve lessons online and tailor them to their classrooms.

Last year, we expanded our commitment to Detroit Area Pre-College Engineering Program, which is recognized as one of the nation's best programs of its kind directed at urban students. It's a true collaboration of parents, teachers, corporations, and universities....

At the elementary level, Ford has recently taken a leadership role in A World in Motion, a program that emphasizes hands-on discovery of science principles, especially for girls and children of color. Developed by the Society of Automotive Engineers in 1991, the program promotes careers in engineering.

40 We will launch another new effort this fall with the Henry Ford Museum, the Fair Lane Learning Institute. This unique institute will look at how rapid social change is affecting education. It will then create innovative and alternative learning environments that support public education.

Of course, it's important to understand that education doesn't stop once you're out of school. It's a lifelong process. . . .

Through the many programs offered under the UAW-Ford Education, Development, and Training Program, employees, spouses and retirees can continue their education in a number of ways—from sharpening technical skills to earning college degrees. Annual college scholarships also are awarded to children of employees and retirees.

Many of these programs are located on-site at the plants themselves and operate—quite literally—as a community college for employees and their families. They are learning centers and through them, our people have access to a well-rounded education that they might not otherwise have.

Conclusion

When the auto industry was establishing itself in Detroit 100 years ago, the advantages of the region were mostly physical and logistical. It was near the sources of raw material—lumber, iron, copper—and near waterways that allowed materials to be transported easily.

45 Those advantages will continue to be useful in the coming decade. But they will not be essential. In the Information Age, worker knowledge—or intellectual capital, as we call it—is what will set companies and communities apart—regardless of geography.

The one sure way to be ready is through education—education of our children in our communities and of our employees at every level of their careers.

In a word, that's the best way to ensure a bright and prosperous future for all of us.

Thank you. I look forward to our panel discussion at the end of the session.

Source: Reprinted by permission from *Vital Speeches of the Day* (15 October 1999): 13–15. Ellipses indicate where material has been omitted from the original printing. Headings and paragraph numbers at five-paragraph intervals have been added.

Although these are usually prepared in advance and revealed at the appropriate time, they may also be created spontaneously during the presentation (as in outlining steps on a chalkboard or dry erase board). A chalkboard or an easel and pad of paper serve beautifully. Ideally, even spontaneous visuals should be created in advance and reproduced from memory or notes during the presentation.

In using printed handout material, carefully plan the timing and manner of distribution. The main thing to guard against is competing with your own handout material—having the audience reading when it should be listening.

An easel is almost essential in displaying poster-board sized pictures, cartoons, charts, maps, scale drawings and other flat materials. Various lettering sets, tracing and template outfits, and graphic supplies purchased in hobby or art supply stores can facilitate the creation of neat visuals, as can many computer software programs.

Exhibits

Visual materials such as demonstrations, displays, dramatizations, models, mockups, dioramas, laboratory equipment, and real objects can be used as exhibits. These are usually shown on a table or stand.

Undoubtedly, the demonstration is one of the best aids in an oral presentation. In fact, at times the entire presentation can be in the form of a demonstration. When performing a demonstration, be sure that all equipment operates flawlessly and that everyone in the audience can see; if practical, allow the audience to participate actively. (The oral presentations on the Winsome dryer, pages 246–248, and on penicillin, pages 486–487, use demonstration, real objects, flat materials, and projected materials.

Projected Materials

Projected materials are those shown on a screen by use of a projector or computer with projection display capability: pictures, slides, videos, transparencies, and computer-generated materials created with a program such as *Freelance, WordPerfect Corel Presentation, AppleWorks, Toolbook, Photoshop, or PowerPoint.* (*PowerPoint* screens are used in the oral description of the Winsome dryer, pages 246–248.) When using projected materials, a pointer or an onscreen pointer is essential, and an assistant is often needed to operate equipment or adjust the lights.

REMINDER

Visuals convey information, supplement and minimize spoken words, and add interest.

PREPARING AND SHOWING VISUALS

Visual materials are most effective when you select the most appropriate kinds of visuals and when you prepare and show them well.

Preparing Visuals

Once you have chosen which kinds of visuals to use, carefully plan their presentation. The following tips will help you.

- Determine the purpose of the visual. Select visuals that will help the audience understand the subject. Adapt them to your overall objective and to your audience.
- Organize the visual. Choose the information and design it for quick visual comprehension.
- Consider the visibility of the aid, paying attention to size, colors, and typography. The size of the visual aid is determined largely by the size of the presentation room and the size of the audience. Visuals should be large enough to be seen by the entire audience.
- Keep the visual simple. Do not include too much information.
- In general, portray only one concept or idea in each visual.
- Make the visual neat and pleasing to the eye. Clean, bold lines and an uncrowded appearance contribute to the visual's attractiveness.
- Select and test needed equipment. If you need equipment to show your visuals—computer, projection display equipment, overhead projector, slide projector, VCR, or screen—select the equipment ahead of time and test it to be sure it is operable. Check the room for locations and types of electrical outlets: these may affect the placement of the visual equipment. You may need a long extension cord. Determining needs and setting up equipment ahead of time allow you to make your presentation in a calm, controlled manner.

Showing Visuals

Showing visual materials should be a smooth, natural part of your presentation. The following suggestions will help you to integrate visuals into your presentations.

- Place the visual so that everyone in the audience can see it.
- Present the visual at precisely the correct time. If you need an assistant, rehearse with the assistant. Showing a visual near the beginning of a presentation often helps you to relax and to establish contact with the audience.
- Face the audience, not the visual, when talking. In using a chalkboard or dry erase board, for instance, be sure to talk to the audience, not the board.
- Keep the visual covered or out of sight until needed. After use, cover or remove the visual, if possible. Exposed drawings, charts, and the like are distracting to the audience.
- Correlate the visual with the verbal explanation. Make the relationship of visual and spoken words explicit.
- When pointing, use the arm and hand next to the visual, rather than reaching across the body. Point with the index finger, with the other fingers loosely curled under the thumb; keep the palm of the hand toward the audience.
- Use a pointer as needed, but don't make it a plaything.

Appropriately used, visuals can decidedly enhance an oral presentation. However, visuals should not be a substitute for the speaker, a prop, or a camouflage for the speaker's inadequacies. Further, unless the visuals are the focus of the presentation, they should not overshadow it.

EVALUATING ORAL PRESENTATIONS

To evaluate an oral presentation, use a criteria chart such as the evaluation form on page 497. The evaluative criteria are listed under two headings—Delivery and Content & Organization; a third heading is added for Overall Effectiveness. The audience are to evaluate one another on each criterion, using this scale:

4 = Outstanding
3 = Good
2 = Fair
1 = Needs improvement
0 = Unacceptable

Then the total number of points for each speaker is tabulated.

The evaluation procedure can be simplified, if desired, by using only the Overall Effectiveness criterion. The highest number of points for a speaker would then be 4 and the lowest 0.

The primary purpose of evaluation by the audience is twofold: (1) to encourage listening and attentive viewing and (2) to apply evaluative criteria to presentations.

EVALUATION OF ORAL PRESENTATIONS

(See page 496 for directions)

Course and Section ______________ Date ______________ Evaluated by (Name) ______________

Students' Names																							
DELIVERY																							
Forceful introduction																							
Poise																							
Eye contact																							
Sticking to mode of delivery																							
Zest (enthusiasm)																							
Voice control																							
Acceptable pronunciation and grammar																							
Avoidance of mannerisms																							
Ease in showing visuals																							
Clear-cut closing																							
Sticking to specified length																							
CONTENT & ORGANIZATION																							
Stating of main points at outset																							
Development of main points																							
Needed repetition and transitions																							
Effective kinds and sizes of visuals																							
OVERALL EFFECTIVENESS																							
TOTAL POINTS																							

4 = Outstanding
3 = Good
2 = Fair
1 = Needs improvement
0 = Unacceptable

COMMENTS:

GENERAL PRINCIPLES for Oral Communication

- Planning and delivering an oral presentation requires different considerations from those in written presentations.
- The speaker should use a mode of presentation appropriate for the occasion.
- The speaker must analyze the audience, their needs, and their expectations and then plan accordingly.
- The speaker must follow ethical practices and have respect for the power of the spoken word.
- The speaker should provide proof and evidence, exhibit freedom from bias, and avoid manipulating the audience.
- In preparing a presentation, the speaker must research the topic, organize the material, and rehearse.
- In the delivery of the oral presentation, the speaker should make sure that the audience can see and hear clearly, make eye contact with the audience, and avoid mannerisms.
- The speaker should choose visuals that complement the presentation.
- Evaluating an oral presentation requires considering the purpose and audience, the delivery, and the content and organization.

CHAPTER SUMMARY

ORAL COMMUNICATION

While most people communicate orally without hesitation in a one-on-one conversation, many people feel uncomfortable in addressing a group. This discomfort can be eased by approaching the preparation and delivery of an oral presentation in a systematic manner.

Although speaking and writing have much in common, speaking differs markedly in the level of diction, amount of repetition, kind of transitions, and kind and size of visuals. Further, speaking, whether informal, quasiformal, or formal, requires special attention to tone and to face-to-face interaction with an audience.

The purpose of oral communication may be to entertain, to persuade, or to inform. In the workplace, particularly in formal presentations, the primary purpose is to persuade or inform.

An effective speaker prepares the text and visuals carefully, avoids mannerisms, makes eye contact with the audience, substantiates ideas with proof and evidence, speaks clearly and distinctly, and stays within the time limits.

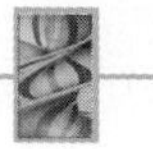

ACTIVITIES

INDIVIDUAL AND COLLABORATIVE ACTIVITIES

16.1. Select three advertisements from magazines and bring them to class. To what needs, wants, and desires do the advertisements appeal? For each one, decide whether the major appeal is emotional or rational. In what ways do the analyses of the advertisements relate to the preparation of an oral presentation, specifically to audience analysis and logical reasoning?

16.2. Review the speech by Peter J. Prestillo, "Preparing Our Workplace for the Next Millennium." Answer these questions:

a. In what ways is the speaker mindful of his audience?
b. How does the introduction present the organization of the speech?
c. In what ways are transitions from one point to another used?
d. In what ways does the speaker present evidence and proof to substantiate his points?
e. How does the speaker bring his presentation to a close?
f. Does the speaker have sufficient knowledge and understanding to make this speech? Support your answer.
g. What does the speaker mean by the phrase, "a potentially devastating shortage of talent" (paragraph 5)?
h. In paragraph 10, the speaker states, "In fact, today's hourly workers at Ford are required to demonstrate a high level of literacy and must understand statistics, metrics and elementary Boolian algebra, among other subjects." What is meant by "hourly workers"? What is implied in the speaker's reference to skills rather than to level of education (high school diploma, associate degree, baccalaureate degree)?
i. Study paragraph 16. List the qualities the speaker specifies. Is this list complete? Support your answer.
j. In what ways can workplace employers encourage continuing education for their workers?

For each oral presentation

a. Adapt and complete a Project Plan Sheet for Oral Presentations, pages 506–507.
b. Prepare the oral presentation by following the steps on pages 481–483.
c. Give the presentation by following the suggestions on pages 488–489.
d. Evaluate yourself and ask your audience to evaluate you, using the Evaluation of Oral Presentations Form on page 508.

16.3. Oral Presentations
As directed by your instructor, prepare one or more of the following for oral presentation.

1. A procedure (set of instructions or process explanation). Review Chapter 9.
2. A description. Review Chapter 10.
3. An extended definition. Review Chapter 11.
4. A report. Review Chapter 13.
5. A proposal. Review Chapter 14.

READING

16.4. As you read the following article, note the ways in which speakers undermine their presentations.

How to Sabotage a Presentation

HUGH HAY-ROE

Not long ago I attended an oil-patch conference at which most of the speakers used either 35mm slides or page-size transparencies for an overhead projector. Only a few of the speakers took advantage of new digital imaging technology to help put across their messages. Most handicapped themselves with feeble delivery and mediocre visual aids. Just for the heck of it, I took notes on the most effective techniques for torpedoing a presentation: There were more than a dozen.

VISUAL AID PROBLEMS

The Searchlight Effect

Keeping an audience awake is a great idea, but not by blinding them with transparencies that have a clear background and weak, spidery text (use boldface wherever feasible). Colored transparencies (yellow, blue, green, red) are so widely available that there's no excuse for giving an audience the equivalent of snow blindness.

Snow Blindness Revisited

Another presenter dazzled the audience with yellow letters on a white background, and then, for variety, gave us black letters on a dark blue background. Anyone wearing glasses that darken or lighten automatically must have loved that presentation!

Mafia Firepower

One speaker, obviously proud of her ammunition, showed a slide with 12 bullets on it. That's a bit much, even using the "build" technique (adding successive bullets, one at a time), which is easy to do with digital imaging and a good graphics program. Two slides with six bullets each would be a good model, and three slides with four bullets each would be still better.

Avant Garde Art

Software applications such as *Freelance* and *PowerPoint* allow you to show a vast amount of statistical information graphically, but it's a mistake to cram it all on the same graph, each in glorious Technicolor™. The audience will quickly quit trying to untangle the multiplicity of curves.

Backward, March!

One speaker who took advantage of two screens and dual projections managed to do so backwards, starting with the right-hand screen and then proceeding to the left. Because we read from left to right, audiences are more comfortable when the slide progression goes in the same direction. When you are going to leave a reference slide on one screen for an extended time, use the left-hand screen: The audience can then progress easily from the known to the unknown (the new slide on the right). For the same reason, the speaker should, if possible, stand to the left of the screen(s), or at least in between two screens—not on the far right.

Binoculars, Anyone?

Quite a few speakers used a minute font size—completely illegible from the back of the auditorium. On overheads, thirty-point boldface fonts (preferably sans serif) work well. The test of readability of 35 mm slides is simple: You should be able to hold the slide up to the light and read the text with the naked eye. Using upper and lower case text improves readability; all caps should be reserved for headings.

Elegant Wallpaper

Today's computer graphics programs provide all sorts of nifty capabilities, some of which should be resolutely avoided in slidemaking. Several speakers evidently felt obliged to liven up their slides with backgrounds that looked like the wallpaper in a Victorian bordello. Dark marble or wood-grain backgrounds are distracting enough with bold white text; with black text, slides may be unreadable. One presenter even used a floral pattern as background for a graph—a great way to camouflage the essentials.

The CPA Syndrome

It's true that most people take in complex information better by the eye than by the ear, but that's no reason to sandbag an audience with a massive spreadsheet containing—apart from the column and row headings—nothing but large numbers in a small font. Such a table is probably the worst type of visual aid.

OTHER PROBLEMS

By no means were all the fumbles in this event limited to the visual aids, of course; presenters dropped the ball in other ways as well.

The Art Connoisseur

Even in a large auditorium, audiences appreciate eye contact with the speaker, particularly if they are sitting near the front. But some speakers in this presentation kept their backs turned to the audience in order to admire, and speak to, their visual aids.

Toys for Speakers

Jingling keys or a handful of pocket change is distracting enough, but two nervous presenters at this conference played with the laser pointer. Inducing the audience to chase the ruby dot around the walls and ceiling is a great way to keep them from focusing on the message. When a laser pointer is not in use, it should be turned off.

The Solemn Reader

Other speakers, even when they had visuals to cue themselves, read every word from a detailed script, without even glancing at the visuals or the audience. The favored voice for reading technical material, a solemn monotone, was ideal for putting people to sleep.

The Speed Demon

What do you do when you have too much material for the allotted time? The answer should be to grit your teeth and reduce the length of the presentation. A couple of presenters thought they had a better solution, though: they galloped through the text, reading at a furious pace. This led to a lack of vocal energy and variety. They droned—and fast. Audiences hate that.

The Unsuspecting Mute

Two other speakers wasted time (and tried the audience's patience) by failing to check the lavaliere microphone. Each got well started before discovering that it was not functioning and had to start over again.

The Mystery Storyteller

Compared with readers, listeners are at a big disadvantage. They can't jump ahead to find the punch line (the key idea) of a technical presentation. If they haven't read an abstract before the talk, they are totally dependent on the speaker to orient them before the tricky details start to unfold. Listeners who don't receive the orientation may become confused or bored; either way, they will likely tune out the speaker. It's surprising how many technical presenters are willing to risk that.

Considering that technical writers often find themselves performing as technical speakers, it seems like a fine idea to get some experience behind the lectern (which countless misinformed souls refer to as a "podium"—something that presenters stand on). People who communicate well in writing may have a big advantage when the time comes to prepare a speech: Often all they need is practice. Shortly after joining Toastmasters International in 1976, I found myself wondering why I had waited so long to get this useful experience.

Source: Written by Hugh Hay-Roe and reprinted with permission from *Intercom* (February 1999: 22–23), the magazine of the Society for Technical Communication, Arlington, VA, U.S.A.

QUESTIONS FOR DISCUSSION

a. What do you know about the meaning of digital imaging technology? Do you use a digital camera? Do you use a scanner?
b. Discuss the fourteen negative techniques presented in the body of the article.
c. How can you rephrase each negative technique into a positive statement?
d. What value do you see for preparing and using a guide of 14 tips for effective delivery and excellent visual aids? Would such a guide help speakers avoid "feeble delivery and mediocre visual aids" mentioned in the opening paragraph?

CASE STUDY

16.5. As you read the following conversation, consider how instruction in giving oral presentations might have helped Bella avoid her current situation.

Big Talk

Jules LaFontaine is astonished to see his friend and colleague Bella Strosser weeping silently in her office. He dashes through the door and closes it quietly.

"Bella, whatever is wrong?" he asks. "Can I help?"

"Oh, nobody can help me now," sobs Bella. "I've lost a major client for ToyTronics."

Jules is astonished. "How can this be? You're the best designer we have!"

"I had to make a speech. I blew it. I just didn't know what to do," admits Bella. "Do you think I'm in trouble?"

"Tell me about it," Jules says. "I'll do what I can."

Bella dries her eyes and blows her nose. "It all started last Wednesday. The head of Marketing called me up and asked me to discuss safety design with a group of buyers from Michigan. He said they were concerned about our use of fireproof and nontoxic materials in our toddler line. He said I'd know what to say." Her voice quivers, and the tears begin. "Well, I just didn't."

"What do you mean?" Jules asks.

"I had a good talk planned and all," snivels Bella. "I started in about EPA regulations and our testing reports, but they didn't seem interested. In fact, most of them walked out after about seven minutes. And I was just doing what they wanted!"

"Hold it, a minute, Bella! You're a design engineer. Were you explaining all that technical stuff from EPA to them? And how long did Marketing say you had to talk?"

"Well, sure, Jules. That's what I had. And they needed our lab results, too. Anyway, that's what Marketing said. They didn't tell me anything about how long I had to talk. They just said to make it good. And I didn't!"

"Maybe not," says Jules thoughtfully. "Those people aren't engineers. They don't care about statistics and legal language and technical terms. They just want to be sure that kids won't choke or burn when they play with our toys. Had you thought about that?"

"Well, no," says Bella. "I thought they'd be impressed with the technical stuff. Besides, I'm scared of speaking. Who'd listen to me, anyway?"

"Oh, come on, Bella. You're a mom. And besides, you know what kids like as well as the facts. Couldn't you explain it that way?"

"Well, now that you mention it, I guess I could. ToyTronics really does design the best and the safest. My kids play with our toys. Why didn't I think of that?"

QUESTIONS FOR DISCUSSION

a. Is Bella right to blame herself for the failure of her talk?
b. What should the head of Marketing have told Bella about her audience?

c. Why did Bella's oral presentation fail?
d. What should Bella have asked about her speech assignment?
e. What kinds of visuals could Bella have included?
f. What sorts of information about product safety would have been helpful to Bella's audience?
g. How can Bella better manage her fear of speaking?
h. What advice do you have for Bella?
i. What advice do you have for ToyTronics to better prepare speakers?

PROJECT PLAN SHEET

FOR ORAL PRESENTATIONS

Audience

- Who will listen to the presentation?
- How will hearers use the information?
- How will your audience guide your communication choices?

Purpose

- What is the purpose of the presentation?
- What need will the presentation meet? What problem can it help to solve?

Subject

- What is the speech's subject matter?
- How technical should the discussion of the subject matter be?
- Do you have sufficient information to complete the subject? If not, what sources or people can help you to locate the additional information?
- What title can clearly identify the speech's subject and purpose?

Author

- Will the presentation be a collaborative or an individual effort?
- If the presentation is collaborative, what are the responsibilities of each team member?
- How can the developers evaluate the success of the completed presentation?

Project Design and Specifications

- Are there models for organization or formats for speeches?
- In what medium (live, tape, interactive television, other) will the speech be presented?
- Are there special features the presentation should have?
- Will the presentation require graphics or other visuals? If so, what kinds and for what purpose?
- What information design features can best help the presentation's audience?

Due Date

- What is the final deadline for the presentation?
- How long will the speech take to plan, research, draft, revise, and rehearse?
- What is the timeline for different stages in preparing the speech?

NOTE: You may need to photocopy this form or make one like it. Ask your instructor for directions.

EVALUATION OF ORAL PRESENTATIONS
(See page 496 for directions)

Course and Section ______________ Date ______________ Evaluated by (Name) ______________

Students' Names																							
DELIVERY																							
Forceful introduction																							
Poise																							
Eye contact																							
Sticking to mode of delivery																							
Zest (enthusiasm)																							
Voice control																							
Acceptable pronunciation and grammar																							
Avoidance of mannerisms																							
Ease in showing visuals																							
Clear-cut closing																							
Sticking to specified length																							
CONTENT & ORGANIZATION																							
Stating of main points at outset																							
Development of main points																							
Needed repetition and transitions																							
Effective kinds and sizes of visuals																							
OVERALL EFFECTIVENESS																							
TOTAL POINTS																							

4 = Outstanding
3 = Good
2 = Fair
1 = Needs improvement
0 = Unacceptable

COMMENTS:

The Technical Communicator's Guide to Research

INTRODUCTION

"The Technical Communicator's Guide to Research" provides information to help you with each step of the research process:

- to determine the kinds of research information you need for academic and workplace projects
- to locate and preview sources efficiently
- to evaluate the information you find
- to acknowledge the borrowed materials you use

Since such a wealth of information is available and because new information appears so quickly, especially on the World Wide Web, research strategy and guides to information are more important for technical communicators than ever before.

The Research Process offers practical advice for approaching research projects. It shows how to plan research by defining what you know about your subject and what you need to learn. By using research questions, you can explore sources and break your research into brief, open-ended explorations. This section also discusses the use of keyword searching and reference guides, provides practical strategies for note taking, and suggests ways to preview and evaluate your source material.

Guides to Information Sources presents the many kinds of reference tools, both in print and in other media, that you can use to locate current information in your research area.

Documentation explains the importance of acknowledging the sources of research material used in academic and workplace writing. This section also discusses three types of documentation—notational, reference, and parenthetical—and provides in-depth discussion and examples of the MLA (Modern Language Association) and APA (American Psychological Association) styles.

The Student Research Report, by Sandra Smith, shows how one technical writing student developed her research project. This section traces each step of a research report project: from the assignment and a completed Project Plan Sheet to a progress report and the completed research report. The student author then discusses her finished project, "Peripheral Neuropathy: A Guide for Diabetics," including her planning and research processes.

The Research Process

INTRODUCTION

Research is the process of finding the best information in the most efficient way. For example, you might want to find the right price for a used car you'd like to buy, or you might want to learn about the history of a company where you have applied to work. Research becomes even more important when it is part of large workplace projects, such as proposals, procedures, or bids. Whether you locate information in a library, on the Web, or by conducting an interview, the information you report should always be accurate and complete.

The way you search and the materials you use will be determined by the purpose of your project. You may use print and electronic materials or you may interview an expert. You may use an online catalog, or you may use print indexes to locate journal articles that are not available in your local library but that are available in electronic form on the Internet. Perhaps you will use an online index to journals and find that some of the articles are available online while others from the same index are available only in print. Use all of the research tools available to you to locate the best information for your project.

This section explains the research process and the materials and techniques that can help you plan and work through the process efficiently and thoroughly.

WHEN DO COMMUNICATORS NEED RESEARCH?

For some projects, you will already have the information that will meet your audience's needs. However, because workplace products, situations, and equipment are constantly changing, you may need to update and supplement the information you already have. For example, if you propose a new word-processing system for your office, you will need to find out:

- How the current word processing system is used
- Its strengths and shortcomings
- The compatibility of the current word processing program with office software and hardware
- How other word processing programs will meet office needs
- How efficiently a new program will convert word processed materials to newer formats
- Training, documentation, and technical support available for the new system
- The effect of future upgrades on the new system

Your research on this subject might include interviews with employees who use the current system and with their managers. Library research might provide reviews of word processing programs and user manuals for some of the programs. Through the Web, you might find more reviews of the word processing programs you are considering, as well as discussions on listservs and newsgroups devoted to this topic. On the Web, you can also visit the sites of companies that develop and sell word processors.

DEFINING RESEARCH QUESTIONS

In the process of research, you seek reliable answers to questions designed to fulfill the purpose of your research project. As you develop your Project Plan Sheet, you'll define your audience, purpose, subject matter, and the requirements of your communication project. Once you have completed your Project Plan Sheet, you need to plan your research by developing a series of research questions that you will answer by locating reliable information.

Research questions should be flexible enough to evolve but still provide a solid connection with a project's purpose and audience needs. While it may be exciting to discover entirely new areas of research and unusual pieces of information, be sure that the information you search for remains directly connected to your original subject and Project Plan Sheet.

The way you research a question and the way you present your answer are shaped by four factors outside the question itself: your audience, purpose, scope, and prior knowledge.

Audience

Before you begin research, consider the audience for your project. How you explain a subject depends on the needs of the audience and the way they will use the information. If you were a rock collector, for example, you might talk to a cub scout troop about the environmental clues that lead collectors to the rocks they want and about the excitement of the search. For a regional jewelers' convention, you might talk about the danger and difficulty of finding and harvesting precious gemstones. The geology society would expect yet a different kind of information. All of your research may be about rocks, but the way you explain rocks, your choice of language, and the depth of information you provide will depend on your audience.

Purpose

The purpose of research, like the purpose of any communication project, is guided by your audience's need for and use of the information you will provide. Audiences' needs will vary, even for the same information. For example, a report on toxic waste could help a company producing hazardous chemicals to find new solutions for toxic waste disposal. The same information could also be used in lawsuits against the chemical company.

Scope

Broad or narrow, the scope of your research will depend on your audience and purpose. For example, you might be asked to report on cost overruns in your company's plants. The scope of this report would be broad and general. On the other hand, you might be asked to report on overruns in a particular plant, including seasonal variations in cost overruns, the difference in cost overruns by shift, and the size of cost overruns by customer. The scope of that report would be narrower and in depth.

Prior Knowledge

When you begin planning and organizing a research project, you will have ideas and perhaps some knowledge about the subject. What you already know will determine how much you need to learn through research. Whatever your topic may be, start by evaluating what you already know. Then determine how much additional information you will need to address the research question effectively. You'll find examples of research questions and an explanation of how they were developed in the interview with a student researcher on pages 579–581.

WORKING WITH LIBRARIANS

Librarians have professional expertise in research and can help you locate useful sources, such as encyclopedias; indexes to magazines, newspapers, and journals; and current guides to online information. If a research resource is electronic, in print, or on film, librarians probably know about it. If you begin your research by working with a librarian, you can save time and more easily locate current guides to information.

Discuss with the librarian specific requirements of your research, your level of expertise on the subject, and the anticipated level and needs of your audience. This discussion will help the librarian to focus on the sources most likely to lead you to the information you need. As you conduct your research, your needs for information may expand. You may need to find more in-depth information or investigate related areas. A librarian can help you as your research evolves.

BEGINNING THE SEARCH: EARLY STEPS

You have several ways to begin your search for sources and information, including:

- Consulting an in-depth, authoritative overview in a specialized encyclopedia.
- Reading recent journal or magazine articles.
- Using a computer with access to one or more indexes to periodicals. Experiment with different keywords that describe the topic. The rapid feedback on what works and what doesn't allows you to quickly explore various aspects of a topic, providing an overview of the kinds of information available.
- Surfing the Web. Because anyone can post anything on the Web, there is no ready filter to sift inaccurate or biased information from reliable information. Web surfing is more useful if done later in the research process, when you have enough knowledge of the topic to evaluate what you find.

You can use any of the methods described below to narrow your research question or to begin collecting information. Or you may want to start your work by using the resources in your local library.

RECORDING AND TRACING YOUR STEPS

Research begins with a question, but as you work, the research question may evolve as you discover new leads. A promising new lead may suggest issues you hadn't considered, and your project may take a new direction.

The best way to keep track of where you are going in your research is to record where you've been. Later in your research, your records can help you find your way back to data you located. From the beginning, keep records of the questions you are asking, the keywords (see pages 517–519) that describe your subject, and the subject headings you used. As you move from one research tool to another, some of those words will recur. Using these terms will help you to focus your search and conduct it more efficiently.

It is important to keep consistent records during your research process, so you need an easy-to-use system to keep track of all of the sources you locate. You might write your notes on cards. By using cards, you can record information in a format that is easy to sort by author, title, or subject. Then you can easily locate the cards you need as you write your research project.

Of course, there are many methods for storing and organizing your source information, including a database or word processing files. Whatever method you decide on, make sure that the bibliographical source information you record is complete and that your records are easy to use. The bibliographical information will be important as you locate your research sources to find additional information. Bibliographical information will also be important as you give credit by documenting information in your finished project.

The list below describes the information to record for different kinds of sources.

Books

- Author or authors
- Editor(s)
- Title of the book (For a book with chapters by different authors, keep a record of the author and title of the chapters you are using.)
- Publisher, place of publication, and date published (or copyright date)
- If the book is in electronic or nonprint format, list the format (audiotape, Web site, or other source) in which the book appears.

Journals and Magazines

- Author or authors (if listed)
- Title of the article
- Title of the magazine

- Volume and issue number
- Date of issue (sometimes a month or season)
- Page numbers on which the article appears
- If the article is in electronic format, the Web address (URL) and the date on which you accessed it
- For full-text online databases, such as Lexis-Nexis Academic Universe, Dow-Jones, or Expanded Academic Index, the date you found the database and the keywords or subject heading you used
- For data on CD-ROM, the dates covered by the disk

Newspapers

- Author or authors (if listed)
- Title of the article
- Title of the newspaper
- Volume and issue number
- Date of issue (sometimes a month or season)
- Page numbers on which the article appears
- If you used a print or microfilm issue of a newspaper, the section of the paper (it might be a number or letter) and the column where the article starts (Sec. C, p6 col 4, for example).

Web Pages

- Web address (URL)
- Author (if listed)
- Sponsor or host of the page
- Date you accessed the page
- Date on which the page was last updated (listed at the bottom of the page)

Interviews

- Interviewee's full name and title
- Place, date, and time of interview
- Records of follow-up letters, e-mails, or telephone calls that added information to the interview

E-mail, Chat Room Discussions, and Listservs

- Since these sources change very quickly, make a complete copy of the material by printing it or saving it to a disk
- Record all available information, including the names of people participating.
- Make your bibliographical description reflect the context in which the conversation occurred, not just the e-mail you plan to use.

Visuals

- Make a clear copy of any visual you wish to use.
- Record the source in which the visual appears and the exact location of the visual in the source.

The specific format you must use to record your sources in your research project will be determined by the style guide your instructor or supervisor requires. These style guides vary by academic discipline and career. You'll find several style guides and examples of different formats on pages 548–556.

USING THE LIBRARY BOOK CATALOG

A library selects, organizes, and stores information in a variety of formats to enable researchers to locate that information. Most library catalogs are now online, which makes searching a library's book holdings fast and easy.

The following entry points into the catalog are all useful; the key is finding the appropriate search terms. If your first choice does not yield the results you want, use these strategies to locate a different entry point.

- **Author**—Use this search category if you know the names of authors who have written on your topic.
- **Title**—Use this search category when you know the titles of books.
- **Subject**—Use this search category when you know standard terminology for your topic.

For author, title, or subject searching, you must be precise. When you have a particular author in mind, or you are sure of the title of a book, you can easily locate the source. But if you are unsure, the keyword search gives you an option. Enter the author's last name, for example, and scan through the results until you find the author you are looking for.

Keyword Searching

Keyword searching can help you narrow subject headings efficiently. Up to three precisely worded subject headings are assigned to each book title. Authorized subject headings appear in *The Library of Congress Subject Headings*. A subject keyword search gives you wider search choices. If the keyword you use appears anywhere in a subject heading, you'll find it. However, you may also get many results unrelated to your particular topic.

- **Author keyword**—Use this search category when you remember part, but not all, of an author's name.
- **Title keyword**—Use this search category when you know the keywords that should appear in a book's title.
- **Subject keyword**—Use this search category when you are unsure of a subject's standard terminology. A subject keyword search is more flexible

than a subject search since the keyword does not have to be the first word in the heading.

To search efficiently with keywords, imagine the titles of a few perfect books on your topic. Pick the three or four most important words that would appear in the titles. Then use these words to conduct a search. If the results are not quite what you need, try subtracting a word, adding a word, or changing a word. In only a few seconds you can see if you've found a good approach.

Here are some titles and the keywords used to locate them. The title keywords are **office AND landscape:**

AUTHOR:
Clarke & Rapuano
TITLE:
The office of Clarke and Rapuano, Inc. (consulting engineers and landscape architects)
PUBLISHED:
New York 1972
DESCRIPTION:
(78) p. (chiefly illus.) 31 cm.
SUBJECTS:
Landscape architecture
Environmental design

AUTHOR:
Palmer, Alvin E., 1935-
TITLE:
Planning the office landscape / Alvin E. Palmer, M. Susan Lewis
PUBLISHED:
New York: McGraw-Hill, © 1977
DESCRIPTION:
xiv, 188 p.: ill.; 24 cm.
NOTES:
Includes indexes.
Bibliography: p. 181.
SUBJECTS:
Office layout
Buildings—Environmental engineering
OTHER AUTHORS:
Lewis, M. Susan, 1947-

AUTHOR:
Fink, Eli Edgar.
TITLE:
A lawyer's landscape . . . the view from his office / by Eli Edgar Fink
PUBLISHED:
Tucson: Lawyers and Judges Publishing Co., 1976
DESCRIPTION:
ix, 86 p.; 24 cm.
SUBJECTS:
Humorous poetry
Practice of law—United States

When you find a book that is exactly what you need, examine the subject headings. Write down the subject headings, and perform searches on promising new subject terms. You may also use the same subject headings later as you search for periodical articles and information on the Web.

As you examine the results of your keyword searches, note the author, title, call number, library location, subject headings, and any other information relevant to your subject. Compare the subject headings you found, and decide which ones to try first. Then do a subject search with one of those headings. Usually, you will do that by going back to the online catalog's menu, and choosing subject instead of keyword. See what comes up. Write down the call number of each source you want, and record the other information you will need.

Before you even look at the books you have selected, you will have gathered some valuable information. This is what you have to work with:

- A variety of terms that are used to describe your topic.
- An overview of the quantity of material published in books on your topic.
- The names of people who write in the field you are researching. This is the first step in identifying experts.
- The call numbers that are used for your topic. You can scan the shelves in those call number areas.

The process of using keywords for searching is so valuable that if your library's catalog does not offer keyword searching, you might want to check the catalog of another library that offers it. You can then check your library's catalog to find books you have located. Books not available locally can be borrowed through interlibrary loan. However, it may take anywhere from a few days to a few weeks for requested books to arrive.

SEARCHING: ADDITIONAL RESOURCES

After completing your preliminary research, you know more about your topic. You can begin searching for other kinds of information.

Periodical Indexes

Periodical indexes list and describe articles in magazines and journals in many of the same ways in which a library's book catalog lists and describes books. If you need to find articles from the 1960s or even the 1890s, printed indexes can help you. These print indexes generally organize listings by author, title, or subject. Some also have abstracts, brief summaries of articles. Most print indexes are now available in electronic format. The electronic version of an index will provide more ways to search, including the keyword method.

Most indexes specialize by subject. For example, you will find titles such as *Business Periodicals Index*, *Social Science Index*, and *Chemical Abstracts*. The title of a specialized index will usually reflect its subject area. General indexes include articles from both journals and magazines in a variety of subject areas. For example, *Expanded Academic Index*, *Periodical Abstracts*, and *Readers' Guide Abstracts* are general indexes. These general indexes are good starting points for researching subjects about which you know little. They can also provide an overview of available source material. Indexes are available in nearly every subject area. You'll find a list of representative print and electronic indexes on pages 532–537. However, this list is only a small sample of the many current indexes available.

Other Research Materials

You may find other research materials useful, depending on your subject. These sources may include annual reports, government documents, and other kinds of data. Most often, you will find references to these materials during the research process. You'll find examples of guides to such sources on pages 541–544.

INTERVIEWING A SOURCE

Sometimes interviewing someone active in your subject area is the best way to gather information. Below are guidelines for planning and conducting a successful interview.

1. Identify what you need to know by consulting your previous research.
2. Find a source who is knowledgeable. Ask for recommendations from people who work in the field, or use your research to identify likely interviewees.
3. Request the interview, briefly explaining your interest, purpose, and the topics to be covered in the interview. You can initiate the request with a phone call, a letter, or an e-mail. You can also conduct the interview in person, on the telephone, or via e-mail. In any case, allow yourself time to set up the interview, plan it, and incorporate the interview information into your project.
4. Plan the interview. Establish a specific time and length for the interview, always respecting the needs of the person you are interviewing. If you will

interview in person or on the telephone, request permission to tape the interview so that you can be sure that your notes will be accurate.

5. Prepare yourself. Conduct research on the subject you need to learn about and the person you are interviewing so that your questions will lead to good answers. Be sure to send copies of these questions to the person you are interviewing before the interview.

Before the interview, ask yourself:

- What is your purpose in conducting the interview?
- What specifics do you want to learn from the interviewee?
- What questions should you ask, and how should you ask them? Your questions should be clear, encouraging a specific subject in the answer. Your questions should also be open, not suggesting the answers you wish to hear. While your interview may lead you to additional questions, organizing a list of questions will guarantee that you cover the subjects you need to learn about.
- How can you best prepare? Learn as much as you can about the subject and the person you plan to interview. In this way, you can take advantage of the interviewee's expertise and be ready to learn about new ideas.

On the day of the interview:

- Check your recording equipment (if you have permission to use it) before the interview.
- Conduct the interview on time, and stay within the time limits you have arranged with your interviewee.
- Prepare a list of interview questions for your own use. Provide plenty of room for notes after each question.

During the interview:

- Introduce yourself. Briefly explain the purpose of the interview. If you want to tape the interview, get permission from the interviewee.
- Keep the interview on track. If the person you are interviewing wanders off the subject, listen long enough to determine whether the answer is going to be useful. If not, interrupt gently. You might say "I see . . . but what I meant to ask was . . ." and repeat the question, perhaps phrasing it differently. Or you might say, "That's really interesting, but I'd really like to focus on. . . ."
- If an answer suggests relevant questions that you had not considered, ask them.
- If you do not tape the interview, take careful notes of your interviewee's comments. Check your notes with the interviewee if you are unsure of their accuracy.

- Be aware of the time. You need time to ask all of the questions you planned. You must also respect the time of the person who has agreed to the interview.

After the interview:

- As soon as possible, organize and elaborate on the notes you took during the interview.
- Promptly write and mail a thank-you note to your interviewee.*

Keep all your notes and lists of questions and answers as well as the tape if you recorded the interview session. Be sure you record the interviewee's full name, title, and when and where the interview was conducted. Keep notes on any follow-up letters or phone calls that added information to the interview.

PREVIEWING SOURCES: USING ABSTRACTS AND SKIMMING

As you locate promising sources, consider how you are going to review the information you've found. Many research materials provide helpful resources for previewing information. For example, a book will provide a preface or introduction that states the scope, purpose, and subject of the work. If a book has an index and appendices, you can use them to locate specific aspects of your topic. The table of contents, which lists a book's chapter titles, and sometimes major headings, will show you the organization and the parts of the book's discussion. Skim chapters for an overview, focusing on the areas that seem to be most helpful for your research. If a book has a bibliography, it may list authors whose names and works you have seen before. Bibliographies can provide leads on experts and sources that you may wish to pursue.

In shorter works, such as journals and reports, headings and subheadings establish the order and key ideas of discussion. Some sources also provide a summary, which can give you a brief overview of the discussion. Many indexes publish abstracts, informational summaries that briefly cover the main ideas of an article. If the abstract provides the information you need, you can use the abstract as your source. (See Chapter 12, pages 295–302, for more information about summaries.)

As you preview sources, note the results on your source descriptions. If parts of sources look promising, note the page locations of useful material.

TAKING NOTES

As you begin to review the sources that can provide the information you need, you will need to take accurate notes. Specifically, your notes should help you

*The suggestions above are used by permission of J. D. Hughey, A.W. Johnson, and B. H. Harper: Speech Communication. Oklahoma State University, Stillwater OK 1990. <http://scip3.okstate.edu/2713/tb07.htm>

answer the research questions you refined during your search. You do not have to document or attribute to others notes on your own ideas or conclusions and information that appears in many sources. However, in your finished project, you must document or acknowledge the source of any specialized information you use. You must document information that is not widely known or not discussed in many sources. You must also document the words, views, conclusions, or opinions of others, whether this information is published or not. By documenting your source material, you respect the ideas or intellectual property of others. For a detailed discussion of intellectual property and documentation conventions, see pages 546–548.

Quoting and Paraphrasing

No matter how you plan to use information in your research project, whatever you record in your notes should be accurate and clear, with a specific reference to the information's exact location in the original source. In taking notes, you can **paraphrase** a source, or you can **quote** it.

When you paraphrase a passage, you restate its ideas in your own words. Even though you have not borrowed another person's words, it is still important to acknowledge the ideas you use. This means that your notes should accurately reflect the ideas you paraphrase and provide a specific location for these ideas in the original source. When you quote, however, you must record the original, word for word, exactly as the quotation appears in the original source, enclosing the quoted passage in quotation marks. (See pages 643–644 in Appendix 2 for the use of quotation marks.) Examples of a quotation and a paraphrase of the same source appear below.

QUOTATION:
"Although early discussions of hypnosis date from the notes of 12th Century scholars, controlled experimentation did not begin until the 19th Century, by which time hypnotism had captivated the popular imagination" (Carter 171).

PARAPHRASE
Hypnotism was discussed in the 12th century, but scientific research of the subject began in the 19th century, when the subject was popularized (Carter 171).

A few conventions allow you to adapt the quotations you record in your notes. If you delete material from an original quotation, you must indicate the omission with three ellipsis dots. If you end a sentence with an ellipsis, the fourth dot will indicate the period at the end of the sentence. (See Appendix 2, pages 647–648.) When it is necessary to insert information in a quotation to clarify the quoted idea, enclose the inserted word or phrase in brackets. (See Appendix 2, page 646.) An example of a quotation with ellipses and brackets follows on page 524.

"The [technical] communicator must analyze . . . audience carefully in order to be able to select the information, style, and medium that can best help . . . information users."

When you use brackets and ellipses, be sure that they do not alter or distort the meaning of the source you are quoting.

In general, quote information that presents the opinions of others, the words of experts in the field, or phrasing that you consider especially concise, clear, or accurate. If a paraphrase will better fit the language of your project, you can always rephrase the quotation in your own words later on. However, since both paraphrases and quotations reflect borrowed information, you must respect your sources by documenting them in your finished project. In addition, since you will be relying on your notes as you write, be sure that your notes, quoted or paraphrased, exactly reflect the ideas of the original authors.

Organizing Your Notes

It is helpful to use the same organizing system for your notes that you use to record your sources. In this way, you can easily locate the sources for every note and arrange notes and sources in the order in which you will use them as you write. Every note should include the following information:

- Keyword or subject of the note
- Note information
- Type of note (quotation, paraphrase, your own ideas)
- Source (author, title, page number)

Your earlier source descriptions will be helpful as you document information later on. An example of a note card appears below.

Subject: Early history of Dell Computer Corporation

"Michael Dell's initiative as a student entrepreneur supported his education and gave him the ideas that encouraged him to begin his company. The same attention . . . to quality and to the needs and interests of customers that made Michael's early business successful has given Dell Computer Corporation the success the company enjoys today."

Quotation

Source: Barry Gray, "The Dell Story," page 26.

The source description for the note appears below.

Author: Barry Gray
Title of article: "The Dell Story"
Title of magazine: *Texas Commerce*
Volume and Issue Number: 7:3
Date of issue: December 1999
Page numbers: 23–29, 77–81

EVALUATING SOURCES

As you conduct your research, you must evaluate your sources. Early on, you should be able to identify information that is not suitable for your project. You will also be able to weed out the obviously biased material that provides opinions without facts to substantiate them. You will then need to evaluate the remaining sources for the following:

- **Author's credentials.** Who wrote the source you have located? What are the author's credentials? You can expect to see the name of an expert cited in other sources or listed in bibliographies of works on your subject.
- **Author's bias.** Does the work reflect the author's loyalties, opinions, or interests? For example, is the author an employee of a company with an interest in the subject? Does the author state opinions as fact?
- **Advertiser's bias.** What is the interest of the publisher of an article, brochure, or Web page? It is helpful to examine where articles are published. For example, an article on the environmental impact of household cleaners published in a popular magazine may or may not be based on research and provide references. However, a journal article on the same subject will be more solidly researched and will provide a bibliography. You may need to use both articles, but be aware of the underlying differences in credibility.
- **Accuracy.** How can you determine the credibility of contradictory sources? Find more sources on the subject, and evaluate what you find. Does most information support one side or another? Is there an ongoing controversy, with good points on each side? Your finished project should reflect all of the credible views so that your report is complete, accurate, and free from any bias of your own.
- **Timeliness.** Is the information you locate sufficiently current? If you are discussing current technology, for example, you'll need to locate the very latest information. However, if you are researching a subject with a broader scope, such as a career area, you may find that other criteria, such as completeness and expertise, will be more helpful in selecting materials.
- **Appearance.** Do a source's appearance and writing reflect hasty preparation? Whether the material is in print or online, look for accurately worded headings, logical organization, clear writing, and correct grammar and spelling. Careless preparation may reflect the quality of information a source provides.

CONCLUDING YOUR RESEARCH

As you compile notes in answer to research questions and check research information against your Project Plan Sheet, you'll reach a point at which you are ready to stop searching, reading, and taking notes and start preparing to draft your project. Ask yourself these questions:

- Have you learned the answers to the research questions you defined?
- Have you answered any new questions that evolved during research?

- Are all of the research questions and the answers you located directly related to the subject and purpose of your project?
- Does the information you located meet your reader's needs and serve your project's purpose?
- Are you satisfied with the quality and completeness of your research information?
- Are your notes and source descriptions complete?

If the answer to all of these questions is yes, you are ready to arrange your notes so that they are in the sequence you'll use as you draft from your organization plan. Your source descriptions for each note will be helpful as you document any quotations, paraphrases, or other information. You will also use the notes and source descriptions to check your draft for accuracy as you revise.

After you have completed your project, review your notes and sources and consider the sources you have located. You'll find that each research project will introduce you to new sources, new search techniques, and new ideas. By reviewing and keeping a record of your work, you will be able to retrace your steps in future searches, and the next research process you begin will be faster, smoother, and more effective.

GENERAL PRINCIPLES for the Research Process

- Researchers should begin with an honest appraisal of what they know about the research subject. If they know very little, they need to begin with a review of background information.
- Effective researchers use a variety of research tools and strategies. By using a variety of sources, they can represent many viewpoints and present a complete discussion to the project's audience.
- References and keywords provide useful research connections. An encyclopedia may refer to three or four key people in the research area, and a book catalog can help locate what has been written by or about them. Periodical indexes can also locate information. Related keywords can provide connections between ideas related to a research subject.
- It is important to keep accurate and complete records and notes throughout the research process. Accurate records allow researchers to locate and document sources. Accurate notes provide the source material needed in the written project.
- When sources disagree, researchers must check a variety of sources to determine what is most credible. Information about discrepant viewpoints should appear in the finished project.
- Research is a process of discovery. As researchers learn more about a subject, they may find new leads to ideas, developing the project in ways they had not considered.

CHAPTER SUMMARY

THE RESEARCH PROCESS

Like the writing process, the research process begins with a definition of a project's audience and purpose. From there, the researcher must assess how much he or she knows about the subject in order to develop research questions, locate research sources, and plan a search strategy. A librarian can make other helpful suggestions before the search begins.

Researchers should use every possible research tool to locate the best available source materials. Research tools include keywords, book catalogs, and print and online indexes. Interviews with experts may provide information not available in other sources.

During the course of planning and research, good notes are essential. Such notes allow researchers to locate sources and to record source material for use and citation in the research project. As research proceeds, it is also important to evaluate the credibility of information located during the search.

Research should provide a complete, accurate, and thorough view of the best available information, so both complete searches and careful evaluation are important for researchers. However, only by starting with a well-defined understanding of a research project's audience and purpose and an efficient plan can a researcher be sure of excellent results.

Guides to Information Sources

INTRODUCTION

As you begin a research project, you'll find that guides to information, in both print and nonprint media, can direct you quickly to the materials your project requires. This section discusses a few such guides, both in the library and on the Internet. Print and online materials are usually located using a systematic, organized search of available sources. To begin research, you must first know how to use information sources efficiently.

You can locate information by using:

- Library catalogs
- Print guides
- Electronic guides
- Public information

You can use these resources to compile a list of the print and nonprint materials—books, periodicals, people, Web sites, companies—that can provide quality information for your research project.

USING THE LIBRARY

Every library contains a vast wealth of information and guides to information, both print materials—books, pamphlets, newspapers, and other periodicals and nonprint materials—tapes, films, microforms, computer disks, compact discs, and online services. But this information is useless to you unless you know how to locate what you need.

The Public Access Catalog (PAC)

The public access catalog (PAC) provides a complete listing of a library's holdings. Once available only on 3- by 5-inch cards, the PAC is increasingly being made available in electronic forms such as a computer output microform (COM) catalog and an online computer system, known as an online public access catalog (OPAC).

In a COM system, the information contained in the public access catalog is input and stored in a computer database to which new acquisitions are routinely added. Periodically, a computer tape of the entire catalog is retrieved and reproduced on either microfiche or microfilm. Library users can quickly access PAC information from stations in the library. If a library has an OPAC system, users can access this information online from distant locations.

The public access catalog entries, called bib (bibliographic) records, are arranged alphabetically according to the first important word in the entry heading (first line). Most items listed in the public access catalog have a minimum of three entries: author, title, and one or more subject entries. The heading of each bib record is determined by the type of entry: author, title, or subject. Only the heading for each entry differs; the basic bibliographic information for each entry is the same.

A typical entry in the public access catalog contains the following information:

1. Heading—subject, author, or title
2. Author (or editor) listed first on the book
3. Complete title of the book
4. Listing of all authors (or editors)
5. City of publication (the state is also given for less well-known cities)
6. Publishing company
7. Year of publication
8. Number of introductory pages
9. Number of text pages
10. Illustrations
11. Height in centimeters (a centimeter is 0.4 inch)
12. International Standard Book Number (ISBN)
13. Headings by which the book is cataloged
14. Call number (the designation for classifying and shelving the book in the library)

Subject Entries from the Public Access Catalog

As you begin looking in the public access catalog, you can use subject entries if you do not know the names of authors and titles. Suppose you are doing research for a report entitled "Medical Applications of the Laser." If you do not know any titles or authors on the subject, first look under "L" for "Laser" and "Lasers." Look under related subject headings also. Some of your most important information might be cataloged under subjects such as "Medicine" or "Surgery" or "Physics."

Using the Online Public Access Catalog (OPAC)

An online computer system (or online public access catalog) makes millions of items of information stored in a central computer available quickly. A terminal links the user to the computer. The online public access catalog (OPAC) provides highly efficient access as you search by the traditional types of entry (subject, author, title) as well as by the following:

- Publication date
- Publisher
- International Standard Book Number (ISBN)
- Language (such as English, Spanish, or French)
- Format (such as films, filmstrips, slides, or compact discs)

On some systems you may search by keyword, which means that you can search the information in each bibliographic record in the computer for a meaningful word, term, or concept. Searching by keyword is particularly valuable if you want to know everything available by or about a particular person, such as Albert Einstein, or if you are not sure of the correct subject heading to use. For more information about how to use keyword searching, see pages 517–519.

Many online public access catalogs integrate library information with OPAC entries. Such library information may include circulation, academic reserves, requests for materials, acquisitions, serials, and bibliographic interfaces. When this information is integrated in an OPAC system, the description of each item includes where the item is located, if it is checked out, if it has been placed on reserve, or if it is on order.

OPACs allow multiple users to access the system simultaneously. OPACs also provide remote access from home or office and they can be used as gateways to search other electronic information retrieval systems, such as some of the periodical indexes that will be discussed later. It is even possible to download and print OPAC information.

Library of Congress Subject Headings

Particularly useful in looking up subject entries in library catalogs is the *Library of Congress Subject Headings* (and its supplements). This book gives subject descriptors used by the Library of Congress, and subsequently by most other libraries.

Periodicals Holdings List

The periodicals holdings list is a catalog of all the periodicals in a library. The alphabetical list gives the name of each magazine and newspaper and the dates of the available issues. The list may also indicate whether the issues are bound or unbound, the format, and where in the library the periodicals are located. When you find titles of articles that seem usable in a periodicals index, you need to know if the library has the specified magazines. The periodicals holdings list will give you this information.

The periodicals holdings list may be in various forms, such as a drawer of 3- by 5-inch cards, an index file, a typed sheet, a microform, or a computer printout, or it may be available online. Some libraries combine the periodicals holdings list with the public access catalog.

Nonprint Collections

Most libraries have a department that supervises collections of audiovisual materials. The librarian who manages this collection can explain how these materials, which can include audiotape, videotapes, disks, CD-ROMs, films, filmstrips, and slides, are cataloged. Nonprint materials may be listed alphabetically in the public access catalog or in a separate catalog.

The Vertical File

Most libraries keep print materials other than books and magazines in a vertical file. Pamphlets, booklets, bulletins, clippings, and other miscellaneous unbound materials are kept in this collection. These materials are filed or cataloged by subject. The vertical file and the *Vertical File Index,* which lists current pamphlets, may be valuable sources of information.

GUIDES TO INFORMATION: INDEXES

Indexes are to periodicals (magazines, newspapers, and journals) what the public access catalog is to books. By consulting indexes to periodicals, you can locate and perhaps even preview articles without looking through dozens of magazines. At the beginning of any index are directions for use, a key to abbreviations, and a list of periodicals indexed.

A great deal of valuable research material on technical or business subjects is likely to come from magazines. For one thing, much information in magazines is never published in book form. In addition, information published in magazines is more likely to be current than that in books because books take far longer to write and publish.

Computerized Periodical Index Systems

Many libraries provide electronic systems for finding periodical materials on specific subjects. The researcher enters the subject words, makes selections from the options that appear on the screen, and chooses which citations to print and look up. Many Internet services provide free access to the same kinds of indexes.

Some indexes provide abstracts of the indexed articles. The abstracts may be printed and used as sources if the original article is unavailable. See Chapter 12 for more information on summaries. Other systems have access to a database of the complete text of the cited material. The complete text, which may be available online, on compact disc (CD), or in microform, may be retrieved, read on the screen or on a reader/printer, and printed if needed.

Many electronic systems use rolling indexes. As current citations and abstracts are added, the space on the system is filled and the older entries are removed. These older entries are often retained in a backfile database available online, on CD-ROM, or in microform. Backfiles can be excellent sources for historical information. To provide additional access to electronic periodical indexes, some libraries are networking computers so that several terminals may access the same electronic index simultaneously. Often, several indexes are available on a local area network (LAN), so research in more than one periodical index may be accomplished at one computer terminal. Some indexes are available on tapes that may be loaded with the online public access catalog (OPAC) and accessed through the same computer terminals as the library's OPAC.

General Periodicals Index. Particularly helpful to students is the *General Periodicals Index*. It comes in four editions: Select, Library, Research I, and Research II. The Research II edition indexes approximately 1,600 general interest and scholarly publications in subject areas such as social sciences, general sciences, humanities, business, management, economics, and current affairs. Updated monthly, it indexes data for the current four years, as well as for the current six months of two newspapers, *The New York Times* and *The Wall Street Journal*. It also contains full text coverage of over 300 of the periodicals indexed.

The following excerpt is from another electronic periodical index, *Magazine Index Plus*.

> InfoTrac $ Magazine Index Plus 1991 – Sep 1994
> Heading: LASER BEAMS
> -Usage
> 5. Catch a ride on a laser beam. (pulsed-detonated
> engines) by Tim Stevens il v242 Industry Week May 17
> '93 p45(1)
> 69C0446
> ABSTRACT / HEADINGS

The entry (one of nine entries under the heading LASER BEAMS—Usage) gives this information.

> "Catch a Ride on a Laser Beam": title of the article
> (Pulsed-detonated engines): additional information about the subject
> Tim Stevens: author
> il: illustrations included in the article
> 242: volume number
> *Industry Week:* magazine title
> May 17, 1993: date
> p45(1): page number and length of the article

The location code 69C0446, near the end of the entry, indicates that the article is available in the Magazine Collection, which contains the full text of selected magazines on microfilm cartridges. The article is on frame 446 of cartridge 69C. ABSTRACT/HEADINGS indicates that a brief summary of the article is available on the index and all the headings under which the article is listed are given. The headings often suggest other terms to use for locating additional citations on a topic.

Readers' Guide to Periodical Literature. The *Readers' Guide* is a general index of almost 250 leading popular magazines published from 1900 to the present. It is published monthly with cumulative issues every three months and a bound cumulation each year. The *Readers' Guide* is also available in CD-ROM and online formats. This cumulation saves the researcher from having to look in so many different issues and keeps the index up to date.

The main body of the *Readers' Guide* is a listing, by subject and author, of periodical articles. Each entry gives the title of the article, the author (if known), the name of the magazine, the volume, the page number, and additional notations for such items as bibliography, illustration, or portrait. Following the main body of the index is an author listing of citations to book reviews.

The excerpt below from the August 1999 issue of the *Readers' Guide to Periodical Literature* is from the listing for the subject "Lasers."

The first entry under "Medical Use" gives this information.

"Curing Snoring": title of article

[use of lasers to trim tissue obstructing air flow]: additional information about the subject

R. Robinson: author

il: illustrations

American Health: magazine title

12: volume number

13: page number

December 1998: date

Readers' Guide Abstracts. The *Readers' Guide Abstracts* each year adds about 60,000 abstracts of the articles indexed in the *Readers' Guide* and of articles in *The New York Times.* Each abstract is a concise summary that presents the major points, facts, and opinions of the original article. *Abstracts* is updated monthly in CD-ROM format and twice a week online. Indexing coverage is January 1983 to the present, and abstracting coverage is September 1984 to the present.

Similar in format to *Readers' Guide Abstracts* are the *Wilson Abstracts* (Business, Applied Science and Technology, General Science, Art, Humanities, and Social Sciences), NewsBank's *Index to Periodicals,* the *ProQuest Abstracts* (Newspaper Abstracts, Periodical Abstracts, ABI/INFORM), and *Health Index.*

Applied Science & Technology Index. The *Applied Science & Technology Index* indexes nearly 400 English-language periodicals in the fields of aeronautics and space science, atmospheric sciences, chemistry, computer technology and applications, construction industry, energy resources and research engineering, engineering, fire and fire prevention, food and food industry, geology, machinery, mathematics, metallurgy, mineralogy, oceanography, petroleum and gas, physics, plastics, textile industry and fabrics, transportation, and other industrial and mechanical arts. The main body of the *Index* lists subject entries to periodical articles. Additionally, this index provides an author listing of citations to book reviews and a product review section. The *Index* is issued monthly except in July and has an annual cumulation. The arrangement of entries is similar to that in the *Readers' Guide to Periodical Literature*. *Applied Science & Technology Index* is available in print, CD-ROM, and online formats.

Business Periodicals Index. This index covers over 350 English-language periodicals in the fields of accounting, advertising and marketing, agriculture, banking, building, chemical industry, communications, computer technology and applications, drug and cosmetic industries, economics, electronics, finance and investments, industrial relations, insurance, international business, management, personnel administration, occupational health and safety, paper and pulp industries, petroleum and gas industries, printing and publishing, public relations, public utilities, real estate, regulation of industry, retailing, taxation, transportation, and other specific businesses, industries, and trades. The main body of the *Index* lists subject entries to business periodical articles. Additionally, there is an author listing of citations to book reviews. The *Business Periodicals Index,* available in print, CD-ROM, and online formats, is published monthly except in August and has quarterly and annual cumulations. It is similar in format and arrangement to the *Readers' Guide* and the *Applied Science & Technology Index.*

Indexes to Professional Journals

In addition to the *General Periodicals Index, Readers' Guide to Periodical Literature, Applied Science & Technology Index,* and *Business Periodicals Index,* the following indexes are examples of the many specialized guides available. Indexes to periodicals are available in a variety of formats, including print, microfilm, CD-ROM, and online. Some indexes are available in several formats. The library you use may or may not have all available formats of the indexes to which it subscribes.

- *Biological and Agricultural Index.* Since 1916; formerly *Agricultural Index.* Available in print, CD-ROM, and online.
- *Business Index.* Since 1979. Available in print, CD-ROM, and online.
- *Cumulative Index to Nursing and Allied Health Literature.* Since 1956; formerly *Cumulative Index to Nursing Literature.* Available in print, tape, CD-ROM, and microform.
- *Engineering Index Monthly.* Since 1884. Available in print, online, CD-ROM, and microform.
- *General Science Index.* Since 1980. Available in print, CD-ROM, and online.
- *Hospital and Health Administration Index.* Since 1945; formerly *Hospital Literature Index.* Available in print, online, and microform.
- *PAIS International in Print.* Since 1915; formerly *Public Affairs Information Service Bulletin.* Available in print, CD-ROM, and online.

- *Social Sciences Index*. Since 1974; formerly part of *Social Sciences and Humanities Index,* 1965–1974, and of *International Index,* 1907–1965. Available in print, tape, and CD-ROM.

Indexes to Newspapers

Newspaper indexes include:

- *Christian Science Monitor Index*. Since 1960, monthly with annual cumulations; formerly *Index to the Christian Science Monitor*. Available in print, CD-ROM, and online.
- *The Times Index*. Since 1906, monthly with annual cumulation. Note: This is an index of the *London Times*. Available in print, microform, and CD-ROM.
- *National Newspaper Index*. Covering *The New York Times, The Christian Science Monitor, The Wall Street Journal,* the *Los Angeles Times,* and *The Washington Post*; since 1979, monthly, "rolling" cumulation index. Covers the most current four years of data. Available in print, CD-ROM, and online.
- *The New York Times Index*. Since 1851, semimonthly, annual cumulation. Available in print, CD-ROM, and online.
- *Newspaper Abstracts*. This index has two editions, Complete and National. The Complete Edition covers *The New York Times, The Wall Street Journal, The Christian Science Monitor, USA Today,* and five regional newspapers. Since 1989, monthly cumulative index. Includes backfiles (1985–1988) for some papers. Available in print and CD-ROM.
- *The Wall Street Journal Index*. Since 1950, monthly, annual volumes. Available in print, CD-ROM, disks, and online.

The New York Times Index is especially useful. Because of its wide scope and relative completeness, the *Times Index* provides a wealth of information. A brief abstract of the news story is included with each entry, allowing a researcher seeking a single fact, such as the date of an event or a person's name, to find all that is needed in the *Index*. In addition, since all material is dated, the *Times Index* serves as an entry into other, unindexed newspapers and magazines. *The New York Times Index* is arranged in dictionary form with cross-references to names and related topics.

Indexes to Government Publications

The U.S. government prints all kinds of books, reports, pamphlets, and periodicals for readers at every level of expertise. The best-known index to government documents is the *Monthly Catalog of United States Government Publications,* published by the U.S. Superintendent of Documents. Each monthly issue lists the documents published that month.

U.S. Government Books is a selective list of government publications of interest to the general public. Also available from the U.S. government are some

175 subject bibliographies of government publications. In *Bibliographies Index,* subjects of bibliographies range from business to education, health, science, and history.

Government Reports Announcements and Index, a guide to reports from research, is published twice a month. It has an annual, cumulative index titled *Government Reports Annual Index. Government Reference Books* is an annotated guide that provides access to important reference works issued by agencies of the United States government. The guide is published every two years; entries are arranged by subject.

Essay Index

Because it can be difficult to locate essays and miscellaneous articles in books, the *Essay and General Literature Index* is a useful guide. It is an index by author, subject, and some titles of essays and articles published in books since 1900. The index is kept up to date by supplements. Subjects covered include social and political sciences, economics, law, education, science, history, the various arts, and literature.

ONLINE COMPUTER SEARCHING

By using an online computer search, researchers can access databases produced by the government, nonprofit organizations, or private companies. To conduct such a search, the researcher establishes contact with the computer containing the database. Then the researcher instructs the computer to search the database for the desired materials.

It is important for researchers to develop search strategies by identifying one or more keywords before going online. For more information on keyword searching, see pages 517–519. A thesaurus or a subject heading list, if available, can be helpful in searching an index. Each index entry has "see" and "see also" terms to help researchers select the proper subject headings to use. A computer search is fast (particularly for searches covering several years), current (some information is available online weeks or months before the printed version), accurate, thorough (more in-depth coverage since there is access to more information), and convenient.

Computer searches may initially require the assistance of a librarian and during busy times may need to be scheduled in advance. In addition, some computer searches require a fee, although the cost of using databases varies greatly. Many searches are free.

Traditionally, database searches have been used for locating citations for journal and newspaper articles on a subject. Now with entire works—periodical articles, books, and even encyclopedias—online, databases may be used for locating detailed financial data and directory listings on companies, statistics, biographical information, information on colleges and universities, and quotations. The information itself is becoming available online, not just the citations to it. The *Gale Directory of Databases,* issued annually in print, CD-ROM, and online, provides a list and description of over 8,500 databases available in many formats.

SIRS: An Example of a Library Resource in Many Formats

A valuable tool for library research is the Social Issues Resources Series (SIRS). SIRS is a selection of articles reprinted from newspapers, magazines, journals, and government publications. The SIRS articles, tables of contents, and indexes are available in print, microfiche, and CD-ROM formats. The articles are organized into volumes each of which covers a different social issue. Among the titles for the volumes are Aging, Alcohol, Communication, Consumerism, Corrections, Defense, Energy, Family, Food, Health, Human Rights, Mental Health, Pollution, Privacy, Sports, Technology, and Women. Each volume contains a minimum of 20 articles and is updated with an annual supplement of another 20 articles. (Each volume eventually contains 100 articles.) The articles are selected to represent various reading levels, differing points of view, and many aspects of the issue.

SIRS Science is a series of five volumes covering Earth Science, Physical Science, Life Science, Medical Science, and Applied Science. Similar to the regular SIRS volumes, these also contain articles selected from various sources. Each volume contains 70 indexed articles in chronological order. The volumes are annual.

The SIRS Critical Issues series contains volumes on current, significant topics. Similar to the regular SIRS volumes, the Critical Issues series began in 1987 with *The AIDS Crisis;* in 1989 *The Atmosphere Crisis* was added.

In 1991, SIRS added the series SIRS Global Perspectives, which contains four volumes: *History, Government, Economics,* and *World Affairs*. Similar to the SIRS Science series, each annual volume contains 70 articles.

Two full-text databases on CD-ROM available from SIRS are *SIRS Researcher* and *SIRS Government Reporter. SIRS Researcher* contains selected full-text articles from over 800 newspapers, magazines, journals, and government publications from 1988 to the present. *SIRS Government Reporter* contains selected full-text documents from federal departments and agencies including tables and charts of census data. Summaries of the documents are included. Both databases are searchable by subject and keyword.

Internet Searching

The Internet is a global computer network of interconnected educational, scientific, business, and governmental networks. Through a computer system, anyone on the Internet can communicate with any other person or group of people who are also connected to it.

A growing variety of resources are now available on the Internet. These include online databases that can provide full-text periodicals, books, and government documents; statistics; business information; and indexes. Also accessible are the online catalogs of many academic and research libraries. Software, pictures, and even sounds can be located on and downloaded from the Internet, although it is important for researchers to respect and acknowledge these sources as they would any other research materials.

Many colleges and businesses provide access to the Internet, and Internet providers have made services affordable for individuals to use at home. In addition to provider fees and sometimes phone line charges, most Internet resources also charge users to search or download information. Some resources are available by subscription and involve fees and passwords or account numbers to access data.

Online Information Resources

The amount of available online materials is so vast that no search engine covers the whole Internet. In fact, the most complete search engines cover less than a third of it. When you find good sites, keep a record of them so that you can find them again. Addresses of Internet resources change frequently. If a site address is not current, you may be able to locate the site by conducting a search using some of the words that describe the site. You can also evaluate the usefulness of a research site by examining its sponsors.

Below is a list of addresses that can help you locate online information in your area of interest.

ACCOUNTING: RUTGERS ACCOUNTING WEB

<http://www.rutgers.edu/accounting>

Sponsored by the Rutgers University Department of Accounting, this Web site provides research resources applicable to education and professional practice, including organizations, network sites, and the Rutgers Accounting Research Center.

BIOLOGY: THE WWW VIRTUAL LIBRARY: BIOSCIENCES

<http://golgi.harvard.edu/BioLinks>

This Harvard University biology site provides links to research materials about nearly every aspect of biology, including biochemistry, evolution, immunology, and other areas of study and research.

CHEMISTRY: THE WWW VIRTUAL LIBRARY: CHEMISTRY

<http://www.biochemistry.ucla.edu>

This site, sponsored by UCLA, provides links through many search engines to information on molecular chemistry, reference information, electronic resources on the sciences, and chemistry resources on the Internet.

COMPUTER SCIENCE: THE WWW VIRTUAL LIBRARY: COMPUTING

<http://src.doc.ic.ac.uk/bySubject/Computing/Overview/html>

This site provides worldwide links to bibliographies, technical reports, and indexes. Sponsored by the Department of Computing in London's Imperial College, the site also provides links to free software and references to print, audio, and multimedia materials.

SOCIAL SCIENCE: THE WWW SOCIAL SCIENCES VIRTUAL LIBRARY

<http://coombs.anu.edu.au>

This site is sponsored by the Research School of Pacific and Asian Studies of Australia's National University in Canberra. The site provides links to online journals, Web searches, and guides to information, with a focus on Asian studies.

These examples represent only a few of the thousands of free Internet research sources available in every field.

The Invisible Web

The "invisible Web" consists of a large portion of the Web that is not picked up by the robots that search engines use to find new sites and can be valuable in helping researchers locate new information. The invisible Web consists of databases, many of them compiled by government agencies and others collected by businesses or individuals and placed on the Internet.

An example of an invisible Web search site is Direct Search:

<http://gwis2.circ.gwu.edu/~gprice/direct.htm>

This inclusive and well-organized site allows you to find and access over 1,000 databases on the following kinds of sources:

- Archives and Library Catalogs
- Bibliographies/Bibliographic Aids
- Books (including the full book text)
- Business/Economics
- Government (U.S. and Foreign)
- Government (U.S. State & City)
- Humanities
- Legal
- News Sources and Serials
- Ready Reference
- Science
- Social Sciences
- Recent Additions to the Collection

Other invisible Web directories are located at:

INFOMINE MULTIPLE DATABASE SEARCH

<http://infomine.ucr.edu/search.phtml>

This UCLA-sponsored database collection (with over 650 search engines) supports academic interests and scholarly sources on subjects including agriculture, biology, medicine, social sciences, mathematics, and humanities. The site provides links to electronic journals.

ALPHASEARCH

<http://www.calvin.edu/library/as/>

Sponsored by Calvin College, AlphaSearch directs users to well-planned sites that cover one idea or one discipline. The site offers keyword searching by academic discipline.

WEBDATA.COM

<http://www.webdata.com/webdata.htm>
Lists topics under the heading "databases." Researchers choose a topic and get a list of databases on the topic, with descriptions of the coverage. This arrangement allows users to preview available resources.

By using the invisible Web, researchers can locate new material efficiently. It is important to make sure that search sites reflect serious research interests in their organization and sponsorship.

REFERENCE WORKS

In addition to locating material, researchers need to consult reference works, reliable sources in many areas of study. Common reference works include encyclopedias, dictionaries, books of statistics, almanacs, bibliographies, handbooks, and yearbooks. Many of these reference works are available in nonprint as well as print formats.

Encyclopedias

Encyclopedias are comprehensive reference works that contain articles on a wide variety of subjects. These reference works may be general or specialized in coverage. General encyclopedias may be a good starting point for locating general information about a subject. *Americana, Britannica, Collier's,* and *World Book* are well-known general encyclopedias that provide articles written for the general reader.

A valuable specialized work is the *McGraw-Hill Encyclopedia of Science and Technology,* an international reference work in twenty volumes (including an index). The articles are arranged in alphabetical order (word by word), covering information for every area of modern science and technology. Each article includes the basic concepts of a subject, a definition, background material, and multiple cross-references.

Specialized encyclopedias are numerous and include: *Encyclopedia of American Forest and Conservation History, Encyclopedia of Psychology, Encyclopedia of Crime and Justice, Encyclopedia of Textiles, Encyclopedia of Banking and Finance, The Encyclopedia of Alcoholism, Encyclopedia of Physics, Encyclopedia of American Business History and Biography,* and *Macmillan Encyclopedia of Architects.*

Dictionaries

Like encyclopedias, dictionaries can be general or specific in coverage. General-use dictionaries include *The American Heritage College Dictionary, Merriam–Webster's Collegiate Dictionary, The Random House Dictionary of the English Language,* and *Webster's New World Dictionary of American English.* Specialized dictionaries exist for subjects such as computer languages, architecture, welding, decorative arts, technical terms, astronomy, U.S. military terms, Christian ethics, and economics.

Statistical References

Statistics can support conclusions, show trends, and validate statements. However, since statistics change frequently, researchers must find and use the most accurate and current statistical information. Up-to-date statistics appear in sources such as almanacs, yearbooks, and encyclopedias as well as recent newspaper and magazine articles. Statistical information on the past, such as wages in the 1930s, may be located in *Historical Statistics of the United States, Colonial Times to 1970.*

Statistics for the United States and other countries in areas such as industry, business, society, and education are found in *Statistics Sources.* Annual updates of statistical data appear in the *Statistical Abstract of the United States* (since 1878) and the *Statistical Yearbook* (since 1949), which covers yearly events for about 150 countries.

Two monthly indexes that provide access to a large area of statistical information are the *American Statistics Index* (ASI) and the *Statistical Reference Index* (SRI). ASI (since 1973) indexes most statistical sources published by the federal government and SRI (since 1980) indexes statistical information in U.S. sources other than those published by the federal government. Both have annual cumulations.

To locate a library's statistical sources, look in the public access catalog under the subject headings "Statistics" and "United States—Statistics." For statistical works about a particular topic, also look under the topic for the subheading "Statistics." For example, you might check "Agriculture—Statistics."

Almanacs

An almanac is an annual publication that provides lists, charts, and tables of specific information. Perhaps the best-known general almanac is *The World Almanac and Book of Facts.* It includes such diverse information as winners of Academy Awards, accidents and deaths on railroads, civilian consumption of major food commodities per person, National Football League champions, notable tall buildings in North American cities, and the U.S. military pay scale. Other almanacs include *Information Please Almanac, The Universal Almanac,* and *Whitaker's Almanac.*

Bibliographies

Bibliographies are lists of sources such as periodicals, books, pamphlets, and nonprint materials. Most bibliographies contain specific citations to information on a subject. However, others, such as the *Encyclopedia of Business Info Sources,* list by subject such types of sources as encyclopedias and dictionaries, handbooks and manuals, periodicals, statistical sources, and almanacs and yearbooks. Bibliographies are particularly useful when researching a topic about which little is known or for which few sources seem to be available.

To find a library's bibliographical holdings on a specific subject, consult the public access catalog. Look under the subject for the subheading "Bibliography," for example, "Business—Bibliography," "Marketing—Bibliography," or "Nursing—Bibliography."

Handbooks

Handbooks provide specialized information for specific fields. The variety of handbook subjects is vast. For example, some handbooks provide information for word processor users, secretaries, electronics engineers, photographers, and construction superintendents. Handbooks exist for the study of suicide, transistors, air conditioning systems design, food additives, simplified television service, and tropical fish—to name a few.

Job seekers interested in information about careers will find the *Occupational Outlook Handbook* quite useful. Published biennially by the U.S. Bureau of Labor Statistics, it describes about 250 occupations. For each occupation, the *Occupational Outlook Handbook* gives the outlook, the number of persons currently employed, salary ranges, educational requirements, and career tracks.

Yearbooks

Yearbooks, as the name suggests, cover events in a given year. An encyclopedia yearbook updates material covered in a basic set—this is an example of a general yearbook. A few of the many specialized yearbooks include the *Yearbook of Agriculture, Yearbook of American and Canadian Churches, Yearbook of Drug Therapy, Yearbook of Astronomy, Yearbook of Sports Medicine*, and *Yearbook of Labor Statistics*.

Guides to Reference Works

For a quick review of what reference materials exist, the following guides are useful. If these guides lead you to materials that are not available in your library, ask the librarian about an interlibrary loan. Remember that interlibrary loans take time.

Sheehy's *Guide to Reference Books* lists approximately 10,000 reference titles of both general reference works and those in the humanities, social sciences, history, and the pure and applied sciences. This guide and its supplements also include valuable information on how to use reference works. The information in Sheehy's *Guide* may be updated by using *American Reference Books Annual* (ARBA), which lists by subject the reference books published each year in the United States. The *Cumulative Book Index* lists books printed in English each year. *Books in Print* is a listing of the majority of the books in print in America.

Guides to Periodicals

Ulrich's International Periodicals Directory includes an alphabetical listing (both subject and title) of in-print periodicals, both American and foreign, and lists some of the works that index the periodicals.

Guides to CD-ROMs

CD-ROMs in Print provides information on CD-ROM products and producers worldwide. The title directory lists titles alphabetically giving title, description, subject areas, provider, and hardware/software requirements. It also includes a subject index.

PUBLICLY AVAILABLE PRINT MATERIALS

Many organizations provide free or inexpensive pamphlets, documents, and reports that can contain useful information, sometimes unpublished or more current than published articles. The U.S. Government Printing Office, the various departments of the United States government, state and local agencies, industries, companies, and professional organizations are just a few of the sources of these materials.

In seeking public information, make your request specific. For instance, a person researching aluminum as a structural building material might be disappointed in the response to a general request to Alcoa for information about aluminum. A request for information about aluminum as a structural building material, however, specifies the subject and thus encourages a more satisfactory response.

When you request public information, consider the timing. Do not assume that all materials will arrive or that the materials will contain research information. You can save time and preview public information by calling toll-free information numbers, making e-mail requests, or reviewing Web page listings of public information provided by organizations.

CHAPTER SUMMARY

GUIDES TO INFORMATION SOURCES

If you learn to use guides to library holdings and library catalogs, are familiar with guides to information, and have experience with different kinds of searching, you'll be able to conduct research in any subject area quickly, efficiently, and confidently.

Documentation

INTRODUCTION

As you draft and revise the project based on your research, you need to give credit for, or document, the information you use. The following discussion explains what you must document for academic and workplace projects. In addition, this section describes types of documentation, lists style guides used in different academic disciplines, and provides specific examples of citations and works cited from two widely used styles.

WHAT AND WHY YOU MUST DOCUMENT

In any research project you must document any borrowed material that you use by recording the source and its location. You must also document the words, views, conclusions, or opinions of others. You must document information not widely known or discussed in many sources. By documenting your sources, you respect the recorded ideas, or intellectual property, of others. Your documentation also shows the ways in which these ideas support your own assertions.

Intellectual property is the recorded form of people's ideas. Whether that property is a book, a visual, or a Web page, copyright law protects the rights of the person who created it. For example, a friend's e-mail, a coworker's memo, or a brochure you picked up at a trade show are all considered copyrighted material. Copyright also covers the way in which information is presented. If you want to use verbal or visual information or the design format, you must properly cite the author of the information, perhaps even getting permission from the author or sponsoring organization.

If you are writing a research report for a class, you will need to identify, or cite, the sources of borrowed information, but you will not have to obtain permission from the copyright holder. This "educational fair use" is built into the copyright law so that students can use published information as part of their education. However, when you are writing for publication or for a company, you must get written permission from the copyright holder. Sometimes you must also pay for the right to use copyrighted material. The Copyright Clearance Center makes it easy for you to obtain copyright clearance for work you wish to publish or to use in workplace projects. With nearly two million copyrighted works in their records, the Copyright Clearance Center can quickly process requests. Their online information site at <http://www.copyright.com> provides more information.

When you use borrowed material—visuals, quotations, paraphrases, original ideas, or unusual data—in a college or workplace communication project, you must acknowledge that use by documenting your sources. When you document, you use a special format to indicate exactly where any borrowed information appears in the text of your project. These in-text citations are presented fully in a list of sources, or works cited at the end of the project. The format you should use to prepare citations in text and in the bibliography at the end of your project will depend on the style required by your project.

TYPES OF DOCUMENTATION SYSTEMS

Although the reason for documentation is the same in all areas—acknowledging borrowed information and stating its exact source—the procedures for citing sources differ in punctuation, capitalization, and sequence of information. In general, documentation systems fall into three categories:

- Notational documentation
- Reference documentation
- Parenthetical documentation

Notational Documentation

Notational references may be footnotes or endnotes. Borrowed material is cited within the text by a superscript numeral that points to footnotes at the bottom of the page or endnotes at the end of the project. Notes are numbered sequentially throughout a report or a chapter. For example, if 17 notes appear in a report, 17 notes will be listed in order, from 1 to 17. Footnotes list the notes at the bottom (or foot) of the page on which the superscript numeral appears, making it easy for the reader to consult a note while reading.

A typical example of a footnote citation (in MLA format) in text and the footnote as it would appear at the bottom of the page of text follows.

> "The rate of the enrollments in two-year colleges is growing at a rate 27% faster than the enrollment rate in colleges and universities over the period from 1997 until 2000."[1]
>
> [1]Robert W. Van Gullik, *Postsecondary Statistical Studies* (Santa Barbara: Demographic Institute, 1999) 137.

Endnotes are listed in sequence on a separate page titled "Endnotes," which appears after the text and before the complete list of works cited. While endnotes are easier to prepare, they create more of an interruption for the reader who must turn to the end of the text.

Today, footnotes and endnotes are used infrequently.

Reference Documentation

In reference documentation, used widely in the sciences, a report provides a numbered list of references at the end of the work. Citations in the text, placed in parentheses or as a superscript after the cited information, refer to the number of each source. The list is numbered in the order in which each citation appears in text. The works cited list is arranged in the order in which each citation appears in the text. Only those sources cited appear in the works cited list. The Council of Biology Editors' *CBE Style Manual* is one of the style guides using a version of reference notation. Although reference documentation is widely used in specialized science fields, it is not common in academic and workplace publications.

Parenthetical Documentation

The most widely used documentation system, parenthetical citation, allows you to cite your sources at any point, even within a sentence. In parenthetical documentation, the source of the information is placed next to the borrowed information, usually using the author's name and the page number. For complete source information, readers can turn to the list of references or works cited at the end of the project. Examples (in MLA format) appear below.

> "The rate of the enrollments in two-year colleges is growing at a rate 27% faster than the enrollment rate in colleges and universities over the period from 1997 until 2000" (Van Gullik 137).
>
> According to Robert W. Van Gullik, "The rate of the enrollments in two-year colleges is growing at a rate 27% faster than the enrollment rate in colleges and universities over the period from 1997 until 2000" (137).

The complete bibliographical description of the cited source will appear in a works cited (MLA) or references (APA) list at the end of the report.

STYLE GUIDES

Documentation manuals, or style guides, explain the type and format authors should use to document their sources. Documentation styles vary from discipline to discipline and even within disciplines. Examples of style guides include the American Mathematical Society's *Manual for Authors of Mathematical Papers*, the Council of Biology Editors' *CBE Style Manual*, the American Chemical Society's *The ACS Style Guide: A Manual for Authors and Editors*, the American Institute of Physics' *AIP Style Manual*, the International Committee of Medical Journal Editors' "Uniform Requirements for Manuscripts Submitted to Biomedical Journals," the *American Medical Association Manual of Style*, and *A Uniform System of Citation* distributed by the Harvard Law Review Association.

Widely used style guides include:

- *The Chicago Manual of Style*. 14th ed. Chicago: U of Chicago P, 1993.
- *The MLA Handbook for Writers of Research Papers*. 5th ed. New York: MLA, 1999. <http://www.mla.org.>
- *Publication Manual of the American Psychological Association*. 4th ed. Washington: American Psychological Assn., 1994. <http://www.apa.org/journals/webref.html>

The MLA Handbook for Writers of Research Papers (*MLA Handbook*) and *Publication Manual of the American Psychological Association* (*APA Manual*) cover the two styles that will be discussed with examples in this section. Each style guide shows how to cite a wide range of sources, providing discussion and examples. Some style guides, such as the MLA and APA, also have Web sites to make it easier for users to learn how to create citations and bibliographic descriptions, especially for new kinds of electronic information. Whatever style guide you use, carefully follow its instructions for in-text citations and works cited entries.

PARENTHETICAL DOCUMENTATION: MLA AND APA

Both the Modern Language Association (MLA) and the American Psychological Association (APA) recommend parenthetical documentation. To use this format to document a citation in text, give the author's last name and the page number. In both MLA and APA formats, a complete description of the reference will also appear in the list of works cited or references at the end of the document.

Below are examples of parenthetical citations in both MLA and APA formats. When an author's name is mentioned in the context of cited information, it is only necessary to identify the page in MLA and the date and page in APA. Note the differences in placement and punctuation between the two documentation styles.

> MLA: According to Jarrold Gould, "Effective international communication calls for an increased awareness of cultural conventions about reading, graphics, and information from authors and translators" (52).

> APA: According to Gould (1999), "Effective international communication calls for an increased awareness of cultural conventions about reading, graphics, and information from authors and translators" (p. 52).

Use the models below when you must identify the author and the source of cited material.

> MLA: "Although many botanists argue that the biodiversity of rain forests is threatened by logging and settlement, governments of affected areas have been slow to respond with policy decisions to protect unique species" (Magree and Miller 195).

> APA: "Although many botanists argue that the biodiversity of rain forests is threatened by logging and settlement, governments of affected areas have been slow to respond with policy decisions to protect unique species" (Magree & Miller, 1998, p. 195).

If you refer to a publication that lists no author, include the first few keywords in the title instead of the author's name in the parenthetical citation. MLA and APA examples appear below.

> MLA: The majority of the employees in fast-food restaurants are 20 years of age or younger ("Fast Food" 6).

APA: The majority of the employees in fast-food restaurants are 20 years of age or younger ("Fast Food," 1999, p. 6). [In APA style the page number may be omitted for paraphrases.]

References to Web pages and other electronic media may need to reflect sections rather than page numbers in the parenthetical reference.

QUOTATION FROM A WEB PAGE
MLA: Ralph Voss states that "the sensitivity of the Australian ecosystem to invasive plants and animals is thoughtfully reflected in the regulations governing transport of animal and agricultural products" (Intro).

APA: Ralph Voss (1999, December 21) states that "the sensitivity of the Australian ecosystem to invasive plants and animals is thoughtfully reflected in the regulations governing transport of animal and agricultural products" (Intro).

Paraphrase in MLA or APA format can include phrasing to indicate their source.

PARAPHRASE
MLA: In technical communication many professionals share the experiences of Megan Little, who in a 1999 e-mail message described the combined effectiveness of her academic training and the support she received from workplace mentors.

APA: In technical communication many professionals share the experiences of Megan Little (1999), who described in an e-mail message the combined effectiveness of her academic training and the support she received from workplace mentors.

WORKS CITED AND REFERENCE LISTS

In MLA and APA documentation, a list of works cited, or references, concludes a research report or publication. This list gives author, title, and publishing information for all source materials referred to in the text.

MLA and APA documentation share some conventions for preparing such a list. For both styles, the items in a works cited list are arranged in alphabetical order by the last name of the first listed author, or, if no author is listed, by the first important word in the title. ("A", "an", and "the" are not considered important words.) The first line in an MLA documentation entry is not indented; each turn line thereafter is indented. The first line in an APA entry is indented; turn lines are not indented. Data from a CD-ROM or a disk should be described in the same way as print information. However, the type of media should also appear after the name of the source (for example: *Times Herald Tribune*, CD-ROM).

Although documentation styles differ, they differ only in such details as capitalization of titles, punctuation, abbreviations, and order of information. All documentation styles give the essential information for locating a particular source. A discussion of MLA and APA conventions for citations (with examples to illustrate each style) follows.

Citation Conventions: MLA

Listings for books, articles, and electronic sources require a particular order and punctuation in an MLA works cited list.

For books:

- last name, first name of author
- title of book
- city of publication
- name of publisher
- year of publication

EXAMPLE

Adams, Michael. *The Glory Days: Colonial Agricultural Practices in Massachusetts Bay.* New York: Putnam, 1998.

For articles in magazines:

- last name, first name of author (if an author is listed)
- title of article
- name of magazine or journal
- date of issue
- page numbers

EXAMPLE

Southworth, Eliza."Gender and Authorship: A Review of Historical Sources." *Publishing Review* Sept. 1999: 17–26.

For electronic media:

- last name, first name of author (if an author is listed)
- title of document
- date of publication
- source
- date of access
- URL

EXAMPLE

Watling, Bill. "Discrimination Online." Communication and Gender Studies Information Page. 17 May 1999. 25 Sept. 1999 <http://web.gen/clark.edu/www/html>.

Examples of MLA Citations

Below are examples of MLA citations for different kinds of references: books, periodicals, and nonprint materials.

Books

Books with one author:

Campion, John. *Squaring the Circle*. Berkeley: Ecotropic, 1999.

Books with two or three authors:

Silverman, Jay, Elaine Hughes, and Diana Roberts Wienbroer. *Rules of Thumb: A Guide for Writers*. 4th ed. New York: McGraw, 1999.

Books with three or more authors:

Lay, Mary M., et al. *Technical Communication*. 2nd ed. Boston: Irwin McGraw–Hill, 2000.

Book with corporate author:

American Cancer Society. *Art, Rage, Us: Art and Writing by Women with Breast Cancer*. San Francisco: Chronicle, 1998.

Book listed by editor:

Garay, Mary Sue, and Stephen A. Bernhardt, eds. *Expanding Literacies: English Teaching and the New Workplace*. Albany: SUNY, 1998.

Government document:

United States. National Committee on Global Change Research. *Global Environmental Change: Research Strategies for the Next Decade*. Washington: National Academy, 1999.

A work in an anthology:

Connors, Robert J. "Landmark Essay: The Rise of Technical Writing Instruction in America." *Three Keys to the Past: The History of Technical Communication*. Ed. Teresa C. Kynell and Michael G. Moran. Stamford: Ablex, 1999. 173–95.

Encyclopedia listing:

"Parallel Port." *The Computer Desktop Encyclopedia*. 1999 ed.

Articles

Article in a journal:

Moore, Patrick. "When Persuasion Fails: Coping with Power Struggles." *Technical Communication* 46 (1999): 351–59.

Article in a magazine:

Meisler, Stanley. "Splendors of Topkapi, Palace of the Ottomans." *Smithsonian* Feb. 2000: 114–22.

Newspaper article:

Jayson, Sharon. "Building for the Future." *Austin American Statesman* 6 Feb. 2000: A1, A19.

Nonprint Media

Filmstrip, film, or videotape:

Technical Occupations. Videocassette. Delphi Productions, 1998.

Computer program:

Microsoft Word 2000. Computer software. PC with Pentium 75 MHZ. 20 MB RAM 146 MB hard disk space. Microsoft, 2000.

Web site:

Stubbs, Chelsea. *Gender and Workplace Roles.* Workplace Studies Page. 21 Jan. 1999. 30 May 1999 <http://web.work/smithville.edu/www/>.

E-mail:

Larkin, Mike. "Plant Parasite Infestations." E-mail to the author. 21 Dec. 1999.

CD-ROM:

Loy, Marc, and Tom Berry. *Java Programming for the Internet.* CD-ROM. Upper Saddle River: Prentice, 1997.

Interview:

Ludlow, Elijah. Personal interview. 3 Jan. 2000.

For MLA citation conventions governing other types of sources, consult the *MLA Handbook for Writers of Research Papers* or the MLA Web page <http://www.mla.org>.

Citation Conventions: APA

Books, articles, and electronic sources require a particular order and punctuation in an APA reference list. Notice the conventions in order, punctuation, and capitalization of titles. The student research report, which appears on pages 565–578, uses APA citation format.

For books:

- last name, first initial of first name of author
- year of publication
- title of book
- city and state of publication
- name of publisher

EXAMPLE

Adams, M. (1998). *The glory days: Colonial agricultural practices in Massachusetts Bay.* New York: Putnam.

For articles in magazines:

- last name, first initial of first name of author (if an author is listed)
- year of issue
- title of article
- name of magazine, volume, number
- page numbers

EXAMPLE

Southworth, E. (1999) Gender and authorship: A review of historical sources. *Publishing Review, 4* (3), 17–26.

For electronic media:

- last name, first initial of first name of author (if an author is listed)
- date of Internet publication
- title of document
- date of access
- URL

EXAMPLE

Watling, B. (1999, May 17). Discrimination online. *Communication and Gender Studies Information Page.* Retrieved September 25, 1999 from the World Wide Web: <http://web.gen/clark.edu/www/html>.

Examples of APA Citations

Below are examples of APA citations for different kinds of references: books, periodicals, and nonprint materials.

Books

Book with one author:

Campion, J. (1999). Squaring the circle. Berkeley: Ecotropic.

Book with two or more authors:

Silverman, J., Hughes, E., & Wienbroer, D. (1999). Rules of thumb: A guide for writers. (4th ed.). New York: McGraw-Hill.

Book with corporate author:

American Cancer Society. (1998). Rage, art, us: Art and writing by women with breast cancer. San Francisco: Chronicle Books.

Book listed by editor:

Garay, M., & Bernhardt, S. (Eds.). (1998). Expanding literacies: English teaching and the new workplace. Albany: SUNY.

Government document:

National Committee on Global Change Research. (1999). Global environmental change: Research strategies for the next decade. Washington, DC: National Academy.

Chapter or essay in an anthology:

Connors, R. (1999). Landmark essay: The rise of technical writing instruction in America. In T. Kynell & M. Moran (Eds.). Three keys to the past: The history of technical communication (pp. 173–195). Stamford, CT: Ablex.

Encyclopedia article:

Parallel port. (1999). The computer desktop encyclopedia (p. 670). New York: Amacom.

Articles

Article in a journal:

Moore, P. When persuasion fails: Coping with power struggles (1999). Technical Communication, 46, 351–359.

Article in a magazine:

Meisler, S. (2000 February). Splendors of Topkapi, palace of the Ottomans. Smithsonian, 30, 114–122.

Newspaper article:

Jayson, S. (2000, February 7). Building for the future. The Austin American Statesman, pp. A1, A19.

Nonprint Media

Filmstrip, film or videotape:

Technical occupations [Videocassette]. (1997). Boulder, CO: Delphi Productions.

Computer program:

Microsoft Word 2000 [Computer software]. (2000). Seattle, WA: Microsoft Corporation.

Specific document on a Web site:

Stubbs, C. (1999, January 21). Gender and workplace roles. Workplace Studies Page. Retrieved May 30, 1999 from the World Wide Web: http://web.gen/clark.edu/www/html>.

E-mail:

In APA style, personal communications (including e-mail, telephone calls, and letters) are cited in text but not listed in the references.

CD-ROM:

Loy, M., & T. Berry. (1997). Java programming for the Internet. [CD-ROM]. Upper Saddle River, NJ: Prentice Hall.

Interview:

In APA style, personal interviews are cited in text but are not listed in the references.

For APA citation conventions governing other types of sources, consult the *Publication Manual of the American Psychological Association,* 4th edition, or visit the APA Web site at <http://www.apa.org/journals/webref.html>.

CHAPTER SUMMARY

DOCUMENTATION

By documenting your sources, you acknowledge the works of others that you have used to support your own ideas. You also make it easy for yourself and others to locate the sources you consulted. Using a particular style guide consistently and correctly helps you to adhere to the conventions of your profession. Your ability to conform to these conventions, like your willingness to credit the ideas of others, demonstrates that you are part of a professional community.

Student Research Report

INTRODUCTION

The following section illustrates the planning, development, and discussion of a student's research report from the original Project Plan Sheet to the finished report. It also includes a progress report, which Sandra Smith, the author, submitted to update her instructor about her work on her project. This progress report provides an overview of Sandra's research, her organization plan, and a bibliography of the sources she had located. Sandra's finished report, "Peripheral Neuropathy: A Guide for Diabetics," appears next, followed by an interview in which Sandra comments on her research project and the importance of research in college and in the workplace.

THE ASSIGNMENT

The assignment that Sandra Smith and her classmates received at the beginning of the semester appears below.

ASSIGNMENT: THE RESEARCH REPORT

The research report is a full-semester project, which you should begin thinking about from the first day of class. Since it counts for 40% of your grade, this report should reflect your best planning, thinking, and writing.

The Project Plan Sheet and progress report are intended to provide you with feedback about your research report as you go about planning for it and working on it. Feel free to meet me [the instructor] to discuss any aspect of this important assignment.

The Project Plan Sheet

You will complete a project plan sheet to present your research report topic for approval before you can begin work on the report itself. The proposal allows you to consider your report topic, audience, and purpose fully before you begin work. You may write on any subject, but the topic must be limited and useful to your readers.

Research

You will need to locate preliminary sources even before you submit your Project Plan Sheet. Your text provides useful introductory information about research skills, describing documentation formats (including the APA format required for this project). A complete list of area research collections is available through the library, as is a list of Web reference services. Please feel free to discuss your ideas and interests with me [the instructor].

The Project Plan Sheet for the research report appears on pages 559–561. In this plan sheet, the writer carefully defined her project's purpose, audience, and subject. She also established a timeline of activities for her research and writing.

PROJECT PLAN SHEET

RESEARCH REPORT

Audience

- Who will read the report?

  Readers will be diabetics, their families, or their employers. Readers will not have medical background or training.

- How will readers use the report?

  Readers will use the report to learn about the relationship between neuropathy and diabetes, to take action to prevent neuropathy, to understand the symptoms of neuropathy, and to seek appropriate treatment.

  Diabetics will read the report to understand and respond appropriately to their medical risk for neuropathy and to seek treatment if they suffer from neuropathy.

- How will your audience guide your communication choices?

  I will need to explain all medical terms and concepts clearly and nontechnically.

Purpose

- What is the purpose of the report?

  The report's purpose is to provide a clear understanding of a potentially disabling but preventable condition so that diabetics can understand what they need to do to prevent it. The report will also help diabetics who suffer from neuropathy understand the condition and available treatments for it. The report will help patients and those who live and work with them to clearly understand neuropathy and respond appropriately to it.

- What need will the report meet? What problem can it help to solve?

  The report can help diabetic patients prevent neuropathy. For diabetics with neuropathy, the report can help them adapt to their condition. The report can also help others understand causes and symptoms of neuropathy.

Subject

- What is the report's subject matter?

  The report will discuss neuropathy as a complication of diabetes.

- How technical should the discussion of the subject matter be?

  The report must be comprehensive, but it must explain medical terms and issues clearly for nontechnical readers.

- Do you have sufficient information to complete the report?

  I have some background understanding of the symptoms of neuropathy, but I will need to conduct research to provide more specifics, especially about causes, symptoms, and treatment.

- What sources or people can help you to locate the best research materials?

  I'll use CD-ROM guides to literature and the Internet. I can also discuss sources with some of my Health Sciences teachers.

- What title can clearly identify the report's subject and purpose?

  I'd like to call the report "Peripheral Neuropathy: A Guide for Diabetics." This title is concise and describes the subject clearly.

Author

- Will the report be a collaborative or an individual effort?

  Individual.

- How can you evaluate the success of the completed report?

  When the report is drafted, I'll ask two diabetics to review the draft to make sure that it answers all of their questions.

Project Design and Specifications

- Are there models for organization or format for the report?

  Yes. My instructor has provided a guide for parts of the report (including title page, table of contents, descriptive summary, and sections of text). APA style is required for citations.

- In what medium will the completed report be presented?

  Print.

- Are there special features the completed report should have?

  Yes. I have described these above.

- Will the report require graphics? If so, what kinds and for what purpose?

  Yes. I will include graphics showing nerve cell components and nerve trunk components.

- What information design features can help the report's audience?

  A strong introduction and careful descriptive headings listed in the table of contents and in the text will guide readers through the sequence and to parts of the report.

Due Date

- What is the final deadline for the completed report?

  I need to submit the report no later than April 24.

- How long will the report take to plan, research, draft, revise, and complete?

  If I start research now, I can be ready to plan and draft in four weeks. This allows me 2½ weeks to revise and to get reader feedback. I'll need to be economical about research, using abstracts and getting advice about the best sources from my instructors. I can also save time by developing clear research questions about my three main research areas: causes, symptoms, and treatments.

- What is the timeline for different stages of the project?

  Research: 3 weeks.

  Outlining: 4 days.

  Drafting: 1 week.

  Review and revision: 2½ weeks.

PROGRESS REPORT

Sandra Smith submitted the following progress report to her instructor to report on her research, explain her plans for the report, and discuss the changes she had decided to make in the original project.

DATE: April 2, 1999
TO: Dr. Katherine Staples
FROM: Sandra Smith
SUBJECT: Progress on Research Report on Neuropathy for Diabetics

RESEARCH REPORT TOPIC

My report will be on neuropathy for diabetics. Neuropathy is a continual tingling sensation in the lower extremes of the legs. The tingling can be mild to extremely painful. The report will go into detail, explaining neuropathy and the causes as related to diabetics. It will describe the symptoms that diabetics could or would be experiencing. It will discuss treatment methods. The report will also give the readers an idea of what to expect during and after treatment.

RESEARCH PERFORMED TO DATE

Information obtained and requested

Journal articles

Health magazines

Health books

Request for pamphlet from Neuropathy Association

I have gathered enough information to write the report. I am waiting on the pamphlet, so I can see if there is any new information on the disease. I used the libraries and some books that I had collected for my mother. I have collected over 20 articles that may be helpful in writing the paper.

WORK TO BE COMPLETED

I am ready to write the rough draft of the report after I receive the pamphlet from the Neuropathy Association. I will then work on the final draft to prepare it for submission.

CHANGES FROM THE ORIGINAL PLAN

The author has altered her original plan so that her finished report will better meet the needs of her readers as defined in her completed Project Plan Sheet.

The only change I am making from my Project Plan Sheet is that I describe neuropathy in less detail than I originally thought necessary. The report is designed to give the readers practical knowledge about the disease, but not so much that they can't read the entire report in one or two sittings. If readers are not afflicted by peripheral neuropathy, I still want them to read the entire report in order to understand the importance of preventing the disease.

REVISIONS IN COST

I have been able to gather all my research on the days that I am already in Austin. This has reduced my gas mileage cost from $25.00 to $5.00. I did take a vacation to work on the research project. I still attended classes, but I used the time after class to do research. I reduced the total cost from $50.00 to $30.00. The cost still doesn't include my time. It is taking a lot of time to gather the information, and it will still take a lot of time to write the paper. However, I don't know how to put a value on the time spent.

ORGANIZATION PLAN

The organization plan provides a useful guide to arranging the information important for readers.

I. Introduction

II. Neuropathy for diabetics
- A. General symptoms
  1. The area of the body that is affected
  2. How symptoms feel

III. Medical overview of neuropathy
- A. Medical description
  1. Short description of the makeup of a nerve that relates to neuropathy
  2. Description of how neuropathy affects the nerves
  3. Description of what actually causes the sensation

IV. Connection
- A. Short description of the sugar in a diabetic's system
- B. Description of the sugar and the nerves that cause neuropathy

V. Diagnosis and treatment
- A. Methods of diagnosis
  1. Physical exam
  2. Electromyography exam
- B. Treatment
  1. Medicine
  2. Pain relief (home remedies)

VI. Long-term expectations
- A. Description of the pattern of neuropathy after treatment
  1. Fluctuations in symptoms
  2. Disappearance of symptoms
- B. Description of care for affected areas to prevent injuries
- C. Support numbers and associations

VII. Summary

ANNOTATED BIBLIOGRAPHY

Complications of diabetics: Nerve damage (neuropathy). (1997). Clinical Research Systems, 170.

I will use the description of the nerve cells from this article.

Dinsmoor, R. (1998). Easing the pain of diabetic neuropathy. Diabetes Self-Management, 15, (4), 37–47.

This article describes the medicines used for the pain sensation of diabetics' neuropathy. It also briefly describes the tests that are used for diagnoses.

Dyck, P. (1996). Nerve growth factor and diabetic neuropathy. The Lancet, 347, (9034), 1044–1046.

This article explains how neuropathy and diabetes affect each other.

Feldman, O. (1998). Foot smarts pitfalls and preservation. Diabetes Self-Management, 15, (6), 52–64.

This article discusses nerve damage to the feet. It also discusses how this affects balance and how it relates to the feet. It gives an address for a good pamphlet for diabetics, which I will pass on in the paper.

Hasselbring, B., & Politzer, B. (1994). Diabetics: Strategies for a full and healthful life. Women's Home Remedies Health Guide, 114–120.

This book gives a good description of the care needed for diabetics' feet. When neuropathy has gone to the extreme, nerve damage does occur, and the person cannot feel his or her feet. Diabetics need to be aware of the simple ways they can protect their feet.

The sources are promising, and the annotations—short notes describing each source—clearly state the content and value of each source.

Lipnick, J. A., & Lee, T. H. (1996). Diabetic neuropathy. American Family Physician, 54, (8), 2478–2487.

This article covers a wide range of subjects that my report will use.

FINAL PROJECT—PERIPHERAL NEUROPATHY: A GUIDE FOR DIABETICS

Sandra's finished research report appears on the following pages.

PERIPHERAL NEUROPATHY: A GUIDE FOR DIABETICS

By
Sandra Smith

Dr. Katherine Staples
Introduction to Technical Writing
Section 6582

TABLE OF CONTENTS

DESCRIPTIVE ABSTRACT

The descriptive abstract provides a clear and accurate overview of the report's subject matter.

Peripheral neuropathy is a common complication for diabetics. This report discusses the overall symptoms and characteristics of neuropathy, and explains how the condition is related to diabetes. The report gives information on the diagnosis, treatment, and prevention of peripheral neuropathy. It also suggests precautions for affected areas to decrease injuries.

INTRODUCTION

The introduction states the report's subject, purpose, and scope, and guides the reader to the list of organizations that can help sufferers of peripheral neuropathy.

Many diabetics are or will be affected by peripheral neuropathy, a nerve disorder that is caused by high blood sugar. Some symptoms of peripheral neuropathy include numbness and tingling in the hands, feet, or legs. This report will help diabetics, or family and friends of diabetics, to become familiar with the symptoms and causes of peripheral neuropathy. It will help them understand how diabetes is connected to peripheral neuropathy, which may help them to prevent the disease. The report discusses the methods used for diagnosis and treatment, including medical intervention and home remedies. It also gives helpful hints on how to protect the affected areas from further injuries. A list of organizations that can provide additional support for diabetics with peripheral neuropathy appears in the Appendix on page 11.

DIABETIC NEUROPATHY

Diabetic neuropathy is nerve damage due to high levels of sugar in the blood. This condition is a common complication for diabetics. There are many forms of neuropathy, which are classified by what nerves and what areas are damaged. The most common form of diabetic neuropathy is polyneuropathy, which is generally known as peripheral neuropathy.

Peripheral neuropathy refers to the damage of nerves that are not part of the central nervous system. These are the nerves that connect the muscles, skin, and internal organs to the brain and the spinal cord. With peripheral neuropathy, several nerves are usually damaged at once, and it affects both sides of the body. Peripheral neuropathy usually affects the sensory nerves. These nerves are responsible for detecting pain, pressure, or changes in temperature. Damage to the sensory nerves, termed sensory neuropathy, can cause painful sensation or loss of sensation in the hands, feet, legs, and arms (What is, 1997). According to the National Institute of Diabetes and Digestive and Kidney Diseases, 30% to 40% of people with diabetes develop sensory neuropathy (Dinsmoor, 1998). Peripheral neuropathy can also affect the motor nerves. This can lead to muscle weakness and wasting. Peripheral and sensory neuropathy will be referred to jointly for the remainder of the report as neuropathy or diabetic neuropathy.

SYMPTOMS

Some diabetics are affected by neuropathy suddenly. Others may become affected gradually over many years. The problem usually starts with weakness, numbness, tenderness, or pain associated with the legs, feet, hands, and arms.

Numbness, Tingling, and Pain

Medical and technical terms are defined carefully throughout the report.

Spontaneous sensations, called paresthesias, may occur. These sensations include numbness, tingling, pins and needles, prickling, burning, coldness, pinching, deep stabs, electric shocks, or buzzing sensations, usually in the legs. The pains may exist continuously, or they may come and go. The pains are usually worse at night and can become severe (Donovan & Latov, 1997).

Weakness in Arms and Legs
Weakness takes place when there is damage to the motor nerves. Leg symptoms include difficulty in walking or running, a sense of heaviness, and stumbling or tiring easily. Muscle cramps are common in the legs. For people with neuropathy, the arms get exhausted when carrying groceries, combing hair, opening jars, and turning doorknobs. The grip in the hands is not as strong as it used to be, so items are dropped frequently. Overall, the body's stamina seems to be lowered because the muscles tire so easily.

Tenderness
The skin becomes very tender. The doctors refer to a "glove and stocking sensation" around the hands, feet, ankle and wrist (Lipnick & Lee, 1996). The hands and feet may be bare, but they feel like they are covered with a pair of stockings or gloves. The simple weight of clothes and bed covers is enough to be frustrating and unbearable.

Absence of Position Sense
Absence of position sense causes unsteadiness and lack of coordination when walking. This is due to the uncertainty of not knowing where feet are due to loss of sensation. Neuropathy causes a change in the way a person walks. The gait may have widened to help with balance, or people may drag their feet. Damage that occurs to the feet, such as cuts and bruises, may go unnoticed, often leading to infection.

PERIPHERAL NERVES

Nerve Description
The basic unit of the peripheral nervous system is the "neuron," or nerve cell, shown in Figure 1. Its job is to carry information by electrical impulse from one part of the body to another. Each nerve is made up of a cell body and a long projection called an "axon" (Donovan & Latov, 1997). The axon carries impulses between the cell body and nerve terminals called "receptors" in the muscles, skin, or internal organs ("Complications of diabetics," 1997).

Both diagrams are introduced with references and discussion in text.

The axons are insulated with a membrane known as the "myelin sheath." The insulation allows the axon to conduct electrical impulses faster and more efficiently. The myelin sheath is produced by another cell type called the "Schwann cell." This cell lies right alongside the myelin sheath ("Complications of diabetics," 1997).

The axons travel in bundles called "nerve trunks"; see Figure 2 (Dyck, 1996). These bundles run throughout the body like wires in an electrical network. They are found in the endoneurium, which also contains the blood vessels that supply nutrients to the nerves.

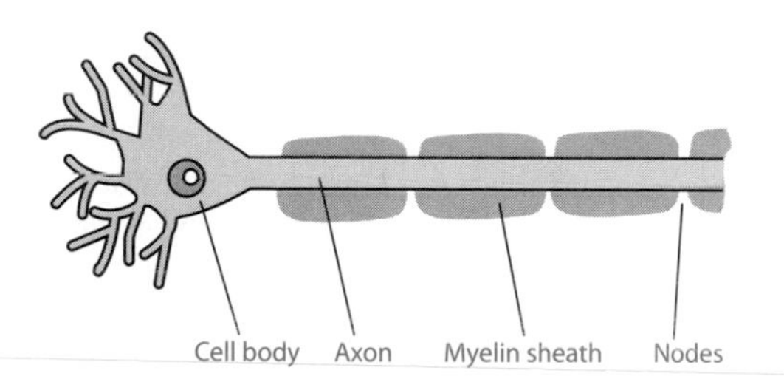

FIGURE 1 Nerve Cell Components
Source: Donovan, M.A. & Latov, N. (1997). Explaining peripheral neuropathy, 2.

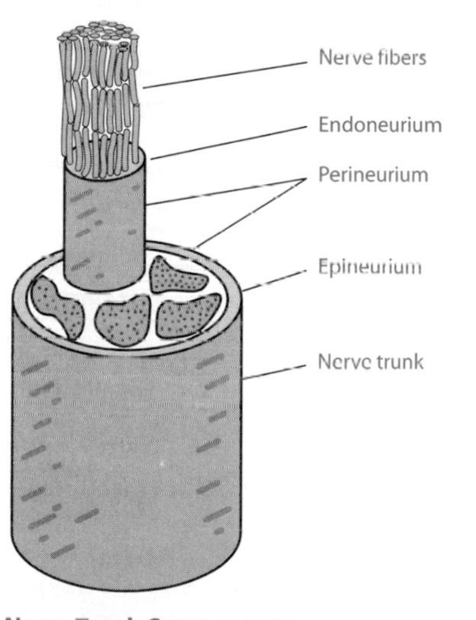

FIGURE 2 Nerve Trunk Components
Source: Donovan, M.A. & Latov, N. (1997). Explaining peripheral neuropathy, 2.

PERIPHERAL NERVE DAMAGE FROM DIABETES

Scientists do not know exactly what causes diabetic neuropathy. They know it is associated with high blood glucose, but they have only theories to explain how neuropathy affects the nerves.

Nerve Fiber Damage

One hypothesis for nerve damage is that chronic high blood sugar levels damage the nerve cell's axon (Donovan & Latov, 1997). The axon is the part of the cell that conducts electrical impulses. Some scientists believe that the chemical balance and energy metabolism within the nerves are changed by the presence of high blood sugar levels (Dinsmoor, 1998). Both of these situations prevent the cell from transmitting electrical impulses properly.

Blood Vessel Damage

The blood vessels supply oxygen and nutrients to the nerve cells. A person with a high blood sugar level has a decreased level of nitric acid in the nerves. The nitric acid is responsible for dilating the blood vessels. When the blood vessels are constricted from the high blood sugar levels, the nerves may not receive their proper nutrients (Cameron & Cotter, 1997). The nerve cells could then become damaged and not operate normally.

Glycosylation Endproducts

Glucose molecules in the bloodstream bind to proteins in the body to form glycosylation endproducts (Cameron & Cotter, 1997). When blood glucose levels are high, molecules are more likely to attach to the proteins on the cells that make up the myelin sheath or nerve fibers (Cameron & Cotter, 1997). This could affect nerve function and cause damage.

DIAGNOSING NEUROPATHY

Doctors who suspect peripheral neuropathy in a patient conduct a physical examination and sometimes laboratory tests to confirm a diagnosis and to determine which nerves are damaged.

Physical Examination

The neurologist takes a careful history of the patient's symptoms. He or she will perform a simple test to see if there is any loss of sensation in the lower extremities. This is done by using a tuning fork, a pin, a cotton swab, or a nylon filament that is placed onto different parts of the foot and lower legs. The neurologist will also visually inspect the condition of the lower extremities. The neurologist will observe the diabetic's walk to check for unsteadiness.

Electromyography

The neurologist will use a technique termed electromyography (EMG) that records the electrical activity in the muscles. This test can identify abnormal nerves and their location, helping to determine the extent of the damage to the nerve axons or to the myelin sheaths (Dyck, 1996). In EMG testing, small needles or conductive patches are inserted into or on the muscle. Electrical impulses are sent along the nerves to a muscle or group of muscles. The electrical impulses are mild and a little surprising, causing the muscles in the leg to contract involuntarily. The machine monitors the contraction pattern of the muscle. The pattern is then compared to a normally functioning muscle. Slower or weaker contractions than normal suggest that there is nerve damage. The extent of the damage determines the length of the EMG test.

PREVENTION

Diabetics are more susceptible to neuropathy the longer they have diabetes. Maintaining a tight control of blood glucose levels is the best way to prevent diabetic neuropathy. The Diabetes Control and Complications Trial showed that tight control of blood sugar levels reduced the risk of developing diabetic neuropathy by 60% (Dinsmoor, 1998). The American Diabetes Association recommends exercising regularly. They also recommend eating foods rich in vitamins and minerals to keep the nerves healthy (Hasselbring & Politzer, 1994). This diet will help reduce the chances of developing diabetic neuropathy.

Smoking narrows the blood vessels and decreases circulation to the feet (Levin, 1998). This will contribute to the ongoing problem of losing the sensation in the lower extremities. Drinking alcohol has a toxic effect on nerves. Both of these practices can increase the chances of developing diabetic neuropathy, so it is best to refrain (Hasselbring & Politzer, 1994).

TREATMENT

There is no proven cure for diabetic neuropathy. Once diabetics are afflicted, neuropathy might continue for the rest of their lives. The goals of treatment are twofold: (1) to eliminate the cause of the disease and (2) to relieve its symptoms (Donovan & Latov, 1997). A number of treatments may help when pain occurs due to neuropathy. Some evidence also suggests that bringing blood sugar levels into a normal range when they've been high can improve or reverse the symptoms of existing diabetic neuropathy (Dinsmoor, 1998). However, symptoms may worsen initially as blood sugar levels are brought under control.

Medicines

Many drugs are used to treat the pain caused by diabetic neuropathy, but no particular drug is guaranteed to work. In most cases, aspirin or acetaminophen offer enough relief. In other cases, stronger drugs may still not be enough to relieve pain (Senueff, 1997).

One of the most effective types of drugs for the pain is a class of antidepressants called tricyclics. Tricyclics work by increasing the brain's level of the neurotransmitter norepinephrine, which is part of the body's natural pain relief system (Lipnik & Lee, 1996). A few common examples of tricyclics include amitriptycyclics (brand name Elavil), nortriptyline (Pamelor), doxepin (Sinequan), imipramine (Tofranil), desipramine (Norpramin), and trazodone (Desyrel) (Dinsmoor, 1998). Smaller doses of the tricyclics are used to treat neuropathy than are needed to treat depression. Even so, tricyclics can still cause a few side effects, such as sedation, constipation, and dizziness.

Another substance that has been reported to alleviate the pain is capsaicin. This substance is derived from capsaicum peppers, which include cayenne peppers, red peppers, African chilies, and tabasco peppers. Capsaicin is available over the counter in salves such as ArthiCare and Zostrix (Dinsmoor, 1998). Capsaicin appears to trigger the release of substance P, which is a chemical transmitter for pain. This leads to a loss of sensation by depleting the nerve terminals of their pain transmitters. Capsaicin can make the pain worse for some individuals, and it can also cause irritation of the eyes, mucous membranes, and skin.

Electrical Stimulation

Some new treatments to relieve the pain caused by neuropathy use electrical stimulation to trigger the body's own natural pain relief mechanisms. These treatments don't have the side effects caused by drugs. Electrical spinal cord stimulation (ESCS) and transcutaneous electrical nerve stimulation (TENS) are the most popular electrical treatments. ESCS involves stimulating the spinal cord by an electrode that is surgically implanted at the base of the spine. The pain relief can last for several hours a day. ESCS was reported to successfully relieve chronic pain in 8 out of 10 people who previously had not responded to conventional drug treatments (Lipnick & Lee, 1996). TENS sends electrical impulses through the skin. Researchers are not sure how TENS works, but 60% to 70% of the people who try it find it helpful (Lipnick & Lee, 1996). Researchers believe that a combination of TENS and drug treatment may have an improved effect on the relief of pain in diabetic neuropathy (Dinsmoor, 1998).

Mind over Matter

Sometimes, pain is more than a person can stand, causing high levels of stress or depression. Relaxation techniques and psychotherapy can help reduce the pain and allow a person to live a better life. These techniques can be self-taught, but many people benefit from formal training. Relaxation training and meditation are a few of the techniques that can relieve the stress. Psychotherapy can help change habits, behaviors, and attitudes that make neuropathy pain worse.

Self-Help Tricks

Since pain is usually worse at night, many diabetics have developed their own methods for relieving the pain long enough for them to fall asleep. Diabetics with painful neuropathy in the feet have found that soaking the feet in warm water right before bedtime can help. Walking around can also help. Diabetics can use a foot cradle over the bed to keep the bed sheets from touching their feet. Muscle aches and pains can be relieved by performing stretching exercises, and even applying analgesic balms such as BenGay. Massaging the affected area is another good way to relieve pain or at least help the mind relax and forget for a while.

PREVENTING DAMAGE TO AFFECTED AREAS

When nerve damage does occur, the feet are usually involved. Diabetes can contribute to foot problems in two ways: (1) It can cause decreased feeling in the feet, so that injuries such as cuts and scrapes may go unnoticed. (2) It can cause decreased circulation to the feet, resulting in a reduced blood supply that may be insufficient to fight infection and heal wounds (Levin, 1998). Here are a few tips that might help with preventing foot injuries.

The writer took special care in her paraphrase of this list. She wanted to reflect the ideas of the original without copying the language.

- Cover the feet with a lotion to prevent dry skin.
- Wash the feet daily with mild soap and dry them thoroughly.
- Avoid slippery stockings or socks with bulky seams.
- Cut toenails straight across to prevent ingrown toenails.
- Never go barefoot.
- Wear shoes that fit well.
- Check inside for foreign objects before putting shoes on.
- To improve circulation, avoid sitting with the legs crossed.
- Don't attempt to remove calluses, corns, or warts at home (Feldman, 1998).

SUMMARY

Diabetics are susceptible to a nerve disorder called neuropathy. Neuropathy causes tingling, numbness, and pain in the legs, ankles, arms, and wrists. The high blood sugar level affecting the nerves causes neuropathy. The diagnosis is determined by a physical examination and by a technique called electromyography. Maintaining tight control of blood glucose levels best prevents neuropathy. There is no cure for neuropathy. The treatment includes drugs and relaxation methods, which help in relieving the pain. The feet need to be protected if affected by neuropathy, due to the loss of sensation.

9

REFERENCES

Cameron, N., & Cotter, M. (1997). Metabolic and vascular factors in the pathogenesis of diabetic neuropathy. Diabetes, 46, (9), S31–S38.

Complications of diabetics: Nerve damage (neuropathy). (1997). Clinical Research Systems, 170.

Dinsmoor, R. (1998). Easing the pain of diabetic neuropathy. Diabetes Self-Management, 15, (4), 37–47.

Donovan, M. A., & Latov, N. (1997). Explaining peripheral neuropathy [Brochure]. New York. The Neuropathy Association, Inc.

Dyck, P. (1996). Nerve growth factor and diabetic neuropathy. The Lancet, 347, (9034), 1044–1046.

Feldman, O. (1998). Foot smarts pitfalls and preservation. Diabetes Self-Management, 15, (6), 52–64.

Hasselbring, B., & Politzer, B. (1994). Diabetes: Strategies for a full and healthful life. Women's Home Remedies Health Guide, 114–120.

Levin, M. Foot care and diabetes [Brochure]. New York. The Neuropathy Association, Inc.

Lipnick, J. A., & Lee, T. H. (1996). Diabetic neuropathy. American Family Physician, 54, (8), 2478–2487.

Seuneff, John A. (1997). Pain medications. Numb toes and aching soles: Coping with peripheral neuropathy. Retrieved October 30, 1997 from the World Wide Web: <http://www.medpress.com>.

What is peripheral neuropathy? (1997). The Neuropathy Association–Overview. Retrieved October 30, 1997 from the World Wide Web: <http://www.neuropathy.org/heuropathy.asp>.

APPENDIX: SUPPORT INFORMATION

Below is a list of organizations that might be of help for diabetics with peripheral neuropathy.

This list is a very helpful report feature.

American Diabetes Association
Diabetes Information Service Center
1660 Duke Street
Alexandria, VA 22314
(800) 232-3472

American Chronic Pain Association, Inc.
P.O. Box 850
Rocklin, CA 95677
(916) 632-0922

Diabetes Self-Management
P.O. Box 52890
Boulder, CO 80322-2890
(800) 234-0923

The Neuropathy Association, Inc.
The Lincoln Building
60 East 42 Street, Suite 942
New York, NY 10165
(800) 247-6968

In the following interview, Sandra Smith's comments about her report explain her choices and activities during her research. Since Sandra is employed full-time while she attends college, she has strong ideas about planning projects and about the importance of research in the workplace.

Q: As you considered topics for your research project, which ones did you find most interesting?

Sandra Smith: Peripheral neuropathy was the only one I really considered. It's a personal concern for me because my mom is a diabetic who suffers from peripheral neuropathy. She couldn't find the reliable medical information she needed.

Q: Were there other topics that you could have written about?

Sandra Smith: There were other topics, but whatever I decided on would have to interest me. One topic could have been the computer field, because that's where I work. I raise emu and ostrich, so that could have been another topic. Whatever topic I chose, it would have to be interesting and useful to other people as well as to me.

Q: How much did you know about peripheral neuropathy before you began your research?

Sandra Smith: I knew that it existed, and I knew the symptoms, and that's about it. I had heard rumors about different treatments, and I needed to learn the facts.

Q: How did you define the purpose and audience for your research project?

Sandra Smith: I wanted to find information to help my mom, but I knew that this information would also help other people. In fact, I've already had several requests for copies of the report.

Q: Do you work as well as attend college?

Sandra Smith: Yes, I work about 50 hours a week at Dell Computer Corporation as a lead for order completion on the manufacturing floor. It's a challenging job. I also commute two hours a day. While I was working on the report, I was taking four college classes, one of them a distance learning section.

Q: How did you find time to conduct research and also to balance your studies, your family, and your work?

Sandra Smith: That required planning. I started with preliminary research with Internet and CD-ROM guides to check sources first so that I knew that they would be worthwhile. I also read widely in my mom's magazines for diabetics, which were very helpful. Of course, I had to find time to care for my emu and ostrich, and I live at home with my family. I found that commuting was a great time for thinking and relaxing. I also used the time between classes and work for study and research. I'd plan and conduct my research one step at a time.

Q: What kinds of sources did you find useful for background information?

Sandra Smith: Since peripheral neuropathy is a relatively new topic, I needed an overview, which I found in materials from the Neuropathy Association. These appeared both in print and on the Internet. The diabetes magazines and publications my mom receives were very helpful for reliable, current, and research-based information about treatments and practical advice.

Q: What kinds of specific research questions did you define?

Sandra Smith: I had several specific questions. I wanted to know if peripheral neuropathy is curable, and about the relationship between blood sugar and peripheral neuropathy, the long-term effects of peripheral neuropathy, and the kinds of reliable treatment available. Of course, one research question led to related issues. For example, once I learned how blood sugar affects nerves, I realized that my audience needed to learn more about the nervous system. I used research questions to build my subject and to plan additional research.

Q: How were you able to determine which sources about this new topic were reliable?

Sandra Smith: I asked my mom's doctor about the Neuropathy Association and about other sources I'd found, especially the magazine for diabetics. He told me that these were all very reliable and current. I wanted to be very careful about locating reliable sources, especially about materials I located on the Internet.

Q: What parts of the report did you find hardest and easiest to write?

Sandra Smith: The easiest part was explaining the symptoms, which I knew firsthand. The hardest part was paraphrasing lists of suggestions for patients. These were very specifically worded, but I wanted to make sure that they fit my report's style and sequence. I had to be very careful about accuracy and clarity as I paraphrased.

Q: In what ways did your experience in researching peripheral neuropathy develop skills that you can apply to your job and to your career goals?

Sandra Smith: On the job, I locate problems, research them, develop possible solutions, and present ideas to my boss as proposals. No matter what the research problem, you need to approach it the same way. My research project has taught me to take a big problem and to break it down into small steps so that each step is approachable.

Q: What advice do you have for students who are planning research projects?

Sandra Smith: Pick a subject that matters—to you and to other people. That's number one. Once you pick your subject, you have to limit it. If you don't define your audience and research goal, you'll waste a lot of time looking up things that aren't helpful, even if they may be interesting.

That's what helped me, anyway. I could have gone on and on about peripheral neuropathy, but I wanted my readers to feel that as they read, they were getting their questions answered clearly. I didn't want my readers getting lost, bored, or confused.

As you first plan a research project, the topic needs to be connected; it needs a primary focus. Then each section of the report needs to be connected to that focus. This way, your research will be efficient and your readers will understand how the parts of your report fit together—and want to read them all.

ACTIVITIES

INDIVIDUAL AND COLLABORATIVE ACTIVITIES

1. Use Internet and print resources to find materials about a company or organization. Then use the information to write a letter to the company expressing your interest in possible employment.
2. Look in current issues of newspapers and magazines to find a story about a legal issue that concerns a specific company, college, or organization. The issue could concern affirmative action, hiring practices, or pollution. Use print and online research to learn about the background of the issue. Apply the issue to the news story in a short written report or an oral presentation.
3. Use the *Occupational Outlook Handbook* to find a career that interests you. Then use the career information you have found to write a job description to be published as an advertisement for employees. Be sure to include the necessary knowledge, skills, and experience required for the job you are advertising.
4. In groups of three or four, develop a research question on major employers in your region. As a class, discuss the effectiveness of each group's research question.

5. Use MLA and APA formats (see pages 552–556) to write a works cited entry and a references entry for the following published title:

   Katherine Staples and Cezar Ornatowski edited an anthology of essays called *Foundations for Teaching Technical Communication: Theory, Practice, and Program Review.* The book was published by Alblex, which is located in Greenwich, CT, in 1997.

6. What do you notice about the differences between the two documentation styles?
7. In her interview, student author Sandra Smith describes the usefulness of research in her career. In groups of three or four, discuss specific ways in which research can be important to your workplace and the kinds of sources you will find most helpful in conducting such research.

READING

8. The following article by Dana R. Johnson originally appeared in *Intercom*, a magazine for technical communicators. Read the selection. In small groups or as a class, respond to the questions for discussion.

Copyright Issues on the Internet

DANA R. JOHNSON

The Internet is overloaded with easily accessible information. As technical communicators, it is important for us to know how to use this information. Many people assume that everything on the Internet is public domain and can be copied at will, but much of this information is protected by copyright. Internet users should assume that a work is covered under copyright protection until it is proven otherwise.

EXPIRATION DATES

Works protected by copyright do not enter the public domain until after their copyrights have expired. The length of copyright protection depends on when the work was created.

The copyright on a work that was published before 1978 will expire seventy-five years from the date of publication. The copyright on a work published by an individual after 1978 will remain under copyright protection until fifty years after the author's death. The copyright on a work published by an employer after 1978 will remain under copyright protection until seventy-five years after publication or one hundred years after creation (whichever comes first).

An employer usually owns the rights to any work created by an employee during the course of employment.

FAIR USE

You can use copyright-protected works without the author's permission if you follow the fair use guidelines. The fair use law is covered in section 107 of the U.S. Copyright Act of 1976.

This law protects individuals who take a small portion of another's work to benefit the public. The fair use law is intended for education, research, criticism, and journalistic purposes that are nonprofit in nature. Making multiple copies of a work or distributing them to others, however, could cross the line from fair use to copyright infringement. A commercial use of the work would also be considered an infringement. There is a detailed description of the fair use law at ***http://www.loc.gov/copyright*** (select "Copyright Law," then "Chapter 1," then "107").

When downloading information off the Internet for personal use, you should first track down the author to ask for permission. If you have trouble contacting the author yourself, the Copyright Clearance Center (CCC) might be a help. The CCC Web site is ***http://www.copyright.com.***

GETTING COPYRIGHT RIGHT

A created work is automatically protected by copyright. But even though it is not required, it is important that authors register their works. Registration gives the author proof that can be used in a federal court against anyone who uses the work without permission (copyright infringement).

An author can register a work by contacting the Copyright Office of the Library of Congress. The copyright owner must submit a $20 filing fee, a registration form (Form TX or Form VA), and a deposit of the work (either a printout of each page of the work or a disk containing a copy of the whole work plus five representative pages in print). Registration forms are available from the Copyright Office Web site at ***http://www.loc.gov/copyright.***

Authors of Internet documents should also protect themselves by including a copyright notice. An example would be the following: "Copyright 1999 by Dana R. Johnson." Although not required, a copyright notice is important. The notice helps protect works from copyright infringement.

LATEST DEVELOPMENT

The Digital Millennium Copyright Act of 1998 was signed by President Clinton on October 28, 1998. This law adds Internet-specific guidelines to the Copyright Act of 1976. To find out what this new law entails, visit ***http://www.loc.gov/copyright.***

Copyright Web Sites

See the following Web sites for detailed information on copyright issues.
http://www.loc.gov/copyright
http://www.fplc.edu/tfield/copynet.htm
http://www.copyright-resources.com
http://www.cetus.org/fairindex.html
http://www.copyright.com
http://law.house.gov/325.htm

Written by Dana R. Johnson and reprinted with permission of *Intercom,* the magazine of the Society for Technical Communication. Arlington, VA, U.S.A.

QUESTIONS FOR DISCUSSION

a. Had you expected Internet publications to be protected in the same way as print publications? Why or why not?
b. Why does the copyright of a work last so long after publication? Who is protected by this law, and in what way?
c. What kinds of commercial use of copyrighted material could you imagine requesting? How would you use such information?
d. "Intellectual property," which an author can claim in publication, can take any form as long as it can be recorded in some way. What kinds of materials appear in a Web site that will not appear in a book or an essay?
e. Individual authors are not the only ones to own copyright to written and recorded materials. Businesses and organizations that pay for work can claim it as their property. What kinds of businesses own copyright on the Web? What kinds of property would businesses claim? Product names? Original artwork and graphics? Text? Music?
f. In what ways can Web resources make it easy and fast for authors to respect the published work of others who have published on the Web?
g. What kinds of legal actions and penalties would be the result of a copyright violation?
h. Copyright offers the owners of intellectual property international protection. In what ways does Web publication make it important

for authors to respect copyright, whether a piece is intended for fair use or commercial use?

i. What kinds of publications do you consider it important to protect legally through the Copyright Office? How would official copyright registration protect the owner of the copyright?

APPENDIX 1

The Search for Employment

CHAPTER GOALS

This chapter:

- Defines the employment search
- Identifies key points in self-assessment
- Lists specific ways to find job leads
- Explores ways to evaluate a job lead
- Discusses transmittal letters and their uses
- Discusses résumés, their elements and design
- Explains how to prepare for a job interview
- Explains how to write follow-up letters
- Examines the role of electronic technology in the employment search process

WHAT IS AN EMPLOYMENT SEARCH?

An employment search is the process a job seeker goes through to locate and establish a cooperative, rewarding, and successful work relationship. Unfortunately, no magic formula exists to ensure such employment. Finding the right job requires effort on your part. You must know your specific skills and goals, the kind of job you seek, acceptable locations, and ways to conduct a successful search.

An excellent way of learning about employment opportunities is talking with people who are currently employed, especially in the kind of job that interests you. Another way of learning about employment opportunities is by reading occupational guides. The *Occupational Outlook Handbook*, available in hard copy and online and published every two years by the U.S. Department of Labor, gives up-to-date information on many fields. The *Encyclopedia of Careers* is another source of basic information.

A U.S. Department of Labor pamphlet, *Tips for Finding the Right Job*, gives advice on determining your job skills and organizing your employment search. Another Department of Labor publication, *Job Search Guide: Strategies for Professionals*, outlines specific steps you can follow to identify employment opportunities and offers information on assessing your skills and interests, conducting the employment search, and networking. These publications are available at state employment service offices or from the U.S. Government Printing Office (telephone 202–512–1800 for prices and ordering information). College career planning and placement offices, as well as career resource libraries, offer counseling, testing, and employment search advice. Any bookstore or library has many of the materials discussed above plus additional information on employment opportunities and procedures. The Internet offers a wide range of free and for-cost services, including career counseling, résumé preparation, occupational profiles, and job databases.

The methods by which companies and organizations find candidates for a job are changing rapidly. In the past, the major method was reading transmittal letters and hard copy résumés. Today companies and organizations have computer-related options for finding candidates for a job: scanning hard copy résumés, establishing and searching databases, and reading electronic résumés.

As you read and study this chapter, you will learn more about the employment search process: self-assessment and career goals, finding and evaluating job leads, applying for a job, writing transmittal letters and résumés, filling out job application forms, interviewing, and writing follow-up letters.

SELF-ASSESSMENT AND CAREER GOALS

An honest self-assessment is a good way to begin an employment search. Look realistically at your skills, aptitudes, and abilities, and think through your attitudes, qualifications, and career goals. Answer questions such as:

- What are my strongest skills?
- What jobs am I qualified for?
- Why should I be hired over someone with similar qualifications?
- Do I prefer to work with people or independently?
- What are my ambitions in life?
- How can my career support my values and interests?
- What are my salary needs?
- Am I willing to relocate?

Next, determine whether you are ready for a specific job by taking advantage of state employment service counseling and testing services or by visiting career counseling and testing centers to identify your occupational aptitudes and interests. Such centers also offer help in choosing and preparing for a career.

Above all, be honest about your capabilities and do not misrepresent yourself. You cannot find what you need unless you are open with yourself and with others.

FINDING JOB LEADS

How do I find a job lead? Where do I look? Who can give me direction? Job seekers frequently ask such questions. Below are suggestions and useful approaches for finding answers.

- **Networking.** Networking means talking with people you know, such as
  - neighbors

- professionals (doctors, lawyers, teachers, office managers, business executives)
- family (immediate and extended)
- schoolmates
- fellow employees (past and present)

You can easily say, "I'm looking for a job as a computer programmer," or "I'm interested in working for WorldCom." Or you can simply indicate an interest in employment without mentioning a specific job or company. Talk to people who have jobs that sound interesting; ask them what they do, how they like their work, and how they got the job.

- **Professional organizations related to your career interest**. Many such organizations have student member rates. Once you join, you will receive publications from that organization, which often include announcements for available positions. For information about trade and professional associations for your occupation, consult reference books such as the *Encyclopedia of Associations*, the *National Directory of Employment Services*, and the *Encyclopedia of Business Information Services*.
- **College placement services**. College placement offices offer a variety of services, including counseling, testing, employment search advice, a career resource library, and workshops. Most college placement services have issues of the *College Placement Annual*, which lists occupational requirements and addresses for over a thousand employers in industry, business, and government. Many colleges and universities also host job fairs at which representatives from a number of companies are available to talk with potential employees. Often, companies will hire college students for internships or coop programs in which the student works for the company part of the year and attends school part of the year. Students who participate in these programs may find companies that will pay their college fees and then hire them full-time after graduation.
- **Private or state public employment agencies.** Private employment agencies typically charge a fee if they locate a job for you, so it is important to find out up front if such a fee is charged, how much it is, and whether you or the employer will pay it. Sometimes the agency requires you to pay a set fee or a percentage of your first month's salary. State public employment agencies have approximately 1,800 local offices to help job seekers and local employers find each other. Look in the local telephone directory in the state government section for "Job Service" or "Employment." State public employment services are free of charge.
- **Headhunting services.** Headhunters are individuals whose companies locate potential employees. The so-called headhunters try to "fit" applicants to jobs and provide guidance to applicants during the employment search process.
- **Searching online**. The Internet offers a variety of information on jobs as well as job research resources and techniques. Typically, job listings are posted by field and discipline, and often you can post your résumé online

or send it to an employer by e-mail. Some Internet services are free; others are not. You might want to try the E-Span Interactive Employment Network, which posts job listings for electronic data-processing professionals, engineering, accounting, and finance, and lists career fairs by region. Ryder Technology, a Fortune 250 company that develops high-tech logistics and transportation systems, sponsors a virtual job fair.

John Sumser, president of Internet Business Network, is quoted in *Money* magazine (May 1997) as estimating that the Internet currently lists some 1.5 million jobs, with 3 million listings projected by May 1998. This statistic indicates the growth trend of Internet job listings. Companies like electronic listing because it is fast, convenient, and inexpensive—and it also draws applicants who are typically well educated and computer literate.

Some sample sites are listed below. **Keep in mind that Internet sites change daily; existing sites are updated regularly, sites disappear, and new sites are created**. You can, however, use search engines such as WebCrawler, Yahoo!, InfoSeek, Magellan, Netscape, and AltaVista to find career-related sites. Use keywords (entry-level jobs, engineering jobs) to direct your search.

- AltaVista Careers <altavista.com> Click on "Careers." One example of a growing number of job-related services on Internet portal sites.
- America's Job Bank <http://www.ajb.dni.us/index.html> Online employment service offering information on approximately 100,000 job openings each week. Service of the U.S. Department of Labor and 1,800 state public employment agencies.
- BridgePath <www.bridgepath.com> Specializes in placing new graduates and young alumni. More than 14,000 job seekers in database. Fill out individual profile; site continually sends you job openings that match your portfolio.
- CareerMosaic <www.careermosaic.com> More than 70,000 job listings, a career resource center, customized job searches, and employer profiles.
- CareerPath.com <http://careerpath.com> Lists approximately 300,000 job listings from 60 newspapers nationwide.
- FedWorld Information Network JOBS Library <http://www.fedworld.gov>
- The Job Bank <http://www.Eajb.dni.US/about-ajb.html> Database of 250,000 jobs, including vacancies in federal agencies.
- Online Career Center <http://www.occ.com> Database of some 250,000 U.S. jobs.
- The Riley Guide <http://www.jobtrak.com/jobguide> Specializes in listings for prospective and recent college graduates.
- Purdue University Placement Service: Job Sites for Job Seekers <http://www.ups.purdue.edu/student/resource.hrm> Reference and resources for career searches. Links to information for entry level workers to professionals.

You can reach industry-specific sites through the Internet Business Network <www.interbiznet.com.eeri>, a network that links job recruiting sites for hundreds of industries ranging from engineering to sports medicine. You can also find field-specific sites for job listings in fields such as health care, computer and high-tech, human resources, criminal justice, academia, and law, and jobs abroad.

Evaluating a Job Lead

Once you identify a job lead, find out as much as you can about the company. Your investigation may lead you to conclude that you don't really want to work for that particular company. More likely, however, your investigation will provide you with useful information for your transmittal letter and résumé, your interview, and later—if you get the job.

Your analysis of the company should include the following:

- **Background.** How large is the company? How old? Who established it? Who owns it? What are the main factors or steps in its development? Is the company financially sound? What profit has the company shown for the past five years? Is the company a subsidiary of a larger operation?
- **Organization and management.** Who are the chief executives? What are the main divisions? What are the geographical locations of the company? Who in the company has the authority to hire for this particular position?
- **Product (or service).** What product does it manufacture or handle? What are its manufacturing processes? What raw materials are used and where do they come from? Who uses the product or service? Is there keen competition? How does the quality of the product or service compare with that of other companies?
- **Personnel.** How many people does the company employ? Where will I fit in? What is the rate of turnover? What is the range of skills required for the total work force? What are the company's policies concerning hiring, sick leave, vacation, overtime, retirement? What kinds of in-service training are provided? What is the salary range? Are there opportunities for advancement? How do people who work for the company feel about their jobs? About the company? Are employees frequently transferred to other geographical locations?

Obtaining answers to these and other pertinent questions may require consulting various sources, such as a local Chamber of Commerce, trade and industrial organizations, government agencies, local newspapers, public and school libraries; arranging informational interviews with company personnel; taking inspection tours of the company; reading company publications such as annual reports; and corresponding with the company.

You might also consult reference directories, available in libraries, which provide basic facts (products, services, number of employees, and so on) about companies, such as:

Dun & Bradstreet's Million Dollar Directory
Standard and Poor's Register of Corporations
Directors and Executives
Moody's Industrial Manual
Thomas Register of American Manufacturers
Ward's Business Directory

You can also learn about a company by reading articles in recent magazines and newspapers. You can find such articles by checking computerized or print indexes, such as:

Business Periodicals Index
Reader's Guide to Periodical Literature
Newspaper Index
New York Times Index
Wall Street Journal Index

Projections on employment and output for some 200 industries are printed every other year by the Bureau of Labor Statistics.

You can also visit a company's Web site by entering the company name in a search engine. Many Internet job sites have links to a company's site. In addition, sites such as **Wall Street Research Net** <www.wsrn.com>, which gives financial data and analyst opinions on some 17,000 companies, and **Business Wire** <www.business wire.com>, which gives corporate profiles on more than 600 firms and links to Dow Jones, Reuters, and 125 other news databases, offer up-to-date news coverage of companies.

REMINDER

If a job lead sounds too good to be true, it may well be. Researching a company or a job opportunity carefully may save you later disappointment or anxiety.

APPLYING FOR A JOB

Once you have identified a plausible job lead, the next step is to apply for the job. You can write a transmittal letter and a résumé to send to the company or individual who hires personnel or you can submit the résumé online. You may be required to complete an application form. The following suggestions will help you with these steps.

Transmittal Letters

Typically, you will write a transmittal letter to accompany every résumé you send and every job application form you complete, unless the form is filled out on site. The transmittal letter introduces you to potential employers and is your opportunity to convince a potential employer that you are a strong candidate for the job. The transmittal letter, typically one page long and in block or modified block letter format, has three sections:

1. **Purpose of the letter.** Why are you writing the letter? Be specific. Identify the job you are applying for, and tell how you found out about it. If you read information in a newspaper, periodical, or brochure, give the name and date; if you learned about the opportunity from an individual, give the person's name. Explain why you want the opportunity. If you have special qualifications, state them.
2. **Background information.** Select information to show that you qualify for the job you seek. Include information to convince the reader that you are the person to hire, and that your skills and knowledge will be an asset to the company that hires you. State that a résumé with more complete information is included and make the proper enclosure notation at the end of the letter.
3. **Request for an interview.** Any firm interested in employing an applicant will want a personal interview. In the closing paragraph of the transmittal letter, therefore, request an interview at the prospective employer's convenience. Should there be restrictions regarding your time, such as classes or work, say so. If distance makes an interview impractical (living in Virginia and applying for a summer job in California), suggest an alternative, such as an interview via telephone, with a local representative, or by sending a tape (audio or video) or a disk. Include information on how and when you can be reached.

Examples of Transmittal Letters

On pages 594–596 are three transmittal letters. The résumé that accompnies the first letter is on page 604.

Résumés

A résumé is an easy-to-read organized listing of details that shows you can perform productively in a specified area of employment. Through written evidence of your skills and knowledge, the résumé provides facts to motivate a potential employer to interview you. A résumé includes details about qualifications such as skills, education and training, work history, experience, personal qualities, and accomplishments. It typically shows these details in an organized list with guide words or headings to identify the content of each block of information.

Keep your résumé concise, since busy people are interested only in necessary information. Also remember that some companies want résumés prepared according to their specifications. Check each company's Web site or call its human resources department to find out if information is available for preferences on résumé preparation and submission.

The purpose of your résumé is to show you are the best candidate for a job. Therefore, select the details that clearly describe you as such a candidate. Plan your résumé—the design, the medium, and the content—accordingly. Select the type of résumé and the content and arrange the information on the

Transmittal Letter (Block Form) for an Applicant Beginning a Career

3621 Bailey Drive
Big Rapids, MI 49307
1 May 2001

Dale Garrett, Office Manager
Carter Insurance Agency
1531 Jayne Avenue
Big Rapids, MI 49307

Dear Mr. Garrett:

Through my data management instructor, Mr. Tom Lewis, I have learned that you have an opening for a data management supervisor. I believe my qualifications make me an excellent choice for the position.

On 31 May I will complete my degree in Data Management. While a student I have worked for Mr. Lewis, Director of Data Management Systems at John Williams Community College, and as a night supervisor in Midwestern Bell Telephone's data management department. Now I am eager to begin a full-time job in my field.

The enclosed résumé gives a brief outline of my work and education. In all the data management-related classes I have taken, I have been ranked as one of the top three students. As a first-year data management major, I received the award for the outstanding first-year student.

My sophomore year I served as Projects Committee Chair for the campus Phi Theta Kappa chapter, an honorary fraternity. Projects included collecting blankets for the homeless, mentoring at-risk students, and recognizing outstanding faculty. I was a member of the college choir for two years. These experiences and activities have helped me learn to work with others and to work as a team member.

I will be happy to supply any additional information about myself. I am available afternoons and Saturdays for an interview and can be reached at (579) 456-2156. I look forward to hearing from you.

Sincerely,

Alice M. Rydel

Alice M. Rydel

Enclosure: Résumé

500 Northside Drive G-1
Clinton, KS 66813
1 October 2001

Box 8249-P
c/o *The Denver Post*
1560 Broadway
Denver, CO 80202

Dear Prospective Employer:

SUBJECT: SECRETARY-ADMINISTRATIVE ASSISTANT POSITION

Your advertisement in the 25 September issue of the *Denver Post* has sparked my interest in becoming a member of your team.

My recent college courses in business plus my years of experience as a secretary qualify me for the position of secretary-administrative assistant. My business communication skills, my work with others, and my computer keyboarding skills enable me to meet any situation with confidence and success. Enclosed is my résumé, which gives details about my educational qualifications and work experience.

Could we discuss the possibility of my joining your company? I can be reached at 719-379-0022 after 3:00 any afternoon; I look forward to hearing from you to schedule an appointment.

Sincerely,

Sheri Rayburn

Sheri Rayburn

Enclosure: Résumé

Transmittal Letter (Block Form) for Applicant with Work Experience

1515 Maria Drive
Jackson, MS 39204
24 September 2001

Ms. Sara Watkins
Director of Pharmacy
Mississippi Baptist Medical Center
1500 North State Street
Jackson, MS 39206

Dear Ms. Watkins:

Through the 22 September 2001 issue of *The Clarion—Ledger,* I learned of your opening for a part-time pharmacy technician in the hospital pharmacy department. Having previously worked as a pharmacy technician, I am very familiar with the responsibilities and duties.

Enclosed is a résumé for your consideration. My education and five years of experience in a hospital pharmacy have prepared me for work as a pharmacy technician. As you will note, as a two-year college honors graduate, I worked a part-time job and participated in many extracurricular activities. In the five years I have worked as a pharmacy technician, I have consistently received performance evaluations between 4.5 and 5 (highest possible rating).

As I pursue a degree in pharmacy, I am eager to continue working part-time as a technician. I would appreciate the opportunity to speak with you in person to further discuss my qualifications, your department objectives, and the contributions I can make to your department.

I am available any afternoon for an interview. I can be reached by phone at (601) 372-5973. I look forward to hearing from you.

Sincerely,

Kristi L. Baker

Kristi L. Baker

Enclosure: Résumé

page to show your qualifications in the most positive way. You may need to prepare more than one résumé if you are seeking more than one type of job.

Information in a Résumé

Certain kinds of information—all true, of course—are expected in a résumé. Below are headings and descriptions of details typically included.

Personal Identification. Personal identification includes your name, address, and telephone number. Do not use nicknames. Other information may include your fax number, pager number, and an e-mail address, if appropriate. Examples of personal identification appear below.

Marlene Knowles

Present Address:	**Permanent Address:**
312B Mary Nelson Hall	12 Ashwood Road
Mississippi College	Florence, SC 29501
Clinton, MS 39056	803.661.6936
601.924.1234	

Tran Ngyn

124 Pike Street	(303) 576-6789 Pager
Colorado Springs, CO 80906	(303) 576-0001 Work

Marabeth Connolly
MC@cis.com

21 Bridge Avenue ≈≈ Brooklyn, NY 11217 ≈≈ 718.636.5432 ≈≈ FAX 718.636.0100

Omit personal data such as age, height, weight, health, marital status, and so on.

Objective or Target. State your immediate objective or target, if you wish.

EXAMPLES

- management position in ready-to-wear apparel
- nuclear engineer in research and development
- hospital administrator
- executive secretary
- radiologist
- research in history and future of international banking
- social worker

From this point, the information in your résumé is your choice. Usually, education and employment are prominently placed in the résumé. List the information that will make a stronger impression on the prospective employer first.

Education. Include the school name, dates of attendance or year of completion, major and minor fields of study, and highest grade completed or degree awarded. If you are still pursuing a degree, give expected graduation date. If you include several levels of education, begin with the most recent education and list in reverse chronology. Generally you can omit high school education if you have a college degree. Give your grade point average if it is 3.5 or above.

Work History or Employment. Include the company, address, dates of employment, position held, and responsibilities. If new to the job market, include volunteer work as well as paid work, part-time and temporary. If you have held a number of jobs, divide the information into sections such as employment related to the objective stated in the résumé and other jobs. You may include only the jobs that relate directly to the objective. Begin with current or most recent employment and list in reverse chronology.

Just listing when and where you worked tells little, however. In addition to dates and job titles, include your specific responsibilities. Military service may be listed as a job; give dates of service, your duty assignment, and your rank at discharge. Here also give specific information about your responsibilities.

Acknowledge gaps in time in the listing of your employment history. For example, you may have left the workforce for a period of time to enter college, to travel for a year, or to care for a family member.

Skills, Capabilities, and Accomplishments. In one or more sections, list any special skills under headings such as Software Used, Research Capabilities, and Personal Achievements. You can also use headings such as Management, Communications, Graphic Design, Research, Writing, Consulting, Sales, Editing, and so on. Include types of software you have used or equipment you handle proficiently and the tasks for which you have used them.

Training, Activities, Memberships, Awards, or Other Specialties. If appropriate for the objective of the résumé, include blocks of information listing training, activities, memberships, awards, or other specialties. For example, list foreign language skills or experience, and school, community, or social activities that demonstrate leadership. Be as accurate as possible. Do not misrepresent your qualifications or your experience.

References. There is no rule for including or omitting references. If you are asked to supply references, select three or four individuals (not relatives), who can speak knowledgeably about your qualifications and character. If you have little or no job experience, you might select former teachers or community leaders. Give the name, occupation, business address, and telephone number of each individual. If you prepare a one-page résumé, you may not have room for references. An option is to state "References supplied upon request." Remember to obtain permission to list a person's name as a reference.

Arranging and Designing Your Résumé

A résumé follows no single standard form. Once you have jotted down facts under the selected topics, you can decide which headings to use and how to arrange the information. You can arrange the sections of your résumé in ways that are most appropriate for the job you are seeking.

Following the personal information, which always goes at the top, select headings and information about your capabilities, accomplishments, education, and work experience that specifically relate to the job you hope to be offered. Most first-time job seekers use a traditional or chronological résumé, which typically begins with Education or Work Experience. Arrange the headings so that the most impressive details appear near the top of the page.

If you are an experienced job-seeker, you may choose a functional or skills résumé, which emphasizes the functions you can perform or skills you have acquired. You might use headings such as Communication, Supervision, Budget, Machines, Personnel, Research, Management, Sales, Project Development, and the like in such a résumé. Following the headings, give a summary of activities that helped you to become competent at performing a function or developing a skill. You may also list specific activities related to each heading. If applicable, give work experience that demonstrates your proficiency in these skills.

Whatever format you choose, pay close attention to the design. The overall appearance of a résumé is critical; it affects the reader's initial response. It could well determine whether your résumé is rejected or considered.

You can create a positive overall appearance by following these guidelines.

- If you use well designed software or a template to create your résumé, you can save time and easily revise your résumé.
- Keep your résumé to one page, unless you have extensive, relevant experience.
- Frame the text with white space at the top, bottom, and sides and between the sections, balancing the text and white space on the page.
- Choose a typeface that is easy to read and use a point size of 10 or 12.
- Use headings.
- Avoid mixing more than two typefaces and three type styles.
- Choose good quality paper in a neutral color such as gray, white, ivory, or a pale beige.
- Check the text for accuracy of spelling and grammar. Do not rely solely on spell check and grammar check features on software. Proofread carefully.
- Print on a laser printer.

Résumé Forms and Style

Résumé style is affected by several factors, including form—will the résumé be print or electronic? Another factor is the arrangement and emphasis of material. Below is a discussion of four résumé forms: print, scannable, online or electronic, and multimedia.

Print Résumés. The print résumé is still the most common form for most job seekers. The print résumé contains blocks of data for work history, education, skills, and capabilities. Within each block of data, the facts are shown in chronological order, beginning with the current or most recent and working back. The print résumé emphasizes abilities through strong action verbs (managed, designed, supervised).

A print résumé should be output from a laser printer on high-quality paper. It can be sent to potential employers through traditional mail delivery, e-mail, or fax.

When you need a résumé, preparing a print résumé first may be a useful approach. This résumé helps you to present your data clearly and forcefully. You can then rearrange the data for other résumé forms as you need them.

Scannable Résumés. As companies receive more and more résumés and as employment search technologies emerge, personnel departments are increasingly scanning résumés to create databases of readily available data on potential employees.

The scannable résumé requires a change in emphasis and in the placement of information on the page. The scannable résumé requires key nouns (manager, designer, supervisor, fluid dynamics, total quality management), even technical jargon, that specifically relate to the career field and the job. Companies search through electronic résumés and job banks, scanning résumés in the database, looking for keywords. Applicants should also consider including a summary statement of their qualifications, being careful to include key nouns.

To prepare a scannable résumé, follow these guidelines:

- Select standard size white or light-colored paper.
- Choose an easily readable font and size, such as 10 or 12 point Times or Helvetica.
- Use a one-column format.
- Use keywords, emphasizing nouns.
- Do not use underlining, italics, graphics, or bold.
- To highlight areas, use asterisks and capital letters.
- Do not fold the résumé. A fold line may cause a misreading when scanned.
- Send originals, produced on a laser printer, not copies.
- Include a separate line of keywords at the bottom of the page.
- Use white space generously. Scanners use it to distinguish sections.
- Keep the résumé simple, clear, and concise, but do not worry about keeping the information to a single page.

Electronic and Online Résumés. Electronic résumés are becoming increasingly popular as companies use the Internet to post jobs and to search for potential employees. Electronic résumés contain essentially the same data as do traditional résumés. A major difference, however, is the use of nouns as keywords, rather than action verbs. Reading job ads helps you to identify keywords used for job classifications that match your qualifications and interests.

Many online services allow you to post your résumé online. Monster Board <http://www.monster.com>, for example, has a database of résumés and allows you to post your résumé using their online résumé form or your own. For a minimal fee, Job Center <http://www.jobcenter.com>, another service, will maintain and try to match your résumé to job offerings for six months. America Online's Career Center has a "virtual agent"—a script to track the kind of job you seek and match it with job offerings.

ResumeMaker Deluxe, a multimedia CD-ROM from Individual Software, helps you post your résumé to the databases of major online résumé banks and search listings of online employment services by location, title, and salary.

A useful Internet connection is JobDirect <www.jobdirect.com>. You can enter your résumé, JobDirect then cross-references your qualifications with current positions listed by participating employers and e-mails the details of each match. You can apply online. The site is updated every 24 hours. Another site is Resumail Network <www.jobs.98.com>. This site downloads its résumé software to you. You can fill it out and upload your résumé to a database of 900 employers advertising jobs on the site. You can also check the jobs yourself.

Be sure a site is reputable. Also, you may want to check the level of confidentiality: Is your permission required before your résumé is released?

Multimedia Résumés. The multimedia résumé is designed to showcase samples of the sender's work, and may include media such as paper, disks, or audio- and videotapes. The multimedia résumé is useful in applying for jobs in which artistic or creative ability is a key component. For example, a person applying for a job as a newspaper photographer submits a portfolio of journalistic photography. A person applying for a job as a rock band guitarist includes sample tapes or CDs of guitar work. If applying for a job with an advertising agency, the applicant submits a portfolio of art work, perhaps on interactive disk with multimedia components. An architect includes sample designs.

Increasingly, job applicants are making their résumés interactive as well as multimedia. Some résumés blend sound, graphics, animation, and text. The résumé with multimedia components, however, is not for everyone. You need a good reason to use multimedia components, and the potential employer or person who receives the résumé must have the equipment and the technical knowledge to review the résumé. The medium for submitting the message is not the most important aspect; the content must be impressive.

Help for Preparing Résumés

Many books on résumés are available in libraries and bookstores, and more information is available electronically. You may find that such resources provide invaluable information for preparing your résumé.

The following Web sites may be helpful:

- Intellimatch Power Resume Builder
  <http://www.intellimatch.com/watson/owaw3.logon>
- Job Bank USA Resume Resources
  <http://www.interbiznet.com/hunt/tools.html>
- The Job Smart Resume Guide
  <http://www.jobsmart.org/tools/résumé/idex.htm>
- Résumé form from Resume Builder on the Monster Listing Service
  <http://www.monster.com/>
- Resumix Creating Your Resume
  <http://www.resumix.com/resume/resumeindex.html>
- The Riley Guide
  <http://www.jobtrak.com/jobguide/index.html>
- Student Center
  <http://www.studentcenter.com>

Résumé templates are also available on the Web and on most word processing software. These templates vary from simple types to more complex patterns. Using Resumix, for example, you enter data on a form, press a button, and see an instantly formatted résumé. Another template, Intellimatch, provides a series of categories to guide you in writing a résumé. Some software packages, such as *Microsoft Word* and *WordPerfect,* for example, provide résumé templates. However, you typically cannot modify the format of these templates.

Résumé writing services will draft a résumé for you and place it directly into company Web sites, usually for less than $100. Some recent versions of word-processing software such as *Microsoft Office 97* and *Corel WordPerfect9* will convert a document to html code for posting on the Web.

Examples of Résumés

On pages 603–606 are four examples of résumés: a traditional chronological résumé, combined functional and chronological résumé, a skills résumé, and a scannable résumé. The combined functional and chronological résumé, page 604, accompanies the transmittal letter on page 594.

Job Application Forms

Many companies provide printed forms for job applicants. The form is actually a detailed data sheet. In completing the form, follow directions, be neat, and answer every question. Some companies suggest that applicants put a dash (-), a zero (0), or NA (not applicable) after a question that does not pertain to them. Doing so shows that you have read the question and not overlooked it. The completed forms of many applicants look very similar. Therefore, it is the accompanying transmittal letter that gives you an opportunity to make the application stand out.

JAMES E. BROWN, JR.
206 Davis Drive
Monroe, LA 71201

Home telephone: (318) 948-7660
JB276@MONROEVM.BITNET

EMPLOYMENT OBJECTIVE

Supervising technicians and engineers in an electronics industry or business with possibility for full management responsibilities

WORK EXPERIENCE

June 1998–Sept. 2001 United States Air Force, Electronics/Communications

- Shift Chief, Long-Haul Transmitter Site. Supervised three technicians in operating 52 transmitters and two microwave systems.
- Team Chief, Group Electronics Engineering Installation Agency. Supervised three technicians in installing weather and communications equipment.
- Technical Writer. Wrote detailed maintenance procedures for electronic equipment manuals.
- Instructor in Electronics Fundamentals. Taught continuous three-month classes of 10 persons.

Summers and Part-Time 1994–1998

- Manager, Campus Apartments
- Disc jockey, radio stations KXOA and KXRQ
- Laboratory Assistant, Broadcast Department
- Technician for theater productions

EDUCATION

B.S. degree in Industrial Management, Louisiana State University, May 1998

CERTIFICATIONS

- NARTE-Certified Electronics Technician
- Top Secret Clearance for work in U.S. Department of Defense

COLLEGE HONORS AND ACTIVITIES

- Member, Society for the Advancement of Management Professionals
- Vice-President, Student Audubon Society
- Dean's List
- Beta Gamma Sigma National Honor Society in Business Administration
- Associated Students Service and Leadership Award
- Listed in *Who's Who in American Colleges and Universities*

Résumé (Combined Functional and Chronological) for an Applicant with Experience as a Part-Time Employee

ALICE M. RYDEL
3621 Bailey Drive
Big Rapids, MI 49307
(601) 456-2156

CAPABILITIES

- Use MacIntosh and IBM computers
- Analyze data to prepare organized financial statistics
- Use Lotus and other spreadsheet programs
- Use automated accounting system to produce monthly statements
- Provide software training
- Manage a staff of twelve
- Keep accurate records of large numbers of accounts

ACCOMPLISHMENTS

- Supervised daily data input in a 12,000-customer billing department.
- Set up a simple but efficient file system for record keeping.
- Managed a computer lab available to 200 students.
- Devised a plan to schedule students for maximum lab use.
- Handled inventory of computer lab and submitted requests for materials, equipment, and maintenance.

EDUCATION

1999-2001 — John Williams Community College
AAS Degree, Computer Programming

1995-1999 — Murrah High School
Diploma, college preparatory and basic business courses

WORK EXPERIENCE

1999-2001
Part-time

John Williams Community College
Big Rapids, MI
Computer Lab Assistant. Responsibilities: Supervised 15 labs per week. Scheduled student lab use. Identified needs and submitted requests for materials, equipment, and maintenance

Midwestern Bell Telephone Company, Billing Department
Big Rapids, MI
Night Supervisor. Responsibilities: Supervised night staff. Kept account records.

HONORS AND AWARDS from John Williams Community College

- Data Management Department Award
- Outstanding First-Year Student in Programming
- Citation for excellence in keyboarding skills

LEADERSHIP ACTIVITIES

Phi Theta Kappa Honorary, Projects Committee Chair
College Choir Tour Schedule Committee Chair

A Skills Résumé

THOMAS D. DAVIS

tddavis@AOL.COM

1045 Drake Place Ellisville, MA 01047 521.363.2371

Employment Objective: Fabrication Superintendent with an opportunity to use my experience and skills in supervision, communication, and fabrication department budgeting.

SKILLS

SUPERVISION:	Managed 20 machinists. Planned work assignments based on priorities. Found solutions to shop production problems.
COMMUNICATION:	Communicated orders to machinists. Wrote monthly reports, such as department reports to an immediate supervisor and reports on budget variances to budget control. Wrote daily reports, such as reports on discrepancies in product conformity.
PERSONNEL:	Interviewed for new personnel. Carried out performance evaluations for raises and promotions.
BUDGET:	Planned budget for a half-million-dollar department budget. Monitored budget spending.
MACHINE SKILLS:	Operated all common machine shop tools, such as lathes, milling machines, grinding machines. Proficient in using related measuring tools and gauges.
EMPLOYMENT:	
1995-Present	Barron Enterprises, Engineering Division, Nye, MA 01047 Fabrication Superintendent, Processing Supervisor
1990-1995	Always Fabrication, Inc., Patterson, IN 47312. Quality Control Checker, Parts Inspector, Layout and Design Assistant
1988-1990	Bickman Manufacturing Company, Cain, IN 47315. Assembly line worker
1986-1988	U.S. Navy, Machinist
EDUCATION:	Cain Community College, A.A., Mechanical Technology
SPECIAL TRAINING:	Indiana Technical Institute • Quality Control with Computers (45 clock hours) • New Materials in Industry (45 clock hours) • Production Planning and Problems (45 clock hours)

A Scannable Résumé

JOHN SMITH
789 Walnut Street
Little Rock, AR 72204
501.569.1001
jsmith3540@aol.com

SUMMARY: Strong background in computer systems, languages, hardware, and software. Advanced experience in graphic design. Knowledge of print production. Basic knowledge of spoken and written Spanish. Excellent communication skills. Experience working as a team member.

SOFTWARE, SYSTEMS, AND LANGUAGE SKILLS: Experience with software QuarkXPress, Windows 95, Microsoft Word, Adobe Photoshop, EXCEL, Aldus Pagemaker. Experience with systems Apple, IBM, and UNIX. Experience with programming languages C++, Cobol, and HTML.

EXPERIENCE: Web page developer, designer, script writer, and programmer. Designed over 30 sites to improve sales for clients. Artist for graphics on a CD package for Lone Hollow, a bluegrass band. Computer lab manager, orienting students on software to be used in class, writing guides for software use, and assisting approximately 160 students daily with PC machines and software.

EMPLOYMENT: Graphic Designer, Laird Public Relations, Dallas, TX, 1/97 to present. Assistant Graphic Designer, The Graphics Corporation, White Settlement, TX, 7/96 to 12/97. Computer Lab Manager, Hinds Community College, Raymond, MS, 8/93 to 6/94.

EDUCATION: University of North Carolina at Greensboro, Greensboro, NC 27412-5001, Graphic Design with a Computer major, Spanish minor, BS degree, 1996. Hinds Community College, Raymond, MS 39154-9799, Commercial Design Technology, AS degree, 1994.

RECOGNITIONS AND AWARDS: Honors Scholarship to Hinds Community College and to University of North Carolina at Greensboro; Outstanding First-Year Graphic Design Technology Student, 1993; 3.6 cumulative grade point average; study abroad one semester in England, Spring 1992; Mexican Exchange Program.

KEYWORDS: graphic designer, programmer, QuarkXPress, Windows 95, Microsoft Word, Adobe Photoshop, EXCEL, Aldus Pagemaker, Apple, IBM, and UNIX, C++, Cobol, HTML, Web page developer, Web page designer, Web page script writer, Spanish speaker and writer, computer lab manager, computer user manual author.

Whether an applicant gets the job is usually a direct result of the interview. Certainly, the information in the transmittal letter and résumé is important, but the *person* behind that information is the real focus. The personal circumstance of the job interview allows the applicant to be more than written data. The impression that the applicant makes is often the deciding factor in getting a job.

Preparing for an Interview

Prepare carefully for an interview. Doing so can reduce your anxiety and nervousness. In preparing for a job interview, analyze yourself. Answer questions such as:

- Why did I apply for this job?
- Do I really want this job?
- Have I applied for jobs with other companies?
- What do I consider my primary qualifications for this job?
- Why should I be hired over someone else with similar qualifications?
- What do I consider my greatest accomplishment?
- Can I take criticism?
- (If you have held other positions) Why did I leave other jobs?
- Do I prefer to work with people or with objects?
- How do I spend my leisure time?
- What are my ambitions?
- What are my salary needs?

Part of preparing for a job interview involves learning as much as you can about the company. Learning something about the person who will interview you also is well worth the effort. Of course, the communication that gained you an interview provided some information about the interviewer, but more information would relieve undue anxiety and help to smooth the way and establish rapport. Just as the salesperson analyzes a prospective customer's wants, needs, and interests before making a big sales effort, so you should analyze the person who will interview you.

Rehearsing for an Interview

With the help of friends, practice an interview. Set up an office situation with desk and chairs. Ask a friend to assume the role of the interviewer and to ask you questions about yourself and questions about the information on your résumé. Ask your friends to comment honestly on the rehearsal. Then swap roles. If possible, tape the interview and play it back for critical analysis. Another way to rehearse is by yourself, in front of a full-length mirror. Dress in the attire you will wear for the interview. Study yourself impartially; go over aloud the points you plan to discuss in the interview. Keep your gestures and facial expressions appropriate. Student Center <http://www.studentcenter.com> offers a virtual interview.

Ordinarily, you cannot know exactly how an interview will be conducted. Thus it is impossible to be prepared for every situation that can arise. It is possible, however, to become acquainted with the usual procedure so that you can more easily adapt to the particular situation. Rehearsals can help you gain self-confidence and organize your thoughts.

Certain rules of etiquette apply in a job interview. For example, dressing appropriately is extremely important. Do not overdress or underdress. Good business manners, if not common sense, demand that you arrive on time, be pleasant and friendly but businesslike, avoid annoying actions (such as chewing gum, tapping your feet, or playing with your jewelry), let the interviewer take the lead, and listen attentively.

Following is a list of job interview tips from the *Occupational Outlook Handbook,* which will be helpful as you plan for and participate in a job interview.

PREPARATION

- Learn about the organization.
- Have a specific job or jobs in mind.
- Review your qualifications for the job.
- Prepare answers to broad questions about yourself.
- Review your résumé.
- Practice an interview with a friend or relative.
- Arrive before the scheduled time of your interview.

PERSONAL APPEARANCE

- Be well groomed.
- Dress appropriately.
- Do not chew gum or smoke.

THE INTERVIEW

- Answer each question concisely.
- Respond promptly.
- Use good manners. Learn the name of your interviewer and shake hands as you meet.
- Use proper English and avoid slang.
- Be cooperative and enthusiastic.
- Ask questions about the position and the organization.
- Thank the interviewer, and follow up with a letter.

TEST (IF EMPLOYER GIVES ONE)

- Listen closely to instructions.
- Read each question carefully.
- Write legibly and clearly.
- Budget your time wisely and don't dwell on one question.

INFORMATION TO BRING TO AN INTERVIEW

- Social Security number.
- Driver's license number.
- Résumé. Although not all employers require applicants to bring a résumé, you should be able to furnish the interviewer with information about your education, training, and previous employment.
- Usually an employer requires three references. Get permission from people before using their names, and make sure they will give you a good reference. Try to avoid using relatives. For each reference, provide the following information: name, address, telephone number, and job title.

Participating in an Interview

During most of the interview, the interviewer will ask questions to which you respond. Your responses should be frank, brief, and to the point, yet complete. Following is a list of common job interview questions.

COMMON JOB INTERVIEW QUESTIONS

1. What are your long-range career objectives? How will you achieve them?
2. What are your other goals?
3. What would you like to be doing in five years? Ten years?
4. What rewards do you expect from your career?
5. What do you think your salary will be in five years?
6. Why did you choose this career?
7. Which would you prefer: excellent pay or job satisfaction?
8. Describe yourself.
9. How would your best friend describe you?
10. What makes you put forth your greatest effort?
11. How has your education prepared you for this career?
12. Why should I hire you?
13. Tell me your specific qualifications that make you the best candidate for this job.
14. Define success.
15. What do you think it takes to be successful in our company?

16. How can you contribute to our organization?
17. Describe the ideal supervisor-subordinate relationship.
18. What is your proudest accomplishment? Why?
19. If you were hiring a person for this position, what qualities would you look for in an applicant?
20. Do your grades reflect your scholastic achievement?
21. What subject was your favorite? Why?
22. What subject did you dislike? Why?
23. Do you have plans for additional education?
24. What have you learned from your extracurricular activities?
25. Do you work well under pressure?
26. Will you accept part-time employment or job sharing? Why/why not?
27. Tell me what you know about this organization.
28. Why do you want to work for us?
29. What criteria are you using to evaluate this company as a potential employer?
30. Would you relocate if necessary? If not, why not?
31. Are you willing to travel?
32. What major problem have you encountered? How did you resolve it?
33. What have you learned from your mistakes?
34. Describe your greatest strength.
35. Explain your greatest weakness.

Be honest in discussing your qualifications, neither exaggerating nor minimizing them. The interviewer's questions and comments can help you determine the type of employee sought; then you can emphasize your suitable qualifications. For instance, if you apply for a business position and the interviewer mentions that the job requires some customer contact, present qualifications that show you have dealt with many people. That is, emphasize any work, experience, or courses that pertain to direct contact work.

Do not expect all questions to be typical. You should also prepare to answer more difficult questions and describe problem situations as a part of a behavioral interview. Employers want to know how well you organize projects, lead teams, and get along with fellow employees or demanding supervisors, in addition to your technical know-how. You may be asked to discuss problem situations, such as:

- If you were a part of a group that did not want to do something, how would you try to persuade the group?
- Describe how you have handled a situation in which you and a colleague disagreed.
- Explain how you handled a stressful situation.

Nina Munk and Suzanne Oliver writing in *Forbes* (24 March 1997) say:

> The gentle interview is dead. . . . These days companies ask applicants to solve brain teasers and riddles, create art out of paper bags, spend a day acting as managers of make-believe companies and solve complex business problems.

They relate an instance when an interviewer went outside and looked at an applicant's car, reasoning that if the car were clean and neat the employee's work would be clean and neat. They tell of a consulting firm that found the best measure of success is a person's ability to manage a complex schedule; this firm looks favorably at résumés whose writers maintain good grades while holding part-time jobs and participating in extracurricular activities.

To prepare for such questions and analyses, think about ways you can best present your skills and abilities. Think about jobs you have held and your responsibilities at those jobs. Plan illustrative stories to show how you performed. If you have several stories planned, you can likely adapt them to any interview question. These illustrative stories can relate to nonjob experience. For example, a student who was asked to describe a situation in which she showed leadership skills explained how she organized a recycling program on campus. She explained how she handled the publicity, made arrangements for recycling bins to be placed on campus, assigned student representatives to buildings, and delegated responsibility to other students who were responsible for getting the recycled materials to an appropriate buyer and planning how to use the money from the sale.

Employers are not interested in hearing, "I have the best leadership skills in my department." Rather, they want you to show that you are the best by detailing specific experiences. Career counselors suggest using the STAR method to answer questions fully and completely.

S Situation
T Task
A Action
R Result

To answer questions about your qualifications and abilities, first briefly describe situations, the tasks the situations required, your actions, and the results. If you have done a careful company and job analysis, you can use the STAR method to show how your skills, based on prior education and experience, fit the job's requirements.

In the course of the interview, you will likely be asked if you have any questions. Don't be afraid to ask what you need to know concerning the company or the job and its duties (such as employee insurance programs, vacation policy, overtime work, travel).

Watch the interviewer for clues that it is time to end the interview. Express appreciation for the time and courtesies given you, say good-bye, and leave. Lingering or prolonging the interview is usually an annoyance to the interviewer.

At the close of the interview, you may be told whether you have the job; or the interviewer may tell you that a decision will be made within a few days. If the interviewer does not definitely offer you a job or indicate when you will be informed about the job, ask when you may telephone to learn the decision.

Following Up an Interview

Following up an interview reflects good manners and good business. Whether the follow-up is in the form of a telephone call, a letter, or another interview usually depends on the hiring decision. If you got the job, a telephone call or a letter can serve as confirmation of the offer, its responsibilities, and the time for reporting to work. You can express appreciation for the opportunity to become connected with the company. If you did not get the job, thank the interviewer by telephone or by letter for his or her time and courtesy. This kind of goodwill is essential to business success.

If a final decision has not yet been reached concerning your employment, a favorable decision may well hinge on a wisely executed follow-up. If the interview ended with "Keep in touch with us" or "Check with us in a few days," you may want to request another interview. A telephone call or a letter would be less effective, for neither is as forceful in maintaining the good impression that has been made. If the interview ended with "We'll keep your application on file" or "We may need a person with your qualifications a little later," follow up the lead. After a few days, write a letter. Mention the interview, express appreciation, include any additional credentials or emphasize credentials that since the interview seem to be particularly significant, and state your continued interest in the position.

Follow-Up Letters

Frequently, after applying for a job (writing and sending a transmittal letter and résumé, filling out and submitting a job application form) or completing an interview, you will write a follow-up letter. The content of the follow-up letter is determined by events that occur after you apply for a job or complete an interview. The letter is simply a courteous response to events following the application or interview.

Follow-up letters may be written for these purposes:

- To request a response from the prospective employer. You would write this kind of letter if you were promised a response by a certain date but have not received it or if you have sent a résumé for a job you hope will be available. An example of a follow-up letter requesting a response is shown on page 613.
- To thank the interviewer, remind the person of your qualifications, and indicate a desire for a positive response. See the body of the thank–you letter on page 614.
- To reaffirm the acceptance of a job and your appreciation for the offer. The letter may also ask questions that you have thought of since getting the job, mention the date you will begin work, and make a statement about looking forward to working with the company. See the body of the acceptance letter on page 614.
- To refuse a job offer. See the body of the refusal letter on page 615.

2121 Oak Street
Fort Collins, CO 80521
July 19, 2001

Mr. William Hatton, Personnel Director
World Bank
20 East 53rd Street
New York, NY 10022

Dear Mr. Hatton:

On June 10, I was interviewed by Mr. John Salman, your representative, for a place in the World Bank's training program. Mr. Salman told me that I would be notified about my application by July 1. Although it is the middle of July, I have not received any response.

Since I must make certain decisions by August 1, could I please hear from you about my employment possibilities with World Bank?

I am, of course, eager to become an employee of World Bank and hope that I will receive a positive response.

Sincerely,

Jayne Mannos

Jayne Mannos

Body of a Sample Thank–You Letter

The interview with you on Wednesday, June 10, was indeed a pleasant, informative experience. Thank you for making me feel at ease.

After hearing the details about the World Bank's training program and opportunities available to persons who complete the program, I am eager to be an employee of World Bank. My experience working with People's Bank and my associate degree in Banking and Finance Technology provide a sound background for me as a trainee.

I look forward to hearing from you that I have been accepted in the World Bank's employee training program.

Body of a Sample Acceptance Letter

Thank you for your offer to hire me as consulting engineer for Wanner, Clare, and Layshock, Inc. I eagerly accept the offer.

The confidence you expressed in my abilities to help the firm improve workers' safety certainly motivates me to be as effective as possible.

I look forward to beginning work on Tuesday, July 10, ready to demonstrate that I deserve the confidence expressed in me.

Thank you for offering me a place in the World Bank's training program. The opportunities available upon completion are enticing.

Since my interview with you, however, the president of People's Bank, where I have worked part-time while completing my associate degree in banking and finance, has offered me permanent employment. The bank will pay for my continued schooling, allow me to work during the summers, and guarantee me full-time employment upon completing my advanced degree.

By accepting this offer from People's Bank, I can live near my aging parents and help care for them.

I appreciate your interest in me and wish World Bank continued success.

ACTIVITIES

A1.1. Read about your career choice or one of interest in the *Occupational Outlook Handbook*, the *Encyclopedia of Careers*, and other such sources. Find answers to the following:

a. What is the nature of the work?
b. What skills are required?
c. What education is required?
d. What personality traits are desirable?
e. What advancement opportunities are available?
f. What is the beginning salary?
g. What are the projections for job opportunities?
h. Where are jobs available?

Write a brief report of your findings.

A1.2. Using the Internet, locate three jobs that you would qualify to apply for when you have completed your formal education. Write a brief analysis of each job, including advantages and disadvantages of each.

A1.3. Prepare a traditional résumé and a scannable résumé based on your present qualifications.

A1.4. Write a transmittal letter to accompany a résumé.

A1.5. Find a job application form (the form used by the U.S. government is especially thorough) and fill out the form neatly and completely. It is usually wise to write out the information on a separate sheet of paper and then transfer it to the application form.

A1.6. Arrange for an interview with your program adviser, a school counselor, an employment counselor, or some other person knowledgeable in your major field of study. Try to determine the availability of jobs, pay scale, job requirements, promotion opportunities, and other pertinent information for the type of work you are preparing for. Present your findings in an oral report.

A1.7. Write a follow-up letter in which you accept a job offer.

A1.8. Write a follow-up letter in which you refuse a job offer.

APPENDIX 2

The Search for Standard English Usage

How to Use This Appendix

From time to time, writers need answers to questions about writing conventions. Is *affect* or *effect* the right word in this sentence? Do commas go inside or outside of quotation marks? Is it acceptable to use abbreviations in writing—and which ones? And what did the peer reviewer mean about a *dangling modifier* on page 3? This appendix is designed to help you find quick answers to such questions. To help you look up information quickly, the appendix is divided into two sections: Standard English Writing Conventions and Common Problems in Using Standard English.

Standard English Writing Conventions is a quick reference guide to the patterns, rules, and conventions you'll need to use in college and on the job. It discusses capitalization, abbreviations, use of numbers, punctuation, spelling and common symbols. In other words, this chapter of the appendix tells you what you need in order to write correctly. The contents of the chapter Standard English Writing Conventions are listed alphabetically on page 620.

Common Problems in Using Standard English provides a quick means of reviewing the terms and patterns that give writers problems. It discusses parts of speech, common errors in grammar, typical problems with word choice and style, and words commonly confused or misused. The contents of the section Common Problems in Using Standard English are listed on page 658.

To help you become familiar with both parts of the appendix, a brief list of topics from each chapter appears below.

The appendix is easy to read as well as easy to use.

Standard English Writing Conventions: Topics

- Abbreviations
- Capitalization
- Numbers
- Plurals of Nouns
- Punctuation
- Spelling
- Symbols

Common Problems in Using Standard English: Topics

- Adjectives
- Adverbs
- Comma Splice
- Conjunctions
- Parallelism in Sentences
- Placement of Modifiers
- Preposition at End of Sentence
- Pronouns
- Pronoun and Antecedent Agreement
- Run-on, or Fused, Sentence
- Sentence Fragments
- Sexist Language
- Shift in Sentence Focus

- Shifts in Person, Number, and Gender
- Subject and Verb Agreement
- Verbs
- Words Often Confused and Misused

Standard English Writing Conventions

Introduction

> Conventions en: Stnd. writen english: Suche as konsistant punctuaion speling and abbrevns are Important too workplace author's and workplace reader's after all? without such consistancieReeders notice the authors arrors and lackof writting ability not thee authors' ideas. . . . readers' sld n't have too fite Writing too get the Infirmation They kneed.

> Conventions in standard written English—such as consistent punctuation, spelling, and abbreviations—are important to workplace authors and workplace readers. After all, without such consistency, readers notice the author's errors and lack of writing ability, not the author's ideas. Readers should not have to fight writing to get the information they need.

Which is easier to understand—the first or the second paragraph? Undoubtedly the second, because it follows generally accepted practices in the conventions of written English. The second paragraph permits you to concentrate on what is being said. This paragraph is thoughtful of you, the reader, because it uses easily recognized writing conventions so that you can pay immediate attention to ideas, not waste time puzzling over the inconsistent way in which the ideas are written. The consistent use of written conventions, or mechanics, is a matter of convention and of courtesy to readers.

The following are accepted practices concerning abbreviations, capitalization, numbers, plurals of nouns, punctuation, spelling, and symbols. Applying these practices to your writing will help your reader understand what you are trying to communicate. Each item of discussion is listed alphabetically to help you use this section for reference.

Abbreviations

Always consult a recent dictionary for abbreviation forms you are not sure about. Some dictionaries list abbreviations together in a special section; other dictionaries list abbreviated forms as regular entries.

Always acceptable abbreviations

1. Abbreviations generally indicate informality. Nevertheless, there are a few abbreviations always acceptable when used to specify a time or a person, such as a.m. (ante meridiem, before noon), BC (before Christ), AD (anno Domini, in the year of our Lord), BCE (before the Common [or Christian] Era), CE (Common [or Christian] Era), Ms. (combined form of Miss and Mrs.), Mrs. (mistress), Mr. (mister), Dr. (doctor). (See item 6 below.)

   EXAMPLE

   Dr. Ann Meyer and Mrs. James Brown will arrive at 7:30 p.m.

Titles following names

2. Certain titles following a person's name may be abbreviated, such as Jr. (Junior), Sr. (Senior), MD (Doctor of Medicine), SJ (Society of Jesus).

EXAMPLES

Martin Luther King, Jr., was assassinated on April 4, 1968. (The commas before and after *Jr.* are optional.)

George B. Schimmet, MD, signed the report.

Titles preceding names

3. Most titles may be abbreviated when they precede a person's full name, but not when they precede only the last name.

EXAMPLES

Doctor Sykes   Dr. Melba Sykes

Terms with numerals

4. Abbreviate certain terms only when they are used with a numeral.

EXAMPLES

a.m., p.m., B.C., A.D., No. (number), $ (dollars)

Careful writers place *B.C.* after the numeral and *A.D.* before the numeral (for example, 325 B.C., A.D. 597).

ACCEPTABLE

He arrives at 2:30 p.m.
Julius Caesar was killed in 44 B.C.
The book costs $12.40.

UNACCEPTABLE

He arrives this p.m.

Repeated term or title (acronyms)

5. To avoid repeating a term or a title intrusively, abbreviate the term or shorten the title. Write out in full the term or title the first time it appears, following it with the abbreviated or shortened form in parentheses; thereafter, use only the shortened form.

EXAMPLES

Many college students take the Graduate Record Exam (GRE) in their senior year; however, some prefer to take the GRE earlier for practice.

Dynamics of Document Design (*Dynamics*) by Karen A. Schriver is a valuable and current discussion of the subject. Both students and professional communicators praise the usefulness of *Dynamics* in supporting document design decisions.

Periods with abbreviations

6. Generally, place a period after each abbreviation. However, there are many exceptions:

EXCEPTIONS

The abbreviations of organizations, of governmental divisions, and of educational degrees usually do not require periods (or spacing between the letters of the abbreviation).

EXAMPLES

IBM (International Business Machines), DECA (Distributive Education Clubs of America), IRS (Internal Revenue Service), AAS (Associate in Applied Science) degree

Postal Service abbreviations

The U.S. Postal Service abbreviations for states are written without periods and in all uppercase (capital) letters. For a complete list of these abbreviations, see page 435 in Chapter 15 Correspondence.

Roman numerals

Roman numerals in sentences are written without periods.

EXAMPLE

Henry VIII didn't let enemies stand in his way.

Units of measure

Units of measure, with the exception of in. (inch) and at. wt. (atomic weight), are written without periods. See item 9 below.

EXAMPLE

He weighs 186 lb and stands 6 ft tall.

Plural terms

7. Abbreviations of plural terms are written in various ways. Add *s* to some abbreviations to indicate more than one; others do not require the *s*.

Abbreviations adding **s**

EXAMPLES

Figs. (or Figures) 1 and 2
20 vols. (volumes)

Abbreviations without **s**

EXAMPLES

pp. (pages)
ff. (and following)

Lowercase letters

8. Generally use lowercase (noncapital) letters for abbreviations, except for abbreviations of proper nouns.

EXAMPLES

mpg (miles per gallon) Btu (British thermal unit)
c.d. (cash discount) UN (United Nations)

Units of measure as symbols

9. Increasingly, the designations for units of measure are being regarded as symbols rather than as abbreviations. As symbols, the designations have only one form regardless of whether the meaning is singular or plural, and are written without a period.

EXAMPLES

100 kph (kilometers per hour)
50 m (meters)

Capitalization

There are few absolute rules concerning capitalization, and many businesses and publishers develop their own standard rules for capitalization. However, the following are widely accepted capitalization conventions.

Sentences

1. Capitalize the first word of a sentence or a group of words understood as a sentence (except a short parenthetical statement within another).

   EXAMPLE

   After the meeting nobody offered to help rearrange the board room. Not one person.

Quotations

2. Capitalize the first word of a direct quotation.

   EXAMPLE

   Melissa replied, "Tomorrow I begin."

Proper nouns

3. Capitalize proper nouns.

People

Names of people and titles referring to specific persons.

EXAMPLES

Frank Lloyd Wright	Mr. Secretary
Aunt Marian	the Governor

Places

Places (geographic locations, streets), but not directions.

EXAMPLES

Canada	Golden Gate Bridge
Canal Street	the Smoky Mountains
the South	the Red River

Go three blocks south; then turn west.

Groups

Nationalities, organizations, institutions, and members of each.

EXAMPLES

Nigerian	Westminster College
Brazilian	League of Women Voters
A Rotarian	International Imports, Inc.

Calendar divisions

Days of the week, months of the year, and special days, but not seasons of the year.

EXAMPLES

Monday	Halloween	New Year's Day
January	spring	summer

Historic occurrences

Historic events, periods, and documents.

EXAMPLES

World War II	the Industrial Revolution
the Magna Carta	Battle of San Juan Hill

Religions

Religions and religious groups.

EXAMPLES

Judaism the United Methodist Church

Deity

Names of the Deity and personal pronouns referring to the Deity.

EXAMPLES

God	Son of God
Creator	His, Him, Thee, Thy, Thine

Bible

Bible, Scripture, and names of the books of the Bible. (These words are not italicized [underlined] in text.)

EXAMPLE

My favorite book in the Bible is Psalms.

Proper noun derivatives

4. Capitalize derivatives of proper nouns when used in their original sense.

EXAMPLES

Chinese citizen	*but* china pattern
Salk vaccine	*but* pasteurized milk

Pronoun I

5. Capitalize the pronoun *I*.

EXAMPLE

In that moment of fear, I could not say a word.

Titles of publications

6. Capitalize titles of books, chapters in books, magazines, newspapers, articles, essays, poems, plays, stories, musical compositions, paintings, motion pictures, and the like. Ordinarily, do not capitalize articles (*a, an, the*), coordinate conjunctions (*and, or, but*), and prepositions (*by, in, with*) unless they are the first words of the title. It is acceptable to capitalize prepositions of five or more letters (such as *between* or *against*).

EXAMPLES

Omni (magazine)
"Tips on Cutting Firewood" (article in a periodical)
The Phantom of the Opera (musical)

Titles with names

7. Capitalize titles immediately preceding or following proper names.

EXAMPLE

Juan Perez, Professor of Computer Science

Substitute names

8. Capitalize words or titles used in place of the names of particular individuals. However, names denoting kinship are not capitalized when immediately preceded by an article or a possessive.

EXAMPLES

Last week Mother and my grandmother gave a party for Sis.
Jill's dad and her uncle went hunting with Father and Uncle Bob.

Trade names

9. Capitalize trade names.

EXAMPLES

Xerox Hershey bars

Certain words with numerals

10. Capitalize the words Figure, Number, Table, and the like (whether written out or abbreviated) when used with a numeral.

EXAMPLES

See Figure 1. See the accompanying figure.
This is Invoice No. 6143. Check the number on the invoice.

Academic subjects

11. Capitalize academic subjects only if derived from proper nouns (such as those naming a language or a nationality) or if followed by a numeral.

EXAMPLES

English	Spanish
history	History 1113
algebra	Algebra 11

Numbers

Numbers often appear in the text of workplace writing. However, they present a special problem for authors: when to use numerals and when to use words. Below are accepted conventions that provide practical guidelines for addressing the problem.

Numerals (Used as Figures)

Dates, houses, telephone numbers, zip codes, specific amounts, mathematical expressions

1. Use numerals for dates, house numbers, telephone numbers, zip codes, specific amounts, mathematical expressions, and the like.

EXAMPLES

July 30, 1987 or 30 July 1987	857-5969
600 Race Street	10:30 p.m.
Chapter 12, p. 14	61 percent

Decimals

2. Use numerals for numbers expressed in decimals. Include a zero before the decimal point in writing fractions with no whole number (integer).

EXAMPLES

12.006
0.01
0.500 (The zeros following 5 show that accuracy exists to the third decimal place.)

Number 10 and above; three or more words

3. Use numerals for number 10 and above or numbers that require three or more written words.

EXAMPLES

She sold 12 cars in 2 1/2 hours.

Several numbers close together

4. Use numerals for several numbers (including fractions) that occur within a sentence or within related sentences.

EXAMPLES

The recipe calls for 3 cups of sugar, 1/2 teaspoon of salt, 2 sticks of butter, and 1/4 cup of cocoa.
The report for this week shows that our office received 127 telephone calls, 200 letters, 30 personal visits, and 3 faxes.

Adjacent numbers

5. Use numerals for one of two numbers occurring next to each other for clarity in reading; the other number is spelled out (see item 5 in next section).

EXAMPLES

two hundred 8 by 10 picture frames
12 fifty-gallon containers

Numerals (Spelled Out as Words)

Approximate or indefinite numbers

1. Use words for numbers that are approximate or indefinite.

EXAMPLES

two hundred 24 × 36 mats
If I had a million dollars, I'd buy a castle in Ireland.
About five hundred machines were returned because of faulty assembly.

Fractions

2. Use words for fractions.

EXAMPLES

The veneer is one-eighth of an inch thick.
Our club receives three-fourths of the general appropriation.

Below 10

3. Use words for numbers below 10.

EXAMPLE

There are four quarts in a gallon.

Beginning of sentence

4. Use words for a number or related numbers that begin a sentence.

EXAMPLES

Fifty cents is a fair entrance fee.
Sixty percent of the freshmen and seventy percent of the sophomores come from this area.

Note: If using words for a number at the beginning of a sentence is awkward, recast the sentence.

UNACCEPTABLE — 2,175 freshmen are enrolled this semester.

AWKWARD — Two thousand, one hundred and seventy-five freshmen are enrolled this semester.

ACCEPTABLE — This semester 2,175 freshmen are enrolled.

Adjacent numbers

5. Use words for one of two numbers occurring next to each other.

EXAMPLES

50 six-cylinder vehicles
four 3600-pound loads

Repeating a number

6. Except in special instances (such as in order letters or legal documents), it is not necessary to repeat a written-out number by giving the numeral in parentheses.

ACCEPTABLE — The trumpet was invented five years ago.

UNNECESSARY — The trumpet was invented five (5) years ago.

PLURALS OF NOUNS

The English language has been greatly influenced by a number of other languages, including Latin, Greek, and French. Some English nouns still retain plural forms from the original language. Below are patterns of irregular plurals for such nouns.

Most nouns: **-s** or **–es**

1. Most nouns in the English language form their plural by adding *–s* or *–es*. Add *–s* unless the plural adds a syllable when the singular noun ends in *s, ch* (soft), *sh, x,* and *z*.

EXAMPLES

pencil	pencils	mass	masses
desk	desks	church	churches
flower	flowers	leash	leashes
boy	boys	fox	foxes
post	posts	buzz	buzzes

Nouns ending in **y**

2. If a noun ends in *y* preceded by a consonant sound, change the *y* to *i* and add *-es*.

EXAMPLES

history	histories
penny	pennies

For other nouns ending in *y*, add *–s*.

EXAMPLES

monkey	monkeys
valley	valleys

Nouns ending in **f** or **fe**

3. A few nouns ending in *f* or *fe* change the *f* to *v* and add *–es* to form the plural.

EXAMPLES

calf	calves	life	lives	shelf	shelves
elf	elves	loaf	loaves	thief	thieves
half	halves	leaf	leaves	wife	wives
knife	knives	sheaf	sheaves	wolf	wolves

In addition, several nouns may either add *–s* or change the *f* to *v* and add *–es*.

EXAMPLES

beef beefs (slang for "complaints") *or* beeves
scarf scarfs *or* scarves
staff staffs (groups of officers) staffs *or* staves (poles or rods)
wharf wharfs *or* wharves

Nouns ending in **o**

4. Most nouns ending in *o* add *–s* to form the plural. Among the exceptions, which add *–es*, are the following:

EXAMPLES

echo	echoes	potato	potatoes
hero	heroes	tomato	tomatoes
mosquito	mosquitoes	veto	vetoes

Compound nouns

5. Most compound nouns form the plural with a final *–s* or *–es*. A few compounds pluralize by changing the operational part of the compound noun.

EXAMPLES

handful	handfuls	son-in-law	sons-in-law (and other in-law compounds)
go-between	go-betweens		
good-by	good-bys	passer-by	passers-by
court-martial	courts-martial	editor in chief	editors in chief

Internal vowel change

6. A few nouns form the plural by an internal vowel change.

EXAMPLES

foot	feet	mouse	mice
goose	geese	tooth	teeth
louse	lice	woman	women
man	men		

-en plurals

7. A few nouns form the plural by adding *–en* or *–ren.*

EXAMPLES

ox	oxen	
child	children	
brother	brothers	*also* brethren

Foreign plurals

8. Several hundred English nouns, originally foreign, have two acceptable plural forms: the original form and the conventional American English *–s* or *–es* form.

EXAMPLES

memorandum	memoranda *or* memorandums
curriculum	curricula *or* curriculums
index	indices *or* indexes
criterion	criteria *or* criterions

Some foreign nouns always keep their original forms in the plural, as in the following words:

EXAMPLES

crisis	crises	die	dice
analysis	analyses	alumna	alumnae (feminine)
alumnus	alumni	thesis	theses
basis	bases	fabliau	fabliaux

Terms, dates, and titles being discussed

9. Letters of the alphabet, signs, symbols, and words used as a topic of discussion form the plural by adding the apostrophe and *-s*.

EXAMPLES

The i's and e's are not clear.
The sentence has too many and's and but's.

Note: Abbreviations and numbers form the plural regularly, that is, by adding *–s* or *–es*.

EXAMPLES

The three PhDs were born in the 1950s.
The number has two sixes.

Same singular and plural forms

10. Some nouns have the same form in both the singular and the plural. In general, names of fish and of game birds are included in this group.

EXAMPLES

cod	swine	sheep
trout	cattle	quail
deer	species	corps

Two forms

11. Some nouns have two forms, the singular indicating oneness or a mass, and the plural indicating different individuals or varieties within a group.

EXAMPLES

a string of fish	four little fishes
a pocketful of money	money (*or* monies) appropriated by Congress
fresh fruit	fruits from Central America

Only plural forms

12. Some nouns have only plural forms. A noun is considered singular, however, if the meaning is singular.

EXAMPLES

measles (Measles is a contagious disease.)	mumps
economics	news

mathematics	physics
dynamics	molasses

A noun is considered plural if the meaning is plural.

EXAMPLE

scissors (The scissors are sharp).

Punctuation

Consistent punctuation is an important part of the written language. After all, marks of punctuation allow readers to follow and understand sentences and the connections between their parts. Punctuation can indicate pauses and stops, separate and set off sentence elements, indicate questions and exclamations, and emphasize main points while subordinating less important ones.

The following discussion of punctuation is presented in groups: marks used primarily at the end of a sentence (period, question mark, and exclamation point); internal marks that set off and separate (comma, semicolon, colon, dash, and virgule); enclosing marks always used in pairs (quotation marks, parentheses, and brackets); and marks used in the punctuation of individual words and of terms (apostrophe, ellipsis points, hyphen, and italics).

Punctuation Marks Used Primarily at the End of a Sentence

Period (.)

Statement, command, or request

1. Use a period at the end of a sentence (and of words understood as a sentence) that makes a statement, gives a command, or makes a request (except a short parenthetical sentence within another).

EXAMPLES

James Naismith invented the game of basketball. (statement)
Choose a book for me. (command)
No. (understood as a sentence)
Naismith (he was a physical education instructor) wanted to provide indoor exercise and competition for students. (parenthetical sentence: no capital, no period)

See also Sentence Fragments, pages 668–669.

Note: The polite request phrased as a question is followed by a period rather than a question mark.

EXAMPLE

Will you please send me a copy of your latest sales catalog.

Initials and abbreviations

2. Use a period after initials and most abbreviations.

EXAMPLES

Dr. H. H. Wright	p. 31
437 mi.	no. 7

The abbreviations of organizations, or governmental divisions, and of educational degrees usually omit periods.

EXAMPLES

BA (Bachelor of Arts) degree
NFL (National Football League)
FBI (Federal Bureau of Investigation)

The U.S. Postal Service abbreviations for states (see page 435 in Chapter 15. Correspondence for a list of the abbreviations), contractions, parts of names used as a whole, roman numerals in sentences, and units of measure with the exception of in. (inch) and at. wt. (atomic weight) are written without periods. (Designations for units of measure are increasingly being regarded as symbols rather than as abbreviations. See item 9 under Abbreviations, pages 623–624.)

EXAMPLES

Hal C. Johnson IV, who lives 50 miles away in Sacramento, CA, cannot attend.

See also Abbreviations, item 6, pages 622–623.

Outline

3. Use a period after each number and letter symbol in an outline.

EXAMPLES

I.

  A.

  B.

For another example, see page 357.

Decimals

4. Use a period to mark decimals.

EXAMPLES

$10.52
A reading of 1.260 indicates a full charge in a battery; 1.190, a half charge.

Question Mark (?)

Questions

1. Use a question mark at the end of every direct question, including a short parenthetical question within another sentence.

EXAMPLES

Have the blood tests been completed?
When you return (When will that be?), please bring travel reports.

Note: An indirect question is followed by a period.

EXAMPLE

He asked if John were present.

Note: A polite request in question form is usually followed by a period.

EXAMPLE

Will you please close the door.

Uncertainty

2. Use a question mark in parentheses to indicate a question about certainty or accuracy.

EXAMPLE

Chaucer 1343(?)–1400
The spindle should revolve at a slow(?) speed.

Exclamation Point (!)

Sudden or strong emotion or surprise (*Not generally used in workplace writing*)

1. Use an exclamation point after words, phrases, or sentences (including parenthetical expressions) to show sudden or strong emotion or force, or to mark the writer's surprise.

EXAMPLES

What a day!
The computer (!) made a mistake.

Note: The exclamation point can be easily overused, thus causing it to lose its force. Avoid using the exclamation point in place of vivid, specific description.

Internal Marks that Set Off and Separate

Comma (,)

Items in a series

1. Use a comma to separate items in a series. The items may be words, phrases, or clauses.

EXAMPLES

How much do you spend each month for food, housing, clothing, and transportation?

Compound sentence

2. Use a comma to separate independent clauses joined by a coordinate conjunction (*and, but, or, neither, nor,* and sometimes *for, so, yet*). (An independent clause is a group of related words that have a subject and verb and that could stand alone as a sentence).

EXAMPLE

There was a time when few women occupied prominent roles, but today they are leaders in local and national affairs and in many organizations.

Note: Omission of the coordinate conjunction results in a comma splice (comma fault), that is, a comma incorrectly splicing together independent clauses. (See Comma Splice, pages 660–661.) If the clauses are short, the comma may be omitted.

EXAMPLE

I aimed and I fired.

See also Semicolon, item 1, page 639.

Equal adjectives

3. Use a comma to separate two adjectives of equal emphasis and with the same relationship to the noun modified.

EXAMPLE

The philanthropist made a generous, unexpected gift to our college.

Note: If *and* can be substituted for the comma or if the order of the adjectives can be reversed without violating meaning, the adjectives are of equal rank and a comma is needed.

Misreading

4. Use a comma to prevent misreading.

EXAMPLES

Besides Sharon, Ann is the only available programmer.
Ever since, he has met his deadlines.

Number units

5. Use a comma to separate units in a number of four or more digits (except telephone numbers, zip code numbers, house numbers, and the like).

EXAMPLES

2,560,781 7,868 *or* 7868

Note: The comma may be omitted from four-digit numbers.

Nonrestrictive modifiers

6. Use commas to set off nonrestrictive modifiers, that is, modifiers which do not limit or change the basic meaning of the sentence.

EXAMPLES

The Guggenheim Museum, designed by Frank Lloyd Wright, is in New York City. (nonrestrictive modifier; commas needed)

A museum designed by Frank Lloyd Wright is in New York City. (restrictive modifier; no comma needed)

Introductory adverb clause

7. Use a comma to set off an adverb clause at the beginning of a sentence. An adverb clause at the end of a sentence is not set off.

EXAMPLES

When you complete these requirements, you will be eligible for the award.
If provided with the right mentoring and education, a line worker can become an effective supervisor.
A line worker can become an effective supervisor if provided with the right mentoring and education.

Introductory verbal modifier

8. Use a comma to set off a verbal modifier (participle or infinitive) at the beginning of a sentence.

EXAMPLES

Experimenting in the laboratory, Sir Alexander Fleming discovered penicillin. (participle)
To understand the continents better, researchers must investigate the oceans. (infinitive)

Appositives

9. Use commas to set off an appositive. (An appositive is a noun or pronoun that follows another noun or pronoun and renames or explains it.)

EXAMPLES

Joseph Priestley, a theologian and scientist, discovered oxygen.
Opium painkillers, such as heroin and morphine, are narcotics.

Note: The commas are usually omitted if the appositive is a proper noun or is closely connected with the word it explains.

EXAMPLES

My associate Mary lives in Phoenix.
The word *occurred* is often misspelled.

Parenthetical expressions, conjunctive adverbs

10. Use commas to set off a parenthetical expression or a conjunctive adverb.

EXAMPLES

A doughnut, for example, has more calories than an apple.
I was late; however, I did not miss the plane.

(See also Semicolon, item 2, page 639.)

Address, dates

11. Use commas to set off each item after the first in an address or a date.

EXAMPLE

My address will be 1045 Carpenter Street, Columbia, Missouri 65201, after today. (House number and street are considered one item; state and zip code are considered one item.)

EXAMPLE

Thomas Jefferson died on July 4, 1826, at Monticello. (Month and day are considered one item.)

Note: If the day precedes the month or if the day is not given, omit the commas.

EXAMPLE

5 July 2001

Quotations

12. Use commas to set off the quoted words (or similar matter) in a direct quotation.

EXAMPLE

"I am going," Mary responded.

Person or thing addressed

13. Use commas to set off the name of the person or thing addressed.

EXAMPLES

If you can, Ms. Yater, I would prefer that you attend the meeting.
Your appearance has improved since last quarter, Will.

Mild interjections

14. Use commas to set off mild interjections, such as *well, yes, no, oh.*

EXAMPLES

Oh, this is satisfactory.
Yes, I agree.

Inverted name

15. Use commas in an inverted name to set off a surname from a given name.

EXAMPLE

Adams, Lucius C., is the first name on the list.

Title after a name

16. Use commas to set off a title following a name. (Setting off *Junior* or *Senior,* or their abbreviations, following a name is optional.)

EXAMPLES

Patton A. Houlihan, President of Irish Imports, is here.
Gregory McPhail, DDS, and Harvey D. Lott, DVM, were classmates.

Contrasting elements

17. Use commas to set off contrasting elements.

EXAMPLES

The harder we work, the sooner we will finish.
Leif Eriksson, not Columbus, discovered the North American continent.

Elliptical clause

18. Use a comma to indicate understood words in an elliptical clause.

EXAMPLE

Sally was elected president; Ralph, vice president.

Inc., Ltd.

19. Use commas to set off the abbreviation for incorporated or limited from a company name. (Newer companies tend to omit the commas.)

EXAMPLES

Drake Enterprises, Inc., is our major competitor.
I believe that Harrells, Ltd., will answer our request.

Introductory elements

20. Use a comma to introduce a word, phrase, or clause.

EXAMPLES

After the presentation, the discussion was lively.
I told myself, you can do this if you really want to.

(See also Colon, pages 640–641.)

Before the conjunction **for**

21. Use a comma to precede *for* when used as a conjunction to prevent ambiguity.

EXAMPLES

The plant is closed, for the employees are on strike.
The plant has been closed for a week. (no comma; *for* is a preposition, not a conjunction.)

Correspondence

22. Use a comma to follow the salutation and complimentary close in a social letter and usually the complimentary close in a business letter.

EXAMPLES

Dear Mother,	Sincerely,
Dear Lynne,	Yours truly,

Note: The salutation in a business letter is usually followed by a colon; a comma may be used if the writer and the recipient know each other well. Some newer business letter formats omit all punctuation following the salutation and the complimentary close.

Tag question

23. Use commas to set off a tag question (such as *will you, won't you, can you*) from the remainder of the sentence.

EXAMPLE

You will conduct the meeting, won't you?

Absolute phrase

24. Use commas to set off an absolute phrase. An absolute phrase (also called a nominative absolute) consists usually of a participial phrase plus a subject of the participle and has no grammatical connection with the clause to which it is attached.

EXAMPLE

The review completed, we developed the final version of the product.

Semicolon (;)

Independent clauses, no coordinate conjunction

1. Use a semicolon to separate independent clauses not joined by a coordinate conjunction (*and, but, or*).

EXAMPLES

Germany has a number of well-known universities; several of them have been in existence since the Middle Ages.
There are four principal blood types; the most common are O and A.

Note 1: If a comma is used instead of the needed semicolon, the mispunctuation is called a comma splice, or comma fault (see pages 660–661).

Note 2: If the semicolon is omitted, the result is a run-on, or fused, sentence (see pages 667–668).

Short, emphatic clauses may be separated by commas.
I came, I saw, I conquered.

Independent clauses, transitional connectives

2. Use a semicolon to separate independent clauses joined by a transitional connective. Transitional connectives include conjunctive adverbs, such as *also, moreover, nonetheless, then, thus,* and explanatory expressions, such as *for example, in fact, on the other hand.*

EXAMPLES

We have studied the organizational culture of the company; thus, we can consider its international growth more intelligently.
During the Manhattan Project, several famous physicists worked in New Mexico; for example, Oppenheimer and Reiser both worked there.

See also Comma, item 10, pages 636–637.

Certain independent clauses, coordinate conjunction

3. Use a semicolon to separate two independent clauses joined by a coordinate conjunction when the clauses contain internal punctuation or when the clauses are long.

EXAMPLES

The office needs better lighting, a workstation, and a conference table; however, our budget allows for none of these things.
Students in professional majors are usually eager to take electives in areas outside their fields of study; but these students rarely have the time to pursue academic interests outside their majors.

Items in a series

4. Use a semicolon to separate items in a series containing internal punctuation.

EXAMPLES

The three major cities on our sales trip are London, Ontario, Canada; Washington, DC, USA; and Tegucigalpa, Honduras, Central America.
The new officers are Annette Orbison, President; Cochrane Samuels, Vice-President; and Anthony Daniels, Treasurer.

Examples

5. Use a semicolon to separate an independent clause containing a list of examples from the preceding independent clause when the list is introduced by *that is, for example, for instance,* or a similar expression.

EXAMPLES

Many creative thinkers had to overcome severe problems; for example, John Milton and James Joyce had sight problems.

Colon (:)

List of series

1. Use a colon to introduce a list or series of items. An expression such as *the following* or *as follows* often precedes the list.

EXAMPLES

A learner comes to understand responsibility in three ways: by example, by instruction, and by experience.
The principal natural fibers used in the production of textile fabrics are as follows: cotton, wool, silk, and linen.

See also Comma, item 20, page 638.

Explanatory

2. Use a colon to introduce a clause that explains, reinforces, or gives an example of a preceding clause or expression.

EXAMPLE

Until recently, American industry used the European system of linear measure as standard: the common unit of length was the inch.

Emphatic appositive

3. Use a colon to direct attention to an emphatic appositive.

EXAMPLES

We have overlooked the obvious problem: cost.
That leaves one question: when will production begin?

Quotation

4. Use a colon to introduce a long or formal quotation.

EXAMPLE

My argument is based on John Meredith's words: "The attitudes, gestures, and movements of the human body are laughable in exact proportion as that body reminds us of a machine."

Formal greeting

5. Use a colon to follow the formal salutation of a business letter.

EXAMPLES

Dear Ms. Boxwood:
Greenway, Inc.:
Sales Department:

See also Comma, item 22, pages 638–639.

Relationships

6. Use a colon to indicate such relationships as volume and page, ratio, and time.

EXAMPLES

42:81–90 (volume 42, pages 81–90)
x:y
3:1
Genesis 4:8
2:50 a.m.

Dash (—)

The dash generally indicates emphasis or a sudden break in thought. Often the dash is interchangeable with a less strong mark of punctuation. Use a dash for emphasis. Otherwise, use an alternate mark of punctuation (usually a comma, colon, or parenthesis). No spacing should appear before or after the dash inside the sentence.

Sudden change

1. Use a dash to mark a sudden break or shift in thought.

EXAMPLE

The product is—but I'll let you find out for yourself.

Appositive series

2. Use a dash to set off a series of appositives.

EXAMPLE

Three major factors—cost, fabric, and fit—influence the purchase of clothing.

Summarizing clause

3. Use a dash to separate a summarizing clause from a series.

EXAMPLE

Productivity, materials, and overhead—these influenced our net earnings last quarter.

Note: The summarizing clause usually begins with *this, that, these, those,* or *such*.

Emphasis

4. Use a dash to set off material for emphasis.

EXAMPLE

My manager—who is also the Union Director—will conduct the meeting.
Flowery phrases—no matter what the intent—have no place in workplace prose.
I want only one thing from my contract—prompt payment.

Note: Writers often have choices among dashes, parentheses, and commas. Dashes emphasize; parentheses subordinate; and commas imply additional information.

Sign of omission

5. Use a dash to indicate the omission of words or letters.

EXAMPLE

The only letters we have been able to decipher in the old letter are —a—l—z.

Virgule (or "Slash") (/)

Alternatives

1. Use a virgule to identify appropriate alternatives.

EXAMPLE

Identify/define the problems with the test results.

Poetry

2. Use a virgule to separate run-on lines of poetry. For readability, leave a space before and after the virgule.

EXAMPLE

"Friends, Romans, countrymen, lend me your ears. / I come to bury Caesar, not to Praise him."
(Shakespeare, *Julius Caesar*)

Per

3. Use a virgule to represent *per* in abbreviations.

EXAMPLES

12 ft/sec 260 mi/hr

Time

4. Use a single virgule to separate divisions of a period of time.

EXAMPLE

Fiscal year 2000/01

Enclosing Marks Used in Pairs

Quotation Marks (" ")

Direct quotations

1. Use quotation marks to enclose every direct quotation.

EXAMPLE

The American Heritage Dictionary defines *dulcimer* as "a musical instrument with wire strings of graduated lengths stretched over a sound box, played with two padded hammers or by plucking."

Note: Quotations of more than one paragraph have quotation marks at the beginning of each paragraph and at the end of the last paragraph. Long quotations, however, are usually set off by indentation, which eliminates the need for quotation marks.

Titles

2. Use quotation marks to enclose the titles of journal and magazine articles, songs, poems, titles of broadcasted programs, and speeches.

EXAMPLE

Included in the collection is Miller's essay "What's Technical about Technical Writing?"

Note: Titles of serial publications (journals, magazines, or newspapers), books, dramatic presentations, films, ships, trains, and aircraft are italicized (or underlined in handwriting). See item 1 under Italics, page 649.

Different usage level

3. Use quotation marks to distinguish words on a different usage level.

EXAMPLE

The technician described the condition of the inspection site as "totally messed up."

Note: Use quotations to highlight different usage levels sparingly. Since this use of quotation marks can offend readers by seeming to apologize for a word or phrase, omit the quotation marks or select a different phrasing. Do not use quotation marks for emphasis.

Nicknames

4. Use quotation marks to enclose nicknames.

EXAMPLE

Edwin “Happy” Smythe is the current project manager.

Single quotation marks

5. Use single quotation marks (‘ ’) for a quotation within a quotation.

EXAMPLE

“I was concerned,” Officer Markham testified, “when the suspect said that he has ‘had strong words and altercations many times in the past’ with the decedent.”

Own title not quoted

6. Do not enclose the title of your own writing projects in quotation marks (unless your title is a quotation).

EXAMPLE

How to Sharpen a Drill Bit

Quotations with other marks of punctuation

7. Use quotation marks correctly with other marks of punctuation.

Period or comma

The closing quotation mark follows a period or comma.

EXAMPLE

“Enter the remaining data before May 1,” the director said, “in order to meet our new deadlines.”

Colon or semicolon

The closing quotation mark precedes the colon or semicolon.

EXAMPLE

I have just finished D. C. Leonard’s “The Web, the Millenium, and the Digital Evolution of Distance Education”; it is a comprehensive and useful essay.

Question mark, exclamation point, dash

The closing quotation marks precede the question mark, exclamation point, or dash when one of these marks punctuates a sentence of which quoted material is a part. The closing quotation mark follows the question mark, exclamation point, or dash when one of these marks punctuates the quoted material.

EXAMPLES

Did he agree to add what our legal representative calls “the relevant documentation”?
Ms. Smythe asked, “Did you receive the e-mail?”

Parentheses ( )

Additional information

1. Use parentheses to enclose additional information (such as definitions, references, or comments) not directly related to the sentence.

EXAMPLES

The Society for Technical Communication (STC) supports the needs and interests of technical writers and editors.

The qualifying examination (a challenge for graduates) is held at the end of the junior year.

Note: Writers can use parentheses, dashes, and commas to enclose information. Dashes emphasize, parentheses subordinate, and commas imply additional information.

Itemizing

2. Use parentheses to enclose numbers or letters that mark items in a list.

EXAMPLE

Research tells us that engineers write in order to (1) transact in-house business, (2) report findings and propose new projects, and (3) communicate with customers and distant collaborators.

References to other pages and sources

3. Use parentheses to enclose references that direct readers to charts, references, and pages.

EXAMPLE

The current life expectancy for males in the United States is 75 (see Figure 3).

4. Capitalization and punctuation with parentheses.

Capitalization

Do not capitalize the first word of a sentence parenthetically enclosed in another sentence.

EXAMPLE

The table shows the allowable loads on each beam in kips (a *kip* is a measure of 1000 pounds).

Period

Omit the period used as end punctuation in a complete sentence parenthetically enclosed in another sentence.

EXAMPLE

The Old North Church (the name is now Christ Church) is the oldest remaining house of worship in Boston.

Question mark, exclamation point

Use a question mark or exclamation point if parenthetically enclosed information requires it.

EXAMPLE

Pour the foundation below the frost line (what is the frost line depth in this region?) in order to assure a stable foundation.

Separate sentence

If the information enclosed in parentheses is a separate sentence, place the end punctuation inside the closing parenthesis.

EXAMPLE

The existing heating system is adequate for ordinary building uses. (Infrared heaters are available for spotheating.)

Brackets ([ ])

Parentheses within parentheses

1. Use brackets for parentheses within parentheses.

EXAMPLE

Susan M. Jones (a graduate of Normal State [now Illinois State University]) was a winner in this year's technical report competition.

Insertion in quotations

2. Use brackets to insert comments or clarifications inside quotations.

EXAMPLES

"Functional [automotive] design involves efficient and safe operation and pleasing form."
"This decision [Miller v. Adams] governs the sentencing of juvenile offenders."

Sic in quotation

3. Use brackets to enclose the Latin word *sic* (which means *so* or *thus*) to indicate an unusual usage or an error in spelling or grammar in the original quotation.

EXAMPLE

The report stated: "sixty drivers had there [sic] licenses revoked for repeat offenses."

Documentation of sources

4. Use brackets to supply missing or unverified sources in documentation.

EXAMPLE

James D. Arnett, *The History of Proposal Writing.* [Boston: McGuire Institute.] 1997.

Marks Used in the Punctuation of Individual Words and of Terms

Apostrophe (')

In contractions

1. Use an apostrophe to take the place of a letter or letters in a contraction.

EXAMPLES

I'm	(I am)
can't	(cannot)

Singular possessive

2. Use an apostrophe to show the possessive form of singular nouns and of indefinite pronouns.

EXAMPLES

citizen's responsibility	Ron's car
somebody's laptop	everyone's concern

Note: To form the possessive of a singular noun, add the apostrophe + *s*.

EXAMPLES

doctor + ' + s = doctor's (the doctor's advice)
Einstein + ' + s = Einstein's (Einstein's research)

Note: The *s* may be omitted in names ending in *s* or *z*, especially if the name has two or more syllables.

EXAMPLES

James's (*or* James') book
Mr. Gomez's (*or* Mr. Gomez') investigation

Note: Personal pronouns (*his, her, their, our*) do not need an apostrophe because they are already possessive in form.

Plural possessive

3. Use the apostrophe to show the possessive form of plural nouns.

EXAMPLE

Technicians' reports (reports of the technicians)

Note: To show plural possessive, first form the plural. If the plural noun ends in *s*, add only an apostrophe. If the plural does not end in an *s*, add an apostrophe + *s*.

Certain plurals

4. Use the apostrophe to form the plural of letters used as words.

EXAMPLE

Try to use fewer and's, and eliminate the I's from the report.

Ellipsis Points (. . .)

Omission inside quotations

1. Use ellipsis points (plural form: ellipses) to indicate that words have been left out of a quotation. Three dots show that words have been left out at the beginning or inside of a quoted sentence. Four dots shows an omission at the end of a quoted sentence. (The fourth dot is the end punctuation of the sentence.)

EXAMPLES

"The average American family spent about $7,000 on food . . . in 1998."

"The adoption of standard time in North America stems from the railroad's search for a means to regularize chaotic schedules. . . . In November, 1883, rail companies agreed to establish zones for each 15 degrees of longitude, with uniform time throughout each zone."

Hyphen (-)

Word division

1. Use a hyphen to separate parts of a word divided at the end of a line. Divide words only between syllables. A careful writer avoids dividing a word if the division creates confusion for readers.

Compound numbers; fractions

2. Use a hyphen to separate parts of compound numbers and fractions written out in words.

EXAMPLES

seventy-four people   twenty-two vehicles
one-eighth of an inch   one-sixteenth-inch thickness

Compound adjectives

3. Use a hyphen to separate parts of compound adjectives when they precede the word they modify

EXAMPLES

an eighteenth-century paper mill   40-hour week

Compound nouns

4. Use a hyphen to separate parts of many compound nouns.

EXAMPLES

brother-in-law   U-turn
kilowatt-hour   proof-of-purchase

Note: Many compound nouns (such as notebook and blueprint) are written as single words. Others (such as card table and steam engine) are written as two words without a hyphen. If you are unsure about hyphenation, look up the term in a dictionary.

Compound verbs

5. Use a hyphen to separate parts of a compound verb.

EXAMPLES

brake-test   oven-temper

Numbers or dates

6. Use a hyphen to separate parts of inclusive numbers or dates.

EXAMPLES

pages 73-79   the years 1997-99

Prefixes

7. Use a hyphen to separate parts of some words whose prefix is separated from the word's main stem.

EXAMPLES

ex-president	self-respect
semi-retired	pre-Glasnost

Italics (Underlining)

Italics (*like these words*) are used in print. The equivalent in typescript or handwriting is underlining.

Titles

1. Italicize (or underline) titles of books, serial publications (such as journals and newspapers), long dramatic works, motion pictures, ships, trains, and aircraft.

EXAMPLE

I reviewed the index for *Science* and *Scientific American.*

Note: Use quotation marks to enclose titles of articles, book chapters, songs, broadcasts, and speeches. (See item 2 under Quotation Marks.)

Note: Do not italicize the titles of sacred writings (the Koran, the Bible), editions, or series.

Terms

2. Italicize terms, letters, or figures when they are referred to as such.

EXAMPLES

Spellcheckers make no distinction between *too* and *to.*
Small type makes it hard to distinguish between *n* and *m.*

Foreign terms

3. Italicize words and phrases considered foreign.

EXAMPLES

The study concerned the *vive la difference* in a suburban area of Detroit.
This item is included gratis with your order. ("Gratis" is not considered a foreign word.)

Note: Check a dictionary to determine if terms are considered foreign.

Emphasis

4. Italicize a word or phrase for special emphasis. Use this italicization sparingly for effect, and be sure that the emphasis does not confuse the meaning of the term you italicize.

EXAMPLE

My final response to the proposal is *no.*

SPELLING

Because of the influence of other languages, spelling in English is often irregular and difficult to learn. However, it *can* be mastered. Consider some of the following suggestions.

Suggestions for Improved Spelling

1. Keep a spelling list of words you regularly misspell. Review the list often, dropping words you can spell correctly and adding new words that give you difficulty.
2. Practice difficult words from your spelling list. One effective method is to write words on small cards for study and review while commuting or during breaks.
3. Use your dictionary, the poor speller's best friend. If you are so unsure of a word's spelling that you can't look up a word, try using a thesaurus or a special dictionary for poor spellers.
4. Use (but don't rely exclusively on) a spelling checker. These programs, which come with nearly all word processing systems, indicate misspelled words and offer correctly spelled alternatives. However, few spelling checkers include proper nouns or words commonly confused and misused (see pages 675–682). If a correctly spelled word exists, a spelling checker will approve it—even though it may not be the word you want! For example, a spelling checker would approve this sentence: "After eating to mulch pizza, I gut quiet sic." (This sentence should read: "After eating too much pizza, I got quite sick.")
5. Proofread your writing very carefully. Examine every word. If a word doesn't look right, look it up in a dictionary. This practice will support your good spelling habits.
6. Remember that some regional pronunciations of English words may not reflect correct spelling. Consider:

   sophmore for *sophomore*
   prompness for *promptness*

   Pronunciation is also misleading for words not of English origin. Consider these examples:

   pneumonia, potpourri, antique, tsunami, xylophone

7. Since many spelling errors are a violation of conventional practices for spelling *ei* and *ie* words and of spelling changes when suffixes are added, you can study lists in the following sections for help.

   Above all, don't give up!

Using *ie* and *ei* Words

The following jingle sums up most of the guides for correct *ie* and *ei* usage:

Use *i* before *e,*
Except after *c,*
Or when sounded like *a,*
As in *neighbor* and *weigh.*

- Generally use *ie* when the sound is a long *e* after any letter except *c.*

EXAMPLES

believe chief

grief / niece
piece / relieve

- Generally use *ei* after *c*.

EXAMPLES

deceive receive receipt

Note: An exception occurs when the combination of letters *cie* is sounded *sh*; in such instances, *c* is followed by *ie*.

EXAMPLES

sufficient efficient conscience

- Generally use *ei* when the sound is *a*.

EXAMPLES

neighbor freight sleigh
weigh reign vein

Spelling Changes When Affixes Are Added

An *affix* is a letter or syllable added either at the beginning or at the end of a word to change its meaning. The addition of affixes, whether *prefix* (added to the beginning of a word) or *suffix* (added to the end of a word), often involves spelling changes.

Prefixes

A prefix is a syllable added to the beginning of a word. One prefix may be spelled in several different ways, usually depending on the beginning letter of the base word. For example, *com, con, cor,* and *co* are all spellings of a prefix meaning "together, with." They are used to form such words as *commit, collect,* and *correspond.*

Common Prefixes Following are common prefixes and examples showing how they are added to base words. The meaning of the prefix is in parentheses.

- **ad** (*to, toward*). In adding the prefix *ad* to a base, the *d* often is changed to the same letter as the beginning letter of the base.

EXAMPLES

ad + breviate = abbreviate
ad + commodate = accommodate

- **com** (*together, with*). The spelling is *com* unless the base word begins with *l* or *r,* then the spelling is *col* and *cor,* respectively.

EXAMPLES

com + mit = commit
com + lect = collect
com + respond = correspond

- **de** (*down, off, away*). This prefix is often incorrectly written *di*. Note the correct spellings of words using this prefix.

EXAMPLES

describe desire
despair destroyed

- **dis** (*apart, from, not*). The prefix *dis* is usually added unchanged to the base word.

EXAMPLES

dis + trust = distrust
dis + satisfied = dissatisfied

- **in** (*not*). The consonant *n* often changes to agree with the beginning letter of the base word.

EXAMPLES

in + reverent = irreverent
in + legible = illegible

Note: The *n* may change to *m*.

EXAMPLES

in + partial = impartial
in + mortal = immortal

- **sub** (*under*). The consonant *b* sometimes changes to agree with the beginning letter of the base word.

EXAMPLES

sub + marine = submarine
sub + let = sublet
sub + fix = suffix
sub + realistic = surrealistic

- **un** (*not*). This is added unchanged.

EXAMPLES

un + able = unable
un + fair = unfair

Suffixes

A suffix is a letter or syllable added to the end of a word. One suffix may be spelled in several different ways, such as *ance* and *ence*. Also, the base word may require a change in form when a suffix is added. Because of these possibilities, adding suffixes often causes spelling difficulties.

Common Suffixes Learning the following suffixes and the spelling of the exemplary words will improve your vocabulary and spelling. The meaning of the suffix is in parentheses.

- **able, ible** (*capable of being*). Adding this suffix to a base word, usually a verb or a noun, forms an adjective.

EXAMPLES

rely—reliable
consider—considerable
separate—separable
read—readable
laugh—laughable
advise—advisable
commend—commendable
sense—sensible
horror—horrible
terror—terrible
destruction—destructible
reduce—reducible
digestion—digestible
comprehension—comprehensible

- **ance, ence** (*act, quality, state of*). Adding this suffix to a base word, usually a verb, forms a noun.

EXAMPLES

appear—appearance
resist—resistance
assist—assistance
attend—attendance
exist—existence
prefer—preference
insist—insistence
correspond—correspondence

Other nouns using **ance, ence** include

EXAMPLES

ignorance
brilliance
significance
importance
abundance
performance
guidance
experience
intelligence
audience
convenience
independence
competence
conscience

- **ary, ery** (*related to, connected with*). Adding this suffix to base words forms nouns and adjectives.

EXAMPLES

boundary
vocabulary
dictionary
library
customary
gallery
cemetery
millinery

- **efy, ify** (*to make, to become*). Adding this suffix forms verbs.

EXAMPLES

liquify
stupefy
putrefy
classify
ratify
testify
falsify
justify

- **ize, ise, yze** (*to cause to be, to become, to make conform with*). These suffixes are verb endings and are all pronounced the same way.

EXAMPLES

recognize
familiarize
generalize
emphasize
realize
revise
advertise
exercise
supervise
analyze
paralyze
modernize
criticize

Note, also, that some nouns end in *ise*.

EXAMPLES

exercise
merchandise
enterprise
franchise

- **ly** (*in a specified manner, like, characteristic of*). Adding this suffix to a base noun forms an adjective; adding *ly* to a base adjective forms an adverb. Generally, *ly* is added to the base word with no change in spelling.

EXAMPLES

monthly
heavenly
earthly
randomly
surely
softly
annually
clearly

If the base word ends in *ic,* usually add *ally.*

EXAMPLES

critically
basically
drastically
automatically

An exception is *public—publicly.*

- **ous** (*full of*). Adding this suffix to a base noun forms an adjective.

EXAMPLES

courageous
dangerous
hazardous
marvelous
outrageous
humorous
advantageous
adventurous
grievous
mischievous
beauteous
bounteous

- Other suffixes include:

-ant (ent, er, or, ian) meaning *one who* or *pertaining to*

-ion (tion, ation, ment) meaning *action, state of, result*

-ish meaning *like a*

-less meaning *without*

-ship meaning *skill, state, quality, office*

Final Letters

Final letters of words often require change before certain suffixes can be added.

- Final **e.** Generally keep a final silent *e* before a suffix beginning with a consonant, but drop it before a suffix beginning with a vowel.

EXAMPLES

use	useful	write	writing
love	lovely	hire	hiring

EXCEPTIONS

true	truly	due	duly	argue	argument

Note: In adding *ing* to some words ending in *e*, retain the *e* to avoid confusion with another word.

EXAMPLES

dye	dyeing	die	dying
singe	singeing	sing	singing

- Final **ce** and **ge.** Retain the *e* when adding *able* to keep the *c* or *g* soft. If the *e* were dropped, the *c* would have a *k* sound in pronunciation and the *g* a hard sound. For example, the word *change* retains the *e* when *able* is added: *changeable.*
- Final **ie.** Before adding *ing*, drop the *e* and change the *i* to *y* to avoid doubling the *i*.

EXAMPLES

tie	tying	lie	lying

- Final **y.** To add suffixes to words ending in a final *y* preceded by a consonant, change the *y* to *i* before adding the suffix. In words ending in *y* preceded by a vowel, the *y* remains unchanged before the suffix.

EXAMPLES

try	tries	survey	surveying

- Final consonants. Double the final consonant before adding a suffix beginning with a vowel if the word is one syllable or if the word is stressed on the last syllable.

hop	hopping	hopped	occur	occurring	occurred
plan	planning	planned	refer	referring	referred
stop	stopping	stopped	forget	forgetting	forgotten

- In adding suffixes to some words, the stress shifts from the last syllable of the base word to the first syllable. When the stress is on the first syllable, do *not* double the final consonant.

EXAMPLES

prefer	preference	confer	conference
refer	reference	defer	deference

-Ceed, -Sede, and -Cede Words

The base words *-ceed, -sede,* and *-cede* sound the same when they are pronounced. However, they cannot correctly be interchanged in spelling.

- **-ceed:** Three words, all verbs, end in *-ceed.*

  proceed succeed exceed

- **-sede:** The only word ending in *-sede* is *supersede.*
- **-cede:** All other words, excluding the four named above, ending in this sound are spelled *-cede.*

EXAMPLES

recede	secede	accede
concede	intercede	precede

Symbols

Symbols are often used in tables, charts, figures, drawings, and diagrams. Although they are not generally used in the text of work for nontechnical readers, symbols appear in material written for more specialized audiences.

Symbols cannot be discussed as definitely as abbreviations, capitalization, and numbers because no one group of symbols is common to all specialized areas. Most organizations and subject groups—medicine and pharmacy, mathematics, commerce and finance, engineering technologies—have their own symbols and practices for the use of these symbols. A person who is a part of any such profession must learn these symbols and the accepted usage practices. Technical dictionaries are the most reliable sources for specialized technical and scientific symbols. However, the following are examples of more common symbols.

- **Medicine and Pharmacy**

EXAMPLES

℞ take (Latin, *recipe*). Used at the beginning of a prescription
℥ ounce
ʒ dram
s̄ write (Latin, *signā*). On a prescription indicates the directions to be printed on the medicine label

- **Mathematics**

EXAMPLES

+ plus	> is greater
− minus	= equals
× times	∫ integral
÷ divide	∠ angle

- **Commerce and Finance**

  EXAMPLES

  # number, as in #7; or pounds, as in 50#
  £ pound sterling, as in British currency
  @ at, as in 10 @ 1¢ each

- **Engineering Technologies**
  Specialized fields, such as electronics, hydraulics, welding, and technical drawing, use standard symbols for different areas. These symbols are usually determined by the American Standards Association.

  For example, one source has nine pages of symbols used by electricians. There are resistor and capacitor symbols, contact and push-button symbols, motor and generator symbols, architectural plans symbols, transformer symbols, and switch and circuit breaker symbols.

  EXAMPLES

⊣⊢	⏚	— —		
contacts-N.O. (normally open)	ground	conductor	squirrel-cage induction motor	ceiling outlet

  Another example is a technical drawing text that has eight pages of symbols including topographic symbols; railway engineering symbols; American Standard piping symbols; heating, ventilating, and ductwork symbols; and plumbing symbols.

  EXAMPLES

⊙ county seat	++ flanged joint
---- national or state line	hot water tank
wood—with the grain	⊗ exit outlet

- **Nonspecialized Areas**

  Commonly recognized and used symbols include

% percent	′ and ″ feet and inches
° degree	**$** and **¢** dollars and cents
& and	

See also units of measure as symbols, item 9 under Abbreviations, pages 623–624.
Most dictionaries include a section on common signs and symbols.

Common Problems in Using Standard English

Introduction

Spoken English varies widely between regions and constantly changes, readily adapting to every situation and every speaker's needs and feelings. The range of any person's informal spoken language is as broad as a lifetime's history of people, places, and situations.

In the workplace, however, the range of interaction between individuals is necessarily limited and often more formal than casual. Written communication reflects a far greater consistency than informal spoken language. Although the style and tone of written English vary according to a document's purpose, readers, and occasion, effective written language in the workplace, as in the academy, is tied to accepted patterns of grammatical correctness and accepted usage, or word choice. In the workplace, correct and consistent writing becomes a democratic means of communication, and a workplace author's ability to recognize, select, and conform to standard writing practices allows the author's ideas to be seen (and judged) on their own merit. Errors and inconsistencies detract from a workplace communicator's credibility in a formal speech presentation as much as in a written document.

Because making informed communication choices requires an awareness of conventions and of potential problem areas, this part of the appendix provides an easy-to-use topic discussion of common grammatical practice and of widely recognized usage. It identifies each potential problem, briefly explaining the problem and providing a pattern to correct it. To make the appendix easy to use for reference and for study, the discussion is arranged alphabetically by topic.

Adjectives

Adjectives provide descriptive information about nouns or pronouns.

1. Adjectives show degrees of comparison in quality, quantity, and manner by affixing *-er* and *-est* to the positive form or by adding *more* and *most* or *less* and *least* to the positive form.

   Degrees of Comparison

POSITIVE	tall	beautiful	wise
COMPARATIVE	taller	more beautiful	less wise
SUPERLATIVE	tallest	most beautiful	least wise

   The comparative degree is used when speaking of two things and the superlative degree is the form used when speaking of three or more things. One-syllable words generally add *–er* and *–est,* whereas words of two or more syllables generally require *more* and *most.* To show degrees of inferiority, *less* and *least* are added.

   - A double comparison, such as *most beautifulest,* should not be used; either *-est* or *most* makes the superlative degree, not both.
   - Adjectives that indicate absolute qualities or conditions (such as *dead, round,* or *perfect*) cannot be compared.

   Thus: *dead, more nearly dead, most nearly dead* (not *deader, deadest*).

2. Adjective clauses are punctuated according to the purpose they serve. If they are essential to the meaning of the sentence (an essential or restrictive clause), they

are not set off by commas. If they simply give additional information (a nonessential clause), they are set off from the rest of the sentence. Adjective clauses beginning with *that* are essential and are not set off (with one exception: when *that* is used as a substitute for *which* to avoid repetition).

> Robert Oppenheimer is the physicist *who led the research team.* (essential to meaning)
>
> Laboratory equipment is usually arranged in ways *that support the sequence of an experiment.* (essential to meaning)
>
> My ideas about the importance of written communication, *which are different from most of my friends' ideas,* were formulated mainly through workplace experience. (gives additional information)

Adverbs

Adverbs provide descriptive information about verbs, adjectives, or adverbs.

- An adjective should not be used when an adverb is needed.

INCORRECT	The intern did *good* on her first report. (adverb needed)
CORRECT	The intern did *well* on her first report.
INCORRECT	For a team to operate *smooth* and *efficient,* all members must cooperate. (adverbs needed)
CORRECT	For a team to operate *smoothly* and *efficiently,* all members must cooperate.

- Introductory adverb clauses are generally followed by a comma. Adverb clauses at the end of a sentence are not set off. (See Comma, page 636.)

> *When the investigation is completed,* the company will make its decision.
>
> The company will make its decision *when the investigation is completed.*

Comma Splice

The comma splice occurs when a comma is used to connect, or splice together, two sentences. The comma splice, a common error in sentence punctuation, is also called the comma fault. (See also the discussion of Run-on, or Fused, Sentence, pages 667–668.)

> Mahogany is a tropical American timber tree, its wood turns reddish brown at maturity.

The comma splice in the above sentence may be corrected by several methods.

- Replacing the comma with a period

> Mahogany is a tropical American timber tree. Its wood turns reddish brown at maturity.

- Replacing the comma with a semicolon

  Mahogany is a tropical American timber tree; its wood turns reddish brown at maturity.

- Adding a coordinate conjunction, such as *and*

  Mahogany is a tropical American timber tree, *and* its wood turns reddish brown at maturity.

- Recasting the sentence

  Mahogany is a tropical American timber tree whose wood turns reddish brown at maturity.

Conjunctions

A conjunction connects words, phrases, or clauses. Conjunctions may be divided into two general classes: coordinate conjunctions and subordinate conjunctions.

1. Coordinate conjunctions—such as *and, but, or*—connect words, phrases, or clauses of equal rank.

   I bought nails, brads, staples, *and* screws. (connects nouns in a series)

   Nancy *or* Elaine will serve on the review committee.(connects nouns)

   - Conjunctive adverbs—such as *however, moreover, therefore, consequently, nevertheless*—are a kind of coordinating conjunction. They link the independent clause in which they occur to the preceding independent clause. The clause they introduce is grammatically independent, but it depends on the preceding clause for complete meaning. Note that a semicolon separates the independent clauses (see Semicolon, pages 639–640) and that the conjunctive adverb is usually set off with commas (see Comma, item 10, page 636).

     The report was researched and written by technical communication students; *therefore*, other students should be interested in reviewing it.

     Jane was called away on assignment; *consequently*, she was unable to participate in the planning session.

   - Correlative conjunctions—*either . . . or, neither . . . nor, both . . . and, not only . . . but also*—are a kind of coordinating conjunction. Correlative conjunctions are used in pairs to connect words, phrases, and clauses of equal rank.

     The report was *not only* well written *but also* thoughtfully designed.

   - The units joined by coordinate conjunctions should be the same *grammatically*. (See also Parallelism in Sentences, pages 662–663.)

CONFUSING	Our department ships two kinds of drills—manual *and* electricity. (adjective and noun)
REVISED	Our department ships two kinds of drills—manual *and* electric. (adjective and adjective)
CONFUSING	I *either* will go today *or* tomorrow. (verb and adverb)

REVISED I will go *either* today *or* tomorrow. (adverb and adverb)

- The units joined by coordinate conjunctions should be the same *logically.* (See also Parallelism in Sentences, below.)

CONFUSING I can't decide whether I want to be a lab technician, a nurse, *or* respiratory therapy. (two people and a field of study)

REVISED I can't decide whether I want to be a lab technician, a nurse, *or* a respiratory therapist. (three people)

REVISED I can't decide whether I want to study lab technology, nursing, *or* respiratory therapy. (three fields of study)

2. Subordinate conjunctions—such as *when, since, because, although, as, as if*—introduce subordinate clauses.

Because the voltmeter was broken, we could not test the circuit.

The manager gave the team members time for research *since the primary purpose of the project was to come up with new ideas.*

Parallelism in Sentences

Parallel structure involves getting like ideas into like constructions. A coordinate conjunction, for example, joins ideas that must be stated in the same grammatical form. Other examples: an adjective should be parallel with an adjective, a verb with a verb, an adverb clause with an adverb clause, and an infinitive phrase with an infinitive phrase. Parallel structure in grammar helps to make parallel meaning clear.

A *computer,* a *table,* and a *filing cabinet* were delivered today. (nouns in parallel structure)

Whether you accept the outcome of the experiment or *whether I accept it* depends on our individual interpretation of the facts. (dependent clauses in parallel structure)

Failure to express all the ideas in the same grammatical form results in faulty parallelism.

FAULTY This study should help the new employee *learn* skills and *to be knowledgeable* about company policy. (verb and infinitive phrase)

REVISED This study should help the new employee *learn* skills and *become* knowledgeable about company policy. (verbs in parallel structure)

FAULTY Three qualities of tungsten steel alloys are *strength, ductility,* and *they have to be tough* (noun, noun, and independent clause)

REVISED	Three qualities of tungsten steel alloys are *strength, ductility,* and *toughness.* (nouns in parallel structure)
FAULTY	The best lighting in an office can be obtained *if windows are placed in the north side, if tables are placed so that the light comes over the employee's left shoulder,* and *by painting the ceiling a very light color.* (dependent clauses, dependent clause, phrase)
REVISED	The best lighting in an office can be obtained *if windows are placed in the north side, if tables are placed so that the light comes over the employee's left shoulder,* and *if the ceiling is painted a very light color.* (dependent clauses in parallel structure)

Placement of Modifiers

Modifiers are words, phrases, or clauses, either adjective or adverb, that limit or restrict other words in the sentence. Careless construction and placement of these modifiers may cause problems such as dangling modifiers, dangling elliptical clauses, misplaced modifiers, and squinting modifiers.

Dangling Modifiers

Dangling modifiers or dangling phrases occur when the word the phrase should modify is hidden within the sentence or is missing.

WORD HIDDEN IN SENTENCE	Holding the wrench tightly, the pipe was adjusted by the plumber. (Implies that the pipe was holding the wrench tightly)
REVISION	Holding the wrench tightly, the plumber adjusted the pipe.
WORD MISSING	By placing a thermometer under the tongue for approximately three minutes, a fever can be detected (Implies that a fever places a thermometer under the tongue)
REVISION	By placing a thermometer under the tongue for approximately three minutes, anyone can tell if a person has a fever.

- Dangling modifiers may be corrected by either rewriting the sentence so that the word modified by the phrase immediately follows the phrase (this word is usually the subject of the sentence) or rewriting the sentence by changing the phrase to a dependent clause.

DANGLING MODIFIER	*Driving down Main Street,* the city auditorium came into view.
SENTENCE REWRITTEN TO CLARIFY WORD MODIFIED	Driving down Main Street, I saw the city auditorium.

SENTENCE REWRITTEN WITH A DEPENDENT CLAUSE	As I was driving down Main Street, the city auditorium came into view.

Dangling Elliptical Clauses

In elliptical clauses some words are understood rather than stated. For example, the dependent clause in the following sentence is elliptical: *When measuring the temperature of a conductor, you must use the Celsius scale;* the subject and part of the verb are omitted. The understood subject of an elliptical clause is the same as the subject of the sentence. This is true in the example: *When (you are) measuring the temperature of a conductor, you must use the Celsius scale.* If the understood subject of the elliptical clause is not the same as the subject of the sentence, the clause is a dangling clause.

DANGLING CLAUSE	*When using an electric saw,* safety glasses should be worn.
REVISION	When using an electric saw, wear safety glasses.

- Dangling elliptical clauses may be corrected by either including within the clause the missing words or rewriting the main sentence so that the stated subject of the sentence and the understood subject of the clause will be the same.

DANGLING CLAUSE	*After changing the starter switch,* the car still would not start.
SENTENCE REVISED: MISSING WORDS INCLUDED	After *the mechanic* changed the starter switch, the car still would not start.
SENTENCE REVISED: UNDERSTOOD SUBJECT AND STATED SUBJECT THE SAME	After changing the starter switch, *the mechanic* still could not start the car.

Misplaced Modifiers

Modifiers must be placed near the word or words modified. If the modifier is correctly placed, there should be no confusion. If the modifier is incorrectly placed, the intended meaning of the sentence may not be clear.

MISPLACED MODIFIER	The machinist placed the work to be machined in the drill press vise *called the workpiece.*
CORRECTLY PLACED MODIFIER	The machinist placed the work to be machined, *called the workpiece,* in the drill press vise.
MISPLACED MODIFIER	Where are the shirts for children *with snaps?*
CORRECTLY PLACED MODIFIER	Where are the children's shirts *with snaps?*

Squinting Modifiers

A modifier should clearly limit or restrict *one* sentence element. If a modifier is so placed within a sentence that it can be taken to limit or restrict either of two elements, the modifier is squinting; that is, the reader cannot tell which way the modifier is looking.

SQUINTING MODIFIER — When the intern began summer work *for the first time* she was expected to follow orders promptly and exactly. (The modifying phrase could belong to the clause that precedes it or to the clause that follows it.)

- Punctuation may solve the problem. The sentence might be written in either of the following ways, depending on the intended meaning.

REVISION — When the intern began summer work *for the first time,* she was expected to follow orders promptly and exactly.

REVISION — When the student began summer work, *for the first time* she was expected to follow orders promptly and exactly.

- Squinting modifiers often are corrected by shifting their position in the sentence. Sentence meaning determines placement of the modifier.

SQUINTING MODIFIER — The employees were advised *when it was midmorning* the new break schedule would go into effect.

REVISION — *When it was midmorning,* the employees were advised that the new break schedule would go into effect.

REVISION — The employees were advised that the new break schedule would go into effect *when it was midmorning.*

- Particularly difficult for some writers is the correct placement of *only, almost,* and *nearly.* These words generally should be placed immediately before the word they modify since changing position of these words within a sentence changes the meaning of the sentence.

In the reorganization of the district the member of Congress *nearly* lost a hundred voters. (didn't lose any voters)

In the reorganization of the district the member of Congress lost *nearly* a hundred voters. (lost almost one hundred voters)

The customer *only* wanted to buy soldering wire. (emphasizes *wanted*)

The customer wanted to buy *only* soldering wire. (emphasizes *soldering wire*)

Preposition at End of Sentence

- A preposition may end a sentence when any other placement of the preposition would result in a clumsy, unnatural sentence.

UNNATURAL — Sex is a topic *about* which many people think.

NATURAL — Sex is a topic which many people think *about.*

Pronouns

- *Myself* should not be used in the place of *I* or *me*.

INCORRECT	The manager or *myself* will be present.
CORRECT	The manager or *I* will be present.
INCORRECT	Reserve the conference room for the supervisor and *myself*.
CORRECT	Reserve the conference room for the supervisor and *me*.

- *Themself, theirselves,* and *hisself* should never be used; they are not acceptable forms.

CORRECT	The members of the team *themselves* voted not to go.
CORRECT	Henry cut *himself*.

Pronoun and Antecedent Agreement

Pronouns take the place of or refer to nouns. They provide a good way to economize in writing. For example, writing a paper on Anthony van Leeuwenhoek's pioneer work in developing microscopes, you could use the pronoun forms *he, his,* and *him* to refer to Leeuwenhoek rather than repeat his name numerous times. Since pronouns take the place of or refer to nouns, there must be number (singular or plural) agreement between the pronoun and its antecedent (the word the pronoun stands for or refers to).

- A singular antecedent requires a singular pronoun; a plural antecedent requires a plural pronoun.

NONAGREEMENT	Gritty ink *erasers* should be avoided because *it* invariably damages the working surface of paper.
AGREEMENT	Gritty ink *erasers* should be avoided because *they* invariably damage the working surface of paper.

- Two or more subjects joined by *and* must be referred to by a plural pronoun.

  The *business manager* and the *accountant* will plan *their* budget.

- Two or more singular subjects joined by *or* or *nor* must be referred to by singular pronouns.

  Laticia or *Margo* will give *her* report first.
  Neither *Conway* nor *Freeman* will give *his* report.

- Partitive nouns, such as *part, half, some, rest,* and *most* may be singular or plural. The number is determined by a phrase following the noun.

  Most of the *applicants* asked that *their* assessments be mailed.

Most of the *coffee* was stored in *its own* container.

- The indefinite pronouns *each, every, everyone* and *everybody, nobody, either, neither, one, anyone* and *anybody,* and *somebody* are singular and thus have singular pronouns referring to them.

  Each of the technicians has indicated *she* wishes *her* own cubicle.

  Neither of the machines has *its* motor repaired.

- The indefinite pronouns *both, many, several,* and *few* are plural; these forms require plural pronouns.

  Several realized *their* failure was the result of not preparing.

  Both felt that nothing could save *them.*

- Still other indefinite pronouns—*most, some, all, none, any,* and *more*—may be either singular or plural. Usually a phrase following the indefinite pronoun will reveal whether the pronoun is singular or plural in meaning.

  Most of the *patients* were complimentary of *their* nurses.

  Most of the *money* was returned to *its* owner.

Run-On, or Fused, Sentence

The run-on, or fused, sentence occurs when two sentences are written with no punctuation to separate them. (See also Comma Splice, pages 660–661.)

> The evaluation of the Computer Assisted Drafting software is encouraging we expect to install the program in an efficient and profitable manner.

Obviously the sentence above needs punctuation to make it understandable. Several methods could be used to make the sentence clearer.

- *Period* between "encouraging" and "we"

  The evaluation of the Computer Assisted Drafting software is encouraging. We expect to use the program efficiently and profitably.

- *Semicolon* between "encouraging" and "we"

  The evaluation of the Computer Assisted Drafting software is encouraging; we expect to use the program efficiently and profitably.

- *Comma* plus *and* between "encouraging" and "we"

  The evaluation of the Computer Assisted Drafting software is encouraging, and we expect to use the program efficiently and profitably.

- *Recasting the sentence.*

Since the evaluation of the Computer Assisted Drafting software is encouraging, we expect to use the program efficiently and profitably.

See also Punctuation, especially usage of the period, the comma, and the semicolon, pages 632, 634–635, and 639–640.

SENTENCE FRAGMENTS

A group of words containing a subject and a verb and standing alone as an independent group of words is a sentence. If a group of words lacks a subject or a verb or cannot stand alone as an independent group of words, the group of words is called a sentence fragment. Sentence fragments generally result from two reasons.

- A noun (subject) followed by a dependent clause or phrase is written as a sentence. The omitted unit is the verb.

FRAGMENT	The engineer's scale, which is graduated in the decimal system.
FRAGMENT	Colors that are opposite.

To make the sentence fragment into an acceptable sentence, add the verb and any modifiers needed to complete the meaning.

SENTENCE	The engineer's scale, which is graduated in the decimal system, is often called the decimal scale.
SENTENCE	Colors that are opposite on the color circle are used in a complementary color scheme.

- Dependent clauses or phrases are written as complete sentences. These are introductory clauses and phrases that require an independent clause.

FRAGMENT	During the Renaissance while some pencil tracings were made from a drawing placed beneath the tracing paper.
FRAGMENT	Another development that has promoted the recognition of the management.

To make the fragments into sentences, add the independent clause.

SENTENCE	During the Renaissance while some pencil tracings are made from a drawing placed underneath the tracing paper, most drawings later were made directly on pencil tracing paper, cloth, vellum, or bond paper.
SENTENCE	Another development that has promoted the recognition of management is the separation of ownership and management.

Acceptable Sentence Fragments

Types of fragments, often called nonsentences, are acceptable in some informal situations, but less frequently in workplace writing.

- Warnings

  Danger!

- Emphasis

  Open the window. Right now.

- Transition between ideas in a paragraph or composition

  Now to the next point of the argument.

- Dialogue

  "How many different meals have been planned for next month?"

  "About ten."

Sexist Language

Sexist language, also called gendered language, presents a special problem in the workplace. It's important to treat customers and employees with respect and equality. Sexist language, however, seems to prefer, or single out, one gender over the other. The choice of *man* over *humanity*, for example, or the use of a term such as *fireman* rather than *firefighter* suggests bias and gender assumptions, even though none may exist. To avoid language that suggests bias, it's important to avoid such language.

The following discussion describes common workplace problems with sexist language and suggests unobtrusive and inoffensive ways to avoid them.

Occupational Titles

Today women are employed in every American workplace, in every occupation, and at every organizational level. As a result, it's important to use occupational titles that are gender free. Here are examples of such gendered and gender-free titles.

SEXIST	GENDER FREE
waiter, waitress	server
stewardess	flight attendant
policeman	police officer
fireman	firefighter
salesman	sales representative
chairman	chair, chairperson

These gender-free titles have come to describe the duties of those who occupy them even more accurately than the older, gendered terms.

It's also important to avoid associating gender with roles that have been traditionally occupied by one gender or the other. For example, it's wise to avoid pronouns that suggest that all nurses are female or that all doctors are male. Here are some ways to avoid the problem.

SEXIST

Each teacher should submit her lesson plans at the end of the week.

A nurse must record medications accurately for all of her patients.

The farmer is responsible for inoculating his livestock.

GENDER FREE

Each teacher should submit lesson plans at the end of the week.
A nurse must record each patient's medications accurately.
Farmers are responsible for inoculating their livestock.

Personal Pronouns English provides us with only two third-person singular pronouns that refer to individuals: *he* and *she*. To avoid automatically using *he*, try using the plural (*they*), or use *he or she*. In this way, you can avoid singling out one gender over another.

PROBLEM	A pilot should always log his flights into his charts.
SOLUTION	Pilots should always log their flights into their charts.
SOLUTION	A pilot should always log a flight into his or her charts.

If you are writing procedures, it's easy to avoid sexist language by writing directly to your reader. In addition to allowing you to avoid sexist language, this practice allows you to address your reader directly, explaining clearly how to perform each step.

PROBLEM	The clerk should key shift data directly into the system.
SOLUTION	Key your shift data directly into the system.

Parallel Treatment in Language and on the Job All any of us can ask for on the job is equal treatment in hiring, evaluation, and promotion. After all, such parallel treatment is fair, avoiding stereotyping and bias, and recognizing and rewarding performance consistently. Since language is a form of workplace behavior, it's important to treat men and women in parallel ways.

- If you need to refer to the gender of employees, do so in ways that reflect equal treatment for men and women.

  SEXIST

  The team consisted of two female pilots, one navigator, and three mechanics.

  PARALLEL

  The team consisted of two female pilots, one male navigator, and three male mechanics.

- Avoid clichés that single out one gender or another; instead use gender-free choices.

SEXIST	GENDER FREE
our man in Madrid	our representative in Madrid
man hours	work hours
man a job	staff (or fill) a job

Whatever the job may be and whoever the workplace partners, doing the work the best way means working together cooperatively and respectfully. By avoiding sexist language on the job, you avoid needlessly seeming to single out individuals.

Shift in Sentence Focus

A writer may begin a sentence that expresses one thought but, somehow, by the end of the sentence, the writer has unintentionally shifted the sentence's focus.

SHIFT IN FOCUS	There are several preparatory steps in spray painting a house, which is easy to mess up if you are not careful.

The focus in the first half of the sentence is on preparatory steps in spray painting a house; the focus in the second half is on how easy it is to mess up when spray painting a house. Somehow, in the middle of the sentence, the writers shifted the focus of the sentence. The result is a sentence that is confusing to the reader.

To correct such a sentence, the writer thinks through the intended purpose and emphasis of the sentence and reviews preceding and subsequent ones. The poorly written sentence might be corrected in two ways:

- Divide into two sentences

REVISION	There are several preparatory steps in spray painting a house. Failure to follow these steps may result in a messed up paint job.

- Recast as one sentence

REVISION	Following several preparatory steps in spray painting a house will help to keep you from messing up.
REVISION	To help assure satisfactory results when spray painting a house, follow these preparatory steps.

Shift in Person, Number, and Gender

Pronouns must agree with antecedents in person, number, and gender.

Avoid shifting from one person to another, from one number to another, or from one gender to another—as illustrated in the examples below. Such shifting creates confusing, awkward constructions.

SHIFT IN PERSON	When *I* was just an apprentice welder, the supervisor expected *you* to know all the welding processes. (*I* is first person; *you* is second person)
REVISION	When *I* was just an apprentice welder, the supervisor expected *me* to know all the welding processes.
SHIFT IN NUMBER	*Each* person in the space center control room watched *his or her* dial anxiously. *They* thought the countdown to "zero" would never come. (*Each* and *his or her* are singular pronouns; *they* is plural)

REVISION	*Each* person in the space center control room watched *his or her* dial anxiously and thought the countdown to "zero" would never come.
SHIFT IN GENDER	The mewing of my cat told me *it* was hungry, so I gave *her* some food. (*It* is neuter; *her* is feminine)
REVISION	The mewing of my cat told me *she* was hungry, so I gave *her* some food.

Subject and Verb Agreement

A subject and verb agree in person and number. Person denotes person speaking (first person), person spoken to (second person), and person spoken of (third person). The person used determines the verb form that follows.

Present Tense

FIRST PERSON	I go	we go	I am	we are
SECOND PERSON	you go	you go	you are	you are
THIRD PERSON	he goes	they go	he is	they are
IMPERSONAL	one goes		one is	

- A subject and verb agree in number; that is, a plural subject requires a plural verb and a singular subject requires a singular verb.

  The *symbol* for hydrogen *is* H. (singular subject; singular verb)

  Comic *strips are* often vignettes of real-life situations. (plural subject; plural verb)

- A compound subject requires a plural verb.

  Employees and *employers work* together to formulate policies.

- The pronoun *you* as subject, whether singular or plural in meaning, takes a plural verb.

  You were assigned the duties of staff nurse on fourth floor.

  Since *you are* new technicians, *you are* to be paid weekly.

- In a sentence containing both a positive and a negative subject, the verb agrees with the positive.

  The *employees,* not the *manager, were asked* to give their opinions regarding working conditions. (Positive subject plural; verb plural)

- If two subjects are joined by *or* or *nor,* the verb agrees with the nearer subject.

  Graphs or a *diagram aids* the interpretation of statistical reports. (subject nearer verb singular; verb singular)

A *graph* or *diagrams aid* the interpretation of this statistical report. (subject nearer verb plural; verb plural)

- A word that is plural in form but names a single object or idea requires a singular verb.

The *United States has changed* from an agricultural to a technical economy.

Twenty-five *dollars was offered* for any usable suggestion.

Six *inches is* the length of the narrow rule.

- The term *the number* generally takes a singular verb; *a number* takes a plural verb.

The number of snow tires purchased each year in Canada *has decreased.*

A number of modern inventions *are* the product of the accumulation of vast storehouses of smaller, minor discoveries.

- When certain subjects are followed by an *of* phrase, the number (singular or plural) of the subject is determined by the number of the object of the *of* phrase. These subjects include fractions, percentages, and the words *all, more, most, some,* and *part.*

Some of his *discussion was* irrelevant. (object of *of* phrase, *discussion*, is singular; verb is singular)

Some of the *problems* in a workplace environment *require* careful analysis by both management and employees. (object of *of* phrase, *problems,* is plural; verb is plural)

- Elements that come between the subject and the verb ordinarily do not affect subject–verb agreement.

One of the numbers *is* difficult to represent because of the great number of zeros necessary.

Such *factors* as temperature, available food, age of organism, or nature of the suspending medium *influence* the swimming speed of a given organism.

- Occasionally, the subject may follow the verb, especially in sentences beginning with the expletives *there* or *here;* however, such word order does not change the subject–verb agreement. (The expletives *there* and *here,* in their usual meaning, are never subjects.)

There *are* two common temperature *scales:* Fahrenheit and Celsius.

There *is* a great *deal* of difference in the counseling techniques used by contemporary ministers.

Here *are* several *types* of film.

- The introductory phrase *It is* or *It was* is always singular, regardless of what follows.

  It is these problems that overwhelm me.

  It was the officers who met.

- Relative pronouns (*who, whom, whose, which, that*) may require a singular or a plural verb, depending on the antecedent of the relative pronoun.

  She is one of those people *who keep* calm in an emergency. (antecedent of *who* is *people,* a plural form; verb plural)

  The earliest meeting *that is* held is at 11:00. (antecedent of *that* is *meeting,* a singular form; verb singular)

Verbs

Verbs are words that reflect action, being, or becoming expressed in time (or tense).

- Use the verb tense that accurately conveys the sequence of events. Do not shift needlessly from one tense to another. Since verb tenses tell the reader when the action is happening, inconsistent verb tenses confuse the reader.

SHIFTED VERBS (CONFUSING)	As time *passed,* technology *becomes* more complex. (*passed* is past tense verb; *becomes* is present tense verb)

Since the two indicated actions occur at the same time, the verbs should be in the same tense.

CONSISTENT VERBS	As time *passed,* technology *became* more complex. (*passed, became* are past tense verbs)
CONSISTENT VERBS	As time *passes,* technology *becomes* more complex. (*passes, becomes* are present tense verbs)
SHIFTED VERBS (CONFUSING)	While I *was installing* appliances last summer, *I had learned* that customers appreciate promptness. (*was installing* is past tense (progressive) verb; *had learned* is past perfect tense verb)

The verb phrase *was installing* indicates a past action in progress while the verb phrase *had learned* indicates an action completed prior to some stated past time. To indicate that both actions were in the past, the sentence might be written as follows:

CONSISTENT VERBS	While I *was installing* appliances last summer, I *learned* that customers appreciate promptness. (*was installing, learned* are past tense verbs)

Active and Passive Voice

Verbs have two voices, active and passive.

1. The active voice indicates action done by the subject. The active voice is forceful and emphatic.

   Roentgen *won* the Nobel Prize for his discovery of X rays.

   Meat slightly marbled with fat *tastes* better.

   The scientist *develops* theories.

2. The passive voice indicates action done to the subject. The recipient of the action receives more emphasis than the doer of the action. A passive-voice verb is always at least two words, a form of the verb *to be* and the past participle (third principal part) of the main verb.

   The surgery *was performed* by Dr. Petronia Vanetti.

   Logarithm tables *are found* in later editions.

- Choose the voice of the verb that permits the desired emphasis.

ACTIVE VOICE	The ballistics experts *examined* the results of the tests. (emphasis on *ballistics experts*)
PASSIVE VOICE	The results of the tests *were examined* by the ballistics experts. (emphasis on *results of the tests*)
ACTIVE VOICE	Juan *gave* the report. (emphasis on *Juan*)
PASSIVE VOICE	The report *was given* by Juan. (emphasis on *report*)

- Avoid needlessly shifting from the active to the passive voice.

NEEDLESS SHIFT IN VOICE (CONFUSING)	Management and labor representatives *discussed* the pay raise, but no decision *was reached.* (*discussed* is active voice; *was reached* is passive voice)
CONSISTENT VOICE	Management and labor representatives *discussed* the pay raise, but they *reached* no decision. (*discussed, reached* are active voice)

Words Often Confused and Misused

Here is a list of frequently misused words often confused by sound and meaning, with suggestions for their standard use in workplace writing.

a, an *A* is used before words beginning with a consonant sound; *an* is used before words with a vowel sound. (*Remember:* Consider sound, not spelling.)

EXAMPLES	This is *a* banana.	The patient needs *a* unit of blood.
	This is *an* orange.	I'll call in *an* hour.

accept, except *Accept* means "to take an object or idea offered" or "to agree to something"; *except* means "to leave out" or "excluding."

EXAMPLES Please *accept* this gift.
Everyone *except* Joe may leave.

access, excess *Access* is a noun meaning "way of approach" or "admittance"; *excess* means "greater amount than required or expected."

EXAMPLES The visitors were not allowed *access* to the laboratory.
Her income is in *excess* of $50,000.

ad *Ad* is a shortcut to writing *advertisement.* In formal writing, however, write out the full word. In informal writing and speech, such abbreviated forms as *ad, auto, phone, photo,* and *TV* may be acceptable.

advice, advise *Advice* is a noun meaning "opinion given," "suggestions"; *advise* is a verb meaning "to suggest," "to recommend."

EXAMPLES We accepted the lawyer's *advice.*
The lawyer *advised* us to drop the charges.

affect, effect *Affect* as a verb means "to influence" or "to pretend"; as a noun, it is a psychological term meaning "feeling" or "emotion." *Effect* as a verb means "to make something happen"; as a noun, it means "result" or "consequence."

EXAMPLES The colors used in a home may *affect* the prospective buyer's decision to purchase.
The new technique will *effect* change in the entire procedure.

aggravate, irritate *Aggravate* means "to make worse or more severe." Avoid using *aggravate* to mean "to irritate" or "to vex," except perhaps in informal writing or speech.

ain't A contraction for *am not, are not, has not, have not.* The form is regarded as nonstandard; careful speakers and writers do not use it.

all ready, already *All ready* means that everyone is prepared or that something is completely prepared; *already* means "completed" or "happened earlier."

EXAMPLES We are *all ready* to go.
It is *already* dark.

all right, alright Use *all right.* In time, *alright* may become accepted, but for now *all right* is generally preferred.

EXAMPLE Your choice is *all right* with me.

all together, altogether *All together* means "united"; *altogether* means "entirely."

EXAMPLES We will meet *all together* at the clubhouse.
There is *altogether* too much noise in the hospital area.

a lot, alot, allot *A lot* is written as two separate words and is a colloquial term meaning "a large amount"; *alot* is a common misspelling for *a lot; allot* is a verb meaning "to give a certain amount."

amount, number *Amount* refers to mass or quantity; *number* refers to items, objects, or ideas that can be counted individually.

EXAMPLES The *amount* of money for clothing is limited.
A large *number* of people are enrolled in the class.

and/or This pairing of coordinate conjunctions indicates appropriate alternatives. Avoid using *and/or* if it misleads or confuses the reader, or if it indicates imprecise thinking. In the sentence, "Jones requests vacation leave for Monday–Wednesday and/or Wednesday–Friday," the request is not clear because the reader does not know whether three days' leave or five days' leave is requested. Use *and* and *or* to mean exactly what you want to say.

EXAMPLES She is a qualified lecturer *and* consultant.
For our vacation we will go to London *or* Madrid *or* both.

See also Virgule, page 642.

angel, angle *Angel* means "a supernatural being"; *angle* means "corner" or "point of view." Be careful not to overuse *angle* meaning "point of view." Use *point of view, aspect,* and the like.

as if, like *As if* is a subordinate conjunction; it should be followed by a subject-verb relationship to form a dependent clause. *Like* is a preposition; in formal writing *like* should be followed by a noun or a pronoun as its object.

EXAMPLES He reacted to the suggestion *as if* he never heard of it.
Pines, *like* cedars, do not have leaves.

as regards, in regard to Avoid these wordy phrases. Use *about* or *concerning* instead.

at this point in time Avoid this wordy phrase. Use *now.*

average, mean, median *Average* is the quotient obtained by dividing the sum of the quantities by the number of quantities. For example, for scores of 70, 75, 80, 82, 100, the *average* is 81.4. *Mean* may be the simple average, or it may be the value midway between the lowest and the highest quantity (in the scores above, 85). *Median* is the middle number (in the scores above, 80).

balance, remainder *Balance* as used in banking, accounting, and weighing means "equality between the totals of two sides." *Remainder* means "what is left over."

EXAMPLES The company's bank *balance* continues to grow.
Our shift will work overtime the *remainder* of the week.

being that, being as how Avoid using either of these wordy phrases. Use *since* or *because*.

EXAMPLE *Because* the bridge is closed, we will have to ride the ferry.

beside, besides *Beside* means "alongside," "by the side of," or "not part of"; *besides* means "furthermore," or "in addition."

EXAMPLES The tree stands *beside* the walk.
Besides the cost there is a handling charge.

bi-, semi- *Bi-* is a prefix meaning "two" and *semi-* is a prefix meaning "half" or "occurring twice within a period of time."

EXAMPLES Production quotas are reviewed *biweekly.*
(every two weeks)
The board of directors meets *semiannually.*
(every half year, or twice a year)

brake, break *Brake* is a noun meaning "an instrument to stop something"; *break* is a verb meaning "to smash," "to cause to fall apart."

EXAMPLES The mechanic relined the *brakes* on the truck.
If the vase is dropped, it will *break.*

capital, capitol *Capital* means "major city of a state or nation," "wealth," or, as an adjective, "chief" or "main." *Capitol* means "building that houses the legislature"; when written with a capital *C,* it usually means the legislative building in Washington.

EXAMPLES Jefferson City is the *capital* of Missouri.
Our company has a large *capital* investment in preferred stocks.
The *capitol* is located on Third Avenue.

cite, sight, site *Cite* is a verb meaning "to refer to"; *sight* is a noun meaning "view" or "spectacle"; *site* is a noun meaning "location."

EXAMPLES *Cite* a reference in the text to support your theory.
Because of poor *sight,* he has to wear glasses.
This is the building *site* for our new home.

coarse, course *Coarse* is an adjective meaning "rough," "harsh," or "vulgar"; *course* is a noun meaning "a way," "a direction."

EXAMPLES The sandpaper is too *coarse* for this wood.
The creek follows a winding *course* to the river.

consensus *Consensus* means "a general agreement of opinion." Therefore, *consensus of opinion* is repetitious. Avoid the expression.

contact *Contact* is overused as a verb, especially in workplace usage. Consider using verbs with more exact meanings, such as *write to, telephone, talk with, inform, advise,* or *ask.*

continual, continuous *Continual* means "often repeated"; *continuous* means "uninterrupted" or "unbroken."

EXAMPLES The conference has had *continual* interruptions.

The rain fell in a *continuous* downpour for an hour.

could of The correct form is *could have*. This error occurs because "could of" sounds like the contraction *could've*.

EXAMPLE We *could have* (could've) completed the work on time.

council, counsel, consul *Council* is a noun meaning "a group of people appointed or elected to serve in an advisory or legislative capacity." *Counsel* as a noun means "advice" or "attorney"; as a verb, it means "to advise." *Consul* is a noun naming the official representing a country in a foreign nation.

EXAMPLES The club has four members on its *council.*
The *counsel* for the defense advised him to testify.
The *consul* from Switzerland was invited to our international tea.

criteria, criterion *Criteria* is the plural of *criterion,* meaning "a standard on which a judgment is made." Although *criteria* is the preferred plural, *criterions* is also acceptable.

EXAMPLES This is the *criterion* that the trainee did not understand.
On these *criteria,* the proposals will be evaluated.

device, devise *Device* is a noun meaning "a contrivance," "an appliance," "a scheme"; *devise* is a verb meaning "to invent."

EXAMPLES This *device* will help prevent pollution in our waterways.
We need to *devise* a safe method for drilling offshore oil wells.

discreet, discrete *Discreet* means "showing good judgment in conduct and especially in speech." *Discrete* means "consisting of distinct, separate, or unconnected elements."

EXAMPLES The administrative assistant was *discreet* in her remarks.
Discrete electronic circuitry was the standard until the advent of integrated circuits.

dual, duel *Dual* means "double." *Duel* as a noun means a fight or contest between two people; as a verb, it means "to fight."

EXAMPLES The car has a *dual* exhaust.
He was shot in a *duel.*

each and every Use one or the other. *Each and every* is a wordy and repetitious way to say *each* or *every.*

EXAMPLE *Each* person should make a contribution.

except for the fact that Avoid using this wordy and awkward phrase.

fact, the fact that Use *that.*

EXAMPLE *That* he was late is indisputable.

had ought, hadn't ought Avoid using these phrases. Use *ought* or *should.*

EXAMPLE He *should not* speak so loudly.

imply, infer *Imply* means to "express indirectly." *Infer* means to "arrive at a conclusion by reasoning from evidence."

EXAMPLES The supervisor *implied* that additional workers would be laid off.
I *inferred* from her comments that I would not be one of them.

in case, in case of, in case that Avoid using this overworked phrase. Use *if.*

in many instances Wordy. Use *frequently* or *often.*

irregardless Though you hear this double negative and see it in print, it is nonstandard. Use *regardless* instead.

its, it's *Its* is the possessive form; *it's* is the contraction for *it is.* The simplest way to avoid confusing these two forms is to think *it is* when writing *it's.*

EXAMPLES The tennis team won *its* match.
It's time for lunch.

lay, lie *Lay* means "to put down" or "place." Forms are *lay, laid,* and *laying.* It is a transitive verb; thus it denotes action going to an object or to the subject.

EXAMPLES *Lay* the books on the table.
We *laid* the floor tile yesterday.

Lie means "to recline" or "rest." Forms are *lie, lay, lain,* and *lying.* It is an intransitive verb; thus it is never followed by a direct object.

EXAMPLES The wrench *lay* in the tool case.
Lying in the tool case was the wrench.

lend, loan *Lend* is a verb. *Loan* is used as a noun or a verb; however, many careful writers use it only as a noun.

EXAMPLES Please *lend* me that diskette.
The bank approved the *loan.*

loose, lose *Loose* means "to release," "to set free," "unattached," or "not securely fastened"; *lose* means "to suffer a loss."

EXAMPLE We turned the horses *loose* in the pasture, but we locked the gate so that we wouldn't *lose* them.

might of, ought to of, must of, would of Nonstandard usage. *Of* should be *have.* See *could of.*

off of Omit *of.* Use *off.*

on account of Use *because.*

one and the same Wordy and repetitious. Use *the same.*

passed, past *Passed* identifies an action and is used as a verb; *past* means "earlier" and is used as a modifier.

EXAMPLES He *passed* us going 80 miles an hour.
In the *past,* bills were sent out each month.

personal, personnel *Personal* is an adjective, meaning "private," "pertaining to the person"; *personnel* is a noun meaning "body of persons employed."

EXAMPLES Please do not open my *personal* mail.
He is in charge of hiring new *personnel.*

plain, plane *Plain* is an adjective meaning "simple," "without decoration"; *plane* is a noun meaning "airplane," "tool," or "type of surface."

EXAMPLES The *plain* decor of the room created a pleasing effect.
We worked all day checking the engine in the *plane.*

principal, principle *Principal* means "highest," "main," or "head"; *principle* means "belief," "rule of conduct," or "fundamental truth."

EXAMPLES The school has a new *principal.*
The refusal to take a bribe was a matter of *principle.*

quiet, quite *Quiet* is an adjective meaning "silent" or "free from noise"; *quite* is an adverb meaning "completely" or "wholly."

EXAMPLES Please be *quiet* in the library.
It's been *quite* a while since I've seen him.

raise, rise *Raise* means "to push up." Forms are *raise, raised, raising.* It is a transitive verb; thus it denotes action going to an object or to the subject.

EXAMPLES *Raise* the window.
The technician *raised* the impedance of the circuit 300 ohms.

Rise means "to go up" or "ascend." Forms are *rise, rose, risen.* It is an intransitive verb and thus it is never followed by a direct object.

EXAMPLES Prices *rise* for several reasons.
The sun *rose* at 6:09 this morning.

rarely ever, seldom ever Avoid using these phrases. Use *rarely* or *seldom.*

respectfully, respectively *Respectfully* means "in a respectful manner"; *respectively* means "in the specified order."

EXAMPLES I *respectfully* explained my objection.
The capitals of Libya, Iceland, and Tasmania are Tripoli, Reykjavík, and Hobart, *respectively.*

sense, since *Sense* means "ability to understand"; *since* is a preposition meaning "until now," an adverb meaning "from then until now," and a conjunction meaning "because."

EXAMPLES At least he has a *sense* of humor.
I have been on duty *since* yesterday.

set, sit *Set* means "to put down" or "place." Its basic form does not change: *set, set, setting.* It is a transitive verb; thus it denotes action going to an object or to the subject.

EXAMPLES Please *set* the test tubes on the table.
The electrician *set* the breaker yesterday.

Sit means "to rest in an upright position." Forms are *sit, sat, sitting.* It is an intransitive verb and thus it is never followed by a direct object.

EXAMPLES Please *sit* here.
The electrician *sat* down when the job was finished.

stationary, stationery *Stationary* is an adjective meaning "fixed"; *stationery* is a noun meaning "paper used in letter writing."

EXAMPLES The workbench is *stationary.*
The school's *stationery* is purchased through our firm.

their, there, they're *Their is* a possessive pronoun; *there* is an adverb of place; *they're* is a contraction for *they are.*

EXAMPLES *Their* band is in the parade.
There goes the parade.
They're in the parade.

till, until Either word may be used.

to, too, two *To* is a preposition; *too* is an adverb telling "How much?"; *two* is a numeral.

EXAMPLES We traveled *to* Cincinnati.
The program is *too* difficult.
The translation requires *two* disks.

try and *Try to* is generally preferred.

type, type of In writing, use *type of.*

used to could Use *formerly was able* or *used to be able.*

where . . . at Omit the *at.* Write "Where is the library" (not "Where is the library at?")

who's, whose *Who's* is the contraction for *who is; whose* is the possessive form of *who.*

EXAMPLES *Who's* on the telephone?
Whose coat is this?

would of The correct form is *would have.* This error occurs because *would of* sounds like the contraction *would've.*

EXAMPLE He *would have* (would've) come if he had not been ill.

CREDITS

Figure 5.1: "Using Life Vests." Reprinted with the permission of Safeair, Inc., Olympia, Washington.

Figure 5.2: Dipu Bhattacharya, Klugy: The Next Generation cartoon. Copyright © Dipu Bhattacharya. Reprinted with the permission of *The Austin Communicator*.

Figure 5.11: "Proven design for a three-way, conventionally built fireplace." Reprinted with the permission of Donley Brothers Co.

William Horton, excerpt from "Pictures Please—Presenting Information Visually" from C. M. Barnum and S. Carliner, *Techniques for Technical Communicators*. Copyright © 1994 by Allyn & Bacon. Reprinted with permission.

Henrietta N. Shirk, "The Impact of New Technologies on Technical Communication" from K. Staples and C. Ornatowski (eds.), *Foundations for Teaching Technical Communication: Theory, Practice, and Program Design*. Copyright © 1997. Reprinted with the permission of JAI Press.

Frances Stiles, "What Is Petty Cash and How Can I Use It?" Reprinted with the permission of Frances Stiles.

"How to Measure Your New Garage Door." Reprinted with the permission of Clopay Building Products Company.

Whirlpool Corporation, "Water and Ice Dispensers (on some models)" from *Use and Care Guide* for side-by-side refrigerator. Reprinted with the permission of Whirlpool Corporation.

Ruth F. Craven and Constance J. Hirnle, "Handwashing" from *Fundamentals of Nursing*. Copyright © 1996. Reprinted with the permission of Lippincott-Raven Publishers.

"Your Car's Engine." Reprinted with the permission of Quaker State Corporation.

"JSF119-611 Primer" from *Code One* (April 1999). Reprinted with the permission of Eric Hehs, editor, *Code One* magazine.

Carl Sagan, excerpt from "The Shores of the Cosmic Ocean" from *Cosmos* (New York: Random House, 1980). Copyright © 1980 by Carl Sagan. Reprinted with the permission of the Estate of Carl Sagan.

Kathy Judge, "Construction of the Pointe Shoe." Reprinted with the permission of Kathy Judge.

"About Outlets (Receptacles)" from *How-to Booklet #3: Switches and Outlets* (Habitat for Humanity). Reprinted with the permission of Creative Homeowner®.

Excerpt from *The World Book Encyclopedia.* Copyright © 1999 by World Book, Inc. Reprinted with the permission of the publisher. www.worldbook.com

Volvo Cars of North America, "Through Thick and Thin." Reprinted with the permission of Volvo Cars of North America.

Hinds Community College Purchasing Department, "Instructions to Bidders." Reprinted with the permission of the Purchasing Department, Hinds Community College, Raymond, Mississippi.

Kim Whitman, "A Personal Portrait of the Rodrigues Fruit Bat" from *Bats,* Vol. 16, No. 4 (Winter 1998). Reprinted with the permission of Bat Conservation International (BCI). For more information about bats, *Bats* magazine, or membership in BCI, please visit the BCI web site at www.batcon.org or write or call: Bat Conservation International, P.O. Box 162603, Austin, Texas 78716, (512) 327-9721. Basic membership, which includes a one-year subscription to *Bats* magazine, is $30.

"Rodrigues Fruit Bat or Rodrigues Flying Fox." Reprinted with the permission of the Philadelphia Zoo and R. N. Sloane, Docent.

"Felling" from *Operator's Safety Manual* (ANSI B 175.1, 1985). Reprinted with the permission of Husqvarna Sweden.

Graphic of Roundup Herbicide Sprayer. Reprinted with the permission of the Fountainhead Company.

Nikon advertisement, "Nikon Coolpix 700." Reprinted with the permission of Nikon.

Toro Company, "Toro® Wheel Horse® Classics—modern tractors with a generation of proven performance." Reprinted with the permission of the Toro Company.

Hanover Consumer Cooperative Society, Inc., "Broccoli raab." Reprinted with the permission of Hanover Consumer Cooperative Society, Inc.

Nevada Barr, excerpt from *Blind Descent.* Copyright © 1998 by Nevada Barr. Reprinted with the permission of Penguin Putnam, Inc.

Excerpt from "Demystify Electric Gadgets, Gizmos" from *Farm Journal* (February 1999). Reprinted with the permission of *Farm Journal.*

Genie Company, "Safety Information: Overview of Potential Hazards." Reprinted with the permission of the Genie Company.

"About . . . Being an Executor" was produced by MetLife's Consumer Education Center and reviewed by the Division for Public Education of the American Bar Association and the Legal Services Corporation.

Overview of M. Tennesen (1999), "Before You Play Doctor" from *Health* 13 (1): 100–103. Reprinted with the permission of Cheril Bible.

Descriptive abstracts of Jean-Pierre Lasota, "Unmasking Black Holes," and Stephen Wroe, "Killer Kangaroos and Other Murderous Marsupials," from *Scientific American* (May 1999). Reprinted with the permission of *Scientific American.*

Descriptive summary of Patricia Flint, Melanie Lord Van Slyke, Doreen Starke-Meyerring, and Aimee Thompson, "Going Online: Helping Technical Communicators Help Translators," from *Technical Communication* (May 1999). Reprinted with permission from *Technical Communication,* the journal of the Society for Technical Communication, Arlington, VA.

Kathy Judge, descriptive summary of "Martial Arts: Teaching the Whole Person." Reprinted with the permission of Kathy Judge.

Abstract of Javier Cid-Ruzafa, Laura E. Caulfield, Yolanda Barron, and Sheila K. West (1999), "Nutrient intakes and adequacy among an older population on the eastern shore of Maryland: The Salisbury Eye Evaluation" from *Journal of the American Dietetic Association* 99: 564–571. Reprinted with the permission of the *Journal of the American Dietetic Association.*

Michael Murrey, summary of "The Ecological Effects of Lead Shot: A Report for Duck Hunters." Reprinted with the permission of Michael Murrey.

Texas Sunset Advisory Committee, "Executive Summary" from *1998 Sunset Advisory Commission Staff Report: Texas Commission on Human Rights.* Reprinted with the permission of the Texas Sunset Advisory Committee.

Janis Ramey, "How to Write a Useful Abstract" from *Intercom* (May 1997). Reprinted with permission from *Intercom,* the magazine of the Society for Technical Communication, Arlington, VA.

"Field Report: Clubhouse—Hinds Community College." Reprinted with the permission of Hinds Community College.

"Proposal for Hinds Community College Honors Football Concession Stand." Reprinted with the permission of Hinds Community College Honors Institute.

Jessica Barlow, "A Hospital Social Work Department." Reprinted with the permission of Jessica Barlow.

Kathy Judge, "Progress on Research Report on Karate as a Physically and Psychologically Integrated Art." Reprinted with the permission of Kathy Judge.

Debbie Davis, "My College Progress Report from Mississippi State University and Hinds Community College: Purpose, Progress and Projections." Reprinted with the permission of Debbie Davis.

James David Wells, "Feasibility Report of Available Compact Sport Pickup Trucks." Reprinted with the permission of James David Wells.

Town of Utica, Mississippi, "1998 Drinking Water Quality Report." As found in the *Hinds County Gazette* (September 30, 1999). Reprinted with the permission of Mary Ann Keith, Editor, *Hinds County Gazette.*

Proposal (Austin Tree Specialists). Reprinted by permission.

Town of Utica, Mississippi, "Request for Proposals for Rehab Inspectors and Asbestos Inspectors." Reprinted with the permission of Mary Ann Keith, editor, *Hinds County Gazette.*

The Kotmeister Foundation, request for proposals for grant funding. Reprinted by permission.

Brian Westbrook, "Internet Home Page Structure and Content Recommendations for Park Primary School." Reprinted with the permission of Brian Westbrook.

Edmond H. Weiss, "Why Your Last Technical Proposal Failed" from *Intercom* (January 1988). Reprinted with the permission of the author and *Intercom,* the magazine of the Society for Technical Communication, Arlington, VA.

Hinds Community College, Department of English, fax cover sheet. Reprinted with the permission of Hinds Community College.

Gastrointestinal Partnerships, fax transmittal form. Reprinted by permission.

Paul Bodmer, e-mail message (June 17, 1999). Reprinted by permission.

Tillman Marshall Marine Company, business letter. Reprinted by permission.

Women in Insurance, business letter. Reprinted by permission.

Southern Educators Life Insurance Company, business letter. Reprinted by permission.

State of Mississippi, Department of Economic and Community Development, two business letters. Reprinted with the permission of the State of Mississippi Department of Economic and Community Development.

Kathy Judge, claim letter. Reprinted with the permission of Kathy Judge.

Mississippi Committee for the Prevention of Child Abuse, Inc., transmittal letter for the *Mississippi Youth Court Desk Book 1999.* Reprinted with the permission of the Mississippi Committee for Prevention of Child Abuse, Inc.

Nell Luter Floyd, "Technology and Ethics in the Workplace: The Ethical Impact of New Technologies on Workers" from *The* [Jackson, MS] *Clarion-Ledger* (August 2, 1998). Reprinted with the permission of *The Clarion-Ledger.*

American Society of Chartered Life Underwriters and Chartered Financial Consultants and Ethics Officer Association, "Ethics in the Computer Age: A Matter of Degree" [survey results]. Reprinted by permission.

Peter J. Prestillo, "The War for Talent" (delivered to the University of Michigan's Management Briefing Seminar, Traverse City, MI, August 4, 1999) from *Vital Speeches of the Day* (October 15, 1999). Reprinted by permission.

Hugh Hay-Roe, "How to Sabotage a Presentation" from *Intercom* (February 1999). Reprinted with the permission of *Intercom,* the magazine of the Society for Technical Communication, Arlington, VA.

Sandra Smith, progress report and "Peripheral Neuropathy: A Guide for Diabetics." Reprinted with permission of Sandra Smith.

INDEX

COMMUNICATION PROJECT PLAN SHEET

Audience

- Who will read or hear the communication project?
- How will readers use the project?
- How will your audience guide your communication choices?

Purpose

- What is the purpose of the communication project?
- What need will the project meet? What problem can it help to solve?

Subject

- What is the communication project's subject matter?
- How technical should the discussion of the subject matter be?
- Do you have sufficient information to complete the subject? If not, what sources or people can help you to locate additional information?
- What title can clearly identify the project's subject and purpose?

Author

- Will the project be a collaborative or an individual effort?
- If the project is collaborative, what are the responsibilities of each team member?
- How can the project developer(s) evaluate the success of the completed project?

Project Design and Specifications

- Are there models for organization or format for the communication project?
- In what medium will the completed project be presented?
- Are there special features the completed project should have?
- Will the project require graphics or other visuals? If so, what kinds and for what purpose?
- What information design features can best help the project's audience?

Due Date

- What is the final deadline for the communication project?
- How long will the project take to plan, research, draft, revise, and complete?
- What is the timeline for different stages of the project?